Ethics in Psychiatry

INTERNATIONAL LIBRARY OF ETHICS, LAW, AND THE NEW MEDICINE

Founding Editors

DAVID C. THOMASMA†
DAVID N. WEISSTUB, *Université de Montréal, Canada*
THOMASINE KIMBROUGH KUSHNER, *University of California, Berkeley, U.S.A.*

Editor

DAVID N. WEISSTUB, *Université de Montréal, Canada*

Editorial Board

VOLUME 45

For further volumes:
http://www.springer.com/series/6224

Hanfried Helmchen · Norman Sartorius
Editors

Ethics in Psychiatry

European Contributions

 Springer

Editors
Prof. Dr. Hanfried Helmchen
Charitè – University Medicine Berlin
Dept. Psychiatry & Psychotherapy, CBF
Eschenallee 3
14050 Berlin
Germany
hanfried.helmchen@charite.de

Prof. Dr. Norman Sartorius
chemin Colladon 14
1209 Geneve
Switzerland
normansartorius@bluewin.ch

ISSN 1567-8008
ISBN 978-90-481-8720-1   e-ISBN 978-90-481-8721-8
DOI 10.1007/978-90-481-8721-8
Springer Dordrecht Heidelberg London New York

Library of Congress Control Number: 2010931105

Printed on acid-free paper

Springer is part of Springer Science+Business Media (www.springer.com)

Acknowledgement

During the past 2 years we had many inspiring and rewarding contacts with Professor David Weisstub, the Series-Editor of the International Library of Ethics, Law, and the New Medicine, the authors of the chapters and the reviewers of the contributions to this book. It gives us great pleasure to thank Professor Weisstub for his continuing collegial support and constructive criticism, the authors for their patience and positive reactions to our recommendations and the reviewers Francesco Amaddeo, George Annas, Paul Appelbaum, Julio Arboleda-Florez, Matthias Berger, Amnon Carmi, Leon Eisenberg, Wolfgang Fleischhacker, Wolfgang Gaebel, Thomas Galert, Abraham Halpern, Claire Henderson, Wolfram Henn, Jinger Hoop, Thomas Kallert, Christian Kopetzki, Hans Lauter, Michael Linden, Mario Maj, Karl Mann, Georg Marckmann, Declan McLoughlin, Hans-Jürgen Möller, Ahmed Okasha, Laura Roberts, John Sadler, Benedetto Saraceno, Anne Slowther, Ryan Spellecy, Giovanni Stanghellini, Alan Stone, Johannes Thome, Sam Tyano, Jochen Vollmann, Siegfried Weyerer, Urban Wiesing, Alison Yung for their reviews of the chapters and their many useful comments. We enjoyed working together with all of them and hope that there will be opportunities to do so in the future.

We thank Jane Helmchen, M.A. for her help in improving the English language of some German authors. We also thank the employees of the publisher Springer, Dordrecht, NL, involved in the production of this book as well as the project manager Sangeetha Satiamurthy from Integra Software Services Pvt. Ltd., Pondicherry 605 008, INDIA, for their advice and engagement in preparing this publication.

Berlin   Hanfried Helmchen
Geneva   Norman Sartorius
July 2010

Contents

Contributors

Ron L.P. Berghmans Department of Health, Ethics and Society, Faculty of Health, Medicine and Life Sciences, Maastricht University, 6200 MD Maastricht, The Netherlands, r.berghmans@hes.unimaas.nl

Deborah Bowman Centre for Medical and Healthcare Education, St George's University of London, London SW17 0RE, UK, dbowman@sgul.ac.uk

Traolach S. Brugha Department of Health Sciences, Leicester General Hospital, University of Leicester, Leicester LE5 4PW, UK, tsb@le.ac.uk

Felicity Callard Health Service and Population Research Department, Institute of Psychiatry, King's College London, PO 34 Service User Research Enterprise, Institute of Psychiatry, De Crespigny, Park Road, London SE5 8AF, UK, felicity.callard@kcl.ac.uk

Dirk Claassen Psychiatrist and Psychotherapist, Former Consultant Psychiatrist and Honorary Senior Lecturer at the East London Mental Health Trust. Psychiatric Practice, D-30161 Hannover, Germany, praxis@drclaassen.com

Elmar Doppelfeld Permanent Working Party of Research Ethics Committees, Köln D-50859, Germany, med.ethik.komm@netcologne.de

Giovanni A. Fava Psychotherapy and Psychosomatics, Department of Psychology, University of Bologna, Bologna 40127, Italy, giovanniandrea.fava@unibo.it

Hanfried Helmchen Department of Psychiatry and Psychotherapy, CBF, Charitè – University Medicine Berlin, D-14050 Berlin, Germany, hanfried.helmchen@charite.de

Joachim Klosterkötter Department of Psychiatry and Psychotherapy, University of Cologne, D-50924 Cologne, Germany, joachim.klosterkoetter@uk-koeln.de

Norbert Konrad Institute of Forensic Psychiatry, Charité – University Medicine, 12203 Berlin, Germany, norbert.konrad@charite.de

Juan José López-Ibor Institute of Psychiatry and Mental Health, San Carlos Hospital, Complutense University, 28035 Madrid, Spain; CIBERSAM (Spanish

Network for Research on Mental Health), Complutense University, 28035 Madrid, Spain, jli@lopez-ibor.com

Maria-Inès López-Ibor CIBERSAM (Spanish Network for Research on Mental Health), Complutense University, 28035 Madrid, Spain; Department of Psychiatry and Medical Psychology, Complutense University, 28035 Madrid, Spain, jli@lopez-ibor.com

Lienhard Maeck Universitäre Psychiatrische Kliniken, CH-4025 Basel, Switzerland, lienhard.maeck@upkbs.ch

Giovanni Maio Institute for Bioethics and History of Medicine, University of Freiburg, D-79104 Freiburg, Germany, maio@ethik.uni-freiburg.de

Arnd T. May Institute of Philosophy, Ruhr Universität, 44780 Bochum, Germany; University Hospital Aachen (UKA), Aachen 52074, Germany, May@medizinethik.de

Roy McClelland School of Medicine, Dentistry & Biomedical Sciences, Queen's University, Belfast, UK, r.j.mcClelland@qub.ac.uk

Nisha Mehta School of Medicine, King's College London, London SE5 8AF, UK, nisha_macmehta@hotmail.com

Howard Meltzer Department of Health Sciences, University of Leicester, Leicester LE1 6TP, UK, hm74@le.ac.uk

Stefan Priebe Unit for Social and Community Psychiatry, Barts and the London School of Medicine and Dentistry, Queen Mary University of London Newham Centre for Mental Health, London E13 8SP, UK, s.priebe@qmul.ac.uk

Peter Propping Institute of Human Genetics, University of Bonn, D-53105 Bonn, Germany, propping@uni-bonn.de

Christian Reimer University of Gießen, D-61231 Bad Nauheim, Germany, ccreimer@t-online.de

Ulrich Rüger University of Göttingen, D-37085 Göttingen, Germany, urueger@gwdg.de

Norman Sartorius Association for the Improvement of Mental Health Programmes, CH-1209 Geneva, Switzerland, sartorius@normansartorius.com

Hans-Martin Sass Kennedy Institute of Ethics, Georgetown University, Washington, 20057 DC, USA; Institute of Philosophy, Ruhr Universität, Bochum 44780, Germany, sass@ethikzentrum.de

Thomas E. Schlaepfer Departments of Psychiatry and Mental Health, The Johns Hopkins University, Baltimore, MD 21210, USA; Department of Psychiatry and Psychotherapy, University Hospital, 53105 Bonn, Germany, schlaepf@jhmi.edu

Bettina Schöne-Seifert Institut für Ethik, Geschichte und Theorie der Medizin, Von-Esmarch-Str. 62, 48149 Münster, Germany, schoeneb@ukmuenster.de

Frauke Schultze-Lutter Research Department, University Hospital of Child and Adolescent Psychiatry Bern (UPD), CH-3000 Bern 60, Bern, Switzerland, frauke.schultze-lutter@kjp.unibe.ch; frauke.schultze-lutter@web.de

Otto W. Steenfeldt-Foss University Health Services of Oslo, Oslo N-0260, Norway, ottost@extern.uio.no

Gabriela Stoppe Universitäre Psychiatrische Kliniken, CH-4025 Basel, Switzerland, gabriela.stoppe@upkbs.ch

Matthis Synofzik Centre for Neurology and Hertie-Institute for Clinical Brain Research, D-72076 Tübingen, Germany, matthis.synofzik@uni-tuebingen.de

George Szmukler Health Service and Population Research Department, King's College London, Institute of Psychiatry, London SE5 8AF, UK, George.Szmukler@iop.kcl.ac.uk

Davinia Talbot Institut für Ethik, Geschichte und Theorie der Medizin, 48149 Münster, Germany, talbot@uni-muenster.de

Ruud ter Meulen Centre for Ethics in Medicine, University of Bristol, Hampton House, Cotham Hill, Bristol BS 6 6AU, UK, r.termeulen@bristol.ac.uk

Graham Thornicroft Health Service and Population Research Department, Institute of Psychiatry, King's College London, London SE5 8AF, UK, graham.thornicroft@kcl.ac.uk

Ambros Uchtenhagen Research Institute for Public Health and Addiction, Zurich University, CH-8005 Zurich, Switzerland, uchtenhagen@isgf.uzh.ch

Christina M. van der Feltz-Cornelis Department of Developmental, Clinical and Crosscultural Psychology, Tilburg University, PO Box 90153, 5000 LE Tilburg, The Netherlands, c.m.vdrfeltz@uvt.nl

Robert Van Voren Global Initiative on Psychiatry, 1282 BG Hilversum, The Netherlands, rvvoren@gip-global.org

Birgit Völlm Institute of Mental Health, Sir Colin Campbell Building, University of Nottingham Innovation Park, Nottingham NG7 2TU, UK, Birgit.Vollm@nottingham.ac.uk

Jochen Vollmann Department of Medical Ethics and the History of Medicine, Ruhr-University Bochum, 44799 Bochum, Germany, jochen.vollmann@ruhr-uni-bochum.de

Introduction

Hanfried Helmchen and Norman Sartorius

The title of this book – Ethics in Psychiatry – expresses our conviction that there is no such thing as 'psychiatric ethics' and that the principles of ethics in general should be used when examining psychiatric research and practice. The subtitle – European contributions – draws attention to the fact that the ethicists, philosophers, lawyers, psychiatrists and historians are all Europeans and that we believe that it is important to make European perspectives of ethical issues better known.

Although Europe – seen from outside – may seem as a culturally homogenous there are many differences between Eastern, Central and Western Europe as well as between the Northern and Southern parts of Europe. Even within countries the cultural differences are often marked – partly because of different history and partly because of migrations that have marked the twentieth century and created sizeable minority groups in several European countries. These differences influence the development and understanding of ethical issues and are reflected in the contributions that compose the volume. We did not ask authors to change their contributions in order to make the chapters follow a particular, impersonal line feeling that the differences between them will illustrate the variety of cultural variations within Europe and indicate the difficulties that might be met by anyone wanting to make the whole continent follow exactly the same lines of ethical consideration. However, this situation can be seen also as a challenge or even a singular chance e.g. to compare prospectively the regionally different legal implementations of the same basic ideas in order to prove which regulations promote best the welfare and well-being of the mentally ill. Furthermore, these differences could give reason to observe, to make aware, and perhaps to control from different perspectives the transformative effects of the biomedical progress on law and ethics.

In addition to cultural influences, personal preferences and ideological convictions may also influence action in the field of psychiatry. Thus, for example, the choice of treatment, the information given to the patient and the evaluation of the results of treatment may be guided by both the theoretical orientation and the clinical

H. Helmchen (✉)
Department of Psychiatry and Psychotherapy, CBF, Charitè – University Medicine Berlin,
Berlin D-14050, Germany
e-mail: hanfried.helmchen@charite.de

experience of the therapist underlining the need to examine the risk benefit ratio for each form of treatment in each individual case taking the preferences of the informed patient into account. The same principle of making decisions dependent on the specific individual must also be central in considering the value of recommendations for guidelines and regulations helping to make ethical choices and concerning matters such as the distinction between need and preference, welfare and well-being, therapy of an illness or enhancement of functioning, persuasion or coercion and help while dying or help to die.

The book is divided into 6 parts. The first part addresses the context in which psychiatry is practised and in which research on issues of importance for psychiatry is carried out. It thus examines the stigmatization of mental illness and the consequences of stigmatization, the legislation relevant to the practice of psychiatry and the arrangements – such as the ethical committees – that have been put in place to ensure that the ethical principles are applied in research and in clinical work. The relatively recent areas of considerable interest – the economic aspects of care for people with mental illness and the conflicts of interest that might arise in the practice of psychiatry and in relevant research are also addressed in this section of the book.

The second part of the book examines four areas of the currently central issues of ethics in relation to health care – informed consent, autonomy of the patient, confidentiality and distributive justice – added to the old medical principles of acting for the welfare of the patient (salus aegroti) and avoiding harm (nil nocere). The challenges of their application in psychiatry are considerable and there is little doubt about the fact that this contributes to the negative image that psychiatry has in many societies. The consensus about the application of ethical principles in psychiatry adopted by the World Psychiatric Association (WPA) – the Declarations of Hawaii and of Madrid – is also discussed in this part of the book.

In the third part of the book a series of contributions examine systematically the ethical aspects of psychiatric assessment and treatment. In the selection of topics we gave priority to areas of most controversy and to areas that are usually not included in texts about ethics – for example the ethical implications of procedures in institutions. They are followed by chapters that address the application of ethical principles in research with persons who are unable to decide for themselves as well as to issues arising in epidemiological and genetic research.

The fourth part of the book discusses ethical issues related to the non-medical uses of psychiatry and focuses on two subjects: the abuse of psychiatry for political purposes and the (ab)use of psychiatry for enhancement purposes. The first of these two topics has torn psychiatry apart in the nineteen sixties and seventies and we felt that despite of the fact that the political abuse of psychiatry seems to have declined in frequency it should be included in the book as an example of an ethical problem whose solution depends mainly on political action rather than only on a change of behaviour of psychiatrists. The second topic on the contrary seems to be gaining in importance in the context of globalization and continuing emphasis on the elevation of personal (and societal) aspirations for success in all spheres of life and might thus remind that new types of ethical problems are likely to emerge in the near future.

The fifth part of the book deals with education about ethics in psychiatry this being one of the possible ways to improve the likelihood that psychiatric research and practice will be beyond ethical reproach. The importance of internalizing the ethical principles and making them an integral part of the value system of medical professionals should be of primary concern for all medical schools: sadly until now this has not happened.

Finally, the Part VI summarizes the major points addressed by the various chapters and draws conclusions based on the material presented in the volume.

The complexity of ethical considerations in relation to psychiatry called for a very careful preparation of the text. The editors have corresponded with the authors about their chapters and given them numerous suggestions. In addition each of the chapters was reviewed by two experts whose opinion was provided to the authors with the expectation that they will finalize their chapter taking the views of the editors and the reviewers into account. In some instances this seemed impossible for the authors and we had therefore had to seek alternative authors. In considering these decisions we have kept in mind that Europe is far from a homogenous entity and that there are very few experts who are familiar with the practice and ethical considerations in all of the more than fifty European countries. In some instances therefore we have accepted chapters that describe the situation in a single country in Europe, as a paradigm illustrating issues that arise in relation to a topic such as that of dementia and end-of-life decisions.

As in any multi-authored book there is a certain amount of overlap between the chapters. This was unavoidable in this book and we have not insisted on removing the overlap – not least also because the chapters, as they now stand, treat their topics comprehensively which would not have been possible if all overlap had been removed. In some instances the overlap is apparent but there is little of it: thus, for example, the chapter on de-institutionalization and the chapter on coercive measures both address the delicate issue of balance between patients' autonomy and their welfare but from different standpoints – once from the point of view of the person responsible for providing service and once from the more abstract, theoretical level.

In some instances the authors wrote vehemently about topics about which they feel passionately and their treatment of the topic is therefore not as comprehensive as might be desired. Thus, for example, the chapter on the political misuse of psychiatry written by an author who has been involved in fighting against it for decades describes the developments in the Soviet Union in fascinating detail and with remarkable insights while neglecting to address the general question of ways of abusing psychiatry for political purposes and its limits towards the systematic neglect of patients' needs and towards the use of psychiatric methods in interrogations concerning criminal activities. We did not feel that we should complement such descriptions with more systematic reviews and believe that these chapters will be well placed to draw attention and engagement of the reader to specific issues.

The psychiatrists' task is to recognize the needs of people suffering from mental illness and to help them by controlling the symptoms of mental disorder, by preventing or reducing the disability related to the disorder, by overcoming their lack or loss of adequate social relationships and by finding ways to facilitate their inclusion into society. In doing so the psychiatrists are confronted with moral problems that they need to recognize and resolve. A central issue in this regard is to find, in each individual case, the right balance between patients' autonomy and the protection against harm related to illness, between the need to empower patients while reducing their disability and between the responsibility of patients themselves and the help that their carers offer. In addition the psychiatrists have to deal with attitudes to mental illness and to psychiatry, with laws and regulations and with their own biases – all of which raises moral questions and ethical issues at the individual and social level.

Fundamental changes of our societies – related, for example, to demographic change, migration, unemployment, social disintegration and conflicts of cultural value systems – as well as the increasing costs of medical care using new techniques and interventions reopen questions about the nature of man and increase the sensitivity for autonomy and dignity of all participants in the treatment process. At the same time, the continuing stigmatization of mental illness and of psychiatry, the gradual acknowledgement of rights of the mentally ill, the changes of the role of the patients (from patient to client) and of therapists (from a father figure to a partner and from an independent professional to a guideline dominated practitioner) modifying the nature of medical care and of the relationship between patients and doctors thus revitalising old and adding new ethical problems. Whether the well established ethical principles and codes are sufficient to solve the new ethical problems – for example the conflicts between the protection and the autonomy of the patient, between individual and common welfare and between the claim of global validity of human rights and culture- bound morals and customs is an open question.

We hope that this book will help in thinking about these issues – in practice, in drafting regulations and guidelines and in the education about ethics in schools of health personnel. We also hope that it will orient research and focus attention on unresolved ethical issues. The achievement of these aims would be the greatest reward for the contributors of this book.

Part I
The Context

Chapter 1
Societal Framework of Psychiatry

Hanfried Helmchen and Norman Sartorius

Contents

Abbreviations

EDEN	European Day-Hospital EvaluatioN
EPSILON	European Psychiatric Services: Inputs Linked to Outcome Domains and Needs
ESEMeD	European Study of the Epidemiology of Mental Disorders
EUROSTAT	European Statistics
IMS	International Measuring System
OECD	Organisation for Economic Co-operation and Development
GP	General Practitioner
WHO	World Health Organisation
SES	Socio-Economic Status

1.1 Society and the Individual

The ideas expressed in the Bill of Rights (1688), the Constitution of the United States of America (1787) and the French Revolution (1789) gave rise to the concept of the constitutional state in which the rights and obligations of the individual towards society – and vice versa – are regulated by law. However, the balance between the rights and obligations of the individuals and the state varies among the countries that have accepted the idea of being a constitutional state. There are

H. Helmchen (✉)
Department of Psychiatry and Psychotherapy, CBF, Charitè – University Medicine Berlin,
Berlin D-14050, Germany
e-mail: hanfried.helmchen@charite.de

H. Helmchen, N. Sartorius (eds.), *Ethics in Psychiatry*, International Library
of Ethics, Law, and the New Medicine 45, DOI 10.1007/978-90-481-8721-8_1,
© Springer Science+Business Media B.V. 2010

states – for example the USA – that emphasize the importance of the individual's freedom and self-reliance which can be harmful for those who are not able to care for themselves such as some of the mentally ill people. On the other side are states with a strong social welfare policy which provide a necessary minimum of social security, sometimes at the expense of individual freedom particularly in societies deformed by totalitarian features as was the former Soviet Union or the German Democratic Republic. Most European countries try to maintain a position between these two extremes, but experience increasing problems in financing such a system because of the demographic change and the progress of scientific medicine entailing higher cost of care, and because people's understanding of the system's subsidiary character fades and the population probably because of that no longer accepts the obligations stemming from the maintenance of strong welfare policies.

Dominating ideas in contemporary Western European societies are the principles of individual autonomy, solidarity and justice. The German philosopher Habermas explained:

> "The more the structures of the living world differentiate the more it can be seen that the increasing autonomy of the 'individuated' single is interwoven with the increasing integration into multifold social dependencies. The more the individuation is going on the more the single subject gets entangled in a closely woven and at the same time more subtle net of reciprocal defencelessness and exposed needs of protection." Thus morals "show the integrity of individuals to advantage by demanding even respect for the dignity of each; however, they also protect to the same extent the intersubjective relations of reciprocal recognition by which the individuals keep themselves as members of a community. Both complementary aspects correspond with the principles of justice and solidarity" (Habermas 1991).

The principle of distributive justice in health care can only be realized through solidarity among members of a society. The importance of solidarity grows with the decrease of resources or with an increase of the needs of care with a stable amount of resources. The principle of solidarity competes with the concept of increasing the common welfare by the powers of the free market. If these concepts prevail the solidarity will weaken and the support for those who are in need will become less available which will lead to lesser care for the mentally ill and an increased moral problem of (mental) health care. With the diminution of the resources provided by the community (previously characterised by solidarity) the psychiatrists obligation to act for the welfare of the mentally ill individual becomes more difficult because they have to face the conflict between their primary obligation to ensure the welfare of the individual and their social obligation to save the limited resources so that other members of community can use them for their welfare.

The traditional focus of ethics in medicine were on ethical problems which arise in medical treatments of the individual, e.g. a conflict between the welfare and autonomy of the patient, to do no harm, or confidentiality. In contrast the ethics of modern public health considers – according to a recent Report from the Nuffield Council on Bioethics:

> "the responsibilities of governments, individuals and others in promoting the health of the population. We conclude that the state has a duty to help everyone lead a healthy life and

reduce inequalities in health. Our 'stewardship model'[1] sets out guiding principles for making decisions about public health policies".[2] Although the report has been attacked by the UK media as a way into a 'nanny state'[3] it is a thoughtful examination of the question: "whose job is it to ensure that we lead a healthy life?", or more specifically "how and when the state should act?".[4]

1.2 Society and the Mentally Ill Individual

The balanced relationship between the individual and his social environment is at risk in the case of mental illness. Mental illness may impair the capacity of individuals to live their lives in accordance with their intentions, objectives, wishes, needs and expectations. It may reduce their chances of using the social context and of adequately tackling its constraints. The social context will also influence the individuals' mental health if it exceeds their capacity to cope with the multitude of stresses resulting from insults (e.g. mobbing, stigmatisation), broken relationships, bad company (e.g. drug abusing friends), overwork (e.g. workaholic), social insecurity (unemployment, homelessness), war persecution and many others. Such social events may influence the causes, the development, the manifestation, the course (remission, recurrence, relapse, chronification) or outcome of mental illness. At least as important for the fate of the mentally ill are attitudes prevailing in a society – such as those leading to exclusion, discrimination and stigmatisation of the mentally ill or those that make it impossible to rejoin society. The conviction that mental illness is untreatable – widely held by the populations of European countries (including mentally ill people) and by many health professionals – can lead to the belief that people struck by mental illness no longer have any value for society and that therefore they should be eliminated.

These ideas found their catastrophic expression during second world war's mercy killings in Germany, in the forced starvation of the mentally ill in mental hospitals in France and other countries, in the forced sterilisation of people with intellectual

---

[1] According to this model 'acceptable public health goals include: (1) reducing the risks of ill health that people are exposed to as a result of other people's actions or behaviours, for example reducing drunk-driving and passive smoking; (2) reducing causes of ill health relating to environmental conditions, such as drinking water safety and housing standards; (3) protecting and promoting of the health of children and other vulnerable people; (4) helping people to overcome addictions and other unhealthy behaviours; (5) ensuring that it is easy for people to lead a healthy life, for example by providing convenient and safe opportunities for exercise; (6) ensuring that people have appropriate access to medical services; (7) reducing unfair health inequalities. At the same time, public health programmes should: (1) not attempt to coerce adults to lead healthy lives; (2) minimise interventions that are introduced without individual consent of those affected, or without procedural justice arrangements (such as democratic decision-making procedures) which provide adequate mandate; (3) seek to minimise interventions that are perceived as unduly intrusive and in conflict with important personal values' (Nuffield Council on Bioethics 2007).

[2] Nuffield Council on Bioethics (2007).

[3] Editorial (2007) The ethics of public health. Lancet 370:1801.

[4] See footnote 3.

disability and in other legally supported ethical transgressions of European societies in relation to their mentally ill members. The notion the mentally ill are of no value to society however has not vanished: it finds its more subtle expression today in the low priority given to mental health programmes in most parts of the world, in the continuous insistence of providing the cheapest possible care to people with mental illness and in other forms of discrimination and stigmatization of the mentally ill.

A rough indication of relevant attitudes of societies may be found in considerable differences of psychiatric care among European countries (Becker and Kilian 2006, de Girolamo et al. 2006). Differences are related to (i) the provision of mental health services, (ii) to their appropriateness to meet the needs of the mentally ill, (iii) to their use and (iv) to the outcome.

> "The number of psychiatrists per 100.000 inhabitants ranged from 3.6 in Spain to 17.5 in Finland, psychiatric beds from 0.4 per 1.000 inhabitants in Italy and Spain to 1.3 in Ireland." (Carta et al. 2004). "In countries such as Switzerland or the United Kingdom in which gerontopsychiatry is institutionalised to a greater extent in the 'medical society' and health politics the service provision for depressed elderly persons seem to be more elaborated and better available" (Bramesfeld 2003). The "levels of unmet need in the population ranging from 3.6% in Northern Ireland (McConnell et al. 2002) to 15% in the Netherlands (Alonso et al. 2007, Bijl and Ravelli 2000)". Studies of representative samples in 6 European countries show actually that need of formal healthcare (Alonso et al. 2007) to respond to clinical (e.g. psychopathological symptoms and side-effects, embarrassing or dangerous behaviour etc.) as well as social needs (e.g. such as personal hygiene, getting meals, household is not provided to almost the half of people who need it etc.).

Since these differences are not only due to methodological problems (Alonso et al. 2002, Kiejna et al. 2002, Kilian et al. 2001, Schene et al. 2000),[5] they are likely to be related to other and probably more important differences in cultural attitudes towards the mentally ill and differences of the priority that European countries give to the development of mental health care. Therefore, a major factor of unmet needs of the mentally ill will be due to societal and attitudinal influences (Alonso et al. 2007, Andrews et al. 2001).

> European cross-country studies found that in general "the care of the mentally ill is rather limited in Europe", the use of psychiatric services is fairly low, especially in Italy (de Girolamo et al. 2006): roughly only one quarter of non-institutionalised people with a mental disorder consulted a formal health service, more than a fifth of them received no treatment: "the factors associated with this limited access and their implications deserve further research" (Alonso et al. 2004). There are, e.g. studies which "show that SES [Socio-Economic Status] influences psychiatric services utilization, however the real factors linking SES and mental health services utilization remain unclear" (Amaddeo and Jones 2007). A mental disorder will be recognised by a GP in only about the half of concerned people (Alonso et al. 2007, Linden et al. 1996), although "the disability rate of patients

---

[5] Which are met in international epidemiological studies of psychiatric services (type, structure, number, access to, use of, efficiency etc) such as a *lack of adequate assessment instruments* some of which have been developed only recently (EPSILON, ESEMeD, EDEN) and some others, *insufficient approaches* (e.g. indicated by low response rates) and *insufficient mental health reporting systems* (up-to-now only global and unspecific data are available gathered by WHO, OECD, EUROSTAT, IMS.

with a well-defined disorder is between 67% and 72.7%" (Linden et al. 1996). Such figures are important because disability may be a determinant of the need of care although
"determining the need of care is a complex process" (Alonso et al. 2007, Andersen 1995).
They are the more important because the majority of people would contact their GP first
(Kovess-Masféty et al. 2007). Ethnic factors seem to play a role (Commander et al. 2003),
e.g. South Asian people consulted their GP significantly more frequently than whites, but
only one half reported disclosing their problem to a GP (Commander et al. 2004).

The fact that people with mental illness do not receive the care that they need can
be related to two groups of factors. The first is that the governments of European
countries (and of most other countries of the world!) do not provide adequate
resources for the psychiatric care and that they provide less for the services for
the mentally ill than for people with other illnesses. The second is that people with
mental illness often do not use services that are available. The first of these facts is
a clear transgression against the principle of distributive justice and a demonstration
of discriminatory tendencies affecting the principle of solidarity. The second fact –
the non-use of services when they are available – is also a transgression against
distributive justice using two different mechanisms. People do not seek psychiatric
care because they are ashamed to do so in a society that stigmatizes mental illness
and people who suffer from them; they are also afraid to use the services because
they are often not providing competent services, i.e. that competence, attitudes and
behaviour of the mental health service personnel will be experienced by mentally
ill patients sometimes and somewhere as insufficient (Schulze 2005) – which is at
least in part a remote effect of the negative selection of personnel entering psychiatry that is itself a stigmatized discipline. The fact that general practitioners are often
not willing or not able to recognize the presence of mental illness in the persons
who contact them for help may also play a role in this respect.

Ethical problems are related to attitudes toward and to concepts as well as systems of psychiatric care, in which the treatment of the individual is embedded.
Thus, e.g. the process of deinstitutionalisation of mentally ill patients faces the
psychiatrist on the individual level with striking a balance between the respect of
the patient's autonomy and his protection. In some countries as in Italy the rash
beginning of deinstitutionalisation led to some individual tragedies, because the
idea of freeing the patients initially overburdened their families and not everywhere
the development of community care was sufficiently supported (de Girolamo et al.
2007, Ernst and Ernst 1992). Or, to give another example, significant differences
in rates of detaining people with mental disorders compulsorily in mental hospitals
(Commander et al. 2003) suggest differences in attitudes and habits of psychiatrists
towards the mentally ill, differences in relating the dignity and autonomy of the
mentally ill to the safety of him as well as to those of the community.

In Italy the process of radical deinstitutionalisation of mentally ill patients
indicated a change in attitudes and concepts in general (Tansella 1986), but its
realisation was hampered by insufficient support from some communities to build
up community mental health services.

Furthermore, the process of deinstitutionalisation seems to be counteracted by a
special form of re-institutionalisation in Europe, i.e. the recent increase of forensic

beds and supported housing, which outweigh in some countries the reduction of beds for psychiatric care. Together with the increase of the general prison population it was suggested that this development may reflect general attitudes to risk containment in society – and "may be more important than changing morbidity and new methods of mental health care delivery" (Priebe et al. 2005). Moreover, it indicates also the persistence of prejudices, stigmatisation and exclusion.

1.3 Conclusion

The purpose of this chapter was to provide a background to the discussion of the elements of the context in which psychiatric care and research take place. It is clear that attitudes held by populations affect the priority given to mental health care and that the principle of distributive justice is therefore not respected in the instance of mental health care. Consequently at least psychiatric policy should adapt the mental health care services more specifically to the needs of the mentally ill, should target these needs more specifically, should improve the capacity of GPs to recognise mental disorders, should motivate the GP to coordinate the treatment of mentally ill patients with the psychiatrist, and should increase the general awareness of the public towards the needs of the mentally ill and the treatments psychiatry has to offer. One of this necessary actions is the international campaign to fight the stigma of mental disorder and of psychiatry (Sartorius 2006).

The chapters that follow will provide specific evidence about the ways in which the society reacts to mental illness by examining the process of stigmatization of mental illness, the impact of the economic imperatives on mental health care, the reflection of the populations' views in legislation, the multiple conflicts of interests besetting action in this field and the way in which the psychiatric profession decided to behave in order to reduce the probability that it will contribute to the transgression of ethical principles.

References

Alonso J, Angermeyer MC, Bernert S, Bruffaerts R, Brugha TS, Bryson H, de Girolamo G, Graaf R, Demyttenaere K, Gasquet I, Haro JM, Katz SJ, Kessler RC, Kovess V, Lépine JP, Ormel J, Polidori G, Russo LJ, Vilagut G, Almansa J, Arbabzadeh-Bouchez S, Autonell J, Bernal M, Buist-Bouwman MA, Codony M, Domingo-Salvany A, Ferrer M, Joo SS, Martínez-Alonso M, Matschinger H, Mazzi F, Morgan Z, Morosini P, Palacín C, Romera B, Taub N, Vollebergh WA (2004) ESEMeD/MHEDEA 2000 Investigators, European Study of the Epidemiology of Mental Disorders (ESEMeD) Project use of mental health services in Europe: results from the European Study of the Epidemiology of Mental Disorders (ESEMeD) project. Acta Psychiatr Scand 109(420):47–54
Alonso J, Codony M, Kovess V, Angermeyer MC, Katz SJ, Haro JM, de Girolamo G, de Graaf R, Demyttenaere K, Vilagut G, Almansa J, Pierre Lépine J, Brugha TS (2007) Population level of unmet need for mental healthcare in Europe. Br J Psychiatry 190:299–306
Alonso J, Ferrer M, Romera B, Vilagut G, Angermeyer M, Bernert S, Brugha TS, Taub N, McColgen Z, de Girolamo G, Polidori G, Mazzi F, De Graaf R, Vollebergh WA,

Buist-Bowman MA, Demyttenaere K, Gasquet I, Haro JM, Palacín C, Autonell J, Katz SJ, Kessler RC, Kovess V, Lépine JP, Arbabzadeh-Bouchez S, Ormel J, Bruffaerts R (2002) The European Study of the Epidemiology of Mental Disorders (ESEMeD/MHEDEA 2000) project: rationale and methods. Int J Methods Psychiatr Res 11(2):55–67

Amaddeo F, Jones J (2007) What is the impact of socio-economic inequalities on the use of mental health services? Epidemiol Psichiatr Soc 16(1):16–19

Andersen RM (1995) Revisiting the behavioral model and access to medical care: does it matter? J Health Soc Behav 36:1

Andrews G, Issakidis C, Carter G (2001) Shortfall in mental health service utilisation. Br J Psychiatry 179:417–425

Becker T, Kilian R (2006) Psychiatric services for people with severe mental illness across western Europe: what can be generalized from current knowledge about differences in provision, costs and outcomes of mental health care? Acta Psychiatr Scand 113(429):9–16

Bijl RV, Ravelli A (2000) Psychiatric morbidity, service use, and need for care in the general population: results of The Netherlands Mental Health Survey and Incidence Study. Am J Public Health 90(4):602–607

Bramesfeld A (2003) Service provision for elderly depressed persons and political and professional awareness for this subject: a comparison of six European countries. Int J Geriatr Psychiatry 18(5):392–401

Carta MG, Kovess V, Hardoy MC, Brugha T, Fryers T, Lehtinen V, Xavier M (2004) Psychosocial wellbeing and psychiatric care in the European Communities: analysis of macro indicators. Soc Psychiatry Psychiatr Epidemiol 39(11):883–892

Commander MJ, ODell SM, Surtees PG, Sashidaran SP (2003) Characteristics of patients and patterns of psychiatric service use in ethnic minorities. Int J Soc Psychiatry 49(3):216–224

Commander MJ, Odell SM, Surtees PG, Sashidharan SP (2004) Care pathways for south Asian and white people with depressive and anxiety disorders in the community. Soc Psychiatry Psychiatr Epidemiol 39(4):259–264

de Girolamo G, Alonso J, Vilagut G (2006) The ESEMeD-WMH project: strenghtening epidemiological research in Europe through the study of variation in prevalence estimates. Epidemiol Psichiatr Soc 15(3):167–173

de Girolamo G, Bassi M, Neri G, Ruggeri M, Santone G, Picardi A (2007) The current state of mental health care in Italy: problems, perspectives, and lessons to learn. Eur Arch Psychiatry Clin Neurosci 257(2):83–91

Ernst K, Ernst C (1992) Der Stand der italenischen Psychiatrie-Reform: das Beispiel der Lombardei im Vergleich zur Schweiz. Nervenarzt 63(11):668–674

Habermas J (1991) Erläuterungen zur Diskursethik. Suhrkamp, Frankfurt/Main, p 229

Kiejna A, Kallert TW, Rymaszewska J (2002) Treatment in psychiatric day hospital in comparison with inpatient wards in different European health care systems–objectives of EDEN project. Psychiatr Pol 36(6 Suppl):361–367 (Polish)

Kilian R, Roick C, Bernert S, Matschinger H, Mory C, Becker T, Angermeyer MC (2001) Instruments for the economical evaluation of psychiatric service systems: methodological foundations of the European standardisation and the German adaptation. Psychiatr Prax 28(Suppl 2):S74–S78

Kovess-Masféty V, Saragoussi D, Sevilla-Dedieu C, Gilbert F, Suchocka A, Arveiller N, Gasquet I, Younes N, Hardy-Bayle MC (2007) What makes people decide who to turn to when faced with a mental health problem? Results from a French survey. BMC Public Health 7(147):188

Linden M, Maier W, Achberger M, Herr R, Helmchen H, Benkert O (1996) Psychiatric diseases and their treatment in general practice in Germany. Results of a World Health Organization (WHO) study. Nervenarzt 67(3):205–215

McConnell P, Bebbington P, McClelland R, Gillespie K, Houghton S (2002) Prevalence of psychiatric disorder and the need for psychiatric care in Northern Ireland: population study in the District of Derry. Br J Psychiatry 181:214–219

Nuffield Council on Bioethics (2007) www.nufieldbioethics.org . Accessed 27 May 2008.

Priebe S, Badesconyi A, Fioritti A, Hansson L, Kilian R, Torres-Gonzales F, Turner T, Wiersma D (2005) Reinstitutionalisation in mental health care: comparison of data on service provision from six European countries. Br Med J 330(7483):123–126

Sartorius N (2006) Lessons from a 10-year global programme against stigma and discrimination because of an illness. Psychol Health Med 11(3):383–388

Schene AH, Koeter M, van Wijngaarden B, Knudsen HC, Leese M, Ruggeri M, White IR, Vázquez-Barquero JL (2000) Methodology of a multi-site reliability study. EPSILON Study 3. European Psychiatric Services: Inputs Linked to Outcome Domains and Needs (EPSILON). Br J Psychiatry 177(39):s15–s20

Schulze B (2005) Stigmatisierungserfahrungen von Betroffenen und Angehörigen: Ergebnisse von Fokusgruppeninterviews. In: Gaebel W, Möller HJ, Rössler W (Hg) Stigma – Diskriminierung – Bewältigung. Der Umgang mit sozialer Ausgrenzung psychisch Kranker. Kohlhammer, Stuttgart, pp 122–144

Tansella M (1986) Community psychiatry without mental hospitals – the Italian experience: a review. J R Soc Med 79(11):664–669

Chapter 2
Stigmatisation of People with Mental Illness and of Psychiatric Institutions

Nisha Mehta and Graham Thornicroft

Contents

Abbreviations

BME   Black Minority and Ethnic
CESCR   Committee on Economic, Social and Cultural Rights
CPT   European Committee for the Prevention of Torture and Inhuman or Degrading Treatment or Punishment
CRPD   Convention on the Rights of Persons with Disabilities
DISC   Discrimination and Stigma Scale
ECHR   European Convention on Human Rights and Fundamental Freedoms 1950
ICCPR   International Covenant on Civil and Political Rights 1976
ICESCR   International Covenant on Economic, Social and Cultural Rights 1976
OHCHR   UN High Commissioner for Human Rights
UDHR   UN Declaration of Human Rights 1948

G. Thornicroft (✉)
Health Service and Population Research Department, Institute of Psychiatry,
King's College London, London SE5 8AF, UK
e-mail: graham.thornicroft@kcl.ac.uk

H. Helmchen, N. Sartorius (eds.), *Ethics in Psychiatry*, International Library
of Ethics, Law, and the New Medicine 45, DOI 10.1007/978-90-481-8721-8_2,
© Springer Science+Business Media B.V. 2010

UN   United Nations
WHO   World Health Organisation

2.1 Introduction

The effects of stigmatisation upon people with mental illness are common and profoundly socially excluding, and so constitute unethical barriers to full social participation. This chapter will therefore discuss the ethical dimension of stigma by defining terms, discussing the existing literature on stigma related to mental illness, considering global patterns of stigma, and examining stigma and human rights within psychiatric institutions.

2.2 Defining Terms

2.2.1 Stigma

Stigma (plural stigmata) has been used to refer to an indelible dot left on the skin after stinging with a sharp instrument, sometimes used to identify vagabonds or slaves (Cannan 1895, Gilman 1982, 1985, Thomas Hobbes of Malmesbury 1657). Recently stigma has come to mean 'any attribute, trait or disorder that marks an individual as being unacceptably different from the 'normal' people with whom he or she routinely interacts, and that elicits some form of community sanction' (Goffman 1963, Hinshaw and Cicchetti 2000, Scambler 1998).

A considerable literature now refers to stigma (Falk 2001, Goffman 1963, Heatherton et al. 2003, Mason 2001, Corrigan 2005a, Wahl 1999, Hayward and Bright 1997, Link et al. 1989, 1997, Link et al. 1999, Pickenhagen and Sartorius 2002, Sartorius and Schulze 2005, Smith 2002). The most complete schema of the component processes of stigmatisation has four key components (Link and Phelan 2001) which are: (i) Labelling, in which personal characteristics, which are signalled or noticed as conveying an important difference. (ii) Stereotyping, which is the linkage of these differences to undesirable characteristics. (iii) Separating, the categorical distinction between the mainstream/normal group and the labelled group as in some respects fundamentally different. (iv) Status loss and discrimination: devaluing, rejecting, and excluding the labelled group. Interestingly, more recently the authors of this model have added a revision to include the emotional reactions of both people who are stigmatised and of the 'stigmatisers', which may accompany each of these stages (Jones et al. 1984, Link et al. 2004).

Stigma can also be seen as an overarching term including three elements:

- problems of knowledge (ignorance or misinformation)
- problems of attitudes (prejudice)
- problems of behaviour (discrimination) (Hinshaw 2007, Link and Phelan 2001, Scambler 1998).

Stigma can produce changes in feelings, attitudes and behaviour for both the person affected (lower self-esteem, poorer self-care, and social withdrawal) and for family members (Kadri et al. 2004, Link et al. 1989, Littlewood et al. 2007, Ritsher and Phelan 2004, Thornicroft 2006, Weiss et al. 2001). Consistent findings have emerged from evaluating stigma in Africa (Alem et al. 1999), Asia (Thara et al. 2003), South America (de Toledo Piza and Blay 2004), in Islamic countries of North Africa and the Near East (Al-Krenawi et al. 2004), and in Europe (Sartorius and Schulze 2005). First, there are few countries, societies or cultures in which people with mental illness are considered to have the same value as people who do not have mental illness, as shown for example lower rates of financial investment in mental health services, which has been described as an aspect of *structural discrimination*. Second, the quality of information that we have is relatively poor, with few comparative studies between countries or over time. Third, there are clear links between popular understandings of mental illness and whether people in mental distress seek help or feel able to disclose their problems (Littlewood 1998). The core experiences of shame (to oneself or to one's family) and blame (from others) are common, although they vary to some extent between cultures. Where comparisons with other conditions have been made, mental illnesses are usually more stigmatised, and indeed this has been called the 'ultimate stigma' (Falk 2001). Finally, the behavioural consequences of stigma (rejection and avoidance) appear to be universal phenomena. Nevertheless, this literature says little about a core issue: how such processes affect the everyday lives of people with mental illness.

Research on stigma and mental illness largely consists of attitude surveys, investigating what people would do in imaginary situations or what they think 'most people' would do, for example, when faced with a neighbour or work colleague with mental illness. This work has emphasised what 'normal' people say rather than the actual experiences of people with mental illness themselves. It also assumes that such statements (usually on knowledge, attitudes or behavioural intentions) are linked with actual behaviour, rather than assessing such behaviour directly. In short, with some clear exceptions, this research has focussed on hypothetical rather than real situations, shorn of emotions and feelings (Crocker et al. 1998), divorced from context (Corrigan et al. 2004), indirectly rather than directly experienced, and without clear implications for how to intervene to reduce social rejection (Rose 2001). In this context, discrimination is understood in this Chapter to mean: 'make an unjust distinction in the treatment of different categories of people, especially on the grounds of race, sex, or age'.

Recently a growing body of qualitative evidence which considers how mental health service users subjectively experience, describe and cope with stigma. This has allowed an enhanced understanding of: the scope and dimensions of stigma; the personal consequences of stigma; mental health service users views on anti-stigma campaign priorities; and the impact of stigma on the family, along with the development of related scales to measure stigma (King et al. 2007).

Understanding stigma is important because it can lead to low rates of help seeking, lack of access to care, under-treatment, material poverty, and to social marginalisation (Thornicroft 2007). These effects can be the consequences of *experienced* (actual) discrimination (for example being unreasonably rejected in a job application), or they can be the consequences of *anticipated* discrimination (for example when an individual does not apply for a job because he or she fully expects to fail in any such application) (Rusch et al. 2005). This distinction between experienced and anticipated discrimination is closely related to what has been described as the difference between 'enacted' and 'felt stigma'. 'Enacted stigma' refers to events of negative discrimination, while 'felt stigma' includes the experience of shame of having a condition, and the fear of encountering 'enacted stigma' (Jacoby 1994), and is associated with lower self-esteem.

2.2.2 Ignorance: The Problem of Knowledge

There is an unprecedented volume of information in the public domain, but the level of accurate knowledge about mental illnesses (sometime called 'mental health literacy') is meagre (Crisp et al. 2005). In a population survey in England, for example, most people (55%) believe that the statement 'someone who cannot be held responsible for his or her own actions' describes a person who is mentally ill (Department of Health). Most (63%) thought that fewer than 10% of the population would experience a mental illness at some time in their lives.

Measures taken to improve public knowledge about depression can be successful, and can reduce the effects of stigmatisation. At the national level, social marketing campaigns have produced positive changes in public attitudes towards people with mental illness, as shown recently in New Zealand and Scotland (Vaughn 2004, Dunion and Gordon 2005). In a campaign in Australia to increase knowledge about depression and its treatment, some states and territories received this intensive, co-ordinated programme, while others did not. In the former, people more often recognised the features of depression, were more likely to support help-seeking for depression, or to accept treatment with counselling and medication (Jorm et al. 2005). Similarly recent evidence comparing trends between Scotland and England in public attitudes towards people with mental illness are consistent with a positive effect of the Scottish 'See Me' anti-stigma campaign (Mehta et al. 2009). A new campaign is now starting in England, entitled 'Time to Change' aiming to fundamentally reduce stigma and discrimination (Henderson and Thornicroft 2009).

2.2.3 Prejudice: The Problem of Negative Attitudes

The term prejudice is used to refer to many social groups which experience disadvantage, for example minority ethnic groups, yet it is employed rarely in relation to people with mental illness. The reactions of a host majority to act with prejudice in rejecting a minority group usually involve not just negative thoughts but also emotion such as anxiety, anger, resentment, hostility, distaste, or disgust. In fact prejudice may more strongly predict discrimination than do stereotypes. Interestingly, there is almost nothing published about emotional reactions to people with mental illness apart from that which describes a fear of violence (Graves et al. 2005). An example of such negative attitudes are the terms used by school students towards people with mental health problems, and in one English study, among 250 such terms used, none were positive and 70% were negative (Rose et al. 2007).

2.2.4 Discrimination: The Problem of Rejecting and Avoidant Behaviour

Attitude and social distance surveys (of unwillingness to have social contact) usually ask either students or members of the general public what they would do in imaginary situations or what they think 'most people' who do, for example, when faced with a neighbour or work colleague with mental illness. Although such research is useful, as discussed above it has emphasised what 'normal' people say rather than the actual experiences of people with mental illnesses themselves and does not assess behaviour and discrimination directly. In short, most work on stigma has been beside the point.

2.2.5 Structural Discrimination

In 1999 the UK government published the Macpherson (Macpherson 1999) report which described the failings of the London Metropolitan Police in handling the investigation of the racially motivated murder of the black teenager, Stephen Lawrence. The racist gang responsible has not been brought to justice. Now widely regarded as a seminal work in the fight against racial discrimination in the UK, the Macpherson report tied police failings to the issue of institutional racism in the police, defined as follows:

> The collective failure of an organisation to provide an appropriate and professional service to people because of their colour, culture, or ethnic origin. It can be seen or detected in processes, attitudes and behaviour which amount to discrimination through unwitting prejudice, ignorance, thoughtlessness and racist stereotyping which disadvantage minority ethnic people.

As a result of this there emerged a notable determination by central government to take the Macpherson recommendations seriously. As a result, the UK government has imposed targets on police authorities requiring the modernisation and diversification of the police to boost numbers of black, minority ethnic (BME) officers. In addition, the Home Office 'Strength in Diversity' agenda is actively pursued within all strands of policing policy development, and the Macpherson recommendations continue to be reproduced in official documents and their implementation monitored in all aspects of policing. The Macpherson ethos has permeated other public institutions, and has been enshrined in legislation, meaning that the identification and elimination of institutional racism as defined by this report is now firmly on the agenda of every UK government department and of every UK public body.

The lessons learned from the successes of this movement may be applied to the concept of 'structural discrimination' identified by Corrigan et al. (2004). The authors argue that an understanding of macro-societal determinants of stigma are just as important as the individual experience of the person with mental illness. 'Structural discrimination' is defined as:

> The policies of private and governmental institutions that intentionally restrict the opportunities of people with mental illness and policies of institutions that yield unintended consequences that hinder the options of people with mental illness (Corrigan et al. 2004).

This definition is similar in many ways to Macpherson's definition of institutional racism. Corrigan et al. argue for further methodological and conceptual work to understand structural discrimination, and this will be of undoubted benefit. Additionally, we argue that by publicising structural discrimination in bold terms, and by relating macro-level analyses to the plight of individuals (in the way that was achieved in the wake of the Lawrence Inquiry), the stigma agenda may achieve similar prominence. People affected by stigma and mental illness are equal if not greater in number than those affected by institutional racism. We can learn important lessons from the successes of the UK race relations struggle. Coupled with the impact of powerful domestic and European anti-discrimination law and proven governmental goodwill and resources (eg the 'See Me' anti-stigma Campaign in Scotland), the European anti-stigma movement has the ingredients to empower governments and institutions to tackle the problem of stigma and mental illness head on.

2.3 Global Patterns of Stigma

Does stigma vary between countries and cultures? The evidence here is stronger, but still frustratingly patchy (Littlewood 2004). Although studies on stigma and mental illness have been carried out in many countries, few have been comparison of two or more places, or have included non-Western nations (Fabrega 1991b).

One study in Africa described attitudes to mentally ill people in rural sites in Ethiopia. Among almost 200 relatives of people with diagnoses of schizophrenia or mood disorders, 75% said that they had experienced stigma due to the presence of mental illness in the family, and a third (37%) wanted to conceal the fact that a relative was ill. Most family members (65%) said that praying was their preferred of treating the condition (Shibre et al. 2001). Among the general population in Ethiopia schizophrenia was judged to be the most severe problem, and talkativeness, aggression and strange behaviour were rated as the most common symptoms of mental illness (Alem et al. 1999).

A survey was conducted in South Africa of over 600 members of the public on their knowledge and attitudes towards people with mental illness (Minde 1976, Stein et al. 1997, Hugo et al. 2003). Different vignettes, portraying depression, schizophrenia, panic disorder or substance misuse were presented to each person. Most thought that these conditions were either related to stress or to a lack of willpower, rather than seeing them as medical disorders (Cheetham and Cheetham 1976). Similar work in Turkey (Ozmen et al. 2004), and in Siberia and Mongolia

(Dietrich et al. 2004) suggests that people in such countries may more ready to make the individual responsible for his or her mental illness and less willing to grant the benefits of the sick role.

Although most of the published work on stigma is by authors in the USA and Canada (Corrigan 2005a, Corrigan et al. 2003, Estroff et al. 2004, Link et al. 2004), there are also a few reports from elsewhere in the Americas and in the Caribbean (Villares and Sartorius 2003). In a review of studies from Argentina, Brazil, Dominica, Mexico, and Nicaragua, mainly from urban sites, a number of common themes emerged. The conditions most often rated as 'mental illnesses' were the psychotic disorder, especially schizophrenia. People with higher levels of education tended to have more favourable attitudes to people with mental illness. Alcoholism was considered to be the most common type of mental disorder. Most people thought that a health professional needs to be consulted by people with mental illnesses (de Toledo Piza and Blay 2004).

Some research has also been published concerning the question of stigma towards mentally ill people in Asian countries and cultures (Fabrega 1991a, Leong and Lau 2001, Ng 1997). Within China (Kleinman and Mechanic 1979), a large scale survey was undertaken of over 600 people with a diagnosis of schizophrenia and over 900 family members (Phillips et al. 2002). Over half of the family members said that stigma had had an important effect on them and their family, and levels of stigmatisation were higher in urban areas and for people who were more highly educated.

Within the field of stigma research it is clear that schizophrenia is the primary focus of interest. It is remarkable that there are almost no studies, for example, on bipolar disorder and stigma. A comparison of attitudes to schizophrenia was undertaken in England and Hong Kong. As predicted, the Chinese respondents expressed more negative attitudes and beliefs about schizophrenia, and preferred a more social model to explain its causation. In both countries most participants, whatever their educational level, showed great ignorance about this condition (Furnham and Chan 2004). This may be why most of population in Hong Kong are very concerned about their mental health and hold rather negative views about mentally ill people (Chou et al. 1996). Less favourable attitudes were common in those with less direct personal contact with people with mental illness (as in most Western studies), and by women (the opposite of what has been found in many Western reports) (Chung et al. 2001).

Although rather less research on stigma has been conducted in India, one study found that among relatives of people with schizophrenia in Chennai (Madras) in Southern India, their main concerns were: effects on marital prospects, fear of rejection by neighbours, and the need to hide the condition from others. Higher levels of stigma were reported by women and by younger people with the condition (Thara and Srinivasan 2000). Women who have mental illness appear to be at a particular disadvantage in India. If they are divorced, sometimes related to concerns about heredity (Raguram et al. 2004), then they often receive no financial support from their former husbands, and they and their families experience intense distress from the additional stigma of being separated or divorced (Thara et al. 2003).

Mental illnesses are seen in Japan to reflect a loss of control, and so are not subject to the force of will power, both of which lead to a sense of shame (Desapriya and Nobutada 2002, Hasui et al. 2000, Sugiura et al. 2000). Although it is tempting to generalise about the degree of stigma in different countries, reality may not allow such simplifications. A comparison of attitudes to mentally ill people in Japan and Bali, for example showed that views towards people with schizophrenia were less favourable in Japan, but that people with depression and obsessive-compulsive disorder were seen to be less acceptable in Bali (Kurihara et al. 2000).

Different countries do tend to share in common a high level of ignorance and misinformation about mental illnesses. A survey of teachers' opinions in Japan and Taiwan showed that relatively few could describe the main features of schizophrenia with any accuracy. The general profile of knowledge, beliefs and attitudes was similar to that found in most Western countries, although the degree of social rejection was somewhat greater in Japan (Kurumatani et al. 2004). In 2002, the Japanese Society of Psychiatry and Neurology at the request of a group of patients' families, reviewed the Japanese name for schizophrenia – 'Seishin Bunretsu Byo' ('mind-split-disease'). The group asked for a new name which was less stigmatising and more accurate, and which represented the change in understanding of schizophrenia from a chronic incurable disease to one in which recovery was possible, and which acknowledged a bio-psycho-social approach to the disease. Accordingly, the name was successfully changed to 'Togo Shitcho Sho' ('Integration Disorder'), which within months was widely accepted amongst Japanese mental health professionals, who indicated that it became easier to explain the diagnosis to patients, establish a therapeutic relationship, improve concordance and reduce stigma (Sato 2009). Moreover, changing of the name of schizophrenia in Japan has led to indications from service users and their families that they feel able to discuss the illness more openly as a result.

What is known from the English language literature of stigma in Islamic communities? Despite earlier indications that the intensity of stigma may be relatively low here (Fabrega 1991a), detailed studies indicate that, on balance, it is no less than we have seen described elsewhere (Al-Krenawi et al. 2000, 2001, Cinnirella and Loewenthal 1999, Karim et al. 2004). A study of family members in Morocco found that 76% had no knowledge about the condition, and many considered it chronic (80%), handicapping (48%), incurable (39%), or linked with sorcery (25%). Most said that they had 'hard lives' because of the diagnosis (Kadri et al. 2004). Turning to religious authority figures is reported to be common in some Moslem countries (Al-Krenawi et al. 2004, Loewenthal et al. 2001). Some studies have found that direct personal contact was not associated with more favourable attitudes to people with mental illness (Arkar and Eker 1992, 1994), especially where behaviour is seen to threaten the social fabric of the community (Coker 2005, Ozmen et al. 2004).

A recent global study used the Discrimination and Stigma Scale (DISC) in a cross-sectional survey in 27 countries using language-equivalent versions of the instrument in face to face interviews between research staff and 732 participants with a clinical diagnosis of schizophrenia (Thornicroft et al. 2009). The most frequently occurring areas of negative experienced discrimination were: making or keeping friends (47%), discrimination by family members (43%), keeping a job (29%); finding a job: (29%), and intimate or sexual relationships (29%). Positive experienced discrimination was rare. Anticipated discrimination was common for: applying for work or training or education (64%); looking for a close relationship (55%), and 72% felt the need to conceal the diagnosis. Anticipated discrimination occurred more often than experienced discrimination. This study suggests that rates of experienced discrimination are relatively high and consistent across countries. For two of the most important domains (work and personal relationships) anticipated discrimination occurs in the absence of experienced discrimination in over a third of participants. This has important implications: disability discrimination laws may not be effective without also developing interventions to reduce anticipated discrimination, for example by enhancing the self-esteem of people with mental illness, so that they will be more likely to apply for jobs.

Prejudice and discrimination by the public against people with mental illness are therefore common, deeply socially damaging, and are a part of more widespread stigmatisation. Stigma against people with mental illness can contribute to negative outcomes as well as perpetuating self-stigmatisation and contributing to low self-esteem. With a growing awareness about such stigma, a number of recent initiatives have been launched in the UK aiming to improve public attitudes. The Royal College of Psychiatrists 'Changing Minds' campaign in England ran between 1998 and 2003. It advertised websites, showed campaign videos in cinemas, distributed leaflets to the general public and healthcare professionals and created reading material for young people for use in the curriculum (Crisp 2004, Crisp et al. 2004, 2005). The Scottish Government 'See Me' campaign (2002-present) has a higher profile, is better funded and more extensive. It aims to deliver specific messages to the Scottish population by using all forms of media as well as cinema advertising, outdoor posters, supporting leaflets in GP surgeries, libraries, prisons, schools, and youth groups It also has a detailed website containing interactive resources and its impact is regularly monitored and progress reported in the public domain (Dunion and Gordon 2005). The investment of public funds in government campaigns is an important step and evidence suggesting that 'See Me' may have had a positive effect on attitudes in Scotland relative to England is encouraging (Mehta et al. 2009).

However, it remains the case that addressing public 'knowledge' and 'attitudes', as discussed above, does not necessarily lead to a change in 'behaviour' and 'discrimination'. This remains an elusive goal and further work is needed to understand the complex relationships between these three elements of stigma and to identify and develop evidence-based tools and interventions with which to tackle discrimination.

2.4 Stigmatisation and Human Rights in Psychiatric Institutions

In many countries people with mental illnesses are treated in unethical ways which prevent them from exercising many of their basic human rights. It is hardly an exaggeration to say that we can estimate the value attached to people in this category quite precisely from seeing how much or how little attention is paid to ensuring that they are treated in fully humane ways (Amnesty International 2000, 2003, Baker 1993).

> All persons have the right to the best available mental heath care, which shall be part of the health and social care system (United Nations).

Countries have responsibilities towards everyone in their jurisdiction under international law. Where there is any conflict with domestic law, international law is superior, provided that it is ratified in the respective country. All of the general international human rights treaties protect the rights of people with mental illness through the principles of equality and non-discrimination, while more specific standards exist in relation to people with mental illness.

A main reference point for international human rights within the United Nations (UN) is the Universal Declaration of Human Rights 1948 (UDHR), which refers to civil, political, economic, social and cultural rights. Civil and political rights, such as the right to liberty, to a fair trial, and to vote, are set out in an internationally binding treaty, the International Covenant on Civil and Political Rights 1976 (ICCPR), which has not been ratified by only seven nations including China (United Nations). Economic, social and cultural rights, such as the rights to an adequate standard of living, the highest attainable standard of physical and mental health, and to education, are described in a second binding treaty, the International Covenant on Economic, Social and Cultural Rights 1976 (ICESCR), which has not been ratified by the USA.

The UN High Commissioner for Human Rights (OHCHR) reports to the UN General Assembly on the implementation of the rights protected by these human rights treaties. Countries which have ratified these binding treaties are then obliged under international law to guarantee to every person on their territory, without discrimination, all the rights enshrined in both (1966a, b, 1991, 1992, 1994, 2006, United Nations 1948)

The body which monitors implementation of the ICESCR is the Committee on Economic, Social and Cultural Rights (CESCR). In a special report explaining how the ICESCR relates specifically to the rights of people with disabilities, the Committee stated:

> The obligation of States parties to the Covenant to promote progressive realisation of the relevant rights to the maximum of their available resources clearly requires Governments to do much more than merely abstain from taking measures which might have a negative impact on persons with disabilities. The obligation in the case of such a vulnerable and disadvantaged group is to take positive action to reduce structural disadvantages and to give appropriate preferential treatment to people with disabilities in order to achieve the objectives of full participation and equality within society for all persons with disabilities. This almost invariably means that additional resources will need to be made available for this purpose and that a wide range of specially tailored measures will be required (United Nations).

In relation to mental illness specifically, the UN Principles for the Protection of Persons with Mental Illness and for the Improvement of Mental Health Care were adopted in 1991, and elaborate the basic rights and freedoms of people with mental illness that must be secured if states are to be in full compliance with the ICESCR. 'The Right to Mental Health' is stated in Article 12 of the ICESCR, which provides the right of everyone to the 'enjoyment of the highest attainable standard of physical and mental health', and identifies some of the measures states should take 'to achieve the full realisation of this right'.

These 'Mental Illness Principles' apply to all people with mental illness, and to all people admitted to psychiatric facilities, whether or not they are diagnosed as having a mental illness. They provide criteria for the determination of mental illness, protection of confidentiality, standards of care, the rights of people in mental health facilities, and the provision of resources. Mental Illness Principle 1 lays down the basic foundation upon which nations' obligations towards people with mental illness are built: that 'all persons with a mental illness, or who are being treated as such persons, shall be treated with humanity and respect for the inherent dignity of the human person', and 'shall have the right to exercise all civil, political, economic, social and cultural rights as recognised in the Universal Declaration of Human Rights, the International Covenant on Economic, Social and Cultural Rights, the International Covenant on Civil and Political Rights and in other relevant instruments'. It also provides that 'all persons have the right to the best available mental health care'. As the United Nations' health agency, the World Health Organisation (WHO) gives substance to the UN's understanding of what is meant by 'the best available mental health care' (World Health Organisation). More recently the Convention on the Rights of Persons with Disabilities (CRPD) has set out an integrated international framework to support the proper acknowledgement of human rights for all people with disabilities (United Nations General Assembly).

Further to these provisions, 46 member states of the Council of Europe are bound or guided by a series of arrangements (Bindman et al. 2003, Kingdon et al. 2004). These include the 1950 European Convention on Human Rights and Fundamental Freedoms (ECHR), and the European Committee for the Prevention of Torture and Inhuman or Degrading Treatment or Punishment (CPT).

Therefore there exist a series of treaties, conventions, covenants, declarations and consensus statements at the national or international level. Despite this, it is possible to synthesise the most common themes and to apply them to any particular country. Table 2.1 shows one example of this approach, as applied to countries within the UK, using 12 core principles (Thornicroft and Szmukler 2005).

The overall ethical motif that lies behind these declarations is the pursuit of justice: that people with mental illness should be treated fairly, and should receive assistance in relation to their needs, and free from unfair discrimination. This also means that people with disabilities from mental illnesses should be treated with parity in comparison to people with disabilities from physical conditions (Corrigan 2005b, Hunt 2005). The framework in Table 2.1 provides practitioners with a useful tool in the application of stigma reduction interventions in psychiatric services. For example, adherence to the 'least restrictive form of care' would favour a move towards deinstitutionalisation, and would suggest that we need to confront rather than avoid transient fluxes in public stigma that may be encountered during the transition phase to community care. 'Dignity' requires us to re-examine

Table 2.1 Fundamental principles relevant to the mental health policies and mental health laws

Principle	England		Scotland 2003	United Nations 1991 (UN)	WHO 2001	World Psychiatric Association 1996 (WPA)
	National Service Framework for Mental Health 1999 (NSFMH)	Social Exclusion Unit 2003 (SEU)				
1. Participation in care	Involve service users		Regard to past and present wishes of person.... full participation		Consumer involvement ... right to information and participation	Person should be accepted as a partner by right in therapeutic process
2. Therapeutic benefit to the individual person	Effective care	Effective care to prevent crises	Importance of providing maximum benefit to person	Right to the best available mental health care. Every person shall have the right to receive such health and social care as is appropriate to his or her health needs ... in the best interest of the person	Efficient treatment	Providing the best therapy available consistent with accepted scientific knowledge. Treatment must always be in the best interest of the Person
3. Choice of acceptable treatments	Acceptable care and choice	Genuine choices	Importance of providing appropriate services to person		Wide range of services	Allow the person to make free and informed decisions

Table 2.1 (continued)

Principle	England		Scotland 2003	United Nations 1991 (UN)	WHO 2001	World Psychiatric Association 1996 (WPA)
	National Service Framework for Mental Health 1999 (NSFMH)	Social Exclusion Unit 2003 (SEU)				
4. Non discrimination	Non discriminatory	Fair access regardless of ethnicity, gender, age or sexuality	Have regard to encouragement of equal opportunities	These principles shall be applied without discrimination of any kind	Equality and non discrimination	Fair and equal treatment of the mentally ill. Discrimination by psychiatrists on the basis of ethnicity or culture, whether directly or by aiding others, is unethical
5. Access	Accessible			Every Person shall have the right to be treated and cared for, as far as possible, in the community in which he or she lives	Local services	

Table 2.1 (continued)

Principle	England		Scotland 2003	United Nations 1991 (UN)	WHO 2001	World Psychiatric Association 1996 (WPA)
	National Service Framework for Mental Health 1999 (NSFMH)	Social Exclusion Unit 2003 (SEU)				
6. Safety	Promote safety			To protect the health or safety of the person concerned or of others, or otherwise to protect public safety, order, health or morals or the fundamental rights and freedoms of others	Physical integrity of service user	
7. Autonomy and empowerment	Independence	Maintain employment		Treatment . . . directed towards preserving and enhancing personal autonomy	Person empowerment, autonomy	Provide the Person with relevant information so as to empower
8. Appropriate family involvement in care		Social and family participation	Have regard to needs and circumstances of person's carer		Partnership with families, involvement of local community	Psychiatrist should consult with the family

Table 2.1 (continued)

Principle	England		Scotland 2003	United Nations 1991 (UN)	WHO 2001	World Psychiatric Association 1996 (WPA)
	National Service Framework for Mental Health 1999 (NSFMH)	Social Exclusion Unit 2003 (SEU)				
9. Dignity				Treated with humanity and respect for the inherent dignity of the human person	Preserve dignity	Psychiatrists to be guided primarily by the respect for persons and concern for their welfare and integrity.... to safeguard the human dignity
10. Least restrictive form of care			Have regard to minimum restriction of the freedom of the person necessary	Every person shall have the right to be treated in the least restrictive environment		Therapeutic interventions that are least restrictive to the freedom of the person
11. Advocacy			Have regard to views of person's named person, carer, guardian, welfare attorney			
12. Capacity				The person whose capacity is at issue shall be entitled to be represented by a counsel		

our approach to service users, from changing our descriptive language where, for example, 'a schizophrenic' becomes 'a person with schizophrenia'. 'Autonomy' requires practitioners to help to empower the patient to live a full and meaning-ful life and to respond in their capacity as an advocate and source of advice and information regarding employment.

2.5 Conclusion

Although there has been a realisation, for almost a century now (Bogardus 1924), that stigma against people with mental illness is both common and severe, little real progress has been made in most countries to ensure social inclusion. While there is some evidence that attitudes, especially towards people with depression, are improving (Dunion and Gordon 2005); Jorm et al. 2005), attitudes in England toward people with mental illnesses as a whole have substantially deteriorated in recent years (Department of Health). The idea of parity with people with other forms of disability, let alone with non-disabled people, is very far from the everyday experience of most people with mental illnesses.

As this chapter demonstrates, the rejection of people with mental illness is both widespread (affecting every aspect of personal, social and work life) (Dear and Wolch 1992, Estroff 1981), and often appears not to be intentional, the totality of such exclusion has been described as structural discrimination (Corrigan 2003, Corrigan et al. 2004, 2005). Just as concerted social and cultural and legal mea-sures have been necessary to begin to reverse discrimination against black and minority ethnic communities in the UK or the USA, so the same determination is now necessary to promote the social inclusion of people with mental illness within society.

To support the pursuit of these goals it is essential to understand the real experi-ences of discrimination by people with mental illness (Chamberlin 2005, Goldman et al. 1995, March 2000, Pinfold et al. 2005, Rogers et al. 1993, Wahl 1999), and their priorities for change. What this amounts to no less than the need to eradicate both direct and structural discrimination against people with mental illness. Such a co-ordinated attack on stigma depends upon recognition that mental illnesses can produce very serious impairments, for example, in concentration or memory. While they can be minimised by a favourable environment, the severity of the primary problems should not be denied or minimised. This will mean that for some peo-ple with mental illness, at least for periods when the symptoms and features of the condition are more prominent, that it is difficult to advocate on their own behalf, and even harder to join with others to campaign for their common interests. Forms of advanced statement or Joint Crisis Plans are examples of such empowerment arrangements (Flood et al. 2006, Henderson et al. 2004).

More broadly, it is clear from every chapter of this book that the social and lob-bying power of people with mental illness and their advocates is weak. Indeed one of the central paradoxes of this book is that while up to three quarters of adults know someone directly who has been affected by a mental illness, we act as if nobody knows anything (Crisp et al. 2005). This is both ignorance and denial on an astonishing scale.

There is an additional fundamental problem: in every country in the world the supply of treatment and care is less or far less than the actual need. Even in the best resourced countries fully half of all people with mental illnesses receive no effective help (Kessler et al. 2005). This fraction is at the root of the problem. Many people do not reveal their distress, even to their closest family members for fear of rejection, in anticipation of being shunned. Where they do mention their difficulties, then family members may well react without understanding or sympathy, and discourage them from embarrassing the family by telling others. And even when people do go to a primary care practitioner, most of the time their mental illness is not recognised, and even less often is it treated properly (Goldberg and Goodyer 2005). The clear need is to provide acceptable care and assistance to people with mental illness or a far greater scale than has ever been done before.

Is mental illness the strongest remaining taboo? If taboo means 'a social or religious custom placing prohibition or restriction on a particular thing or person' (Soanes and Stevenson 2003), then certainly the ways in which many people with mental illness are left in social (Dear and Wolch 1992), material (Estroff 1981), and cultural poverty (Draine et al. 2002a, b, Fanon 1963) suggest that our societies allow, if not to deliberately orchestrate (Morone 1997), a form of 'structural violence' against this category of people (Kelly 2005).

The recognition of both individual and systemic discrimination against people with mental illness is a very recent phenomenon (Corrigan 2005a, Corrigan et al. 2005, Porter 1998, Sartorius and Schulze 2005, Sayce and Curran 2006, Sayce and O'Brian 2004, World Health Organisation 2005, Shorter 1997, Goldman 2000). In the future we may well see that from this recognition grows a civil rights movement dedicated to the liberation of people with mental illness from being marginalised and treated in inhumane and unethical ways (Chamberlin 2005, Hinshaw and Cicchetti 2000, Kingdon et al. 2004, Rose 2001).

References

Al-Krenawi A, Graham JR, Dean YZ, Eltaiba N (2004) Cross-national study of attitudes towards seeking professional help: Jordan, United Arab Emirates (UAE) and Arabs in Israel. Int J Soc Psychiatry 50:102–114
Al-Krenawi A, Graham JR, Kandah J (2000) Gendered utilization differences of mental health services in Jordan. Community Ment Health J 36:501–511
Al-Krenawi A, Graham JR, Ophir M, Kandah J (2001) Ethnic and gender differences in mental health utilization: the case of Muslim Jordanian and Moroccan Jewish Israeli out-patient psychiatric patients. Int J Soc Psychiatry 47:42–54
Alem A, Jacobsson L, Araya M, Kebede D, Kullgren G (1999) How are mental disorders seen and where is help sought in a rural Ethiopian community? A key informant study in Butajira, Ethiopia. Acta Psychiatr Scand Suppl 397:40–47
Amnesty International (2000) Ethical codes and declarations relevant to the health professions. Amnesty International, London

Amnesty International (2003) Mental illness, the neglected quarter: summary report. Amnesty International, Dublin

Arkar H, Eker D (1992) Influence of having a hospitalized mentally ill member in the family on attitudes toward mental patients in Turkey. Soc Psychiatry Psychiatr Epidemiol 27:151–155

Arkar H, Eker D (1994) Effect of psychiatric labels on attitudes toward mental illness in a Turkish sample. Int J Soc Psychiatry 40:205–213

Baker D (1993) Human rights for persons with disabilities. In: Nagler M, Kemp EJ (eds) Perspectives on disability. Health Markets Research, Palo Alto, CA

Bindman J, Maingay S, Szmukler G (2003) The Human Rights Act and mental health legislation. Br J Psychiatry 182:91–94

Bogardus ES (1924) Social distance and its origins. J Appl Sociol 9:216–226

Cannan E (1895) The stigma of pauperism. Econ Rev:380–391

Chamberlin J (2005) User/consumer involvement in mental health service delivery. Epidemiol Psichiatr Soc 14(1):10–14

Cheetham WS, Cheetham RJ (1976) Concepts of mental illness amongst the rural Xhosa people in South Africa. Aust NZ J Psychiatry 10:39–45

Chou KL, Mak KY, Chung PK, Ho K (1996) Attitudes towards mental patients in Hong Kong. Int J Soc Psychiatry 42:213–219

Chung KF, Chen EY, Liu CS (2001) University students' attitudes towards mental patients and psychiatric treatment. Int J Soc Psychiatry 47:63–72

Cinnirella M, Loewenthal KM (1999) Religious and ethnic group influences on beliefs about mental illness: a qualitative interview study. Br J Med Psychol 72(Pt 4):505–524

Coker EM (2005) Selfhood and social distance: toward a cultural understanding of psychiatric stigma in Egypt. Soc Sci Med 61:920–930

Corrigan PW (2003) Towards an integrated, structural model of psychiatric rehabilitation. Psychiatr Rehabil J 26:346–358

Corrigan P (2005a) On the stigma of mental illness. American Psychological Association, Washington, DC.

Corrigan PW (2005b) Mental illness stigma as social injustice: yet another dream to be achieved. In: Corrigan PW (ed) On the stigma of mental illness. Practical strategies for research and social change. American Psychological Press, Washington, DC

Corrigan PW, Markowitz FE, Watson AC (2004) Structural levels of mental illness stigma and discrimination. Schizophr Bull 30:481–491

Corrigan P, Thompson V, Lambert D, Sangster Y, Noel JG, Campbell J (2003) Perceptions of discrimination among persons with serious mental illness. Psychiatr Serv 54:1105–1110

Corrigan PW, Watson AC, Gracia G, Slopen N, Rasinski K, Hall LL (2005) Newspaper stories as measures of structural stigma. Psychiatr Serv 56:551–556

Crisp A (2004) Every family in the land: understanding prejudice and discrimination against people with mental illness. Royal Society of Medicine Press, London.

Crisp AH, Cowan L, Hart D (2004) The college's anti-stigma campaign 1998–2003. Psychiatr Bull 28:133–136

Crisp A, Gelder MG, Goddard E, Meltzer H (2005) Stigmatization of people with mental illnesses: a follow-up study within the Changing minds campaign of the Royal College of Psychiatrists. World Psychiatry 4:106–113

Crocker J, Major B, Steele C (1998) Social stigma. In: Gilbert D, Fiske ST, Lindzey G (eds) The handbook of social psychology. McGraw-Hill, Boston, MA

Dear M, Wolch J (1992) Landscapes of despair. Princeton University Press, Princeton, NJ.

Desapriya EB, Nobutada I (2002) Stigma of mental illness in Japan. Lancet 359:1866

de Toledo Piza PE, Blay SL (2004) Community perception of mental disorders – a systematic review of Latin American and Caribbean studies. Soc Psychiatry Psychiatr Epidemiol 39: 955–961

Dietrich S, Beck M, Bujantugs B, Kenzine D, Matschinger H, Angermeyer MC (2004) The relationship between public causal beliefs and social distance toward mentally ill people. Aust NZ J Psychiatry 38:348–354

Draine J, Salzer MS, Culhane DP, Hadley TR (2002a) Poverty, social problems, and serious mental illness. Psychiatr Serv 53:899

Draine J, Salzer MS, Culhane DP, Hadley TR (2002b) Role of social disadvantage in crime, joblessness, and homelessness among persons with serious mental illness. Psychiatr Serv 53:565–573

Dunion L, Gordon L (2005) Tackling the attitude problem. The achievements to date of Scotland's 'see me' anti-stigma campaign. Ment Health Today (March):22–25

Estroff S (1981) Making it crazy: ethnography of psychiatric clients in an American community. University of California Press, Berkeley, CA.

Estroff SE, Penn DL, Toporek JR (2004) From stigma to discrimination: an analysis of community efforts to reduce the negative consequences of having a psychiatric disorder and label. Schizophr Bull 30:493–509

Fabrega H Jr. (1991a) Psychiatric stigma in non-Western societies. Compr Psychiatry 32:534–551

Fabrega H Jr. (1991b) The culture and history of psychiatric stigma in early modern and modern Western societies: a review of recent literature. Compr Psychiatry 32:97–119

Falk G (2001) Stigma: how we treat outsiders. Prometheus Books, New York, NY.

Fanon F (1963) The wretched of the Earth. Grove Press, New York, NY.

Flood C, Byford S, Henderson C, Leese M, Thornicroft G, Sutherby K, Szmukler G (2006) Joint crisis plans for people with psychosis: economic evaluation of a randomised controlled trial. BMJ 333:729

Furnham A, Chan E (2004) Lay theories of schizophrenia. A cross-cultural comparison of British and Hong Kong Chinese attitudes, attributions and beliefs. Soc Psychiatry Psychiatr Epidemiol 39:543–552

Gilman SL (1982) Seeing the insane. Wiley, New York, NY

Gilman SL (1985) Difference and pathology: stereotypes of sexuality, race and madness. Cornell University Press, Ithaca, NY.

Goffman I (1963) Stigma: notes on the management of spoiled identity. Penguin Books, Harmondsworth.

Goldberg D, Goodyer I (2005) Genesis of common mental disorders. Brunner-Routledge, London.

Goldman HH (2000) Implementing the lessons of mental health service demonstrations: human rights issues. Acta Psychiatr Scand 399:51–54

Goldman HH, Rachuba L, Van Tosh L (1995) Methods of assessing mental health consumers' preferences for housing and support services. Psychiatr Serv 46:169–172

Graves RE, Cassisi JE, Penn DL (2005) Psychophysiological evaluation of stigma towards schizophrenia. Schizophr Res 76:317–327

Hasui C, Sakamoto S, Suguira B, Kitamura T (2000) Stigmatization of mental illness in Japan: images and frequency of encounters with diagnostic categories of mental illness among medical and non-medical university students. J Psychiatry Law 28:253–266

Hayward P, Bright JA (1997) Stigma and mental illness: a review and critique. J Ment Health 6:345–354

Heatherton TF, Kleck RE, Hebl MR, Hull JG (2003) The social psychology of stigma. Guilford Press, New York, NY.

Henderson C, Flood C, Leese M, Thornicroft G, Sutherby K, Szmukler G (2004) Effect of joint crisis plans on use of compulsory treatment in psychiatry: single blind randomised controlled trial. BMJ 329:136–140

Henderson C, Thornicroft G (2009) Time to change: tackling stigma and discrimination against people with mental illness in England. Lancet. in press

Hinshaw S (2007) The mark of shame. Oxford University Press, Oxford.

Hinshaw SP, Cicchetti D (2000) Stigma and mental disorder: conceptions of illness, public attitudes, personal disclosure, and social policy. Dev Psychopathol 12:555–598

Hobbes of Malmesbury, Thomas (1657) Markes of the absurd geometry, rural language, Scottish church politics, and barbarisms of John Wallis professor of geometry and doctor of divinity. Printed for Andrew Cooke, London

Home Office 'Strength in Diversity' Agenda. http://www.homeoffice.gov.uk/documents/cons-strength-in-diverse-170904/. 7 Jan 2009.

Hugo CJ, Boshoff DE, Traut A, Zungu-Dirwayi N, Stein DJ (2003) Community attitudes toward and knowledge of mental illness in South Africa. Soc Psychiatry Psychiatr Epidemiol 38:715–719

Hunt P (2005) Economic, cultural and social rights. Report of the special Rapporteur on the right of everyone to enjoyment of the highest attainable standard of physical and mental health. Commission on Human Rights, 61st Session, Item 10 on the provisional agenda. United Nations Economic and Social Council, New York, NY

Jacoby A (1994) Felt versus enacted stigma: a concept revisited. Evidence from a study of people with epilepsy in remission. Soc Sci Med 38:269–274

Jones E, Farina A, Hastorf A, Markus H, Milller D, Scott R (1984) Social stigma: the psychology of marked to relationships. W.H. Freeman & Co, New York, NY.

Jorm AF, Christensen H, Griffiths KM (2005) The impact of beyondblue: the national depression initiative on the Australian public's recognition of depression and beliefs about treatments. Aust N Z J Psychiatry 39:248–254

Kadri N, Manoudi F, Berrada S, Moussaoui D (2004) Stigma impact on Moroccan families of patients with schizophrenia. Can J Psychiatry 49:625–629

Karim S, Saeed K, Rana MH, Mubbashar MH, Jenkins R (2004) Pakistan mental health country profile. Int Rev Psychiatry 16:83–92

Kelly BD (2005) Structural violence and schizophrenia. Soc Sci Med 61:721–730

Kessler RC, Demler O, Frank RG, Olfson M, Pincus HA, Walters EE, Wang P, Wells KB, Zaslavsky AM (2005) Prevalence and treatment of mental disorders, 1990 to 2003. N Engl J Med 352:2515–2523

King M, Dinos S, Shaw J, Watson R, Stevens S, Passetti F, Weich S, Serfaty M (2007) The stigma scale: development of a standardised measure of the stigma of mental illness. Br J Psychiatry 190:248–254

Kingdon D, Jones R, Lonnqvist J (2004) Protecting the human rights of people with mental disorder: new recommendations emerging from the Council of Europe. Br J Psychiatry 185:277–279

Kleinman A, Mechanic D (1979) Some observations of mental illness and its treatment in the People's Republic of China. J Nerv Ment Dis 167:267–274

Kurihara T, Kato M, Sakamoto S, Reverger R, Kitamura T (2000) Public attitudes towards the mentally ill: a cross-cultural study between Bali and Tokyo. Psychiatry Clin Neurosci 54:547–552

Kurumatani T, Ukawa K, Kawaguchi Y, Miyata S, Suzuki M, Ide H, Seki W, Chikamori E, Hwu HG, Liao SC, Edwards GD, Shinfuku N, Uemoto M (2004) Teachers' knowledge, beliefs and attitudes concerning schizophrenia- a cross-cultural approach in Japan and Taiwan. Soc Psychiatry Psychiatr Epidemiol 39:402–409

Leong FT, Lau AS (2001) Barriers to providing effective mental health services to Asian Americans. Ment Health Serv Res 3:201–214

Link BG, Cullen FT, Struening EL, Shrout PE, Dohrenwend BP (1989) A modified labeling theory approach in the area of mental disorders: an empirical assessment. Am Sociol Rev 54:100–123

Link BG, Phelan JC (2001) Conceptualizing stigma. Ann Rev Sociol 27:363–385

Link BG, Phelan JC, Bresnahan M, Stueve A, Pescosolido BA (1999) Public conceptions of mental illness: labels, causes, dangerousness, and social distance. Am J Public Health 89:1328–1333

Link BG, Struening EL, Rahav M, Phelan JC, Nuttbrock L (1997) On stigma and its consequences: evidence from a longitudinal study of men with dual diagnoses of mental illness and substance abuse. J Health Soc Behav 38:177–190

Link BG, Yang LH, Phelan JC, Collins PY (2004) Measuring mental illness stigma. Schizophr Bull 30:511–541

Littlewood R (1998) Cultural variation in the stigmatisation of mental illness. Lancet 352:1056–1057

Littlewood R (2004) Cultural and national aspects of stigmatisation. In: Crisp AH (ed) Every family in the land. Royal Society of Medicine, London

Littlewood R, Jadhav S, Ryder AG (2007) A cross-national study of the stigmatization of severe psychiatric illness: historical review, methodological considerations and development of the questionnaire. Transcult Psychiatry 44:171–202

Loewenthal KM, Cinnirella M, Evdoka G, Murphy P (2001) Faith conquers all? Beliefs about the role of religious factors in coping with depression among different cultural-religious groups in the UK. Br J Med Psychol 74:293–303

Macpherson W (1999) The Stephen lawrence inquiry: the report of an inquiry by Sir William Macpherson of cluny. Advised by Tom Cook, The right reverend Dr John Sentamu, Dr Richard Stone. Presented to parliament by the secretary of state for the Home Department by Command of Her Majesty, 1999. The Stationery Office.

March DT (2000) Personal accounts of consumer/survivors: insights and implications. J Clin Psychol 56:1447–1457

Mason T (2001) Stigma and social exclusion in healthcare. Routledge, London.

Mehta N, Kassam A, Leese M, Butler G, Thornicroft G (2009) Public attitudes towards people with mental illness in England and Scotland, 1994–2003. Brit J Psychiatry 194(3):278–284

Minde M (1976) History of mental health services in South Africa. Part XIII. The National Council for Mental Health. S Afr Med J 50:1452–1456

Morone JA (1997) Enemies of the people: the moral dimension to public health. J Health Polit Policy Law 22:993–1020

Ng CH (1997) The stigma of mental illness in Asian cultures. Aust NZ J Psychiatry 31:382–390

Ozmen E, Ogel K, Aker T, Sagduyu A, Tamar D, Boratav C (2004) Public attitudes to depression in urban Turkey - the influence of perceptions and causal attributions on social distance towards individuals suffering from depression. Soc Psychiatry Psychiatr Epidemiol 39:1010–1016

Phillips MR, Pearson V, Li F, Xu M, Yang L (2002) Stigma and expressed emotion: a study of people with schizophrenia and their family members in China. Br J Psychiatry 181:488–493

Pickenhagen A, Sartorius N (2002) The WPA Global Programme to reduce stigma and discrimination because of schizophrenia. World Psychiatric Association, Geneva.

Pinfold V, Byrne P, Toulmin H (2005) Challenging stigma and discrimination in communities: a focus group study identifying UK mental health service users' main campaign priorities. Int J Soc Psychiatry 51:128–138

Porter R (1998) Can the stigma of mental illness be changed? Lancet 352:1049–1050

Raguram R, Raghu TM, Vounatsou P, Weiss MG (2004) Schizophrenia and the cultural epidemiology of stigma in Bangalore, India. J Nerv Ment Dis 192:734–744

Ritsher JB, Phelan JC (2004) Internalized stigma predicts erosion of morale among psychiatric outpatients. Psychiatry Res 129:257–265

Rogers A, Pilgrim D, Lacey R (1993) Experiencing psychiatry. Users views of services. MIND, London.

Rose D (2001) Users' voices, the perspectives of mental health service users on community and hospital care. The Sainsbury Centre, London.

Rose D, Thornicroft G, Pinfold V, Kassam A (2007) 250 labels used to stigmatise people with mental illness. BMC Health Serv Res 7:97

Rusch N, Angermeyer MC, Corrigan PW (2005) The stigma of mental illness: concepts, forms, and consequences. Psychiatr Prax 32:221–232

Sartorius N, Schulze H (2005) Reducing the stigma of mental illness. A report from a Global Programme of the World Psychiatric Association. Cambridge University Press, Cambridge.

Sato M (2009) Renaming schizophrenia: a Japanese perspective. World Psychiatry 5:53–55

Sayce L, Curran C (2006) Tackling social exclusion across Europe. In: Knapp M, McDaid D, Mossialos E, Thornicroft G (eds) Mental health policy and practice across Europe. The future direction of mental health care. Open University Press, Milton Keynes

Sayce L, O'Brian N (2004) The future of equality and human rights in Britain-opportunities and risks for disabled people. Disabil Soc 19:663–667

Scambler G (1998) Stigma and disease: changing paradigms. Lancet 352:1054–1055
Shibre T, Negash A, Kullgren G, Kebede D, Alem A, Fekadu A, Fekadu D, Madhin G, Jacobsson
    L (2001) Perception of stigma among family members of individuals with schizophrenia and
    major affective disorders in rural Ethiopia. Soc Psychiatry Psychiatr Epidemiol 36:299–303
Shorter E (1997) A history of psychiatry. Wiley, New York, NY.
Smith M (2002) Stigma. Adv Psychiatr Treat 8:317–325
Soanes C, Stevenson A (2003) Concise oxford English dictionary. Oxford University Press,
    Oxford.
Stein DJ, Wessels C, Van Kradenberg J, Emsley RA (1997) The mental health information centre
    of South Africa: a report of the first 500 calls. Cent Afr J Med 43:244–246
Sugiura T, Sakamoto S, Kijima N, Kitamura F, Kitamura T (2000) Stigmatizing perception of
    mental illness by Japanese students: comparison of different psychiatric disorders. J Nerv Ment
    Dis 188:239–242
Thara R, Kamath S, Kumar S (2003) Women with schizophrenia and broken marriages–doubly
    disadvantaged? Part II: family perspective. Int J Soc Psychiatry 49:233–240
Thara R, Srinivasan TN (2000) How stigmatising is schizophrenia in India? Int. J Soc Psychiatry
    46:135–141
Thornicroft G (2006) Shunned: discrimination against people with mental illness. Oxford
    University Press, Oxford.
Thornicroft G (2007) Most people with mental illness are not treated. Lancet 370:807–808
Thornicroft G, Brohan E, Rose D, Sartorius N, Leese M (2009) Global pattern of experienced and
    anticipated discrimination against people with schizophrenia: a cross-sectional survey. Lancet
    373:408–415
Thornicroft G, Szmukler G (2005) The draft mental health bill in England: without principles.
    Psychiatr Bull 29:244–247
United Nations General Assembly (2006) Convention on the rights of persons with disabilities.
    United Nations, New York, NY
United Nations (1948) Universal declaration of human rights. Adopted and proclaimed by the UN
    General Assembly Resolution 217A (III) of 10 December 1948. United Nations, New York,
    NY
United Nations (1966a) International covenant on civil and political rights. Adopted by the
    UN General Assembly Resolution 2200A (XXI) of 16 December 1966. United Nations
    (http://www.ohchr.org/english/countries/ratification/4.htm), New York, NY
United Nations (1966b) International covenant on economic, social and cultural rights. Adopted by
    UN General Assembly Resolution 2200A (XXI) of 16 Dec 1966. United Nations, New York,
    NY
United Nations (1991) UN Resolution 46/119 on the protection of persons with mental illness and
    the improvement of mental health care, adopted by the General Assembly on 17 Dec 1991.
    United Nations, New York, NY
United Nations (1992) UN Principles for the protection of persons with mental illness and for the
    improvement of mental health care. Adopted by UN General Assembly Resolution 46/119 of
    18 Feb 1992. United Nations, New York, NY
United Nations (1994) United Nations. Persons with disabilities. General comment number 5
    (Eleventh Session 1994). UN Doc E/C 12/1994/13. UN Committee on Economic, Social and
    Cultural Rights. United Nations, New York, NY
Vaughn G (2004) Like minds, like mine. In: Saxena S, Garrison P (eds) Mental health promotion:
    case studies from countries. World Health Organisation, Geneva
Villares C, Sartorius N (2003) Challenging the stigma of schizophrenia. Rev Bras Psiquiatr 25:1–2
Wahl OF (1999) Mental health consumers' experience of stigma. Schizophr Bull 25:
    467–478
Weiss MG, Jadhav S, Raguram R, Vaunatsou P, Littlewood L (2001) Psychiatric stigma across
    cultures: local validation in Bangalore and London. Anthropol Med 8:71–87
World Health Organisation (2005) WHO resource book on mental health, human rights and
    legislation. World Health Organisation, Geneva

Chapter 3
Economical Framework of Psychiatric Care

Christina M. van der Feltz-Cornelis

Contents

C.M. van der Feltz-Cornelis (✉)
Department of Developmental, Clinical and Crosscultural Psychology, Tilburg University,
PO Box 90153, 5000 LE Tilburg, The Netherlands
e-mail: c.m.vdrfeltz@uvt.nl

H. Helmchen, N. Sartorius (eds.), *Ethics in Psychiatry*, International Library
of Ethics, Law, and the New Medicine 45, DOI 10.1007/978-90-481-8721-8_3,
© Springer Science+Business Media B.V. 2010

Abbreviations

APA   American Psychiatric Association
CIOMS   Council of International Organizations for Medical Science
DRGs   Diagnosis Related Groups
DTC   Direct-To-Consumer
EBM   Evidence Based Medicine
GP   General Practitioner
ICER   Incremental Cost-Effectiveness Ratio
iMTA   Institute of Medical Technology Assessment
NICE   National Institute of Clinical Excellence
QALY   Quality Adjusted Life Year

3.1 Introduction

In this chapter, the principles of an ethical framework for the mentally ill and central ethical and professional practices in psychiatric care will be discussed. The economical framework, the principle of equity as well as epidemiological, medical and societal consequences of increased welfare and deceleration of mortality will be described. The role of Evidence Based Medicine and policies concerning cost effectiveness of mental health care will be alluded in terms of the consequences for psychiatric patients and mental health care. Moral implications of managed care and of the influence of pharmaceutical companies and governmental policies will be described. Ethical implications of new developments in mental health care such as E-mental health and collaborative care, new treatment forms in which the autonomy of the patient is enhanced, will be indicated. Finally, possible solutions on micro-, meso- and macrolevel will be suggested.

3.2 Ethical Framework for the Mentally Ill

In the Hippocratic tradition, the two fundamental ethical principles for medicine were to do good and to avoid evil (Primeau 2002). In contemporary Medicine, these ethical principles have been extended into four principles: autonomy of the patient, beneficence and non-maleficence of the physician, and justice, and described as such in the Belmont Report (National Commission for the Protection of Human Subjects of Biomedical and Behavioral Research 1978) (Fig. 3.1).

Fig. 3.1 Four ethical principles (Belmont Report) (National Commission for the Protection of Human Subjects of Biomedical and Behavioral Research 1978)

Autonomy of the patient
Beneficence of the physician
Non-maleficence of the physician
Justice

A main ethical tradition in Western thought relevant to psychiatry is the teleological tradition, expressed in terms of responsibility and consequences of actions and 'moral imagination', or the capacity to empathize with others, is a main aspect of this tradition (Lolas 1996). Such a capacity is needed in every physician, and it has been suggested that it might be especially challenged in the practice of psychiatry as 'psychiatrists can be confronted with patients that do not express the wish for psychiatric treatment although indications are that this might be their real, though unrecognized, need'(Lolas 2006). Another important European tradition of ethical reasoning is the deontological one, in which actions are taken because they are intrinsically good, without taking into account if the net benefit of the action will be resulting in more good over evil; the latter would have been an utilitaristic line of thought. In other words, ethical decisions play a role in everyday psychiatric care – and they are increasingly influenced by its economical framework.

If ethical decisions would be taken that would inevitably result in an impossibility to sustain the resulting medical treatment because of lack of economic means, this would result in non-ethical injustice in the division of means. Therefore, viability and economic sustainability have ethical implications in themselves. However, ethical decisions should be taken considering not only their viability and economic sustainability. They should be based primarily upon sound moral values and their acceptability in a given cultural context. For example, the concept of individual autonomy may be interpreted differently in urbanized Western or non-Western society; in the latter, the extended family or the clan may be of great influence as compared to the more individualistic Western society. Also, the concept of justice may be interpreted differently in different cultures, be it under influence of certain religious convictions or otherwise. Ethical considerations should pay attention to such differences.

Several Medical Associations as well as Psychiatric Associations published a Code of Ethics (American Medical Association 2008, American Psychiatric Association 2001, Canadian Medical Association 1996, Canadian Psychiatric Association 1984) (see Chapter 5) in which the physician–patient relationship is considered to be at the heart of medical and psychiatric practice. In our society, this relationship has been conceptualized as 'a consensual agreement between two autonomous individuals who are free to enter, sustain, or discontinue the relationship unconstrained by discrimination, coercion, or fear of physician abandonment' (Stanford 2008). Central ethical and professional practices in psychiatric care as described by the APA (American Psychiatric Association 2001) are shown in Fig. 3.2.

Confidentiality	Informed Consent
Honesty and Trust	Decision-making capacity
Non-participation in fraud	Involuntary psychiatric treatment

Fig. 3.2 Central ethical and professional practices in psychiatric care (American Psychiatric Association 2001)

3.3 The Economical Framework of Psychiatry

In this post-modern, post-communist, globalized world, high technical medical care becomes potentially available for bigger and bigger populations. Wide scale vaccination programs and better welfare have diminished infant mortality in many countries and people have the possibility to enjoy a long and healthy life more than ever. Also, as a result of better life circumstances and medical care, mortality rates decline. Lynch et al. found that amongst adults aged 20–105 for 1968–1992 in US government data, mortality deceleration increases over time, and this decline in mortality rates over time results in decompression of mortality: in other words, more diseases contribute to the occurrence of mortality than before (Lynch and Brown 2001). As a result, prevalence of welfare related diseases such as Diabetes Mellitus and cardiovascular disease together with an increasingly elderly population put a strain on society, demanding more and more in terms of medical resources (Murray and Lopez 1997). Associated with this deceleration of mortality, mental illnesses have become prominent factors contributing to higher societal burden of disease.

3.3.1 Four Models for Access to Mental Health Care

Entrance to health care has been organised in different ways, providing a role for primary care as *gatekeeper*, or providing other ways of access. Four such main models to provide medical care have been available. (I) One model is a primary care model such as the individual care model in which a Primary Care Practitioner or General Practitioner develops an individual long-term relationship with his patients, and refers them to general or mental health care if needed. (II) Another model, as compared to this primary care as gatekeeper model, is a *service model* in which physicians, often medical specialists, work on a shift basis in clinics, seeing all patients coming in at a certain day, without long-term relationship and without clear gatekeeper mechanism. The former often has aspects of a private practice of which the physician is an employer; the second can be a cooperation of self employed medical specialists, but more often, this is a Medical Service Centre in which physicians are hired as employees and do not have responsibility for the management or organisation. (III) A third model is the model of the Community-based Mental Health Service, provided by National or Regional Health Services. In this model, adopted in Italy and other countries such as the UK, patients can directly access mental health services, also without previous contact with GPs. After the first contact they are usually followed by a multidisciplinary team which is responsible for the pathway of care (Tansella et al. 2006). (IV) A fourth model of access to care is access by the insurance company, that delivers care to their insured patients in need more or less directly, as for example in *managed care*. In managed care, an insurance company provides medical care for employees of a certain company that buys medical care for its employees. In several countries, managed care has been developed with

Table 3.1   Annual cost in billions of dollars (1990 Data) (California Psychiatric Association 1993)

	Cardiovascular diseases	Mental disorders
Direct costs	$85	$67
Indirect costs	$75	$75
Other related costs	$0	$6
Total costs	$160	$148

clear cost-reduction incentives. Indications for treatment are provided by administrative personnel; choice of treatment and physician is not free for the patient, or there may be reduction of personal continuity in patient care. Cost containment or cost reduction and limited access to care are rule; some insurance companies do not provide care for mental disorder or addiction, or they provide unequal allocation of resources, e.g. with regard to antidementive or neuroleptic drugs, that may be effective but expensive, although algorithms for use of atypical antipsychotics have been tested for cost effectiveness and recommended as evidence based in the UK and by the US Joint Commission (Davies et al. 2008, Goren et al. 2008). An illustration may be helpful. According to a report of the National Advisor Mental Health Council (California Psychiatric Association 1993), based on 1990 data, the incidence of mental illness is comparable to cardiovascular disorders, with annual incidence rates of 22% versus 20% of the US adult population in a given year. The direct annual cost of treating all mental disorders is $67 billion, 40% of which is spent on treating the severely mentally ill, who constitute 2.8% of the population. The direct annual cost of treating all cardiovascular disease is $85 billion, as shown in Table 3.1.

According to the Californian Psychiatric Association, given the advances in psychiatric treatments, many of these dollars might be saved by providing access to needed services. The existence of effective treatments is only relevant to those who can obtain them. In managed care, patients with severe mental illness and their families often find that the appropriate treatment is inaccessible because they lack insurance or the coverage is inadequate for mental illness (California Psychiatric Association 1993). Socially weak patients with chronic psychiatric disorders, such as schizophrenia, especially suffer from such restrictions of solidarity financing of psychiatric services. The ethical conclusion is that managed care may violate the autonomy as well as the beneficence of patients particularly of those with mental disorders.

3.3.2 The Birth of Evidence Based Psychiatry

The burden of mental disorder and the need for medical care for these disorders is high. Also, greater possibilities and therefore greater demand for treatment have come about. The question has risen how the costs of treatment could be contained in order to be able to provide as many patients as possible with the medical care they need. This has placed physicians in a position where they had to account for the effectiveness of their treatments, as treatment without clear benefits or cure risked to be no longer financed. In this respect, the case of Osheroff against Chestnut Lodge can be considered a case in point. Osheroff was a medical specialist who suffered from serious recurrent major depressive disorder and applied for treatment in Chestnut Lodge, a private clinic renowned for the quality of its psychodynamic treatment along analytic lines. He was diagnosed not only with depressive disorder, but also with personality disorder, for which he received long-term psychodynamic treatment. He received no psychopharmacological treatment, although his Major Depressive Disorder deteriorated during treatment. The patient developed psychotic symptoms, and suffered serious consequences in the professional and personal realm. After several years, he received treatment with antidepressants elsewhere, and the Major Depressive Disorder that had haunted him for years went into remission. Osheroff filed suit against Chestnut Lodge for damages suffered by him as Chestnut Lodge had failed to offer him appropriate treatment for his Major Depressive Disorder, although they had diagnosed him as suffering from this condition. A lawsuit entailed in which Osheroff claimed that Chestnut Lodge had withheld him evidence based treatment; finally the law suit was settled out of court. The case generated widespread discussion. It was argued that 'This case involved the proposed right of the patient to effective treatment and that treatments whose efficacy has been demonstrated have priority over treatments whose efficacy has not been established.' (Klerman 1990) Although this was disputed (Stone 1990), this case catalyzed the development of Evidence Based Medicine (EBM) in the field of psychiatry. EBM is the conscientious, explicit and thorough application of state of the art use of 'best evidence' for treatment decisions for patients. In clinical practice, this means the integration of individual clinical expertise with external scientific evidence (Buchanan et al. 1997). EBM resulted in formulation of Guidelines for evidence based treatment of many medical diseases as well as mental disorders. However, the Osheroff debate continues to bother psychiatrists, and the criticism, similar to criticism of EBM, goes that 'there is an apparent dichotomy between knowledge derived from a reductionist scientific method, as manifest in evidence-based medicine, and that of a narrative form of knowledge derived from clinical experience, and that scientific knowledge dominates over narrative knowledge in psychiatry, which could lead to annihilation of all forms of knowledge other than science' (Sackett et al. 1996). Also, it is considered a possible pitfall of EBM guidelines that the physician may blindly follow the guideline without discussing treatment options with the patient; however, in the classical description of EBM, discussing treatment options with the patient and choosing treatment with consent of the patient is considered an essential part of the process of EBM. A reaction to the Osheroff case as well as to this criticism of guidelines, is emphasis on the process of negotiating a treatment plan with the patient, and asking his informed consent for that (Malcolm 1986). This was felt to be an important way to improve treatment of patients.

3.3.3 Evidence Based Guidelines

From an economical point of view, EBM and the development of medical guidelines was of great importance for the notion of cost-effectiveness of medical treatment. In the guidelines, only Evidence Based treatment modes were acceptable, which meant that they should be based on research proving that the treatment was more effective than placebo or than treatment as usual; or on consensus of experts in case of lack of such studies. Also, the Guidelines should take the economical perspective into account. An example of such Guideline development is described by the National Institute of Clinical Excellence (NICE). In order to formulate an advice, judgments have to be made both about what is good and bad in the available science (scientific value judgments) and for society (social value judgments). If a new treatment strategy turns out to be more effective but also more expensive than current standard practice, it has to be decided how much health increase can be expected to arise from the increase in costs. This so-called incremental cost effectiveness ratio is generally expressed in the cost per quality adjusted life year (QALY) (Rawlins and Culyer 2004). Based on guidelines developed this way, policies can be developed aimed at making appropriate medical treatments available to the patients needing them. This is done according to the *principle of justice*: patients fulfilling clear cut diagnostic criteria that require this Evidence Based Treatment should receive the treatment at the least possible cost, in order to enable delivery of medical services to the population as a whole.

Sometimes the conclusion that a certain treatment should be made available is not as clear cut as desirable. For example, if a treatment was not only effective but also expensive, should this treatment be provided to patients or not? In order to solve such questions, the incremental cost-effectiveness ratio (ICER) can be used as shown in Fig. 3.3 (Rawlins and Culyer 2004). As indicated by Rawlins and Culyer (2004) treatments with ratios to the left of A would generally be regarded as cost effective. They should therefore be provided if an indication exists; not providing them would be unethical. Treatments with ratios to the right of B would be considered ineffective and expensive and therefore it would not be ethical to provide them, as this would deny other patients with different conditions access to other, more cost effective treatments (Rawlins and Culyer 2004). Figure 3.3 is shown below.

Fig. 3.3 Probability of rejection on grounds of cost effectiveness (Rawlins and Culyer 2004). Figure shown with permission of BMJ Publishing Group

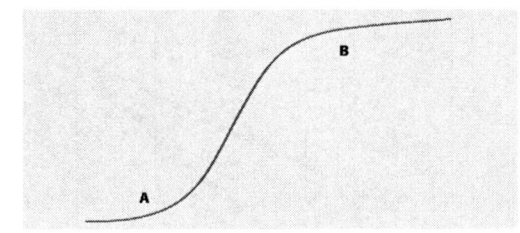

Increasing cost/QALY (log scale)

3.3.4 Policy Decisions Based on Cost Effectiveness

Another way to picture the ICER and related policy decisions of a new over an old treatment option is the cost-effectiveness plane, as shown in Fig. 3.4. This is discussed thoroughly by Klok and Postma (2004). A reflection of this discussion is given here. The cost-effectiveness plane consists of four quadrants: the south east quadrant shows a new treatment that is more effective and less expensive compared to the current treatment; policies aimed at providing this treatment would be ethically sound. The North-West Quadrant shows treatments that are less effective and more costly than current treatment; policies aimed at providing this treatment would be unethical. The North East quadrant shows treatments that are more effective but more costly. Treatments in this quadrant may need ethical consideration of treatment policy. For example, in vitro fertilisation is effective but expensive. As this treatment can provide people great happiness and satisfaction if it results in pregnancy and childbirth, but is very expensive as well, in many countries it is provided up to a certain number of trials in order to give couples a reasonable chance without losing too many resources for medical treatment. A similar case can be made for currently available antidementive drugs, which are evidence based effective, but costly; however, they spare the patient and his family a lot of suffering and thus contribute greatly to their quality of life. The last quadrant is the South West quadrant, in which the new treatment is less effective and less expensive than the old treatment; it would be unethical to choose such a treatment. Unfortunately, sometimes ICERS end up in several quadrants and in such a case it may be complex to distinguish and to describe the desired policies for a new treatment. Figure 3.4 is shown below.

Policy decisions as described above are often guided by price per QALY. QALY, or Quality Adjusted Life Year, is a measure of disease burden, including both the quality and the quantity of a life lived. The QALY is based on the number of years of

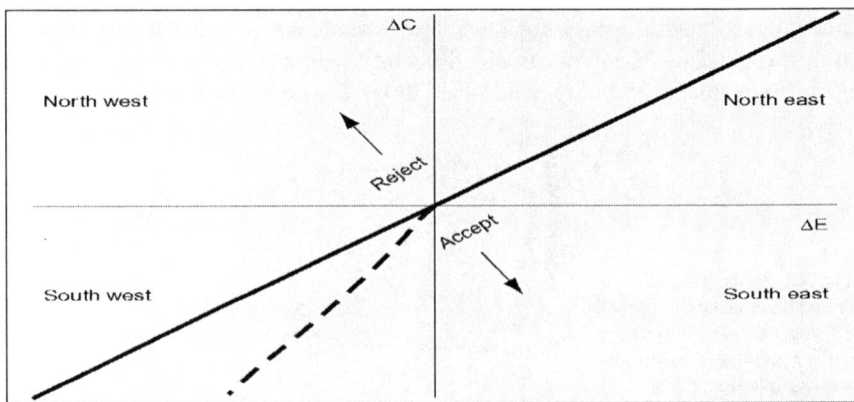

Fig. 3.4 Cost effectiveness plane and quadrants (Klok and Postma 2004). Legenda: $\Delta C = \Delta Cost$; $\Delta E = \Delta Effectiveness$. Figure shown with permission of Authors and Editor

life that would be added by the intervention. Each year in perfect health is assigned the value of 1.0 down to a value of 0.0 for death. If the extra years would not be lived in full health, for example if the patient would lose a limb, or be depressed, then the extra life-years are given a value between 0 and 1 to account for this. If, due to an intervention, a formerly depressed person would gain 1 year without depressive days, the value of this year gained may be balanced by a price of the intervention. A price of €20.000 per QALY per life year gained is considered an acceptable price for the Netherlands; in the UK it is £25,000–£30,000 per QALY and in the USA it is $50,000 per QALY (Klok and Postma 2004, Buitenlands nieuws 2008). Obviously the price per QALY may differ between countries; this is a similar finding as the finding that a statistical life has different values in different continents with ranges from $0.65 million for South Korea to $8.5 million for Japan (Miller 2000). Also, age may play a role in the valuation of a statistical life (Johansson 2006). Laupacis et al. denote several grades of recommendation of treatment interventions depending on the price per QALY, with a range from $20,000 to more than $100,000. They emphasize, however, that issues other than cost effectiveness, such as ethical and political considerations, affect the implementation of a new technology (Laupacis et al. 1992). Therefore, policy decisions based on ICERs will always depend on national and cultural context and thus will be impossible to generalize to a global level.

3.4 Moral Implications and Ethical Pitfalls: Ethical Aspects of Standardisation of Psychiatric Care

3.4.1 Equity Versus Restriction of Resources for the Mentally Ill

Economic and financial influences are driving forces of this change of medicine including psychiatry, and as a result this standardisation of psychiatric care may conflict with its necessary individualisation and of the ethical principles mentioned above. Making policy decisions according to cost effectiveness of application of Evidence Based Guidelines can originate from an ethical principle of the division of means for the benefit of patients according to principles of justice, or so-called *equity of funding*. In this vein, health care systems have been seeking to allocate their limited funds rationally in order to secure equal opportunity of access for equal need. However, in several countries economic considerations led to reduction of beds and all-in price regulations such as in the Diagnosis Related Groups (DRG) System. For example, although it would be reasonable, after many years of refinement of such costing systems, for hospitals to expect that large year to year variations in weighting for certain diagnostic categories would only occur under exceptional circumstances, a review of DRG weights in Australia demonstrated that some DRGs were significantly discounted compared to the previous year, leading to shortage of beds (AMA Victoria 2008). Such policies simply lead to non-availability of required treatment, which is unethical. In such a case, the presumption that providing treatment according to the principles of cost-effectiveness will guarantee patients in need of care access to this care cannot be held, which is not according to the ethical principle of justice. In an Editorial on the legal and ethical impact of psychiatric DRGs, Morgenlander and Greenwald argue that the allocation of health care resources reflects the values of those doing the allocation (see Chapter 2), and unfortunately equity of funding, in particular for mentally ill patients, is not always a cherished value principle (Morgenlander and Greenwald 1985).

3.4.2 Guideline Versus Autonomy of the Patient

Another pitfall is that the use of guidelines diminishes the autonomy of patients; the patient will be offered treatment according to the guideline, whereas guidelines are supposed to take economic aspects into account, in other words to start with the cheapest, or most efficient, kind of treatment. In a recent report of the Institute of Medical Technology Assessment (iMTA) for the Dutch Ministry of Health, such suggestions were made for the use of antidepressants in anxiety and depressive disorder, whereas in the Guidelines for these disorders, the choice was left open to be made based on patient and physician preference (Swan Tan et al. 2008). If economic considerations would lead to enforcement of the cheapest treatment option as first or only treatment option, this may counter the development of patient preferences as the leading principle for treatment in the framework of medical indication.

Furthermore, although there is a tendency that the patient develops features of a self-determined client, such as the patient-general practitioner Mayer who died of pancreatic cancer and published his views on the cost-containing policies concerning his case (Mayer 2008), there are limits to this development. Some patients are able to be critical consumers, but others may be too ill for that. For these patients with serious mental illness, the concept of free will is difficult to sustain in view of neurological or psychiatric limitations, a reality that has been described in terms of neurological determinism (Ledermann 1982). Governmental actions aimed at diminishing options for choice of treatment for patients, as psychiatrists and mental health institutions are forced to work under managed care conditions with limited resources, are grossly unethical in view of the limited possibilities of these patients to alter their fate or to negotiate appropriate care. Such patients may end up at the lowest part of society, as homeless mentally ill, without insurance and without possibility to work and come out of a vicious destructive cycle. Different cultures have different solutions; in several European countries, societal institutions provide basic care for these patients, provide shelter and medical care and help patients improving their financial and social predicament, thus enabling them to resume functioning at a respectable level of society. In non-western societies, the family or clan may play this role. In the USA, such structures do not exist sufficiently and the homeless mentally ill form a substantial societal, medical and personal problem. No indication exists that this situation would have improved since the nineties. Data on the percentage of mentally ill amongst the homeless are somewhat scattered, but on a more regional level, data are available. For example, in the state of Ohio, the prevalence of mental illness in the homeless is reported to be increased (Lazarus and Sharfstein 2002), and in a fact sheet, the National coalition for the homeless suggests several policies to be taken for this group (MHS Cleveland 2008).

3.4.3 Managed Care: Interference in the Patient–Physician Relationship

Another ethical problem is that the psychiatrist used to be an employer of his own medical service, but now the psychiatrist often acts as employee of an insurance or managed care company with back up of the ministry of health care, working according to principles of managed care. This can lead to conflicts. At the very least, this development puts a serious strain on the physician–patient relationship, as the psychiatrist is not only exclusively responsible for his individual patient but also to the insurance financing system. Lazarus et al. give a thorough description of the possible pitfalls in such a situation, and of ways to handle them, which will be summarized here (Lazarus and Sharfstein 2002). They identify the following potential ethical problem areas: confidentiality breach by handling of personal patient information by other, non medical, people than the physician and the medical team (see Chapter 11); conflicts of interest (see Chapter 4) as the interests of the managed care company might interfere with that of the patient; informed consent that might be breached if treatment resources are known to be limited and this might lead to lack of informing the patient that effective treatment is available, but not by the managed care company; and negative enforcement of honesty, which is fundamental to the doctor–patient relationship, but which may lead to non availability of treatment if because of certain information in the patient file, benefits might be denied by utilization managers. These potential ethical problems that interfere with the patient–physician relationship are summarized in Fig. 3.5.

Lazarus et al. emphasize that whereas the psychiatrist–patient relationship was highly personal and private in the past, this has changed dramatically with managed care. In general, their advice is to focus on honesty and confidentiality in this

Fig. 3.5 Ethical problems that interfere with the patient–physician relationship

Confidentiality breach
Conflicts of interest
Breach of informed consent
Negative enforcement of honesty

relationship, and on good informed consent according to ethical standards. On the other hand, if utilization review is intrusive and interfering, if treatment should be provided that is not available in this managed care company, or if the physician is faced with non medical personnel of the company providing services that should be withheld for a physician, the advice is to discuss this with the patient or with the company managers, and if this does not have the desired results, to discontinue participation with the particular company (Lazarus and Sharfstein 2002). This seems ethically sound advice. The question remains to what extent psychiatrists and their patients have the option to curb interference with treatment in such cases.

3.4.4 Marketing Strategies of Pharmaceutical Companies: Interference with Treatment Strategies

Another source of interference with treatment are the marketing strategies of the pharmaceutical industry, such as widening the indications for treatment with certain medication, or trying to influence psychiatrists and residents by pharmaceutical representatives or with industry-sponsored events. Also, attempts are made to influence guideline development or psychiatric research by framing its questions or withholding publications. Governments take measures against such strategies, Medical Journals ask disclosure of connections with such pharmaceutical industries, (Lenzer 2008) and some Psychiatric Associations diminish funding by pharmaceutical companies in an attempt to curb the influence of the pharmaceutical industry (see Chapter 4).

3.4.5 Interference with Treatment: What Can the Psychiatrist Do Against It?

The question rises if this problem of interference with medical treatment by a third party is new to physicians, or old wine in new bottles. In Antiquity, the physician was a slave and dedicated to his owner to provide the best medical care he could give. However, interference and breach of confidentiality would occur then as well. The Roman Emperor Claudius died suddenly in 54 AD. Although there was, and is, some discussion about the cause of death (Marmion and Wiedemann 2002), the general conclusion was that he was poisoned by his wife, Agrippina. However effective her poisoning was, it was not her first attempt; she had to try several times. She had tried to poison his food; however, Claudius was aware of that, and tried to avoid eating food that she gave him. If he would fall ill nevertheless, he managed to survive by eating the figs from a fig tree in his garden. He told his physician, Xenophon; after successive failed poisoning attempts, Agrippina bid the physician and asked an explanation; Xenophon mentioned the figs. Several days later, Claudius fell ill and died. The Empress had put poison on part of the food on her plate, namely on the mushrooms; and she had shared her food, particularly the mushrooms, with

Claudius, refraining from the mushrooms herself. However, this time she had put poison on the figs as well.

Attempts to breach the physician–patient relationship in order to interfere with treatment are of all times and physicians always have had to cope with them. In this time, economical considerations have induced governments, managed care companies, insurance companies and pharmaceutical companies to try to influence allocation of resources for treatment in various ways and for various motives. Ethical motives, such as equity, can play a role, but unfortunately, motives originating from the market position of the companies implied may do so as well. On an ethical level, the ethical standards formulated in the beginning of this chapter should be maintained and in order to do so, they should be subject of study exploring how the ethical standards can be kept in several new situations. In some cases, such as in managed care or in case of marketing activities by pharmaceutical companies, individual psychiatrists can play a role enforcing ethical standards on the micro level by being aware and raising awareness of their patients by ethically sound informed consent procedures; however, supportive policies at the meso and macrolevel are needed as well.

3.5 New Developments

Exploring ethical consequences of new treatment developments in psychiatry is an important aspect of further developing appropriate ethical standards. The increasing acknowledgement of rights and entitlements of the mentally ill changes roles of both the patient and the psychiatrist. The patient changes towards a self determined actor with own preferences for treatment; the role of the psychiatrist has become less paternalistic. This modifies the physician–patient relationship, and raises new ethical challenges.

3.5.1 The Autonomous Patient: New Developments and Ethical Controversies

New treatment models have been developed enhancing this self determined role of the patient, such as *collaborative care*, in which contracting a treatment plan and monitoring progress together with the patient is a primary step in treatment. Within collaborative care the role of the care manager is introduced to assist and manage the patient by giving structured and systematic interventions. A network is formed around the patient with at least two of the three following professionals: a general practitioner (GP), a care manager, who may be a psychiatric nurse, a social worker or a psychologist closely working together with the GP; and a consultant psychiatrist. Progress and outcome of treatment is being monitored and in case of insufficient improvement treatment may be changed according to the principles of *stepped care*. The first step is discussing a treatment plan with the patient and deciding between several effective treatment modes based on patient preference, the so-called contracting phase. Evidence-based treatment options from present guidelines are used for this (Katon et al. 1995, 1999). In this new developing relationship, informed consent is a prerequisite for the treatment model. The physician–patient relationship is extended to a triangle, in which the nurse/care manager plays a monitoring role. From an ethical point of view, this is a simple extrapolation of a long existing model, with an enlarged range of confidentiality.

Another development in line with patient emancipation is the rapid development of *internet self-help treatment*. This is an effective treatment mode for depressive or anxiety symptoms, and for problem drinking (Riper et al. 2008, Spek et al. 2007). From an ethical point of view, prevention of confidentiality breach i.e. by using scrambled internet sites not susceptible for hackers, an ethical protocol providing for protection of patient related emails or data on a server including a quality-assurance program evaluating how well the electronic media rules are being followed (Harty-Golder 2007) and a good definition of the patient–doctor relationship, if there is any, should be a focus of attention here. For the latter, a description of what exactly is medical treatment requiring a physician–patient relationship, what is not, and who monitors what is the case, is in place. More autonomy of the patient also increases the responsibility of the patient, and this may have controversial effects. Cases have been known of patients with serious mental illness asking for prescription medicines by internet doctors, providing misinformation in order to hide suicidal motives, and committing suicide with the medication thus obtained.

Also, in many countries Direct-To-Consumer (DTC) advertising of prescription drugs by pharmaceutical companies is prohibited; in the US and New Zealand, it is not. In the USA, DTC is both ubiquitous and controversial. Research showed that DTC leads to more patient-led prescription of antidepressants by physicians (Kravitz et al. 2005). Although providing DTC can be seen as a positive development in terms of increasing patient autonomy, regulation is needed as pharmaceutical companies may have a clear cut market incentive to advertise their products, more than an ethical incentive such as improving patient autonomy or trying to avert underuse of medication and undertreatment of mental disorder. Moreover, in case of depressive disorder, providing treatment does not necessarily imply providing antidepressants, as Cognitive Behavioral Treatment has been shown to be effective in many cases as well.

3.5.2 *Patient Preference as a Concept Incorporated in Research*

Another new development that takes the patient preference into account, is research evaluating a treatment according to the preference of the patient. This has been introduced as an acceptable and relevant design for randomised clinical trials. Apart from two regular design arms in which a new treatment is compared to an old treatment or

Care As Usual, an extra patient preference arm is taken up in the design, for patients who do not want to be randomized, the so-called patient preference arm (Howard and Thornicroft 2006). Note that in this arm, the influence of the fact that the patient follows his preference in the treatment is tested, rather than the effectiveness of the treatment itself.

3.5.3 The Contribution of the Patient to Society as a Factor: Return to Work as Treatment Result

Another aspect of importance for the economic framework for health care for the mentally ill is the societal relevance of treatment. Generally, effectiveness of treatment is assessed in terms of severity of symptoms or remission of disorder, but there is a new development of assessment in terms of *Return To Work*. Until now, not much research has been done assessing this outcome in treatment of depressive disorder, but recently studies have been set up aimed at this area of research (van der Feltz-Cornelis et al. 2007). Successful treatment options in this area might lead to priority for treatment of patients with disorders that may be compatible with resuming work, such as bipolar disorder, mood disorder, anxiety disorder or somatoform disorder. It might also lead to priority for treatment of employees with mental disorder that are on sick leave by buying in of treatment by their employer. It seems that ethical protocols should be developed for that, as this might interfere with equity of treatment. On the other hand, if keeping people at work increases the societal means that can be put into medical resources, it might be ethical to allocate mental health resources in favour of employees in the workforce needing them.

3.6 Ethical Arguments

Ethical principles for treatment of patients with mental disorder have been well established and evidence based psychiatry as well as the development of guidelines can guarantee a patient that whichever psychiatrist he or she consults, (s)he will have access to evidence based medical care. However, ethical principles and codes have been under pressure by shifting balances between welfare and will of a patient, or between individual and common welfare. Unfortunately the market development that has been introduced in medicine has a tendency to interfere with the physician–patient relationship or with treatment itself, with informed consent procedures, with privacy and with allocation of resources. The influence of pharmaceutical companies has been obvious and has been, and still is, the subject of close surveillance by governments. However, governments have instituted managed care and have interfered with distribution of resources for medical care in a way incompatible with the principle of equity. The stigma (see Chapter 2) that still burdens patients with mental disorder may have an influence here as well. A stringent review of ethical standards on micro-, meso- and macrolevel of economic aspects of psychiatric care is in place.

3.6.1 Micro Level

It has been suggested that ethical reflection in psychiatry should centre on the role of the psychiatrist in, amongst others, clinical practice and scientific research, education of residents in psychiatry, and advocacy of the plight of patients with mental disorders and their family (Lolas 2006). In terms of clinical practice, individual psychiatrists should be alert that there is no infringement on informed consent, honesty and trust, the patients decision-making capacity, or therapeutic boundary-keeping, in their psychiatric practice or in the managed care company in which they are employed. In scientific research, they should ask informed consent as described in the Declaration of Helsinki, in the Belmont Report (National Commission for the Protection of Human Subjects of Biomedical and Behavioral Research 1978) and in the CIOMS Guidelines (Council of International Organizations for Medical Science (CIOMS) 2002) as well as in the Declarations of Hawaii and of Madrid. Also, research into the impact of patient preferences on outcomes of care should be evaluated. Individual psychiatrists should see to it that they stick to ethical guidelines and governmental standards required for relations with pharmaceutical companies as mentioned above.

3.6.2 Meso Level

In the *education of residents* (see Chapter 31) in psychiatry, attention should be paid to new ethical developments (Hauck et al. 2002), to the ethical pitfalls of working in a managed care organisation and how to deal with it (Lazarus and Sharfstein 2002), and to the hazards of working with pharmaceutical industries and how to keep an ethical profile while doing so, and regulations for this should be put down in a guideline for residents' education in a practical manner. Staff of residency training sites should make special efforts to be a role model, following existing codes, regulations and guidelines (Primeau 2002). Also, in the training as a psychiatrist, attention should be paid to the negative influence of stigma on policies concerning the mentally ill, and on strategies to cope with this on individual and organisational level.

Health care managers should be aware of factors influencing equity of care. Health inequalities vary with quality of health services, access to those services, or health profiles of people, and this has consequences for resource allocation. Reducing health inequalities is not necessarily compatible with equity of access, nor with the objective of maximizing health gain. Resources may have to be allocated in an inequal way depending on epidemiological characteristics such as age, gender and income; so-called *differentiated inequality* (Gethmann et al. 2004). The results have profound consequences for approaches towards economic evaluation, the role of clinical guidelines and performance management, as well as for resource allocation methods (Hauck et al. 2002).

Examples can be providing health care for the mentally ill in various settings according to prevalence of the disorder, treatment possibilities and probability of attaining high adherence levels for treatment.

The existence of *Evidence Based Guidelines* implies that they should be implemented as a logical next step. Quality of care can preclude negative health outcomes in the mentally ill by implementation of appropriate care on a low-cost basis. This was established in case of dementia(Banerjee et al. 2007), schizophrenia (Dickey and Normand 2004) and addiction (Nabitz and Walburg 2000), and is under study in mental disorder in the primary care setting in several countries (Beaglehole et al. 2008, Oxman et al. 2006). Depressive disorder (Greenberg et al. 2006) as well as bipolar disorder (Simon et al. 2006) have been found to improve by quality of care programs at relatively low costs compared to the health gain implied.

3.6.3 Macro Level

Advocacy of the plight of patients with mental disorders and their family is needed, and a decisive scrutiny of social values that permit restriction of medical and other resources for the mentally ill.[1] The public and policy makers should become aware of the fact that a *society* is as weak as its weakest members, and that a culture is determined, amongst others, by the way they treat those weakest, such as the mentally ill. In striving towards cost effectiveness of treatments for the mentally ill, first, reflection is needed on the question for what means the resources that become available by this cost reduction will be used. If those resources are acquired by restricting, even on the basis of equity, medical resources for the medically or mentally ill, then the destination for which those resources would be used should be at least as needed as the medical treatments that were not provided to obtain those resources. It would be unethical if this would not be the case. Or, as Dooyeweerd states: 'Economy is the sparing or frugal mode of administering scarce goods, implying an alternative choice of their destination with regard to the satisfaction of different human needs'(Dooyeweerd 1955). Therefore, cost effectiveness should not be an end in itself. For an ethical approach to the economy of psychiatry, not only the principle of equity should be followed, but health profiles should be used to allocate resources or health care as well, and allocation of money or resources should always be related to an ethically sound destination.

A way to approach this would be the requirement of a compulsory ethical statement of insurance companies and managed care companies, in which they should declare their ethical standpoint towards the issues of the physician–patient relationship, confidentiality, beneficence, non maleficence and justice. They should state

---

[1] See Chapter 2.

if they aim towards a just division of care for all patients entitled to such care, including mental health and addiction care; they should explain how they proceed in terms of Confidentiality, Informed Consent, Honesty and Trust, Decision-making capacity, Non-participation in fraud, Involuntary psychiatric treatment and respect for Therapeutic Boundary-keeping. They should explain how they guarantee that patients with mental disorder can expect to be treated according to ethical standards in this respect in their organisation, and how they give psychiatrists working in their organisation the possibility to work according to the ethical standards mentioned above. This ethical statute should be communicated to companies buying care of the insurance or managed care company as well as to psychiatrists working for this company and to patients seeking financing for their medical care, thus enabling them to choose if they want to use the care provided by this company, or if they want to work for this company. Governments should play an enhancing role, requesting from insurance companies by law that they fulfil such ethical standards, and by installing an ethical board controlling these companies.

Users and family organisations could play a role in planning, evaluation and management of services, as well as in research (Rose and Lucas 2007).

On a higher level, a committee could suggest choices of treatment to the government based on cost effectiveness and on ethical principles; such committees already exist in some countries in the field of forensic psychiatry; however, they could be installed for psychiatry as a whole.

Another point of interest for governments should be the consequences of the increasing need for private funding to compensate for decreasing public support for teaching and research in Universities, (Primeau 2002) enhancing the potential influence of pharmaceutical companies. Several regulations are already in place in several countries for pharmaceutical companies and close monitoring of these regulations is in place. However, governments should consider making funds available to perform research and to provide teaching programs for Universities without bonds to pharmaceutical industries.

3.7 Summary

Ethical principles of medicine have been laid down in the Belmont report in the four principles of autonomy of the patient, beneficence and non-maleficence of the physician, and justice. Policy decisions regarding allocation of resources for medical care should be based upon sound scientific and societal values and their acceptability in a given cultural context. Economical considerations have induced governments, managed care companies, insurance companies and pharmaceutical companies to try to influence allocation of resources for treatment in various ways and for various motives. Ethical motives, such as equity, can play a role, but unfortunately, motives originating from the market position of the companies implied may do so as well. The principle of equity has attained general acceptance as epidemiological, medical and societal consequences of increased welfare and deceleration of mortality culminated in a high burden of society with mental disorder and high patient demand for treatment. Evidence Based Medicine and policies concerning cost effectiveness of mental health care have enforced this development. However, this standardisation of psychiatric care has been found to conflict with its necessary individualisation and the ethical principles mentioned above. Health care systems have been seeking to allocate their limited funds rationally in order to secure equal opportunity of access for equal need. However, in several countries economic considerations led to reduction of beds, sometimes using psychiatric arguments for dehospitalisation; in that case, the money saved by reduction of beds should be used for the cost-intensive out-patient long-term care (see Chapter 21); economic considerations also led to all-in price regulations. A stringent review of ethical standards on micro-, meso- and macrolevel of economic aspects of psychiatric care is needed and should take into account, amongst others, new developments enhancing patient autonomy, risks ensuing from this enhanced autonomy, and ethically sound ways to handle the controversy between economic incentives and ethical demands. In some cases, such as in managed care or in case of marketing activities by pharmaceutical companies, individual psychiatrists can play a role enforcing ethical standards on the micro level by being aware and raising awareness of their patients by ethically sound informed consent procedures; however, supportive policies at the mesolevel, i.e. in the education of psychiatry residents, and of the macrolevel, i.e. in regulations for pharmaceutical companies and ethical standards for managed care and insurance companies, are needed as well. The public and policy makers should become aware of the fact that a culture is determined, amongst others, by the way they treat the mentally ill. In striving towards cost effectiveness of treatments for the mentally ill, reflection is needed on the question for what means the resources that become available by this cost reduction will be used. Cost effectiveness should not be an end in itself. For an ethical approach to the economy of psychiatry, not only the principle of equity should be followed, but allocation of money or resources should always be related to an ethically sound destination.

References

AMA Victoria (2008) Equitable funding of Diagnosis Related Groups (DRG) http://www.amavic.com.au/page/About_Us/AMA_Agenda/Current_Issues/Equitable_funding_of_Diagnosis_Related_Groups_DRG. Report. 2008. AMA Victoria, Australia

American Medical Association (2008) Opinion of council on ethical and Judicial Affairs. E-9.9011 Continuing Medical Education. www.ama-assn.org/ama/pub/article/4001-427.html

American Psychiatric Association (2001) The principles of medical ethics with annotations applicable to psychiatry. APA, Washington, DC

Banerjee S, Willis R, Matthews D, Contell F, Chan J, Murray J (2007) Improving the quality of care for mild to moderate dementia: an evaluation of the Croydon Memory Service Model. Int J Geriatr Psychiatry 22:782–788

Beaglehole R, Epping-Jordan J, Patel V, Chopra M, Ebrahim S, Kidd M, Haines A (2008) Improving the prevention and management of chronic disease in low-income and middle-income countries: a priority for primary health care. Lancet 372:940–949

Buchanan RW, Strauss ME, Breier A, Kirkpatrick B, Carpenter WT Jr (1997) Attentional impairments in deficit and nondeficit forms of schizophrenia. Am J Psychiatry 154: 363–370

Buitenlands nieuws (2008) Omdat ik het waard ben. Ned Tijdschr Geneeskd 152:2361

California Psychiatric Association (1993) Psychiatric highlights. Developments and ideas for cost effective psychiatric health care. Effectiveness of Psychiatric treatment. Psychiatric highlights. August 1993, 1:2, http://www.calpsych.org/publications/psychhighlights/psaug93.html

Canadian Medical Association (1996) Code of ethics of the canadian medical association. Can Med Assoc J 115:1176A–1176B

Canadian Psychiatric Association (1984) Amendment to position paper on ethics. Can J Psychiatry 29:634

Council of International Organizations for Medical Science (CIOMS) (2002) International ethical guidelines for biomedical research involving human subjects. CIOMS, Geneva

Davies A, Vardeva K, Loze JY, L'Italien GJ, Sennfalt K, van Baardewijk M (2008) Cost-effectiveness of atypical antipsychotics for the management of schizophrenia in the UK. Curr Med Res Opin 24(11):3275–3285

Dickey B, Normand SL (2004) Towards a model for testing the relationship between quality of care and costs. J Ment Health Policy Econ 7:15–21

Dooyeweerd H (1955) A new critique of theoretical thought, Part II. Paris, Amsterdam, p 66

Gethmann CF, Gerok W, Helmchen H, Henke KD, Mittelstrass J, Schmidt-Assmann E, Stock G, Taupitz J, Thiele F (2004) Gesundheit nach Mass? Eine transdisziplinäre Studie zu den Grundlagen eines dauerhaften Gesundheitssystems. Akademie, Berlin

Goren JL, Parks JJ, Ghinassi FA, Milton CG, Oldham JM, Hernandez P, Chan J, Hermann RC (2008) When is antipsychotic polypharmacy supported by research evidence? Implications for QI. Jt Comm J Qual Patient Saf 34:571–582

Greenberg MD, Pincus HA, Ghinassi FA (2006) Of treatment systems and depression: an overview of quality-improvement opportunities in hospital-based psychiatric care. Harv Rev Psychiatry 14:195–203

Harty-Golder B (2007) The legally tricky e-information age. MLO Med Lab Obs 39:42

Hauck K, Shaw R, Smith PC (2002) Reducing avoidable inequalities in health: a new criterion for setting health care capitation payments. Health Econ 11:667–677

Howard L, Thornicroft G (2006) Patient preference randomised controlled trials in mental health research. Br J Psychiatry 188:303–304

Johansson PO (2006) On the Definition and Estimation of the Value of a 'statistical life'. Working paper. University of Milan, Milan. Fifth Milan European Economy Workshop http://www.economia.unimi.it/index.php?id=45&wp=217&mode=view&L=1

Katon W, Von Korff M, Lin E, Simon GE, Bush T, Robinson P, Russo J (1995) Collaborative management to achieve treatment guidelines. Impact on depression in primary care. JAMA 273:1026–1031

Katon W, Von Korff M, Lin E, Simon G, Walker E, Unutzer J, Bush T, Russo J, Ludman E (1999) Stepped collaborative care for primary care patients with persistent symptoms of depression: a randomized trial. Arch Gen Psychiatry 56:1109–1115

Klerman GL (1990) The psychiatric patient's right to effective treatment: implications of Osheroff v. Chestnut Lodge. Am J Psychiatry 147:409–418

Klok RM, Postma MJ (2004) Four quadrants of the cost-effectiveness plane: some considerations on the south-west quadrant. Editor Pharmacoecon Outcomes Res 4:599–601

Kravitz RL, Epstein RM, Feldman MD, Franz CE, Azari R, Wilkes MS, Hinton L, Franks P (2005) Influence of patients' requests for direct-to-consumer advertised antidepressants: a randomized controlled trial. JAMA 293:1995–2002

Laupacis A, Feeny D, Detsky AS, Tugwell PX (1992) How attractive does a new technology have to be to warrant adoption and utilization? Tentative guidelines for using clinical and economic evaluations. CMAJ 146:473–481

Lazarus JA, Sharfstein SS (2002) Ethics in managed care. Psychiatr Clin North Am 25:561–574

Ledermann EK (1982) Ethics in psychiatry–the patient's freedom and bondage. J Med Ethics 8:191–194

Lenzer J (2008) Review launched after Harvard psychiatrist failed to disclose industry funding. Br Med J 336:1327

Lolas F (1996) Bioethical narratives: toward the construction of social space for moral imagination. Int J Bioethics 7:53–55

Lolas F (2006) Ethics in psychiatry: a framework. World Psychiatry 5:185–187

Lynch SM, Brown JS (2001) Reconsidering mortality compression and deceleration: an alterative model of mortality rates. Demography 38:79–95

Malcolm JG (1986) Treatment choices and informed consent in psychiatry: implications of the Osheroff case for the profession. J Psychiatry Law 14:9–106

Marmion VJ, Wiedemann TE (2002) The death of Claudius. J R Soc Med 95:260–261

Mayer R (2008) Because I'm worth it. Br Med J 337:a1248

MHS Cleveland (2008) Ohio. Report http://www.mhs-inc.org/News2008.asp

Miller TR (2000) Variations between countries in the evalues of statistical life. J Transp Econ Policy 34:169–188

Morgenlander KH, Greenwald DE (1985) Psychiatric DRGs: the legal and ethical impact. QRB Qual Rev Bull 11:175–179

Murray CJ, Lopez AD (1997) Alternative projections of mortality and disability by cause 1990–2020: Global burden of disease study. Lancet 349:1498–1504

Nabitz UW, Walburg JA (2000) Addicted to quality–winning the Dutch Quality Award based on the EFQM Model. Int J Health Care Qual Assur Inc Leadersh Health Serv 13:259–265

National Coalition for the homeless (2008) Mental Illness and homelessness. http://www.nationalhomeless.org/factsheets/Mental_Illness.html. NCH factsheet No. 5, June

National Commission for the Protection of Human Subjects of Biomedical and Behavioral Research (1978) The Belmont Report: ethical principles for the protection of human subjects in research. Report

Oxman TE, Schulberg HC, Greenberg RL, Dietrich AJ, Williams JW Jr., Nutting PA, Bruce ML (2006) A fidelity measure for integrated management of depression in primary care. Med Care 44:1030–1037

Primeau F (2002) Basic principles of ethics for psychiatry. Canadian Psychiatric Association Bulletin, Montreal, QC, October

Rawlins MD, Culyer AJ (2004) National institute for clinical excellence and its value judgments. Br Med J 329:224–227

Riper H, Kramer J, Smit F, Conijn B, Schippers G, Cuijpers P (2008) Web-based self-help for problem drinkers: a pragmatic randomized trial. Addiction 103:218–227

Rose D, Lucas J (2007) The user and survivor movement in Europe. Report Mental Health Policy and Practice across Europe

Sackett DL, Rosenberg WM, Gray JA, Haynes RB, Richardson WS (1996) Evidence based medicine: what it is and what it isn't. Br Med J 312:71–72

Simon GE, Ludman EJ, Bauer MS, Unutzer J, Operskalski B (2006) Long-term effectiveness and cost of a systematic care program for bipolar disorder. Arch Gen Psychiatry 63:500–508

Spek V, Cuijpers P, Nyklicek I, Riper H, Keyzer J, Pop V (2007) Internet-based cognitive behaviour therapy for symptoms of depression and anxiety: a meta-analysis. Psychol Med 37:319–328

Stanford University (2008) Principles of ethics and professionalism in psychiatry. Report www.stanford.edu/group/psylawseminar/Ethics.htm

Stone AA (1990) Law, science, and psychiatric malpractice: a response to Klerman's indictment of psychoanalytic psychiatry. Am J Psychiatry 147:419–427

Swan Tan S, Schawo S, Rutten-van Mölken M, Hakkaart-van Roijen L, Rutten FF (2008) Quick Scan: The role of guidelines in the impact of pharmacotherapy. (In Dutch: De rol van richtlijnen bij de nadere invulling van farmaceutische aanspraak). Institute for Medical Technology Assessment. Erasmus MC, Rotterdam, October

Tansella M, Amaddeo F, Burti L, Lasalvia A, Ruggeri M (2006) Evaluating a community-based mental health service focusing on severe mental illness. The Verona experience. Acta Psychiatr Scand 113:90–94

van der Feltz-Cornelis CM, Meeuwissen JA, de Jong FJ, Hoedeman R, Elfeddali I (2007) Randomised controlled trial of a psychiatric consultation model for treatment of common mental disorder in the occupational health setting. BMC Health Serv Res 7:29

Chapter 4
Conflicts of Interest

Giovanni A. Fava

Contents

G.A. Fava (✉)
Psychotherapy and Psychosomatics, Department of Psychology, University of Bologna,
Bologna 40127, Italy
e-mail: giovanniandrea.fava@unibo.it

Disclosure: The author has received grant support for his studies from the Italian Ministry of Education, University and Research, the Italian National Research Council, the Italian National Institute of Health, the Carisbo Foundation, the Cassa di Risparmio di Cesena Foundation, Telefono Azzurro and Compagnia di S. Paolo. He is editor-in-chief of Psychotherapy and Psychosomatics (Karger, Basel).

H. Helmchen, N. Sartorius (eds.), *Ethics in Psychiatry*, International Library
of Ethics, Law, and the New Medicine 45, DOI 10.1007/978-90-481-8721-8_4,
© Springer Science+Business Media B.V. 2010

Abbreviations

CME	Continuing Medical Education
DSM-IV	Diagnostic and Statistic Manual, 4. Revision, (American Psychiatric Association)
JAMA	Journal of the American Medical Association
NEJM	New England Journal of Medicine
PMA	Professional Medical Associations

4.1 Introduction

The notion of conflicts of interest is widely used but may entail different meanings. Margolis (1979) distinguishes between conflicting interests and conflicts of interest. The former occur in any situation where competing considerations are presumed to be legitimate. Conflicts of interest, on the other hand, are characterized by individual occupying dual roles which should not be performed simultaneously. Because of the potential for abuse, performing both roles at the same time is considered to be inappropriate and may lead to apparent conflicts of interest, potential conflicts of interest and real conflicts of interest. Conflicts of interest may be financial and non-financial. Table 4.1 lists the main sources of financial conflict of interest. Table 4.2 displays the principal issues in non-financial conflicts of interest as outlined by Maj (2008) and Möller (2009).

At times the two types of secondary interest may be intertwined, as in the case of 'special interest groups' (Fava 2001), self-selected academic oligarchies that influence clinical and scientific information. Members of corporate-driven special interest groups, in virtue of their financial power and close ties with other members of the group, have the task of systematically preventing dissemination of data which may be in conflict with their interests.

Table 4.1 Main sources of financial conflicts of interest in medicine

Being a clinician/researcher and:
- an employee of a private firm
- a stockholder
- a member of a company board of directors
- a regular consultant of a private firm
- an occasional consultant of a private firm
- an official speaker of a private firm
- an occasional speaker of a private firm
- getting refunds from a private firm
- recipient of honoraria
- a clinical investigator in a sponsored trial
- recipient of research support from a private firm
- owning a patent

Table 4.2 Main sources of non-financial conflicts of interest in medicine

- Personal recognition
- Career advancement
- Visibility in the media
- Favouring a relative, friend or colleague
- The allegiance to a school of thought
- Political commitment
- Rivalry between experts
- Representative of a certain professional society
- Involvement in specific educational activities

I will describe some of the insights that research on financial conflicts of interest has generated in psychiatry, their implications for psychiatric research and practice, and some strategies which may counteract this phenomenon. I will then examine non-financial conflicts of interest in psychiatry and how they could be properly handled.

4.2 Financial Conflicts of Interest in Psychiatry

In the past decade there has been a considerable amount of research on financial conflicts of interest in medicine and psychiatry.

4.2.1 Prevalence

The first idea of the prevalence of situations of conflict of interest in scientific research came from a landmark study which appeared in the 1990s. Krimsky et al. (1998) analyzed 789 articles written by authors from Massachusetts universities publishing in leading scientific journals in 1992. In one out of three cases, at least one author had a vested interest in research. Krimsky et al. (1998) took a very conservative stand as to what constitutes a financial conflict of interest: owning a patent directly related to the published work; being a major stockholder or executive in a company with commercial interests tied to the research, or serving on the board of directors of such a company. The percentage of cases of conflicts of interest would have greatly increased if consultancies and honoraria had been taken into account. The study clearly showed the extent of corporate presence in scientific publishing, even though it did not necessarily indicate that such conflicts biased the research they referred to.

The same group of researchers addressed the issue of the financial ties with the pharmaceutical industry of the 170 DSM-IV panel members. Ninety five (56%) had one or more associations with companies. The percentage reached 100% of the members of the panels on mood disorders and schizophrenia and was above 80% in anxiety and eating disorders (Cosgrove et al. 2006).

4.2.2 Disclosure

Disclosure has emerged as a first and essential step for dealing with conflicts of interest contamination in science. But, despite journals' policies, it is seldom performed (in less than 1% of medical articles according to a study by Krimsky 2001).

Such disclosure often takes place in the media, instead of coming from the authors or scientific community. The proliferating connections between the psychiatrists and the pharmaceutical industry and the failure to disclose them appropriately have brought the credibility of psychiatry to an unprecedented crisis (Anonymous 2008).

It must be noted that, while disclosure has become standard practice in North American meetings and journals, it has not achieved wide currency in Europe, where the demand for disclosure developed quite late. In most of the cases, disclosure is only a formal and meaningless list of collaborations and it is very difficult for the reader or listener to discern any specific link with the research that is presented.

4.2.3 Scientific Societies and the Drug Industry

Glassman et al. (1999) investigated whether revenues generated from pharmaceutical advertisements in medical journals create potential conflicts of interest for nonprofit physician organizations that own those journals. They found that financial conflicts of interest were substantial, and some prestigious medical organizations, such as those underlying the *Journal of the American Medical Association* (JAMA) and the *New England Journal of Medicine* (NEJM), could be viewed as beholden to the drug industry. Scientific societies may control medical journals and affect editorial policies and the selection of papers. Further, financial ties may also affect the scientific meetings of those societies (selection of topics, speakers, etc...). This is something anyone walking in a major society meeting may easily perceive.

4.2.4 Clinical Practice Guidelines and the Pharmaceutical Industry

Choudhry et al. (2002) examined authors of clinical practice guidelines endorsed by North American and European societies on common adult diseases. Eighty-seven percent of authors had some form of interaction with the pharmaceutical industry (58% had received financial support to perform research and 38% had served as employees or consultants for a pharmaceutical company). In published versions of the 44 clinical practice guidelines, specific declarations regarding the personal financial interactions of individual authors with the pharmaceutical industry were made in only two cases. Cosgrove et al. (2009) have examined, by multimodal screening techniques, the degree and type of financial ties to the pharmaceutical industry held by authors of three Clinical Practice Guidelines of the American Psychiatric Association. Ninety percent of authors had financial ties to companies that manufacture drugs which were identified in the guidelines as recommended therapies. None of the financial associations were disclosed.

4.2.5 Attendance to Drug Sponsored Scientific Events and Prescription of the Sponsor's Medication

A review (Wazana 2000) has outlined how attending sponsored continuing medical education (CME) events and accepting funding for travel or lodging for educational symposia were associated with an increased prescription rate of the sponsor's medication. Attending presentations given by pharmaceutical representative speakers was also associated with nonrational prescribing. Wilkes (2000) commented on the consequences of the interactions:

> Physicians take gifts from drug companies and then spend patients' money to help make the same pharmaceutical industry the most profitable in the world. (. . .) All these behaviors are directly opposed to what patients and society expect from us in return for the privileges that have been bestowed.

And, as the subtitle of the editorial indicates, when trust goes, so does the healing power of physicians.

4.2.6 Pharmaceutical Sponsorship and Clinical Trial

It has been repeatedly reported that studies sponsored by pharmaceutical companies are more likely to have outcomes favourable to the sponsor (Bhandasi et al. 2004, Lexchin et al. 2003). Industry sponsorship also results in restrictions on publication and data sharing and in selective reporting (Melander et al. 2003). Perlis et al. (2005) examined funding sources and authors' financial conflicts of interest in clinical trials published in four leading American journals concerned with psychiatry. Sixty percent were funded from a pharmaceutical industry, and conflict of interest was associated with a greater likelihood of reporting a drug to be superior to placebo. These findings were replicated by Kelly et al. (2006) and Montgomery et al. (2004). Further, Melander et al. (2003) analyzed controlled studies of selective serotonin reuptake inhibitors and found that sponsored studies with favourable results were more often published than negative studies. A very good example of this selective publication is given by the scandal following the finding that a major pharmaceutical company allegedly withheld from the medical community clinical trial findings which indicated that a widely used antidepressant had no beneficial effect in treating adolescents (Kondro 2004). A recent paper provide a good illustration as to how selective publication of antidepressant trials promotes their apparent efficacy. Thirty-seven of the 74 FDA-registered studies which were associated with positive outcome were published and 1 was not, whereas only 3 of the 36 negative studies were published as such (Turner et al. 2008). This casts serious doubts on the representativeness of the drug trials which are included in a meta-analysis (Fava 2006). Further, even systematic reviews require careful critical appraisal. Conflicts of interest may affect this appraisal. Evidence-based medicine thus may be a deceptive instrument of propaganda.

Heres et al. (2006) have analyzed the sources of bias which may limit the validity of head-to-head comparison studies of second-generation antipsychotics, such as equivalent dosages, study entry criteria, statistical analysis, reporting of results and wording of findings. It should be noted, however, that negative studies may not get published also because they may have substantial methodological problems (particularly in head to head comparison) or as a decision of scientific editors who privilege the positive ones (Möller 2008a, b). International data banks of all trials that are performed have been recently established and are a major step for overcoming these problems as to meta-analyses.

4.2.7 Data Ownership

Mello et al. (2005) explored the legal agreements that exist between industry sponsors and academic investigators. In 80% of institutions the sponsor may own the data and in 50% the sponsor may write up the results for publication. There have been many instances in the media about the struggles between clinical researchers and pharmaceutical companies as to the publication and analysis of data. In most of the instances investigators have been quite alone in their battles. In sponsored scientific presentations at meetings, it is a quite common practice that the slides of speakers are reviewed and approved by the sponsor. Research ethics committees should advise the researchers that such restrictions are not compatible with the academic freedom of scientists and that they should have the ownership as well as the only responsibility on their data.

4.2.8 Marketing and Advertising

In 1992 Wilkes et al. assessed the accuracy of scientific data presented in the pharmaceutical advertisements of 10 leading medical journals. Each full-page pharmaceutical advertisement was sent to 3 reviewers. In 30% of the cases, 2 or more reviewers disagreed with the advertiser's claim that the drug was the 'drug of choice'. In 44% of cases the reviewers felt that the advertisement would lead to improper prescribing if a physician had no other information about the drug other than that contained in the advertisements. Spielmans et al. (2008) specifically addressed the issue of psychiatric medications and found that in only half of the cases claims regarding the efficacy of medications were supported by studies.

Marketing has been aggressive, particularly at meetings. In a study of all exhibit booths of pharmaceutical companies at the 2002 American Psychiatric Association (APA) convention, a total of 16 violations of the APA'S own exhibit rules has been found (Lurie et al. 2005). Private companies have set campaigns to shape a favourable climate of opinion for their drugs. These campaigns take the form of commercially strategic clinical trials (which have been defined by Carroll 2008 as 'experimercials'), journal publications that are 'infomercial' and educational activities whose main aim is to sell the participant to the sponsor (Fava 2001). The game is clear: to get as close as possible to universal consumption of a drug, by manipulating evidence and withholding data.

The marketing of gabapentin through misinformation and manipulation is a good example of these strategies in psychiatry (Landefeld and Steinman 2009).

4.2.9 Ghostwriting

A very conservative benchmark for ghostwriting of papers published in biomedical journals has been set at roughly 10% (Moffatt and Elliott 2007). However, this percentage greatly increases in case of drug firm initiated trial and may involve the majority of published papers (Sismondo 2007). In some cases drug firms produce the articles and look for prospective authors who are then paid for signing the papers. Jureidini and McHenry (2009) have illustrated the consequences of ghostwriting in a paper concerned with paroxetine in adolescents.

4.2.10 Implications of Conflicts of Interests

The potential adverse consequence of misleading advertisement, selective publication, biased reports, corporate driven symposia and scientific information, ghostwriting and lack of disclosure are obvious as to prescribing habits of psychiatrists: increased new costs when physicians have been persuaded to prescribe expensive new drugs over equally affective lower-cost drugs or nonpharmaceutical treatments (e.g., psychotherapy), or needless harm or even death because physicians have been persuaded to prescribe products for uses for which they have not been adequately tested or are suitable.

However, there are other serious consequences.

One aspect has to do with the issue of censorship (Fava 2009a). By carefully selecting the literature in a biased direction and offering a manipulated interpretation of clinical trials (including those supported by public sources) key opinion leaders may become 'the role models whose views are to be taken seriously' (Iheanacho 2008). The drug industry may thus take control of scientific societies, clinical practice guidelines and reporting investigations in meetings and journals. Independent investigators, who feel the moral obligation to tell the truth, may object to the manipulation of evidence operated by these special interest groups who thus practice the appropriate amount of retaliation excluding them from symposia, removing them from academic appointments and preventing access to research funding (Borgstein and Watine 2001). Isolation is the ultimate outcome. For a pharmaceutical company delaying or minimizing knowledge of a side effect of a medication has cash value.

Yet, there are more subtle forms of censorship. One has to do with setting a financial threshold for publishing research findings. In recent years, there has been a progressive demand on free availability of resources on the internet and on centralizing medical information. Public access to medical journals means that the authors will have to pay for at least part of the expenses. Publication loaded with conflict of interest would not really have any problem, whereas this will be a major difficulty for unsupported, young investigators. These investigators, because of their questioning that may lead to new discoveries, are the lifeblood of science (Borgstein and Watine 2001). The issue is thus not open access to self selected information, but discrimination of independent sources within information overload (Fava and Guidi 2007).

Another subtle form of censorship is by counteracting undesirable published information with massive doses of propaganda. Noam Chomsky (1997a) has been instrumental in disclosing the link between propaganda and media control. Filtering information (selective perception), engineering opinions, using the public relations industry and marginalizing dissident cultures are the well-known modalities of action. The presentation of the literature on long-term treatment with antidepressant drugs exemplified this strategy (Fava 2002).

Yet, it is deliberate self-censorship which may yield the most dangerous effects. As suggested by a recent survey of journalists (Lee and Chan 2008), it is common and eliminates the need for formal cuts and modifications. The typical example is the intervention of an established investigator in a drug-sponsored symposium. He or she refrains from making promotional statements, leaving the dirty job to someone else in the symposium. However, he or she does not comment on unsubstantiated and commercial statements from other speakers in the panel.

Obviously, financial interests are not the only source of censorship in clinical medicine, and may be substituted or supplemented by cultural, political and ideological issues.

If medical knowledge is the cumulative experience of human history 'a legacy from those who have gone before to those who live today', 'a social possession' (Holman 1976, p.21), suppression of memory (i.e. reliance on the most recent papers) and ignorance of the historical intellectual debate may be another subtle from of censorship. Noam Chomsky (1997b), in an essay on the intellectual climate during the Cold War, reminds us of how the neglect of the history of the disciplines was instrumental to preventing a critical attitude toward the establishment. If we do not know where we came from, we have a very poor idea of where we are going to. The challenge is to preserve pluralism, critical thinking and intellectual freedom in a setting more and more characterized by conformism, political appropriateness and the cult of mediocrity (Fava 2009a).

Another important implication of conflict of interest is the conceptual crisis which afflicts psychiatry nowadays and stems from a narrow and reductionistic concept of science applied to the understanding and treatment of mental disorders (Bursztajn et al. 1981, Fava 2006, Perlin et al. 2008).

If 'medical journals are an extension of the marketing arm of pharmaceutical companies' (Smith 2005) and corporate interests result in self-selecting academic oligarchies (special interest groups) (Fava 2001) one may wonder how can a clinician discern important information ('Is the treatment effective and to whom shall I give it? What harm may I do by using it?') from the massive amount of propaganda delivered by medical journals and meetings. Nierenberg et al. (2008) have illustrated how systematic biases in decision-making induced by pharmaceutical companies occur in clinical psychopharmacology. As Healy (2008) commented 'RCTs originated as efforts to debunk therapeutic claims, but the force field in which medicine is now practiced has transformed them into technologies that mandate action. As a result the evidence originally designed to stop misguided therapeutic band wagons has become the fuel for band wagons. Where the placebo arms of antidepressant, antipsychotic or mood stabilizer trials suggest we should not be using the drugs as readily as we do, the trials of these products, embodied in guidelines, have instead become a means to enforce treatment'.

Therapeutic results are always the result of several ingredients, which may be specific or non-specific (Fava 2009b). Antidepressant drugs are therapeutic tools of modest efficacy in a setting characterized by the clinician's full availability for specific times, the patient's opportunity to ventilate thoughts and feelings, the development of a patient-doctor interaction, and the perception of competent care (Downing and Rickels 1978, Gliedman et al. 1958, Uhlenhuth et al. 1966). Efforts to study the role of non specific factors in determining the response to acute and long-term treatment have always been minimal. In the meanwhile, one may wonder what could be consequences, in terms of economic waste, mortality and morbidity, of clinical management by young generations of psychiatrists nurtured by evidence biased psychiatry, meta-analyses, overselling of drug related ingredients of modest efficacy (Fava 2009b).

4.2.11 Operational Proposals

There have been several attempts to address financial conflicts of interest in medical societies, journals and public organizations. Codes of conflict have been developed, such as those endorsed by American College of Cardiology Foundation and the American Heart Association (ACCF/AHA 2004). Within the psychiatric field, the recommendations of World Federation of Societies of Biological Psychiatry (Helmchen 2004) deserve to be mentioned. It is suggested that the physician evaluates (and refrain from, if appropriate) all relationships between himself/herself and industry, includes in the evaluation possible perceptions of such relationships by patients, colleagues and the public, collaborates with the industry only on the basis of clearly formulated contracts, and discloses all relationships when it is appropriate.

A crucial problem lies in the lack of a definition of substantial conflicts of interest. Are eating a pizza at a drug-sponsored lunch or being a regular consultant to a firm the same thing? Table 4.3 outlines some tentative criteria which are based on Krimsky et al.'s work (1998). The first two situations shown in the table involve the concept of continuity of a relationship with a private firm. Indeed, occasional consultancies, grants for performing an investigation, or receiving honoraria or refunds in specific occasions would not be a source of substantial conflict of interest. The latter two situations depicted in the table indicate major financial sources of bias.

Table 4.3 Criteria for the presence of substantial conflicts of interest of a researcher

The researcher meets at least one of the following:
1. Being an employee of a private firm
2. Being a regular consultant or in the board of directors of a firm
3. Being a stockholder of a firm related to the field of research
4. Owning a patent directly related to the published work

Another issue is the fact that the problem of conflicts of interest has been viewed so far mainly in negative terms: how to limit corporate influence in medical research. There has been little or no emphasis on the fact that the scientific community is draining itself of a reservoir of disinterested experts who can be called upon to advise government policy makers and physicians on the safety and efficacy of treatments, on the hazard of chemicals and on the safety of technology (Krimsky et al. 1998). Do we believe that researchers who opted for not having any form of conflict of interest and, by doing this, gave up financial gains, are of special value? Or do we believe that their opinion is in no way different from that of researchers with substantial conflict of interest and that they are simply a pathetic remnant of the past century?

Yet, the experts who are free of conflict of interest may find increasing difficulties in obtaining appropriate visibility at meetings and in journals and in getting support for their research. It is not that disinterested experts are extinct: it is that they are marginalized by the gatekeepers of corporate interest within public institutions, scientific societies and medical journals (Fava 2001).

As a result, if we believe in the value of independent research and researchers and in the need of preserving and promoting this independence, we must endorse the steps which are outlined in Table 4.4 (Fava 2007, 2008). If a grant agency committee, or a medical journal, or a scientific meeting committee does not include experts with no substantial conflicts of interest, and particularly those who have none, it does not deserve credibility.

For certain positions (e.g. editor-in-chief of a medical journal), the situation should be evaluated on an individual basis. For instance, tie to a single firm, contrary to what is often assumed, allows an easy monitoring of an editor's job (he or she can be excluded from assessing papers dealing with products of that firm), whereas multiple forms of conflict of interest make this control impossible. At times, advertising departments appear to influence editorial decisions. Such influence may

Table 4.4 Lines of support of independent researchers who are free of substantial conflict of interest

1. Priority for obtaining grants from public agencies supported by taxpayer money
2. Priority for scientific societies and medical journals editorship positions
3. Adequate visibility in scientific societies meetings programs
4. Inclusion only of researchers with no substantial conflict of interest in clinical practice guidelines groups
5. Conflict-free investigations and reviews should be emphasized in training and continuing medical education and should have priority in medical journals

be particularly strong if the editor is vulnerable because of his/her conflicts of interest.

Information overload is the key vehicle of pharmaceutical propaganda (Fava and Guidi 2007). A psychiatrist may be overwhelmed by scientific articles, often of redundant nature. He or she may become aware of certain articles because of firms pointing to those, or because they appear in very well-known and distributed journals. Yet this may be very misleading. Conflict-free articles (particularly review papers) and purely subscription-based journals should become the focus of attention of clinicians who have become educated to the issues of conflict of interest.

Only in this context, interventions aimed to getting a better control of conflicts of interest may become successful (Table 4.5). While disclosure has become standard practice in North American meetings and journals, it is still poorly practiced in Europe. It should be emphasized that in psychiatry conflicts of interest may arise not only when there are ties with the pharmaceutical industry, but also when the researchers, for instance, are involved in private schools for training in psychotherapy. Disclosure is the minimal requirement for scientific credibility. It should have a specific time frame (e.g., 3 years). When an endless list of financial ties is provided, it should be clear that it becomes virtually meaningless, unless the potential implications of such ties are described in a note.

Each scientific organization should have a conflicts of interest advisory committee that represents different segments of the organization and that should be a referral point to individual members identifying possible conflicts of interest. Scientific organizations may also request disengagement from corporations that abuse public trust (e.g., false advertising, regulatory fines etc.) and do not allow publications of scientific results. Individual members of a society, not unlike the alternative consumer, can also decline participation in specific meetings or society events, or refuse to pay the dues of the society, or write to the journal which was involved in a specific case of conflict of interest (and the letter should be published, whereas it is seldom done with the excuse of lack of space or by not having a dangerous letter section). Members attending a meeting of their association should be able to rate the quality and the influence of the pharmaceutical industry with appropriate evaluation forms and to manifest their dissent (electronic mail is a powerful instrument for it).

Table 4.5 Steps to addressing financial conflicts of interest in medical research

1. Disclosure should become the rule in all scientific meetings and journals
2. Each scientific organization should have a conflicts of interest advisory committee
3. Individual members of societies and readers of medical journals should express their dissent from presentations and articles biased by conflicts of interest
4. Specific policies for integrity in science by professional societies, universities, granting agencies, pharmaceutical companies
5. Independent review bodies (within each field) for examining the issues concerned with conflicts of interest
6. Educational plan for recognizing conflict of interest and the role of treatment ingredients

The development of specific policies for integrity of agencies and pharmaceutical industries are also important.

The creation of independent review bodies (within each field) for examining the issues concerned with conflicts of interest would be another important step. Ethics research committees may play an important role in setting standards for issues such as data ownership.

Finally, professional training programs (e.g. medical school, residency training, etc.) should teach individuals to recognize conflict of interest situations and increase awareness of biased interpretations of research results and treatment ingredients (Dubovsky and Dubovsky 2007).

When these proposals were first presented as a Forum article in World Psychiatry (Fava 2007) they stirred considerable controversies. Guy Goodwin (2007) charged these suggestions with the risk of sterilizing academic activities from the pharmaceutical industry and denied any loss of intellectual freedom with financial ties to the industry. Eduard Vieta (2007) remarked that conflicts of interest are dimensional, not dichotomous, and they may have ideological, financial, social or academic nature. Wolfgang Fleischhacker (2007) emphasized that we all face conflicts in our professional lives and that the style and tone of the proposal by Fava (2007) were 'regrettable'. Paul and Tohen (2007), who represented the personal views of two distinguished researchers working in the pharmaceutical industry, were more supportive and judged the outlined proposal as 'a good start'.

Very recently, another operational proposal for controlling conflicts of interest in professional medical associations (PMA) in United States has been presented (Rothman et al. 2009). The Authors acknowledged that while many PMA had issued guidelines on conflicts of interest, there was little uniformity among them and new rules were necessary. These new rules involved working toward the following steps:

1. complete ban on pharmaceutical and medical device industry funding of a PMA' general budget, except for income from journal advertising and exhibit hall fees,
2. if funds become available for educational and research purposes they should be truly unrestricted and managed by Committee Members who are completely free of all industry ties. PMA can no longer endorse, facilitate or accept funding for satellite symposia,
3. members of Committee that formulate practice guidelines or outcome measures should have no ties to industry. The commonly expressed concern that such restriction excludes the most qualified individuals can be circumvented by circulating drafts of guidelines widely for comment
4. ban of industry support of PMA's publications, products and affiliated foundations
5. PMA's leaders (president and officers) should be free of any financial conflicts of interest during their tenure

These proposal are a bold and welcome change which is in the best interest of the PMA and the larger society. They appear to be much more restrictive than those entailed by the definition of substantial conflicts of interest and effectively address the problem in medical associations (Table 4.5). They provide a clear appraisal of the failure of procedures based only on disclosure, peer review, and appeal to individual integrity and independence. However, even these very restrictive rules fail to address some of the lines of support of independent researchers outlined in Table 4.4, such as priority for obtaining grants from public agencies and in publishing conflict-free investigations and reviews in medical journals.

4.3 Non Financial Conflicts of Interest

There are several sources of conflicts of interest which are not of financial nature (Table 4.2). They have attracted limited interest in the literature, yet they appear to be quite important.

Mario Maj (2008) has recently summarized several issues concerned with non-financial conflicts of interest. He pointed out that the notion of 'special interest groups', which has been applied to researchers who share similar financial ties (Fava 2001), may also apply to groups characterized by academic and/or political and/or personal ties.

The peer review process, which Horrobin (1990) defined as often resulting in 'suppression of innovation' offers very good examples of these conflicts (Möller 2006, 2009). It has been repeatedly reported that scientists (particularly in basic sciences) who act as reviewers for scientific journals may deliberately delay assessment and ultimately reject papers dealing with the same area of research for having the opportunity of performing the study in their laboratory in the meanwhile. Reviewers may also be particularly critical of papers which do not fit with their view of the scientific area and/or contradict their own findings.

As remarked by Möller (2009), very few journals in psychiatry require a disclosure of non-financial conflicts of interest for reviewers. *Psychotherapy and Psychosomatics* is one of the few: experts are asked to review a paper also in the presence of non-financial conflicts of interest, but to provide a full disclosure to the editor. The *International Journal of Neuropsychopharmacology* requires this disclosure also to submitting authors. This may be particularly important, for instance, in case of psychotherapy papers, when the author is a leader in the approach (Maj 2008). It is, however, rather difficult to set a threshold for disclosure of submitting authors when the source of the problem is purely intellectual. It is an increasingly common practice that the author submits a list of suitable and unsuitable reviewers together with the paper (e.g., *American Journal of Psychiatry, Journal of Clinical Psychiatry*). The suggested reviewers may have considerable academic ties (e.g., authorship, grants, committees, etc.) with the submitting author. *Psychotherapy and Psychosomatics* does not take author's suggestions for reviewers into consideration, but excludes scientists whom the authors has requested not to be involved in the review process. The first two points of Table 4.6 provide steps for addressing peer-review related conflicts of interest in publications.

Table 4.6 Steps to addressing non-financial conflicts of interest in medical research

1. Disclosure of both non-financial and financial conflicts of interest should be performed to the editor by reviewers of a scientific journal
2. The request by the submitting author of excluding specific experts from the review process should be granted by the editor, whereas assignment to suggested reviewers may be discretional
3. Members of a committee in granting agencies and external experts cannot act as reviewers if they share previous publications and research projects or belong to the same institution or are relative of the grant applicant
4. All grant application reviewers should provide full disclosure of both non financial and financial conflicts of interest
5. Increased input from the general public, patients'associations and scientists whose main field is outside of the one which is the target of grant applications
6. Inclusion of people who represent the perspectives of research subjects and who are not academic researchers in Institutional Review Boards
7. Independent review bodies (within each field) for examining the issues concerned with non-financial conflict of interest

The following points are concerned with the work of committees in granting agencies. Here a substantial problem lies in the definition of non-financial conflict of interest. Being co-authors in previous publications and research projects or belonging to the same institution are two common sources of exclusion from acting as a reviewer for the grant applications. In this case committee members should not be present when a relevant application is discussed. Members of special interest groups, who act as editors, reviewers and consultants to medical journals and nonprofit research organizations, tend to particularly reward mediocrity (Fava 2005). Talent, in fact, is frequently associated with independence, which undermines the power structure. Mediocrity assures lifelong commitment and is rewarded by the power structure adequately. Frequently one is impressed by the intellectual mediocrity and poor originality of funded research. Alvan Feinstein (1999) warned against the harmful effects of restrictive ideological beliefs in clinical science, leading to proposals which are characterized by 'trivial variations on tired themes'. He suggested major revisions in the choice of appropriate advisory committees. 'The experts who are intimately familiar with a single domain may not always have the vision or imagination to perceive and evaluate challenges that arise beyond that domain. What is needed is creative thought not just knowledgeable expertise' (Feinstein 1999, p. 466). He thus called for the inclusion of scientists who are not experts in the field and increased input from the public and groups of patients. Similar suggestions have been advanced as to including the perspectives of experimental subjects as to Institutional Review Boards (Levinsky 2002). The role of non-financial conflicts of interest (eagerness to get studies completed and personal benefits from publications and acquisition of grants) in undermining the safety of research subjects has in fact been outlined (Levinsky 2002).

4.4 Conclusions

American efforts to control conflicts of interest have been mainly concerned with disclosure and health industry practices that create conflicts of interest (Brennan et al. 2006). A Physician Payments Sunshine Act has been introduced in the US Congress in 2007 to legally oblige drug and device companies to post in a publicly accessible database all payments of more than $ 500 they make to physicians (Anonymous 2008). A policy proposal has been made (Brennan et al. 2006) as to gifts, even of relatively small items, including meals; payment for attendance at lectures and conferences; continuing medical education activities; payment for travel to meeting or for participation in speakers' bureaus or for time while attending meetings; ghostwriting services; pharmaceutical samples; grants for research projects and payment for consulting relationships. A much tougher proposal has been recently advanced as to Professional Medical Associations (Rothman et al. 2009).

The European proposal which has been described here (Fava 2007, 2008) is based on preserving intellectual freedom by providing room to researchers who opted for not having forms of substantial conflict of interest.

An eminent clinician of the past century, John A. Ryle, summarized the social responsibilities of the physician as follows:

> "The life and work of the physician proceed under the direction of three main influences: the scientific, the humane, and the ethical. Whereas other men of science have, until now, found it possible to pursue their intellectual tasks without reference to human need and without regard for ethical considerations other than those immediately connected with the pursuit of the truth and respect for colleagues, the medical man has carried a far heavier and more complex burden of responsibility. He has had and has now in ever-increasing measure – and in the addition to the consideration which he owes to himself and his dependants – a special duty to his patients, to the community, to his colleagues, and to his science or calling (Ryle 1948, p. 101)."

These are values which Robert G. Petersdorf claimed four decades later, in 1989:

> "We can no longer tolerate the dishonesty, cheating, fraud and conflict of interest that have invaded science and medicine. By choosing these professions we have assumed a trust that is predicated upon integrity. We must not deviate from it" (Petersdorf 1989, p. 123).

Halstead R. Holman, in a paper published in Hospital Practice in 1976 which anticipated some developments in health care of the following decades, observed that

> ... the medical establishment is not primarily engaged in the disinterested pursuit of knowledge into medical practice; rather in significant part it is engaged in special interest advocacy, pursuing and preserving social power. The concept of excellence is a component of the ideological justification of that role (p. 11).

Holman identified a decline in intellectual freedom as a major source of the 'excellence deception', which perpetuates prevailing practices, deflects criticism, and insulates the profession from alternative views and social relations that would illuminate and improve health care.

Preserving intellectual freedom in the setting of proliferating connection between the pharmaceutical industry and the physicians is thus a major ethical challenge of psychiatry today and should be pursued in any possible way.

References

ACCF/AHA (2004) Consensus conference report on professionalism and ethics. Circulation 110:2506–2549

Anonymous (2008) More than one bad apple. Nature 445:835

Bhandasi M, Busse JW, Jackowski D et al (2004) Association between industry funding and statistically significant pro-industry findings in medical and surgical randomized trials. Can Med Ass J 170:477–480

Borgstein J, Watine J (2001) Feudal lords of science and medicine. West J Med 175:139–140

Brennan TA, Rothmann DJ, Blank L et al (2006) Health industry practices that create conflicts of interest. JAMA 295:429–433

Bursztajn HJ, Feinblom RI, Hamm RM, Brodsky A (1981) Medical choices, medical chances: how patients, families, and physicians can cope with uncertainty. Delacorte, New York, NY

Carroll BJ (2008) Antipsychotic drugs for depression? http://hcrenewal.blogspost.com/2008/01/antipsychotic-drugs-for -depression. html

Chomsky N (1997a) Media control: the spectacular achievements of propaganda. Seven Stories, New York, NY

Chomsky N (1997b) The cold war and the university. In: Chomsky N, Katznelson I, Lewontin RC, Montgomery D, Nader L, Ohmann R, Siever R, Wallerstein I, Zinn H (eds) The cold war & the university. The New Press, New York, NY, pp 171–194

Choudhry NK, Stelfox HT, Detsky AS (2002) Relationships between authors and clinical practice guidelines and the pharmaceutical industry. JAMA 287:612–617

Cosgrove L, Bursztajn HJ, Krimsky S et al (2009) Conflicts of interest and disclosure in American Psychiatric Association's clinical practice guidelines. Psychother Psychosom 78:228–232

Cosgrove L, Krimsky S, Vijayaraghvan M et al (2006) Financial ties between DSM-IV panel members and the pharmaceutical industry. Psychother Psychosom 75:154–160

Downing RW, Rickels K (1978) Nonspecific factors and their interaction with psychological treatment in pharmacotherapy. In: Lipton MA, Di Mascio A, Killam KF (eds) Psychopharmacology: a generation of progress. Raven Press, New York, NY, pp 1419–1427

Dubovsky SL, Dubovsky AN (2007) Psychotropic drug prescriber's survival guide. Norton, New York, NY

Fava GA (2001) Conflict of interest and special interest groups. Psychother Psychosom 70:1–5

Fava GA (2002) Long-term treatment with antidepressant drugs: the spectacular achievements of propaganda. Psychother Psychosom 71:127–132

Fava GA (2005) The cult of mediocrity. Psychother Psychosom 74:1–2

Fava GA (2006) The intellectual crisis of psychiatric research. Psychother Psychosom 75:202–208

Fava GA (2007) Financial conflicts of interest in psychiatry. World Psychiatry 6:19–24

Fava GA (2008) Should the drug industry use key opinion leaders? No. BMJ 333:1405

Fava GA (2009a) Preserving intellectual freedom in clinical medicine. Psychother Psychosom 78:1–5

Fava GA (2009b) The decline of pharmaceutical psychiatry and the increasing role of psychological medicine. Psychother Psychosom 78:220–227

Fava GA, Guidi J (2007) Information overload, the patient and the clinician. Psychother Psychosom 76:1–3

Feinstein AR (1999) Basic biomedical science and the destruction of the pathophysiologic bridge from bench to bedside. Am J Med 107:461–467

Fleischhacker WW (2007) Conflicting views on conflicts of interest in medicine. World Psychiatry 6:32–33

Glassman PA, Hunter Hayer J, Nakamura T (1999) Pharmacological advertising revenue and physician organizations: how much is too much? West J Med 171:234–235

Gliedman CH, Nash EH, Huber SD, Stone AR, Frank JD (1958) Reduction of symptoms by pharmacologically inert substances and by short-term psychotherapy. AMA Arch Neurol Psychiatry 79:345–351

Goodwin G (2007) Conflicting interest and doing right. World Psychiatry 6:25–26

Healy D (2008) Irrational healers? Psychother Psychosom 77:198–200

Helmchen H (2004) Ethical implications of relationships between psychiatrist and the pharmaceutical industry. http://www.wfsbp.org/publications.html

Heres S, Davis J, Main K et al (2006) Why olanzapine beats risperidone, risperidone beats quetiapine, and quetiapine beats olanzapine? Am J Psychiatry 163:185–194

Holman HR (1976) The "excellence" deception in medicine. Hospital Practice 11:11–21

Horrobin DF (1990) The philosophical basis of peer review and the suppression of innovation. JAMA 263:1438–1441

Iheanacho I (2008) Paying for the view. BMJ 336:160

Jureidini JN, McHenry LB (2009) Key opinion leaders and pediatric antidepressant overprescribing. Psychother Psychosom 78:197–201

Kelly RE, Cohen LJ, Semple RJ et al (2006) Relationship between drug company funding and outcomes of clinical psychiatric research. Psychol Med 36:1647–1656

Kondro W (2004) Drug company experts advised staff to withhold data about SSRI use. Can Med Ass J 170:783

Krimsky S (2001) Journal policies on conflict of interest: if this is the therapy, what's the disease? Psychother Psychosom 70:115–117

Krimsky S, Rothenberg LS, Stott P et al (1998) Scientific journals and their authors' financial interests: a pilot study. Psychother Psychosom 67:194–201

Landefeld CS, Steinman MA (2009) The Neurontin legacy. N Engl J Med 360:103–106

Lee FLF, Chan JM (2008) Professionalism, political orientation, and perceived self-censorship. Issue Stud 44:205–238

Levinsky NG (2002) Nonfinancial conflicts of interest in research. N Engl J Med 347:759–761

Lexchin J, Bero LA, Djulbegovic B et al (2003) Pharmaceutical industry sponsorship and research outcome and quality. Br Med J 326:1167–1176

Lurie P, Tran T, Wolfe SM, Goodman R (2005) Violations of exhibiting and FDA rules at an American Psychiatric Association annual meeting. J Public Health Policy 26: 389–399

Maj M (2008) Non-financial conflicts of interests in psychiatric research and practice. Br J Psychiatry 193:91–92

Margolis J (1979) Conflict of interest and conflicting interests. In: Beauchamp TL, Bowie NE (eds) Ethical theory and business. Prentice-Hall, Englewood Cliffs, NJ, 361–372

Melander H, Ahlquist-Rastad J, Beermann B (2003) Evidence b(i)ased medicine. BMJ 326: 1171–1175

Mello MM, Clarridge BR, Studdert DM (2005) Academic medical centers' standards for clinical-trial agreements with industry. N Engl J Med 352:2202–2210

Moffatt B, Elliott C (2007) Ghost marketing. Perspect Biol Med 50:18–31

Moller HJ (2006) Ethical aspects of publishing. World J Biol Psychiatry 7:66–69

Moller HJ (2008a) Isn't the efficacy of antidepressant clinically relevant? Eur Arch Psychiatry Clin Neurosci 258:451–455

Moller HJ (2008b) Do effectiveness ("real world") studies on antipsychotics tell us the real truth? Eur Arch Psychiatry Clin Neurosci 258:257–270

Moller HJ (2009) Declaration of conflicts of interest in scientific publications. World J Biol Psychiatry 10:2–5

Montgomery JH, Byerly M, Carmody T et al (2004) An analysis of the effect of funding source in randomized clinical trials of second generation antipsychotics for the treatment of schizophrenia. Controlled Clin Trials 25:598–612

Nierenberg AA, Smoller J, Eidelman P, Wa YP, Tilley CA (2008) Critical thinking about adverse drug effects: lessons for the psychology of risk and medical decision-making for clinical psychopharmacology. Psychother Psychosom 77:201–208

Paul SM, Tohen M (2007) Conflicts of interest and the credibility of psychiatric research. World Psychiatry 6:33–34

Perlin ML, Bursztajn HJ, Gledhill K, Szeli E (2008) Psychiatric ethics and the rights of the persons with mental disabilities in mental institutions and the community. UNESCO Chair in Bioethics, Haifa

Perlis RH, Perlis CS, Wu Y et al (2005) Industry sponsorship and financial conflict of interest in the reporting of clinical trials in psychiatry. Am J Psychiatry 162:1957–1960

Petersdorf RG (1989) A matter of integrity. Acad Med 64:119–123

Rothman DJ, MvDonald WJ, Berkowitz CD et al (2009) Professional medical associations and their relationships with industry. A proposal for controlling conflict of interest. JAMA 301:1367–1372

Ryle JA (1948) Changing disciplines. Lectures on the history, method and motives of social pathology. Oxford University Press, London

Sismondo S (2007) Ghost management. PLOS Med 4(9):e286

Smith R (2005) Medical journals are an extension of the marketing arm of pharmaceutical companies. PLoS Med 2:e138

Spielmans GI, Thielges SA, Dent AL et al (2008) The accuracy of psychiatric medication advertisements in medical journals. J Nerv Ment Dis 196:267–273

Turner EH, Matthews AM, Linardatos E et al (2008) Selective publication of antidepressant trials and its influence on apparent efficacy. N Engl J Med 358:252–260

Uhlenhuth EN, Rickels K, Fisher S et al (1966) Drug, doctor's verbal attitude and clinic setting in the symptomatic response to pharmacotherapy. Psychopharmacologia 9:392–418

Vieta E (2007) Psychiatry: from interest in conflicts to conflicts of interest. World Psychiatry 6:27–29

Wazana A (2000) Physicians and the pharmaceutical industry. Is a gift ever just a gift? JAMA 283:373–380

Wilkes MS (2000) Conflict, what conflict? When trust goes, so does the healing power of physicians. West J Med 172:6–8

Wilkes MS, Doblin BH, Shapiro MF (1992) Pharmaceutical advertisements in leading medical journals. Experts' assessments. Ann Intern Med 116:912–919

Chapter 5
Between Legislation and Bioethics: The European Convention on Human Rights and Biomedicine

Felicity Callard

Contents

F. Callard (✉)
Health Service and Population Research Department, Institute of Psychiatry, King's College London, PO 34 Service User Research Enterprise, Institute of Psychiatry, De Crespigny Park Road, London, SE5 8AF, UK
e-mail: felicity.callard@kcl.ac.uk

H. Helmchen, N. Sartorius (eds.), *Ethics in Psychiatry*, International Library of Ethics, Law, and the New Medicine 45, DOI 10.1007/978-90-481-8721-8_5,
© Springer Science+Business Media B.V. 2010

Abbreviations

CAHBI	Council of Europe Ad Hoc Committee of Experts on Bioethics
CDBI	Council of Europe Steering Committee on Bioethics
CIOMS	Council for International Organizations of Medical Sciences
CPT	Council of Europe European Committee for the Prevention of Torture and Inhuman or Degrading Treatment or Punishment
GCP	Good Clinical Practice
ICH	International Conference on Harmonisation (of Technical Requirements for Registration of Pharmaceuticals for Human Use)
ICH–GCP	ICH Harmonised Tripartite Consolidated Guideline for GCP

5.1 Locating the Convention on Human Rights and Biomedicine

This chapter explores the role that regional legislation plays in framing both human rights and ethical principles in psychiatry by considering the Council of Europe's Convention on Human Rights and Biomedicine.[1] The Convention contributes to an emergent legislative, regulatory and discursive formation, which is characterized by its alloy of human rights and bioethics. The proportions of each component within the alloy, as well as the nature of the amalgamation, are subject to ongoing, lively debate.

The Convention on Biomedicine has been described as 'one of the most important bioethics texts from the point of view of international policy and law' (Hottois 2000, p. 133). As well as setting out the rights of patients and of research participants, it engages with topics (e.g. predictive genetic testing and research on embryos in vitro) that are at the inflammatory centre of bioethical debate. It also specifies that Parties to the Convention engage in 'appropriate public discussion' regarding developments in biomedicine. It thereby preoccupies itself with three important domains of relevance to ethics in psychiatry: psychiatry's manner of engaging with its patients, with those on whom it depends to conduct research, and with the publics that surround it.

In this chapter, I explore how the Convention on Biomedicine stages the intersection of human rights legislation, bioethics and biomedicine. Each is, I argue, being re-configured in relation to one another. An analysis of the Convention allows us, then, to consider the impact of biomedicine on several categories central to human rights and/or bioethics – those of patient, doctor, person, science, and research participant. I read the Convention as a document in tension. On the one hand, much of its focus is on the regulation of medical encounters (whether clinical or research focused); this is unsurprising given the Convention's title. On the other, the

---

[1] In full, the *Convention for the Protection of Human Rights and Dignity of the Human Being with Regard to the Application of Biology and Medicine: Convention on Human Rights and Biomedicine*, CETS No. 164; hereafter referred to as the Convention on Biomedicine). See http://conventions.coe.int/treaty/EN/Treaties/Html/164.htm

Convention at moments bears witness to new pressures on those categories on which it is grounded; it thereby brings to visibility some of the difficulties that surface when addressing biomedical matters through a legislative framework.

The chapter focuses on particular features of the Convention and their implications for psychiatry. Towards the end of the chapter, I contextualize the Convention in light of broader debates concerning human rights, bioethics and biomedicine. In so doing, I indicate how legislation both responds to changes in scientific and medical practices, and attempts to solidify the frameworks one might use to describe, as well as to regulate, this shifting terrain. I argue that the Convention on Biomedicine is central to the emergence of a particular politico-ethico-legal regime. This regime attempts to align the domains of human rights and bioethics through the regulation and formalization of the relationship between doctor and patient, and between researcher and research participant.

Psychiatry, we should recall, is a significant actor as regards both human rights legislation and biomedicine. A growing body of scholarship in law, the social sciences and the health sciences addresses the role of legislation in advancing and protecting the rights and entitlements of those with psychiatric diagnoses (Bartlett et al. 2007, Bernardet et al. 2002, Gostin 2001, Gostin and Gable 2004, Hale 2007, Lewis 2002, Niveau 2004, Parker 2007). That psychiatry is the only branch of medicine to use coercion means that it has often been at the forefront of legislative debates over medical interventions. At the kernel of many of the instruments addressing such interventions lies the spectre – sometimes addressed explicitly, sometimes not – of the involuntarily detained and/or treated psychiatric patient. In relation to biomedicine, too, psychiatry is in the vanguard of new technologies, methods and paradigms.[2] Notably, biological psychiatry – both in research and clinical practice – no longer plays itself out across the old, familiar dyad of doctor and patient. Its widening scope has the potential to subsume many more of us within the category of 'patienthood'. That individuals are increasingly positioned as susceptible to a disorder through models of genetic risk has installed what one might call the category of pre-patienthood (Rose 2007). At the same time, we are witnessing the increasing capaciousness of certain diagnostic categories, and the move towards using dimensional spectra rather than discrete categories. One could say, then, that the category of the patient is being fundamentally reworked, given that the divide that we liked to imagine might hold between disorder and normality – and between mental illness and mental health – is no longer so clear. Since the Convention on Biomedicine is often termed 'a patient rights treaty' (e.g. Katholieke Universiteit Leuven/Centre for Biomedical Ethics and Law 2008), it behoves us to interrogate who, exactly, is the patient called forth when her rights are granted protection.

There has also been a shift in how the interaction between patient and doctor is imagined (if not performed in practice). The challenges to a paternalistic

---

[2] Consider, for example, pharmacogenetics/genomics, non-Mendelian models of genetic risk and pathogenesis, genome-wide association studies, the use of animal models within behavioural genetics, and the development of biobanks and epidemiological databases.

model of medicine have come from a broad, politically disparate set of forces: these include the service user and patient rights movements; the changing organization of healthcare (which, through the logic of choice and the marketplace, is increasingly positioning the patient as the customer);[3] and the growth of accounting and accountability within healthcare (along with their many techniques of regulation and surveillance).

At the same time, biomedical research frequently demands a different kind of engagement from what it used to call its 'research subjects'. There is growing acknowledgement of the need for research involving 'vulnerable' patients, and hence pressure to widen the pool from which research participants are drawn. And there is increasing emphasis on acquiring biological data from large numbers of persons who are not – or not yet – patients. The development of biobanks, for example, are projects dependent on engaging *citizens* (Tutton 2007) rather than on recruiting patients. Research that demands the synthesis of genetic data has provoked extensive debate regarding their acquisition, storage, use and confidentiality. And, not least, there is a growing move towards engaging individuals in research as *participants* and *collaborators* rather than as *subjects* (Corrigan and Tutton 2006). The involvement of service users in research demands the elaboration of new models to describe the relationships between clinicians, researchers and recipients of healthcare. In light of these various pressures, the category of the medical research participant can no longer adequately be defined as split between 'the consenting patient' and 'the healthy control'.

5.2 Context of and Background to the Convention on Human Rights and Biomedicine

The Convention on Biomedicine is the outcome of a complex, lengthy process overseen by the Council of Europe. This process commenced in 1985, when the ad hoc Committee of Experts on Bioethics (CAHBI) was constituted to advise on legal and bioethical issues arising from rapid transformations in the biomedical sciences, and ended on 4 April 1997, when the Convention opened for signature in Oviedo, Spain.[4] At the time that this chapter went to press (April 2010), it had been signed and ratified by 26 States (the most recent being Montenegro in March 2010), and signed (but not ratified) by an additional 8 States. Two notable eligible States that have not signed are Germany and the United Kingdom.[5]

---

[3] We should not forget that the presence of coercion within psychiatry acts as a forceful counterweight to the logic of choice and that of the autonomous consumer.

[4] The Convention entered into force on 1 December 1999, when Spain became the fifth State Party to ratify.

[5] There are very different reasons lying behind these decisions not to sign. The United Kingdom finds certain articles of the Convention in conflict with its national legislation – most notably Article 18, which prohibits the creation of human embryos for research purposes. Germany had expressed grave concern throughout the drafting process of the Convention, particularly in relation

The Convention has significant scope – addressing the care provided by the full range of healthcare professionals, and the regulation of medical research *tout court*. But it in no way specifically addresses psychiatry, and in particular barely discusses the vexed questions of involuntary detention and treatment. Many of the Convention's articles are indebted to preceding conventions, codes and guidelines. The Convention, in turn, contributes to a complex web of norms, legislation, declarations and recommendations of relevance to psychiatric practice and research, and applicable in the European region. These include:

- *European Convention on Human Rights* (1950) and the extensive judgements of the European Court of Human Rights, most notably in relation to involuntary detention and treatment of psychiatric patients.
- *European Committee for the Prevention of Torture and Inhuman or Degrading Treatment or Punishment (CPT)*: Set up in 1987, this Committee ensures a proactive, non-judicial mechanism for the protection of human rights alongside the reactive, judicial mechanism provided by the European Court of Human Rights.
- *International Covenant on Civil and Political Rights* (1966): Article 7 specifies that 'No one shall be subjected to torture or to cruel, inhuman or degrading treatment or punishment. In particular, no one shall be subjected without his free consent to medical or scientific experimentation'. (This Article, we should note, implies the prohibition of *all* research in which a person is unable freely to consent.)
- Council for International Organizations of Medical Sciences (CIOMS) *International Ethical Guidelines for Biomedical Research Involving Human Subjects* (2002).
- World Medical Association. *Declaration of Helsinki* (1964, last amended in Seoul in October 2008).
- Committee of Ministers of the Council of Europe. *Recommendation No. R (99) 4 concerning the Legal Protection of Incapable Adults*.
- European Union. *Directive 2001/20/EC* on Good Clinical Practice in Clinical Trials.
- International Conference on Harmonisation of Technical Requirements for Registration of Pharmaceuticals for Human Use. *ICH Harmonised Tripartite Consolidated Guideline for GCP (ICH–GCP)*, adopted in 1996. The *ICH–GCP* is not an international law treaty, but an international ethical and scientific quality standard that each regulatory authority adopts as one of its own guidelines.
- World Health Organization, Regional Office for Europe. *Promotion of the rights of patients* (World Health Organization Regional Office for Europe 1995).
- World Health Organization European Region. *Mental Health Declaration for Europe: facing the challenges, building solutions*. Responsibilities of the

---

to articles concerning research involving those unable to give consent. Germany's opposition to the Convention has a complex history, and demands an understanding of the very particular place that bioethics holds within Germany.

Ministers of Health of the Member States include: 'enforc[ing] mental health policy and legislation that sets standards for mental health activities and uphold[ing] human rights'.

The Council of Europe has engaged with topics germane to biomedicine and bioethics since the 1980s. Non-binding recommendations to Member States have addressed the use of medical data, genetic screening diagnosis and counseling, as well as psychiatry and human rights.[6]

The Convention imposes both negative obligations and positive duties on States Parties. It provides *a set of minimum standards* for the protection of the patient and research participant: it explicitly provides States with the opportunity to allow a wider measure of protection with regard to the application of biology and medicine than that stipulated within the Convention (Article 27). Notably, ratification of the Convention obliges the State to implement legislation that will ensure that its national laws conform to the Convention's principles. But the Convention also allows national law to restrict many of the granted rights (Article 26), including that of free, informed consent to medical treatment: this carries important implications for the intractable psychiatric question of coercive treatment.

5.3 Content of the Convention on Human Rights and Biomedicine

There are 14 chapters within the Convention containing a total of 38 articles. The Explanatory Report to the Convention gives more specific guidance on how the Convention ought to be both interpreted and implemented.[7] There are currently three additional protocols to the Convention that are binding on the States ratifying them, and one that opened for signature in November 2008:

- *Additional Protocol on the Prohibition of Cloning Human Beings* (opened for signature 12 January 1998; entered into force 1 March 2001)
- *Additional Protocol on Transplantation of Organs and Tissues of Human Origin* (opened for signature 24 January 2002; entered into force 1 May 2006)
- *Additional Protocol concerning Biomedical Research* (opened for signature 25 January 2005; entered into force 1 September 2007). This protocol addresses all research involving a person, excepting epidemiological research and that using biological material
- Additional Protocol on Genetic Testing for Health Purposes (opened for signature 27 November 2008).

---

[6] For example, *Recommendation No. R (81) 1* on regulations for automated medical data banks; *Recommendation No. R (83) 10* on the protection of personal data used for scientific research and statistics; *Recommendation No. R (97) 5* on the use of medical data; *Recommendation No. R (90) 13* on genetic screening diagnosis and counselling; and *Recommendation (1235) 1* on psychiatry and human rights.

[7] See http://conventions.coe.int/Treaty/EN/Reports/Html/164.htm

There are a number of comprehensive accounts of the history and content of the Convention (see in particular Andorno 2005, de Wachter 1997, Roscam Abbing 1996, 1998). I shall not therefore cover the full scope of the Convention, but foreground some of its components that have implications for ethics in psychiatry.[8]

5.3.1 Article 3: Equitable Access to Health Care

In acknowledging both health needs and available resources, the article requires Parties to 'take appropriate measures with a view to providing . . . equitable access to health care of appropriate quality'. The Explanatory Report clarifies that '"equitable" means the absence of unjustified discrimination', and is not synonymous 'with absolute equality'.[9] The Article does not create an individual right on which a person may rely in legal proceedings against the State.

The article bends to the demands of pragmatism: it concedes the constraints of available resources, and imposes an obligation on States 'to use their best endeavours' to ensure the aim of equitable access to health care. Hottois (2000, p. 138; italics in original) has lamented the article's lack of a clear definition of the minimum requirements of health care policy that *'from the ethical point of view, and on behalf of human rights and human dignity, all societies should fulfil'*. The Explanatory Report specifies only that 'care must be of a fitting standard in the light of scientific progress'.

The Article concerns access to healthcare rather than the 'right to health'. Paul Hunt, UN Special Rapporteur on the Right to the Highest Attainable Standard of Health from 2002–2008, is developing an analytical framework concerning the right to health (founded upon General comment 14 of the UN Committee on Economic, Social and Cultural Rights) (Hunt and Mesquita 2006). Here, health is understood not simply as 'the absence of disease', and certainly not solely in relation to health care: it includes rights in relation to the underlying *determinants* of health. Given what we know about the power of socio-economic determinants to protect or undermine mental health, the 'right to health', rather than a right concerning access to healthcare, would be a more potent framework through which to address the positive rights of those with psychiatric diagnoses. Several voices (e.g. Farmer and Campos 2004) are calling for bioethics to turn from its traditional focus on the

---

[8] I am leaving to one side those parts of the Convention that focus on specific biomedical clinical and research practices (e.g. predictive genetic testing, interventions on the genome). Article 18, which addresses research on embryos in vitro, has been subject to vociferous debate and has probably received most commentary in print (e.g. Braake 2004, Hansen 2004, Reuter 2000, Walin 2007).

[9] Unjustified discrimination would include, I contend, the hostile and/or exclusionary responses that people with particular psychiatric diagnoses (e.g. personality disorders) often receive from mental health services, as well as the ignoring or under-treating of the physical health care needs of those with psychiatric diagnoses.

doctor–patient relationship towards a deeper interrogation of health inequalities and health determinants. That the Convention restricts its focus to the question of access to healthcare, then, endorses one vision of the intersection of human rights, bioethics and biomedicine over other possible visions.

5.3.2 Article 5: General Rule re Consent

The article specifies that any intervention in the health field must operate with free and informed consent to it, and subsequent to the provision of 'appropriate information as to the purpose and nature of the intervention as well as on its consequences and risks'. Interventions cover *all* medical acts – for the purpose of preventive care, diagnosis, treatment, rehabilitation or research.

The Explanatory Report stresses patients' 'autonomy in their relationship with health care professionals', and observes that informed consent 'restrains the paternalist approaches which might ignore the wish of the patient'. But there is a fly in the ointment as regards the implementation of free and informed consent in psychiatry. If the patient's wish entails disagreement with the health professional, that wish can end up transmogrifying into evidence of a patient's incapacity. As Peter Bartlett, professor of mental health law, has argued: 'In practice, the concern of users regarding capacity tests is that people are found capable if and only if they agree with the views of their physician' (Bartlett 2003, p. 341).

5.3.3 Article 6: Protection of Persons Not Able to Consent

Excepting very specific circumstances, a person without the capacity to consent can be subject to an intervention only for 'his or her direct benefit'. Additional specifications detail the position of minors and of adults unable to give full and valid consent because of mental incapacity. In both cases, there is an emphasis on acquiring the minor's or adult's *opinion* (even though authorization must be acquired from the representative or person/body provided for by law). There are provisions that allow the minor's opinion to be taken increasingly into account, even up to the point that 'the consent of a minor should be necessary, or at least sufficient for some interventions'.

The legal position of minors as regards medical interventions differs substantially in countries within the Council of Europe (see Buchner and Hart 2008 on Germany; Sprumont and Gytis 2005 on Estonia, Lithuania and Latvia; Stultiëns et al. 2007 on the EU Member States that have ratified the Convention [and which include Estonia and Lithuania]). For example, Cyprus, Greece and Slovakia specify that the age at which health care decisions may be taken autonomously is 18 years (legal majority); the Czech Republic allows evaluation on a case by case basis according to the age and majority of the minor and the seriousness of the intervention (Stultiëns et al. 2007).

The Convention does not adjudicate on whether capacity to consent ought to be based on actual capacity (at that moment and in relation to the proposed intervention) or on legal incapacity. Recommendation No. R (99) 4 on Principles Concerning the Legal Protection of Incapable Adults (adopted by the Committee of Ministers on 23 February 1999), which was drafted subsequent to the Convention, clarified that actual capacity ought to be favoured wherever possible (see Jansen 2000).

5.3.4 Article 7: Protection of Persons Who Have a Mental Disorder

Article 7 permits coercive treatment in situations in which not treating a person with 'a mental disorder of a serious nature' for that disorder would be likely to result in serious harm to his or her health. But Article 26 allows restrictions on the exercise of the Convention's rights 'in the interest of public safety, for the prevention of crime, for the protection of public health or for the protection of the rights and freedom of others'.[10] The Explanatory Report further explicates that a person who may be a possible source of serious harm *to others* owing to his/her mental disorder may be subjected to confinement or treatment *without his/her consent*. Johan Legemaate, professor of health law, has argued strongly, however, that the primary concern of the Convention on Biomedicine is *with the patient's health*: '[l]egislation in which considerations regarding the protection of others prevail over the patient's health interests contravenes the [Convention's] provisions and intentions' (Legemaate 2005, p. 130).

5.3.5 Article 9: Previously Expressed Wishes

This article specifies that if a patient is not able to express his/her wishes at the time of the intervention, previously expressed wishes 'shall be taken into account'. This phrase carefully manoeuvres around the complex and diverse legal status of living wills and advance directives.

The Explanatory Report notes that there are occasions when such previously expressed wishes should *not* necessarily be followed, for example, when a long time has elapsed between their expression and the intervention and 'science has since progressed'. This statement gives some cause for alarm as regards psychiatry. To adjudicate that science 'has progressed' demands evidence. In a recent systematic review of patients' perspectives on electroconvulsive therapy, at least one third of patients reported persistent memory loss. This countered the claim made at that time by the Royal College of Psychiatrists (the professional body for psychiatrists in the United Kingdom and the Republic of Ireland) that over 80% of patients were satisfied with electroconvulsive therapy and that memory loss was

---

[10] Cf. Tannsjo (2004), who appears to discount Article 26.

not clinically important. The authors argued that their review demonstrated that the Royal College's statement – which would have been widely regarded as representing scientific consensus on this intervention – was unfounded (Rose et al. 2003). In this case, the orthodox position regarding 'the progression of science' – one in which electroconvulsive therapy is deemed both effective and satisfactory – is shown to be both partial and misleading. Determining what 'progression in science' means in psychiatry is not a straightforward procedure.

5.3.6 Article 10: Private Life and Right to Information

This Article applies the more general right to privacy set out in the European Convention on Human Rights to the specific field of personal health information. A person's 'right to know' covers all health information. This is of clear relevance to ethical practice in psychiatry: diagnoses are frequently withheld from patients, and such diagnoses can have profound implications – not least in terms of their potential for stigma, discrimination and social exclusion (Sartorius 2002). That the diagnostic classificatory systems used by psychiatry are frequently contested both within and beyond medicine means that the collection of 'health information' within psychiatry cannot be an unmarked, neutral practice.

Article 10 also contains the 'right not to know' about certain aspects of one's health (see Fig. 5.1). This reflects the new complexity that genetic research has brought to questions of autonomy, privacy and information. How ought a person to respond to the offer of a genetic test that reveals susceptibility to a genetic condition? Does she – and/or the establishment proffering the genetic test – have an ethical duty to pass any knowledge resulting from genetic tests to her kin? The status of 'not knowing' or 'not wishing to know' has provoked rich debate. Roberto Andorno has strongly defended the right not to know one's genetic status on the grounds that it marks an enhancement rather than diminution of autonomy (Andorno 2004). Graeme Laurie shifts from the language of rights to that of *interest*. He argues that individuals are entitled to enjoy psychological privacy – and it is this that can be pierced by unsolicited disclosures (e.g. about one's health status). On his account, some protection for psychological integrity is desirable, because of the interest that individuals might have in not knowing (Laurie 2002). The sociologists Michel Callon and Vololona Rabeharisoa move the debate into different

---

Bulgaria

In Bulgaria, while the patient has a right not to know, this right is circumscribed: Article 92(2) of the Health Act specifies that it may be applied only in relation to the disease in connection with which the patient has sought medical aid (Goffin et al. 2007).

Fig. 5.1 The 'right not to know' in Bulgaria

terrain entirely, in their analysis of one individual's refusal to engage with knowledge about genetic transmission. They interpret the individual's lack of engagement as a positive refusal of the free-choice, autonomous and responsible liberal subject position. This refusal marks the individual's affirmation of a set of attachments and obligations – an ethics, in other words – different from the one underpinning genetics (Callon and Rabeharisoa 2004).

Article 10 indicates a moment in the Convention in which it is particularly evident that discussions regarding biomedicine are placing pressure on formulations concerning autonomy, medical information, and the responsibilities of the health professional. While Article 10 is couched in the language of rights, debate over the substance of the article move between the overlapping but analytically distinctive languages of rights, interests, bioethics, and ethics.

5.3.7 Articles 15–17: Scientific Research

While Article 15 issues a strong statement that biological and medical research shall be carried out freely, there are firm constraints placed on research involving persons (see Fig. 5.2).[11] These include:

- That no alternative of comparable effectiveness to research on humans exists
- That the risks occurred by the person are not disproportionate to the potential benefits of the research
- That the research has been approved by a competent body involving multidisciplinary review of ethical acceptability
- That persons involved in the research have been informed of their rights
- That necessary consent has been given and has been documented. This consent can be freely withdrawn at any time.

Finland

Medical research involving minors is regulated by the Medical Research Act of 1999 (No. 488 of Statutes, 9 April 1999), which was drafted specifically with the Convention in mind.
Lötjönen (2008) argues that the Act widens the coverage of the requirement of minimal risk (the Finnish Act uses 'slight risk') to *all* medical research, not simply research where there is no direct benefit to the participant:

Minors may be research subjects only where it is not possible to obtain the same scientific results using other research subjects and where the risk of harming or distressing the research subject is only very slight. (Section 8 of the Medical Research Act, unofficial translation by the Ministry of Social Affairs and Health http://www.finlex.fi/en/laki/kaannokset/1999/en19990488.pdf)

Fig. 5.2 Country-based legislation as regards those unable to consent to research

---

[11] The extensive commentary regarding this article includes Lötjönen (2005, 2006), Welie and Berghmans (2006).

There are additional restrictions regarding persons unable to consent to research:

- The results of the research have the potential to produce real and direct benefit to the person's health
- That it is not possible to conduct research of comparable effectiveness with individuals capable of giving consent
- Necessary authorisation has been given in writing
- The person concerned does not object.

In exceptional circumstances, research that does not have the potential to produce results of direct benefit to the health of the research participant may be authorised. Such research must 'entail only minimal risk and minimal burden'. This requirement imposes an absolute criterion that acknowledges both the objective ('risk') and subjective ('burden').

The EU Directive regarding clinical trials appears slightly more capacious in what it will allow. In the case of research where there is no potential benefit, the Directive specifies – both in relation to research involving minors and incapacitated adults – that clinical trials must be designed 'to minimise pain, discomfort, fear and other foreseeable risk in relation to the disease and developmental stage'. (It does specify, however, that the threshold for risk and degree of distress be specifically defined and constantly monitored.) The differences between the Convention and the Directive bear witness to the intractable difficulty of adequately protecting research participants at the same time as not stopping biomedical research in its tracks.

Welie and Berghmans (2006) express reservations over whether research can ever meet such a requirement. The Act allows 'non-therapeutic' research involving those who lack mental capability in cases where research could not take place without the participation of such persons. But this demands that the 'risk associated with participation is negligible and the inconveniences minimal' (Article 4). Welie and Berghmans argue that the imperative of proportionality implies that 'even the slightest risk or discomfort for subjects would imply inadmissibility of the research if not compensated for by some expected advantage for the subjects' (p. 73).

5.3.8 Article 28: Public Debate

States Parties must ensure that there is 'appropriate public discussion' of 'the fundamental questions' raised by developments in biology and medicine. The Explanatory Report notes that the purpose of this article is to create public awareness, and that society's views must be ascertained as regards any problems.

Such formulations largely remain, I argue, within what Brian Wynne has termed the 'deficit model' of public understanding of science (Wynne 1991). This model characterises the relation between science and its public through assuming that citizens' ambivalence or resistance to science and technology can be explained by their lack of knowledge. What is required, then, is that the public's awareness and knowledge be improved (with the concomitant expectation that this will reduce their ambivalence or resistance). Wynne and many others have subjected the deficit model to sustained critique, objecting to its linear and paternalistic tenor. They lament that the model's problematisation of the relationship between science and the public actually translates into the problematisation solely of the *latter term* ('the public').

While the title of article 28 refers to 'debate', the article's teeth are blunted by the very act of signing and ratifying the Convention. For this ratification demands that States Parties endorse particular positions regarding biomedicine – including, in light of article 28, a particular position regarding the relationship between science and its publics! – and thereby removes the possibility of truly open debate on such matters (Borst-Eilers 2005). Public 'debate' as prescribed by the article actually implies, I contend, a much weaker, one-way 'informing' of the public.

5.4 Influence of the Convention on Legislative Mechanisms

5.4.1 Country-Based Patient Rights Legislation

The Convention on Biomedicine has acted as an engine for legislative transformation across the European region. Ianeva has argued that the ratification of international treaties regarding genetics and biomedicine within Eastern European countries 'is the fasted [sic] way to regulate those matters and is becoming the venue of choice' (Ianeva 2006, p. 2). The Convention has force not simply because it is a binding instrument on those States Parties ratifying it: it also acts as a hortatory device that is influential in shaping norms, expectations, and discourse across a wider geographical span.

The Convention has, unsurprisingly, been a spur to comparative health law research (Manuel et al. 1999, Nys 2008). Most notable in this regard is the research being conducted by the Centre for Biomedical Ethics and Law of the Catholic University of Leuven, Belgium, in collaboration with EuroGentest (see http://europatientrights.eu). This research involves an examination of patient rights legislation in EU Member States (e.g. Goffin et al. 2008, Nys et al. 2007).[12] This Centre's research has developed the following categorisation of types of patient rights legislation (see Figs. 5.3 and 5.4):

- *special/split*: 'special' denotes States where patient rights are contained in specific legislation, in comparison with 'split' States where they appear in a variety of laws.

---

[12] This Centre is issuing a series of *European Ethical-Legal Papers*. These evaluate the efforts and progress of individual Member States of the European Union to promote and protect the rights of patients and users of health services in relation to the Convention on Biomedicine (for details, see Katholieke Universiteit Leuven/Centre for Biomedical Ethics and Law 2009).

- *horizontal/vertical:*

    - *Horizontal ('civil law') approach*: patient rights are well-defined and action-able against specified parties. The patient has a right of appeal to a Court or similar authority if they are not respected. If violation occurs, compensation and/or sanction can be imposed.
    - *Vertical ('public law') approach*: Patients have 'quasi-legal rights', in that they have no avenue for direct action against the healthcare provider. Instead, obli-gations are placed on doctors and other healthcare providers (e.g. in legally binding codes of medical deontology).

- *nominate/innominate*

    - *Nominate* implies a specific regulation of the treatment contract between doc-tor and patient within the civil code, whereby the 'treatment contract' can be regarded as a special case of a general 'contract for services'.
    - *Innominate* describes a general acceptance of the contractual nature of the rights of patients and duties of physicians within jurisprudence/legal literature, without a specific, nominate contract of services.

5.4.2 The Challenge of Harmonization

The Convention was explicitly designed to assist in 'harmonis[ing] existing stan-dards' (Explanatory Report, p. 2). In this, it contributes to the harmonising tenden-cies of law from the Council of Europe and the European Union. But we ought not to assume that a harmonising *tendency* results in de facto legislative harmonisation. The research by the Centre for Biomedical Ethics and Law/ EuroGentest indicates the diversity of legislative approaches and specific responses within States Parties to the Convention's requirements surrounding patient rights.

Such legislative variability has also been made visible by the Eunomia project (European Evaluation of Coercion in Psychiatry and Harmonisation of Best Clinical Practice). Dressing and Salize found huge variety in the legal regulations used across the European Union for detaining those with mental disorders (Dressing and Salize 2004). They argued that while a variety of legal reforms had been undertaken in rela-tion to patients' rights vis-à-vis compulsory admission, 'there is no overall common approach' to safeguarding these rights.

Sjef Gevers, professor of health law, in reflecting on the influence of both Strasbourg and Brussels on health law, has argued that 'If there is anything sur-prising, it is rather how tenaciously local national health law sometimes still is' (Gevers 2008, p. 264). Both Gevers (2008) and Legemaate (2005) have documented several contributors to the diversity in approaches to informed consent, confidential-ity and the legal protection of psychiatric patients: the differences between common and civil law approaches; the differences in political as well as health care struc-tures; and cultural and national norms. In addition, while we are beginning to know more about the specificities of legislation in States within the Council of Europe, we know relatively little about the effectiveness in daily practice of these legislative

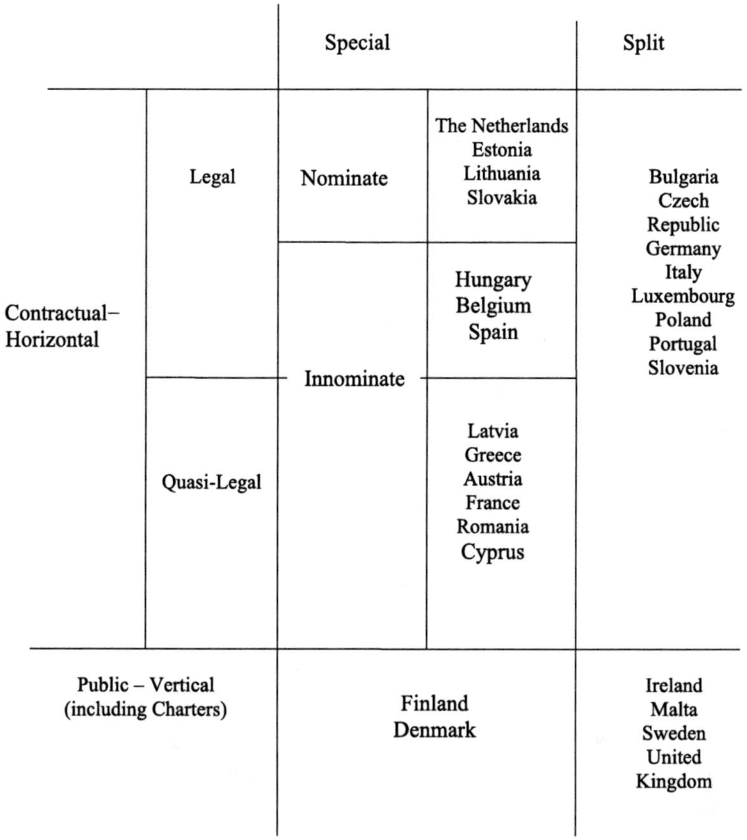

			Special	Split
Contractual–Horizontal	Legal	Nominate	The Netherlands Estonia Lithuania Slovakia	Bulgaria Czech Republic Germany Italy Luxembourg Poland Portugal Slovenia
	Quasi-Legal	Innominate	Hungary Belgium Spain	
			Latvia Greece Austria France Romania Cyprus	
Public – Vertical (including Charters)			Finland Denmark	Ireland Malta Sweden United Kingdom

Fig. 5.3 Typology of patient rights legislation (from http://europatientrights.eu [Katholieke Universiteit Leuven/Centre for Biomedical Ethics and Law]; permission for use granted by Professor Nys])

Netherlands

The Netherlands was the first European country to introduce (in 1994) a specific regulation of the treatment contract between doctor and patient in its civil code. Legislation in Estonia, Lithuania and Slovakia is greatly indebted to this model.

Finland

The Act on the Status and Rights of Patients (No. 785/1992) was the first patient rights act: it regulates the right to information, the right to self-determination, and the maintenance of patient records (Goffin et al. 2008).

Fig. 5.4 Influential country-based legislative approaches to patient rights legislation

Croatia

Croatia ratified the Convention in July 2003. In November 2004, it passed the *Act on the Protection of Patients' Rights* (Official Gazette no. 169/04) to align its legislation with the Convention.

Babić-Bosanac and Džakula (2006), Rušinović-Sunara and Finka (2008) and Rušinović-Sunara (personal communication 28 January 2009) provide a range of factors to explain why the introduction of legal mechanisms in Croatia has not resulted in sufficient protection or respect being given to patients' rights:

- The persistence of the traditional medical-paternalistic approach between doctors and patients (on the part of patients as well as doctors)

- Insufficient informing and education of health professionals re patients' rights

- Insufficient education of the public re their rights as patients; or, people are insufficiently prepared/able to claim their rights when they have been violated

- No standards or norms to set in action or evaluate protection (and improvement) of patients' rights in relation to specific health care practices

- Systemic, market-oriented changes in the healthcare and insurance system

- Absence of independent patients' rights representatives

- Inadequacies in the legislation itself

Fig. 5.5 A country-based case study of difficulties in implementing patient rights legislation

provisions (Roscam Abbing 2004) (Though see Fig. 5.5 for a country-based case study of difficulties in implementing patient rights legislation).

5.4.3 European Court of Human Rights

It is important to emphasize that it is the European Court of Human Rights, rather than the Convention on Biomedicine, that defines the contours of what are arguably the most pressing ethical questions in psychiatry, namely those relating to the involuntary detention and treatment of psychiatric patients. One must turn to Article 5 of the European Convention on Human Rights, rather than to the Convention of Biomedicine, for elucidation of when the deprivation of liberty can be justified.

While the Convention on Biomedicine does not provide persons with a right of petition to the European Court of Human Rights, Article 29 allows the European Court to provide advisory opinions on legal questions concerning interpretation of the Convention. In addition, infringements of rights within the Convention on Biomedicine may be considered under the European Convention on Human Rights (if rights contained in the latter Convention are also considered to have been violated). The case of *Glass v. UK* (Application No. 61827/00; judgement made 9 March 2004), for example, concerned an alleged breach under Article 8 of the European Court of Human Rights. The Court invoked the Convention on

Biomedicine (citing Articles 5 and 9). This is particularly interesting given that the United Kingdom has not even signed the Convention on Biomedicine. Judit Sándor, in analysing references to this Convention in court, suggests that it is currently mentioned not as a legal norm but rather as a general standard: 'The assessment of the text has been uncertain as to whether it is ethical or legal in nature, yet it seems to have become indispensable to refer to it in cases having bioethical implications' (see Sándor 2008, p. 27). She points, then, to the interpretative instability currently characterizing this instance of a new mode of combining rights and ethics.

5.5 The Broader Terrain

In this final section, I return to my claim that the Convention contributes to a new politico-legislative-ethical regime that is attempting to align bioethics and human rights. I do so by posing, briefly, three questions that are addressed to the point of intersection between human rights, bioethics and biomedicine.

5.5.1 What Is the Relationship Between Human Rights and Bioethics?

The Convention explicitly stages itself as a human rights document: the Explanatory Document notes that the Convention on Biomedicine and the Convention for the Protection of Human Rights and Fundamental Freedoms (1950) 'share not only the same underlying approach but also many ethical principles and legal concepts'. But during the drafting period, the Convention on Biomedicine was known as the 'Bioethics Convention'. The report documenting preparatory work on the Convention gives a fascinating account of the turn away from the term 'bioethics', noting explicitly that it had 'negative connotations' in the German-speaking countries and that it 'did not adequately emphasize the legal (apart from ethical) nature of the provisions in the text' (Steering Committee on Bioethics (CDBI) 2000, p. 5). Undoubtedly, the diverse ethical positions held by those involved in the drafting of the Convention meant that the document had to yield to the demands of consensus. '[L]aw with its internationally at least basically accepted notions and terms would shape the text', Reuter argues, 'to a much greater extent than ethics with its evident plurality' (Reuter 2000, p. 187). But while the term 'bioethics' disappears at the level of the title, it does not disappear at the level of content.

The Convention on Biomedicine therefore needs to be set in the context of a larger, spirited debate concerning the emergence and/or very possibility of international bioethics.[13] A number of writers have addressed the divergence between

---

[13] This has been most evident in relation to the heated discussions concerning the Universal Declaration on Bioethics and Human Rights (UDBHR), which was adopted by UNESCO in October 2005.

human rights and bioethics approaches to similar topics (e.g. the edited collection of Freeman 2008). Richard Ashcroft has noted the conceptual difficulty of establishing the relationship between bioethics and human rights: does bioethics provide human rights with a philosophical foundation, or does it operate in 'applied form' through its ability to address discrete and concrete issues (Ashcroft 2008)? In particular, Ashcroft points to the divergence between the positioning of human rights as foundational within human rights discourse, in comparison with the tendency for bioethicists to ground human rights through other abstract categories such as autonomy.

This poses challenges for a Convention that attempts to draw human rights and bioethics into alignment. The Convention explicitly turns to the categories of dignity and identity to anchor the protection of rights and fundamental freedoms with regard to the application of biology and medicine. But both dignity and identity are subject to substantial contestation in bioethical debates. It would take me too far afield to address the complex discussions surrounding dignity (e.g. Macklin 2003, Salako 2008, Schmidt 2007). Suffice it to say here that at the crux of many of these debates is whether and how biomedical practices conflict with or undermine 'human dignity'. That the Convention grounds itself *ab initio* in this concept, at the same time as it attempts to adjudicate on biomedicine, might seem to beg one of the most fundamental bioethical questions. Likewise, the language of personhood that grounds human rights sits awkwardly alongside the Convention's preoccupation with non-person-like entities (such as the embryo, the genome and – indeed – biomedicine itself) (see Ashcroft 2008). How, successful, then, is the Convention in its amalgamated address both to persons with human rights and to such abstract categories as the interests of science or of society?[14]

The Convention, beyond its dependence on the language of dignity, is inflected throughout with the value of autonomy (Article 5 regarding consent, for example, 'makes clear patients' autonomy in their relationship with health care professionals'). But while autonomy is frequently invoked – often as an indisputable good – in bioethical debates, it is also facing pressure on a number of fronts. We have already seen differences in opinion as to whether autonomy is the appropriate category through which to ground a patient's 'wish not to know' health information. Let us briefly consider two additional examples.

The first concerns ethical frameworks through which we might approach and regulate human genetic research. Bartha Knoppers and Ruth Chadwick, bioethical and legal experts, have been arguing that 'ethics as usual' is not sufficient as regards biomedical and biotechnological practices. They call for a move away from *autonomy* as the ultimate arbiter shaping ethical deliberation (at least in bioethical debates in the Global North), and towards an acknowledgement of the importance of 'the participatory approach'. Population-based genetic research, and the participatory approach it requires, points, they argue, to the emergence of values of

---

[14] Article 2 specifies that 'The interests and welfare of the human being shall prevail over the sole interest of society or science'.

'reciprocity, mutuality, solidarity, citizenry and universality'. Such values counter the traditional endorsement of individual choice and of ever more stringent data privacy requirements (Knoppers and Chadwick 2005).

The second concerns informed consent. Neil Manson and Onora O'Neill, in an important monograph entitled *Rethinking Informed Consent in Bioethics*, have challenged the model of informed consent grounded in individual autonomy. They move away from a prioritization of individual autonomy, arguing that the demand that consent be fully explicit and fully specific is unrealistic, and hence that such consent is rarely adequately secured. On their account, the model of individual autonomy buttresses the misplaced assumption that the transmission or disclosure to the patient or research participant of more – and more – information always supports better-informed choice. 'An excessive emphasis on individual choice', they argue, 'does not ensure sufficient attention to the full range of ethical, legal and professional obligations' central to biomedical practice (Manson and O'Neill 2007, p. 191).

5.5.2 Can Rights Undermine Ethics?

The Convention's Explanatory Report notes in relation to Article 4 that:

> An important element of these duties [of the doctor] is the respect of the rights of the patient. The latter creates and increases mutual trust. The therapeutic alliance will be strengthened if the rights of the patient are fully respected.

But ought we to rest content with the assumption that 'rights discourse' necessarily increases mutual trust? While Henrietta Roscam Abbing has argued that the strengthening of the patient's role via patient rights legislation has meant that 'the doctor–patient relationship has become marked by mutual trust, with respect for each other's position and responsibilities' (Roscam Abbing 2004, p. 11), others have maintained that precisely the opposite can occur. The emphasis on rights, duties and sanctions can, it is argued, result both in less trust and in less ethical behaviour: healthcare professionals can become more preoccupied with fulfilling procedural and technical requirements (concerning consent, the dissemination of information, etc.) than on engaging in an ethics of care (Mol 2008, Sperling 2008).

Of central concern is the tendency for legislative and governance arrangements to narrow the range of vision to compliance with the law and thereby to occlude engagement with ethical considerations. These considerations range from ethical conduct in the practice of healthcare itself, to broader arguments concerning health inequalities, the conduct and geographical location of clinical trials, and the use of, and people's access to, various therapies. In addition, an increase in legislative and governance arrangements often goes hand in hand with the 'marketisation' of healthcare – and 'marketisation' has of course been subject to sustained critique, on the grounds of both ethics and social justice.

Psychiatry undoubtedly adds complexity to each of these issues. While debates in other medical fields have turned, for example, to inequities as regards access to pharmacological therapies, the use of pharmacological treatment in psychiatry is far more contested. Often, indeed, ethical discussions centre on the *overuse* of particular kinds of pharmacological treatment. More generally, the possibility, as well as the frequent reality, of coercive practices intruding into the relationship between the psychiatric professional and the patient mean that debates over how best to establish trust, and enhance the therapeutic alliance, are inevitably more vexed. Indeed, in psychiatric contexts in which human rights violations are common, to worry over whether the discourse of patients rights discourse works in tandem with, or perpendicular to, trust and the therapeutic alliance, might seem a luxury too far.

5.5.3 What Is the Relationship Between Biomedicine and Society?

The Explanatory Report of the Convention notes 'the ambivalent nature' of many of the advances in medicine and science:

> The scientists and practitioners behind them have worthy aims and often attain them. But some of the known or alleged developments of their work are taking or could potentially take a dangerous turn, as a result of a distortion of the original objectives. Science, with its new complexity and extensive ramifications, thus presents a dark side or a bright side according to how it is used.

This statement makes a familiar distinction between the unsullied aims of science and its at times corrupted applications. But biomedicine, as a number of scholars within philosophy and science and technology studies have argued, does not allow us to make this divide between the 'pure' search for knowledge and its subsequent application. This is not least because biomedicine acquires knowledge by *interfering with* its object, which means that we cannot position moral and legal issues as pertaining only to the application phase, but must see them as embedded within the very practice of scientific research (Honnefelder 2005, p. 15).

The Convention, then, at once acknowledges that biomedicine poses new questions vis-à-vis the practices of regulation that might govern medical intervention, and yet remains constrained by its reliance on what I regard as an inadequate model through which to understand the relationship between science and society. Biomedicine, through its ability to intervene at the level of our organic nature, has the ability to transform the nature of our human subjecthood. Such transformations are, indeed, arguably already in the process of shifting normative conceptions of selfhood, of agency and of human freedoms. This transformational potential throws its shadow over the Convention as a whole. The very existence of the Convention derives, on the one hand, from the acknowledgement of the challenges that biomedicine is posing. But the language of the Convention – in its cleaving to familiar models of the patient, of his or her autonomy, and of science and its relationship to society – seems at the same time reluctant to engage with them.

5.6 Conclusion

Human rights have been one of the battle cries taken up by the mental health service user movement and by those campaigning for the civil rights of those subjected to psychiatric detainment and/or treatment. Many have regarded the discourse of ethics as insufficiently robust to combat deleterious psychiatric practices. But the language of rights within the Convention on Biomedicine is indebted to other forces beyond those of human rights and civil rights in health movements. Those forces include those consolidating consumerist models of healthcare, as well as regulatory frameworks designed to lubricate an increasingly globalised circuit of pharmaceutical and other biotechnological/biomedical research. There is, concurrently, growing recognition of how the reach and inequitable distribution of biotechnology renders national law insufficient as regards appropriate legislative control over such powerful practices as multi-country clinical trials and the extraction of value from biobanks. To interrogate the ethical implications of the codification and regulation of a range of relationships through the language of rights is no easy matter.

The Convention on Biomedicine contributes, then, to the consolidation of a range of tendencies within both healthcare and medical research practice. These include the logic of medical accountability; of a contractual relationship between doctor/healthcare facility and patient; and of medical professionalism that is framed through the promise of information transmission that will result in patient choice. This contractual relationship is both occasioned by, and subject to stress from, the changing shape of biomedicine. For while rights are called upon to protect the patient and the research participant, biomedicine is in the process of transforming the meaning of those categories, and of reconfiguring the nature of the ties that link individuals to one another, to their own tissue and genetic data, to healthcare professionals, and to biotechnologies such as genetic databases. These transformations are occasioning both calls for greater protection through well-established bioethical and legal categories (e.g. models of informed consent and data privacy), and demands for new ethico-technological frameworks through which to address this new biomedical terrain (e.g. Knoppers and Chadwick's call for an ethics of responsibility and participation; Manson and O'Neill's new model of informed consent).

The Convention is not solely a vehicle for the harmonisation of patients' rights legislation. It also offers a harmonizing framework to legislators, patients, doctors, healthcare policy makers, and the public. This is a framework both for the forging of norms, standards and principles, and for the concretization of what those very categories (law, 'patienthood', healthcare, bioethics, the public, biomedicine) at the heart of bioethical and legislative interrogations of biomedicine might mean. The Convention attempts to unite legal models and bioethical principles through regulating the practices of healthcare professionals and researchers, and through offering patients and research participants empowerment via the exercise of autonomous decision-making, the protection of privacy, and the practice of informed consent.

How ought we to interpret the Convention's contributions to ethics in psychiatry? It is apposite to recall the statement by David Callahan, co-founder of the Hastings Center (the US research institution dedicated to bioethics), that '[law] may be the best institution we have, but it is a poor substitute for moral consensus and public debate on ethics' (Callahan 1996, quoted in Sperling 2008). It would be foolhardy to jib at the protections afforded by the Conventions on Human Rights and Biomedicine – not least because of the numerous and continuing human rights violations experienced by many people with psychiatric diagnoses across the European region both in the context of clinical care and medical research. Nonetheless, it is important to reflect, in a book on ethics in psychiatry, on how the Convention shapes the available terrain for debate as regards both human rights and bioethics, and what might be occluded within its framework.

Acknowledgments The author acknowledges financial support from the National Institute for Health Research (NIHR) Specialist Biomedical Research Centre for Mental Health award to South London and Maudsley NHS Foundation Trust (SLaM) and the Institute of Psychiatry, King's College London. She thanks Dr Dula Rušinović-Sunara of the Croatian Association of Patients' Rights for providing her with material on the Association's attempts to reform patients' rights legislation in Croatia.

References

Andorno R (2004) The right not to know: an autonomy based approach. J Med Ethics 30:435–439
Andorno R (2005) The Oviedo convention: a European legal framework at the intersection of human rights and health law. J Int Biotech Law 2:133–143
Ashcroft R (2008) The troubled relationship between bioethics and human rights. In: Freeman MDA (ed) Law and bioethics. Oxford University Press, Oxford, pp 31–51
Babić-Bosanac S, Džakula A (2006) Patients' rights in the Republic of Croatia. Eur J Health Law 13:399–411
Bartlett P (2003) The test of compulsion in mental health law: capacity, therapeutic benefit and dangerousness as possible criteria. Med Law Rev 11:326–352
Bartlett P, Lewis O, Thorold O (2007) Mental disability and the European convention on human rights. Martinus Nijhoff, Leiden/Boston
Bernardet P, Douraki T, Vaillant C (2002) Psychiatrie, droits de l'homme et défense des usagers en Europe. Editions Erès, Ramonville Saint-Agne
Borst-Eilers E (2005) The role of public debate and politics in the implementation of the convention. In: Gevers J, Hondius E, Hubben J (eds) Health law, human rights and the biomedicine convention. Martinus Nijhoff, Leiden/Boston, pp 247–254
Braake TA (2004) The Dutch 2002 Embryos act and the convention on human rights and biomedicine: some issues. Eur J Health Law 11(2):139–151
Buchner B, Hart D (2008) Research with minors in Germany. Eur J Health Law 15:127–134
Callahan D (1996) Escaping from legalism: is it possible? Hastings Center Report 26(Nov-Dec), 34–35
Callon M, Rabeharisoa V (2004) Gino's lesson on humanity: genetics, mutual entanglements and the sociologist's role. Econ Soc 33(1):1–27
Corrigan O, Tutton R (2006) What's in a name? Subjects, volunteers, participants and activists in clinical research. Clin Ethics 1:101–104
de Wachter MA (1997) The European convention on bioethics. Hastings Cent Rep 27(1):13–23
Dressing H, Salize HJ (2004) Compulsory admission of mentally ill patients in European Union Member States. Soc Psychiatry Psychiatr Epidemiol 39:797–803
Farmer P, Campos NG (2004) New malaise: bioethics and human rights in the global era. J Law Med Ethics 32:243–251
Freeman M (ed) (2008) Law and bioethics. Oxford University Press, Oxford

Gevers S (2008) Health law in Europe: from the present to the future. Eur J Health Law 15:261–272

Goffin T, Borry P, Dierickx K, Nys H (2008) Why eight EU Member States signed, but not yet ratified the Convention for Human rights and biomedicine. Health Policy 86(2–3): 222–233

Goffin T, Zinovieva D, Borry P, Dierickx K, Nys H (2007) Patient rights in the EU – Bulgaria. European Ethical-Legal Papers (Vol. No 8). Leuven

Gostin LO (2001) Beyond moral claims: a human rights approach in mental health. Camb Q Healthc Ethics 10:264–274

Gostin L, Gable L (2004) The human rights of persons with mental disabilities: a global perspective on the application of human rights principles to mental health. MD Law Rev 63(1):20–121

Hale B (2007) Justice and equality in mental health law: the European experience. Int J Law Psychiatry 30:18–28

Hansen B (2004) Embryonic stem cell research: terminological ambiguity may lead to legal obscurity. Med Law 23(1):19–28

Honnefelder L (2005) Science law and ethics: the Biomedicine Convention as an ethico–legal response to current scientific challenges. In: J Gevers, Hondius E, Hubben J (eds) Health law, human rights and the biomedicine convention. Martinus Nijhoff, Leiden/Boston, pp 13–22

Hottois G (2000) A philosophical and critical analysis of the European convention of bioethics. J Med Philos 25(2):133–146

Hunt P, Mesquita J (2006) Mental disabilities and the human right to the highest attainable standard of health. Hum Rights Q 28:332–356

Ianeva E (2006) Biopolitics in Eastern Europe: specific political and legal challenges. Paper presented at the Connecting Civil Society – Implementing Basic Values Conference, Berlin, March 17–19. Retrieved 2 February, 2009, from http://www.boell.de/alt/downloads_uk/Ianeva_Panel1.pdf

Jansen S (2000) Recommendation No R (99) 4 of the Committee of Ministers to member states on principles concerning the legal protection of incapable adults, an introduction in particular to part V Interventions in the health field. Eur J Health Law 7:333–347

Katholieke Universiteit Leuven/Centre for Biomedical Ethics and Law (2008) Facts: Biomedicine Convention. Retrieved 1 February 2009, from http://europatientrights.eu/biomedicine_convention/biomedicine_convention_facts.html

Katholieke Universiteit Leuven/Centre for Biomedical Ethics and Law (2009) European Ethical-Legal Papers. Retrieved 2 February, 2009, from http://www.kuleuven.be/cbmer/page.php?LAN=E&ID=383&TID=0&FILE=subject&PAGE=1

Knoppers BM, Chadwick R (2005) Human genetic research: emerging trends in ethics. Nat Rev Genet 6(1):75–79

Laurie G (2002) Genetic privacy: a challenge to medico-legal norms. Cambridge University Press, Cambridge

Legemaate J (2005) Psychiatry and human rights. In: Gevers J, Hondius E, Hubben J (eds) Health law, human rights and the biomedicine convention. Martinus Nijhoff, Leiden/Boston, pp 119–130

Lewis O (2002) Protecting the rights of people with mental disabilities: the European Convention on Human Rights. Eur J Health Law 9:293–320

Lötjönen S (2005) Research on human subjects. In: Gevers J, Hondius E, Hubben J (eds) Health law, human rights and the biomedicine convention. Martinus Nijhoff, Leiden/Boston, pp 175–190

Lötjönen S (2006) Medical research on patients with dementia – the role of advance directives in European legal instruments. Eur J Health Law 13:235–261

Lötjönen S (2008) Medical research on minors in Finland. Eur J Health Law 15:135–144

Macklin R (2003) Dignity is a useless concept. BMJ 327:1419–1420

Manson NC, O'Neill O (2007) Rethinking informed consent in bioethics. Cambridge University Press, Cambridge

Manuel C, Hairion D, Auquier P, Reviron D, Giocanti D, Terriou P et al (1999) Is the legislation of European states in keeping with the recent convention on human rights and biomedicine? Eur J Health Law 6(1):55–69

Mol A (2008) The logic of care: health and the problem of patient choice. Routledge, London

Niveau G (2004) Preventing human rights abuses in psychiatric establishments: the work of the CPT. Eur Psychiatry 19(3):146–154

Nys H (2008) The biomedicine convention as an object and stimulus for comparative research in the European Journal of Health Law. Eur J Health Law 15:273–283

Nys H, Stultiëns L, Borry P, Goffin T, Dierickx K (2007) Patient rights in EU Member States after the ratification of the convention on human rights and biomedicine. Health Policy 83: 223–235

Parker C (2007) Developing mental health policy: a human rights perspective. In: Knapp M, McDaid D, Mossialos E, Thornicroft G (eds) Mental health policy and practice across Europe: the future direction of mental health care. Open University Press, Maidenhead, pp 308–335

Reuter L (2000) Human is what is born of a human: personhood, rationality, and an European convention. J Med Philos 25(2):181–194

Roscam Abbing HDC (1996) Human rights and medicine: a Council of Europe convention. Eur J Health Law 3(3):201–205

Roscam Abbing HDC (1998) The Convention on human rights and biomedicine: an appraisal of the council of Europe convention. Euj J Health Law 5(4):377–387

Roscam Abbing HDC (2004) Rights of patients in the European context, ten years and after. Eur J Health Law 11:7–15

Rose N (2007) The politics of life itself: biomedicine, power, and subjectivity in the twenty-first century. Princeton University Press, Princeton, NJ

Rose D, Wykes T, Leese M, Bindman J, Fleischmann P (2003) Patients' perspectives on electroconvulsive therapy: systematic review. BMJ 326:1363–1366

Rušinović-Sunara D, Finka D (2008) Reflections on the development of health care and patients' rights in Croatia. Med Law 27(2):357–364

Salako SE (2008) The Council of Europe Convention on Human Rights and Biomedicine: a new look at international biomedical law and ethics. Med Law 27(2):339–356

Sándor J (2008) Human rights and bioethics: competitors or allies? The role of international law in shaping the contours of a new discipline. Med Law 27:15–28

Sartorius N (2002) Iatrogenic stigma of mental illness. BMJ 324:1470–1471

Schmidt H (2007) Whose dignity? Resolving ambiguities in the scope of 'human dignity' in the universal declaration on bioethics and human rights. J Med Ethics 33(10):578–584

Sperling D (2008) Law and bioethics: a rights-based relationship and its troubling implications. In: Freeman M (ed) Law and bioethics. Current legal issues 2008, vol 11. Oxford University Press, Oxford, pp 52–78

Sprumont D, Gytis A (2005) The importance of national laws in the implementation of European legislation on biomedical research. Eur J Health Law 11:245–267

Steering Committee on Bioethics (CDBI)(2000) Convention on the protection of human rights and dignity of the human being with regard to the application of biology and medicine: convention on human rights and biomedicine (ETS No 164) Preparatory work on the Convention. Strasbourg: Council of Europe

Stultiëns L, Goffin T, Borry P, Dierickx K, Nys H (2007) Minors and informed consent: a comparative approach. Eur J Health Law 14(1):21–46

Tannsjo T (2004) The convention on human rights and biomedicine and the use of coercion in psychiatry. J Med Ethics 30:430–434

Tutton R (2007) Constructing participation in genetic databases: citizenship, governance, and ambivalence. Sci Technol Human Values 32(2):172–195

Walin L (2007) Ambiguity of the embryo protection in the human rights and biomedicine convention: experiences from the Nordic countries. Eur J Health Law 14(2):131–148

Welie SPK, Berghmans RLP (2006) Inclusion of patients with severe mental illness in clinical trials: issues and recommendations surrounding informed consent. CNS Drugs 20(1):67–83

World Health Organization Regional Office for Europe (1995) Promotion of the rights of patients in Europe: proceedings of a WHO consultation. Kluwer Law International, The Hague/London

Wynne B (1991) Knowledges in context. Sci Technol Human Values 16(1):111–121

Chapter 6
Ethics Committees for Clinical Research – The West-European Paradigm

Elmar Doppelfeld

Contents

Abbreviations

CIOMS	Council for International Organizations of Medical Sciences
DFG	Deutsche Forschungsgemeinschaft
IRB	Institutional Review Board
NIH	National Institutes of Health
REC	Research Ethics Committee...
SOP	Standard Operation Procedures
WMA	World Medical Association

6.1 Introduction

The term 'Ethics Committee' covers bodies of different responsibility and legal competence. There are ethics committees dealing with general questions in medicine and healthcare, others are more linked to the situation in hospitals and to

E. Doppelfeld (✉)
Permanent Working Party of Research Ethics Committees, Köln D-50859, Germany
e-mail: med.ethik.komm@netcologne.de

H. Helmchen, N. Sartorius (eds.), *Ethics in Psychiatry*, International Library
of Ethics, Law, and the New Medicine 45, DOI 10.1007/978-90-481-8721-8_6,
© Springer Science+Business Media B.V. 2010

questions on individual treatment (see Chapter 7). In contrast to these group other ethics committees are exclusively entrusted with the ethical examination of research projects. For the latter type the denomination 'research ethics committee' abbreviated as 'RECs', becomes more and more familiar. Research ethics committees are the exclusive focus of the following contribution.

It is widely accepted that Henry Beechers publication on ethics in medical research in 1966 has been the decisive initiative for establishing bodies to examine medical research projects (Beecher 1966). The author presented in this publication the results of his critical review of 100 scientific contributions published in a well reputed American medical journal. He classified 22 projects as ethically disputable, among these 12 as ethically questionable.

As a consequence the National Institutes of Health (NIH) introduced as a condition for federal grants for medical research projects the assessment of a project by a body established at the study site (Federal Register 1978). These 'Institutional Review Boards' (IRB) should only approve a project under condition that the protection of research participants was safeguarded.

Research foundations in other States adopted this procedure like the 'Deutsche Forschungsgemeinschaft (DFG)' in Germany. The DFG prompted the establishment of IRBs in 1973 at the universities of Ulm and of Göttingen. So in contrast to a widely spread assumption research ethics committees have not been introduced by the medical profession but on request of authorities or similar bodies. However the medical profession has without any doubt the merit to promulgate the new instrument of protection in the research field.

The revised Declaration of Helsinki (Tokyo 1975) of the World Medical Association (WMA) required that a precise research protocol should be presented to a specifically appointed, independent committee for 'consideration, comment and guidance' (1976). To these classic points of assessment the version 2008 of the Declaration (2008) added 'approval'. In the following decades these committees, now known as 'ethics committees', have been established around the World by different institutions and on different legal basis. Establishing institutions may be universities, faculties of medicine, hospitals, medical associations, States nation wide or regional authorities. The basis for their competence ranges from the code of deontology of the medical profession, in most States legally not binding, to a formal legislation.

6.2 Legal Basis of Research Ethics Committees

The tendency of States in Europe and outside of Europe to introduce a formal legal basis for the work of RECs has grown up predominantly in the last 2 decades. One of the reasons may be that States are more and more aware of their obligation to protect human rights and fundamental freedoms of research participants. Considering that modern medical research is performed by researchers from different professions it seems not to be appropriate to bind all these researchers by the code of deontology of only one profession. For these two reasons the Council of Europe adopted as a legally binding protecting instrument the well known Convention of Oviedo (1997) and its additional protocol concerning biomedical research (2005) (see Chapter 5). The Member States of the Council of Europe are free to sign and to ratify the mentioned instruments. Until now, this Convention has been signed by 34 Member States of the Council of Europe and has been implemented by 24 of these States by ratification into their national legal system. Germany and some other Member States like the United Kingdom have not yet signed and not yet ratified the Convention. Other Member States like The Netherlands or France are preparing the ratification. The legal instruments of the Council of Europe require that all medical research projects are assessed by a REC and thereafter, if requested by national law, are approved by a competent body.

Only to cover clinical drug trials the European Union has introduced a legal regulation by its Directive 2001/20/EC (2001) as a basis for the internal law of its Member States.

In contrast to the treaty regulations of the Council of Europe leaving the decision of adopting a legal instrument to its members, the Member States of the European Union are obliged to implement the provisions of this Directive into their national system. Concerning RECs the Directive requires that Member States establish a system of RECs to assess drug trials following a specific list of items. The Member States adopted this community right in very different ways. In general research ethics committees established at authorities, universities, research institutions or research ministries are legally competent for the assessment of drug trials. Exceptionally research ethics committees at medical associations are charged by this task. All provisions concerning the responsibility of RECs in drug research are regulated by law. The different national regulations may be found in an overview recently published (Wells 2007) It is advised to contact the national authorities and the national representations of RECs, cited in the above mentioned publication, before starting a drug research project in one of the EU-Member States. In general also for research outside the European Union a researcher is well advised to look for the national law, which is the only binding instrument. Declarations or guidelines of Non Governmental Organisations like the World Medical Association or CIOMS are per origin not legally binding. It should be mentioned that the version of the Declaration of Helsinki as adopted in 2008 by the World Medical Association is, seen in a general view, in conformity with legally binding instruments. However to become legally binding a decision of the concerned State is required.

The diversity of regulations can be remarked also on the specific field of RECs. Whereas by the Directive a certain harmonization of duties and responsibilities of RECs for the assessment of clinical drug trials within the EU has been achieved, the regulations concerning the place and obligations of RECs in all other fields of medical research are different from State to State. Therefore a specific common information cannot be given. It is strongly recommended that researchers intending research on all non-drug fields ask the national competent institutions to avoid any misunderstanding.

There is another source of misunderstanding: very often the term 'clinical trial' is used as a synonym for 'drug trial'. It should be kept in mind that the expression 'clinical trial' covers all clinical research. The proportion of clinical drug trials to other clinical trials e.g. in surgery or radiology varies from country to country. It may be that nearly all clinical trials are drug trials or that a certain amount are non drug trials. In Germany e.g. 53% of all trials submitted to RECs are drug trials, whereas 47% are trials in other medical research fields.[1]

The Convention of Oviedo of the Council of Europe and its additional protocol concerning biomedical research as the only international legally binding instruments for medical research contain among others general provisions on the responsibility, obligation, position, assessment conditions for RECs. These conditions could serve as a guideline for Member States of the Council to regulate the work of RECs.[2]

The legal instruments of the Council of Europe and of the European Union have contributed already in the past to harmonize the position, the rights, the obligations and the power of RECs in Western as in Central and Eastern Europe. Main differences are occurring in the affiliation of RECs which are predominantly attached in Western Europe to hospitals or Faculties of Medicine. In other regions of Europe RECs at research institutions, at national or regional legally competent authorities or National Ethics Committees are entitled to assess biomedical research. However regulations are different from State to State and even different from continent to continent. It is not possible to cover all local situations in this contribution. In general ethical principles as laid down e.g. in the Declaration of Helsinki or in the CIOMS guidelines are at least basically respected, but often interpreted on the basis of regional culture and tradition.

Fields to be regulated mostly by national law are in particular independence, composition and the legal competence of RECs in relation to authorities. Furthermore the qualification of members, privacy, liability, fees for the work of the REC and remuneration of its members, procedure in case of conflict of interests and a system of appeal should be addressed by the regulations. Also the working procedure of a REC should be clearly stated by Standard Operation Procedures (SOPs), so that the applicant knows the correct application form for a submission. The time schedule for the assessment by the REC and for the decision must be publicly available.

In most countries several RECs are entitled to assess projects of medical research. Independent from their legal position RECs in a State should be encouraged to build networks for the harmonization of their decisions and for the improvement of their work. In Germany the 'Arbeitskreis Medizinischer Ethik-Kommissionen in der Bundesrepublik Deutschland' (Permanent Working Party of Research Ethics Committees in Germany), established in the year 1983 has contributed to these aims.

---

[1] This calculation is based on the answers to annual questionnaires sent to the members of the Permanent Working Party; the answers are currently published in the Year-Book 'Medizin-Ethik' Deutscher Ärzteverlag, Köln.

[2] Recently the Steering Committee on Bioethics of the Council of Europe (CDBI) instituted a Working Party for the elaboration of a guide for RECs with the aim of harmonisation of their procedure.

There are similar networks in other countries like France, the United Kingdom, Austria or Switzerland. These networks may help to inform foreign researchers on the regulations concerning RECs in a specific State.

6.3 Scope

The main obligation of RECs is to assess projects of medical research. The often used adjective 'biomedical' implies that also medical research using methods of other fields like biology may be covered. Unfortunately there is no unanimously accepted definition neither of medical or of biomedical research (for a more detailed discussion see Helmchen 2002). The Permanent Working Party of German Research Ethics Committees accepted the following definition for the daily work of RECs: 'Research on man is every activity involving a risk for the somatic or psychological integrity of the participant, which is undertaken with the aim to obtain, independently from an individual case, general knowledge in the fields of prevention, diagnosis, treatment or pathophysiology' (Doppelfeld 1990).

This description covers all projects of basic research, of research projects in combination with healthcare, of clinical drug trials and of research using medicinal products in human beings. This German definition is at least for a part in accordance with that of 'clinical trials' as proposed e.g. by the NIH. Depending on national approaches and legislation compassionate use e.g. the use of a not yet licensed drug in individual cases may be accepted as a trial. In Germany this type of healthcare, admitted as 'Heilversuch', is not considered as medical research, but as a step to generate a scientific hypothesis to be proved in a trial. Consequently, in Germany 'Heilversuche' do not belong to the competence of a REC.

The scope of RECs has been enlarged during the last decades by the fields 'identifiable data' and 'research on biological material of human origin'. Again depending on the national situation research projects using data related to a person who can be identified may fall into the competence of a REC. This competence is more and more accepted. In contrast research using anonymized data does not need to be assessed by a REC. However there is an upcoming request that the method used to achieve this anonymization should be shown to and approved by a REC. RECs often require that the person concerned should be asked to give the free informed consent for the anonymization and that he or she should be entitled to introduce restrictions. The necessary information should contain a note that after anonymization an identification of the person concerned will be impossible. However such an identification might be useful if during a research results with relevance for the health of the person concerned have been gained (see Chapters 11 and 26).

Research on removed biological material is often justified by the need to use stored material for projects necessary in the light of new scientific results or of new scientific questions. In clinical practice persons mostly agree in first line to the removal of their tissue for diagnostic purposes.

It is more and more accepted, that the use of stored biological material for research purposes requires a specific consent of the donor. It may not be concluded that a patient implicitly agrees that the tissue, initially removed for diagnostic purposes, may be used for other aims not yet known. National regulations, if ever existing, vary from country to country. Among the international legally binding instruments only the Convention of Oviedo contains in its article 22 a specific provision saying that the use of removed tissue for a purpose other than that for which it has been removed needs a free informed consent. For research using biological material of human origin with its very specific legal provisions to safeguard the rights of the donors the Committee of Ministers of the Council of Europe has adopted in 2006 a recommendation to the Member States (2006). This instrument proposes that all research projects in this field be assessed by a REC.

The classification of that instrument as 'recommendation' and giving it by that a legally non-binding character is an indication for the difficulties to find international legally binding regulations for research using human tissue.

As criteria for the assessment of medical research 'scientific quality', 'accordance with law' and 'ethical acceptability' are mostly accepted. In some countries like Germany the RECs are entitled to prove the fulfilment of these three conditions. In other countries the responsibility of the RECs is restricted to the point 'ethical acceptability', whereas other bodies are charged to examine the scientific quality and the conformity with law. These bodies issue their opinions which may be sent by themselves to the REC or be presented by the applicant to it. In some countries the REC has to accept these statements of the competent bodies, in other countries a REC may ask for a revision by the same or by another body.

The mentioned criteria will be considered in the way as accepted in Germany. It should be kept in mind that the interpretation of the criteria may vary from country to country or even from region to region within a country.

6.4 Responsibilities of Research Ethics Committees

6.4.1 Scientific Quality

The scientific quality is a basic condition for any acceptance of a medical research project. No person may be invited to participate in a project of questionable or missing scientific quality. In addition to this protective attitude in favour of research participants it must be excluded that resources for scientifically unjustified projects are wasted. The scientific quality may be attested by a competent body according to national law. If RECs are entitled to assess the quality they should follow standard operation procedures (SOPs). The scientific examination may be performed by a member of the REC who is competent for the field addressed by the submitted project. Examination by a competent member requires exclusion of any kind of conflict of interests (see Section 2.4). Such conflict is given when that member is involved in the project in any way or if the applicant belongs to its staff. If a conflict of interests exists or if no member of the REC is competent an external expert must be involved in the decision making. Very often this condition is fulfilled by a written external expertise, some RECs prefer the presence of the external expert. The EU-Directive 2001/20/EC requires formally for the assessment of drug trials on minors paediatric experience within the assessing REC or taking external advice in the in the field of paediatrics. In the same manner specific expertise is required for the assessment of drug trials on adults not able to consent.

By the obligation to ask for external expertise if appropriate it can be excluded that during the scientific examination a research field is neglected which is not regularly represented in the membership of a REC.

The scientific examination is based on established parameters like results published in the relevant newest literature, sufficient preclinical research in laboratories and on animals, appropriateness of the envisaged methods in relation to the intended aim of the research and a risk/benefit/ calculation. This calculation of course firstly implies the risk and benefit for the participant (see Chapter 25).

In some countries a benefit for the medical science or the relevance for healthcare has to be included in the calculation. By those provisions a social relation between research and society is established.

It should be underlined that strong or world wide accepted conditions for the risk/benefit/ calculation do not exist. Often a general approach is used: the lower the potential benefit, the lower the acceptable risk.

In a similar way the additional protocol concerning biomedical research of the Council of Europe states that research shall not involve risks and burdens to the participant disproportionate to its potential benefit. This provision allows in clinical research a weighing up of potential direct benefits for a patient against risk and burden to be accepted e.g. in cancer treatment research – a field of research with a potential direct benefit for the participant.

For research without a potential direct benefit for the participant – research on healthy volunteers – the protocol only admits acceptable risk and acceptable burden. The 'acceptability' has to be assessed by the REC. It is of course to the envisaged participant to agree to this assessment. For research without a potential direct benefit on persons not able to consent the protocol only allows minimal risk and minimal burden as absolute limitations (see Chapter 25).

In relation to the scientific quality of research the qualification of the researcher and the quality of the study site shall be assessed. Specific points of this examination are the medical supervision of a research project and the provisions to react to contingencies and emergency situations.

6.4.2 Legal Aspects

A research project of proven scientific quality can only be performed in accordance with relevant national law. Actually national law prevails since all international legally binding instruments regulating research must be implemented into the legal system of States, which sign and ratify those instruments. This is also true for directives of the European Union, which give the binding frame for national law. Different from this legal character regulations of the European Union have the force to bind the Member States directly. Whether legal aspects are examined by a specific body or by a REC depends on national regulations. Relevant legal provisions may be contained in a single 'research law' or in different legal regulations. In nearly all States specific laws for drug research are in force, whereas other fields of research are not covered by specific provisions like in Germany. In Germany legal aspects belong to the competence of REC. These RECs examine whether the general norms of penal and civil law are respected. In addition they look for the accordance of the submitted project with more specific provisions as introduced e.g. by the federal laws on drug or medicinal products, by the federal law for data protection or by other regulations e.g. on radiation protection. A specific attention is paid to the fore-seen compensation of damages in relation to the research project. In Germany such a compensation by a specific insurance is prescribed only for drug research. RECs strongly advise the applicants to seek an insurance for research in other fields. The participant will be informed on the provisions to cover any kind of damage. He or she may than decide to participate even in case where no insurance will be in place to cover damage.

The most important legal and ethical condition for the participation in a research project is the free informed consent of the person concerned or in case of inability to consent the authorization by his or her legal representative (see Chapters 9, 10 and 25). Free informed consent is one of the oldest conditions for participation in medical research. It has been for the first time codified as early as on 29th December 1900 by the decree of the Prussian Minister of Education (1901) and has been taken over by following regulations like the different versions of the Declaration of Helsinki.

It must be safeguarded that a valid free informed consent is given. Such a con-sent is only valid if given on the basis of appropriate, complete and understandable information. The information shall contain the notice that a free informed consent may be refused or may be withdrawn at any time without any disadvantage specifi-cally for the needed medical care. Furthermore the conditions to seek the informed consent must be declared with respect to avoid any undue influence on the invited participant. Information material and the procedure to seek the informed consent must be shown to the REC.

Normally a free informed consent can only be given by a person able to consent. It should be regarded that the terms 'legal incapacity' and 'not able to consent' do not mean the same in all cases. Whereas incapacity covers the legal impossibility to consent, 'not able to consent' sticks to the actual situation. A victim of a traffic accident may formally stay with capacity, but may be in relation to the injuries not be able to consent. Minors are with growing maturity and understanding more and more able to consent, but their legal incapacity rests until to the age foreseen by national law. For research projects on adults able to consent the outlined problem is not relevant.

In contrast a lot of difficulties arise if research is intended on persons not able to consent. The inability to consent may be caused by the age of the participants:

– minors –, or by the specific health situation: – persons suffering from dementia or accident injuries. The inability may disappear with growing maturity of the minor or by recovery from injuries or other diseases. In many situations the inability stays for the rest of life.

It is generally accepted that research on persons not able to consent may be performed if there is no alternative of comparable effectiveness. The research cannot be carried out on persons able to consent with the same results. In addition this research should promise a potential direct benefit for the person concerned. This precondition and the authorization according to national law by a legal representative of the person unable to consent are part of the examination of legal aspects.

Different problems with controversial positions exist on research without a potential direct benefit for the participating person unable to consent. It is accepted that this research is needed to achieve basic knowledge e.g. of the metabolism of a molecule in the newborn before using it as a drug or as part of a drug, of the normal structure of the hip-joint of a newborn before introducing ultrasound as a screening method to detect malformations of that joint. Even if this research is carried out for the benefit of others, a benefit for the individual is not excluded, but in general not expected.

There is a big variety of regulations for this delicate field. Some countries are restrictive like Germany, other countries allow such research, often in accordance with the Convention of Oviedo and its additional protocol concerning biomedical research Both instruments state that this research should be an exception, that protective legal provisions for the participants shall be in force. The scientific quality must be given, alternative research of comparable effectiveness is not possible. As a rather new condition the mentioned instruments introduced that research without a potential direct benefit on persons unable to consent must only entail minimal risk and minimal burden for the person concerned. These absolute limitations may not be altered even if a benefit for participants seems to result. In that case a new study under the perspective 'potential direct benefit' could be appropriate.

6.4.3 Ethical Acceptability

The main task of each REC is the decision on the ethical acceptability of a research project of scientific quality which is in accordance with national law. There may occur discrepancies in the examination of the scientific quality and of the accordance with law. But the ethical assessment can far more be the source of controversial discussions.

The RECs try to overcome this challenge by a multidisciplinary dialog of its members. Philosophers, theologians or lay persons may contribute a lot to the relevant debates during decision making. It should not be forgotten that in such debates the convictions of the participants based on religion, on life experience, on tradition play an important role. These convictions finally lead to the assessment whether a project is 'good' or 'bad', the ultimate result of an ethical analysis.

These convictions of moral conditions may differ from country to country and even from region to region within a country. These differences may explain the encountered different results of the ethical assessment on a research project without scientific or legal problems. The interpretation of commonly accepted principles like autonomy, protection of dignity and identity of the research participant may differ.

6.4.4 Relation to Authorities

In most countries a research project may only carried out with the approval of a competent body. Depending on national law authorities alone or in cooperation with RECs are entitled to approve a research project. In some countries RECs are the only competent bodies. To assure that before any approval is given an ethical assessment has been performed, the Convention of Oviedo and its additional protocol concerning biomedical research require that a research project must only be approved after independent examination of its scientific merit and after multidisciplinary review of its ethical acceptability. The EU-Directive has introduced an independent examination of drug trials by both a REC and a competent national authority. The decisions both of them are requested to start the research. In the wording of the Directive the decision of a REC appears as 'favourable opinion', which is, in a legal sense different, from approval.

6.5 Legal Binding Force of RECs Decisions

Depending on national law the decision of a REC may be legally binding for all fields of medical research or only for some branches. In the EU Member States the 'favourable opinion' is a legally binding condition to start clinical drug trials. Without that favourable opinion any drug research is illegal. National law may prescribe that such a legal binding force of RECs decisions is extended to all medical research. In this understanding the decision is an approval or a disapproval. If this binding character of the approval is only attributed to drug research, the decision of a REC in all other research fields is only an advice to the researcher with no legal obligation to be followed or not. Researchers should be aware of the specific legal force of a decision of a REC in the country where they perform research.

6.6 System of Appeal

The decision of a REC may be not acceptable and seem to the applicant to be unjustified. Therefore, in particular in case the decision of the REC is binding by law, a system of appeal should be in place. The REC cannot work as 'first and last instance'. There are different solutions according to national law. In Germany courts of administration are competent for litigations concerning the decision of a REC. In other countries the body which established the REC may serve as institution to revise those decisions. Some countries have introduced a system of appeal within the group of RECs. An applicant may appeal another REC of the same legal position or appeal a national Ethics Committee legally competent for research. According to national regulations the decision of the second REC or of the national committee may be final or may be furthermore revised by a court or by another body.

References

Additional Protocol to the Convention on Human Rights and Biomedicine Concerning Biomedical Research (2005) Strasbourg, 25. I.2005, CETS No. 195

Beecher H (1966) Ethics and clinical research. N Engl J Med 274:1354–1360

Convention for the Protection of Human Rights and Dignity of the Human Being with regard to the Application of Biology and Medicine: Convention on Human Rights and Biomedicine (1997), Oviedo, 4.IV. 1997. European Treaty Series – No. 164, Council of Europe, Strasbourg

Council of Europe (2006) Recommendation Rec (2006) 4 of the Committee of Ministers to member states on research on biological materials of human origin, 15 March 2006, Strasbourg

Declaration of Helsinki (1976) International source: The World Medical Association, www.wma.net; German source: 3. Die revidierte Deklaration von Helsinki, B Anz. Nr 152 vom 14.August 1976

Declaration of Helsinki (2008) Ethical Principles for Medical Research Involving Human Subjects, WMA General Assembly, Seoul, www.wma.net

DIRECTIVE 2001/20/EC OF THE EUROPEAN PARLIAMENT AND OF THE COUNCIL of 4 April 2001 on the approximation of the laws, regulations and administrative provisions of the Member States relating to the implementation of good clinical practice in the conduct of clinical trials on medicinal products for human use. Official Journal of the European Communities 1.5.2001, L 121–34

Doppelfeld E (1990) Bericht über die 7. Jahresversammlung des Arbeitskreises Medizinischer Ethik-Kommissionen. In: Toellner R (Hrsg) Künstliche Beatmung, Medizin-Ethik 2. Gustav Fischer, Stuttgart, New York, NY, p 201

Federal Register (1978) 11328 (source: E. Deutsch, Medizinrecht, Springer, 1997)

Helmchen H (2002) Biomedizinische Forschung mit einwilligungsunfähigen Erwachsenen. In: Taupitz J (Hrsg) Das Menschenrechtsübereinkommen zur Biomedizin des Europarates – taugliches Vorbild für eine weltweit geltende Regelung? Springer, Berlin, Heidelberg, New York, NY, pp 83–115

National Institutes of Health Glossary of Clinical Trial Terms, http://clinicaltrials.gov/

Prussian Minister of Education (1901) Anweisung an die Vorsteher der Kliniken, Polikliniken und sonstigen Krankenanstalten, Zentralblatt für die Gesamte Unterrichtsverwaltung in Preußen, Berlin, pp 188–189

Wells F (2007) (Editor for EFGCP): The procedure for the ethical review of protocols for clinical research projects in the European Union. Int J Pharm Med 21(1):1–113

Chapter 7
Clinical Ethics Committees and Ethics Consultation in Psychiatry

Jochen Vollmann

Contents

Abbreviations

CEC   Clinical Ethics Committee
CPR   Cardiac-Pulmonary Resuscitation
PEG   Percutaneous Endoscopic Gastrostomy

7.1 Introduction

While research ethics committees in the field of medicine are quite well known and established,[1] clinical ethics committees and clinical ethics consultation are fairly new to many European countries. Research ethics councils have been working with

J. Vollmann (✉)
Department of Medical Ethics and the History of Medicine, Ruhr-University Bochum,
44799 Bochum, Germany
e-mail: jochen.vollmann@ruhr-uni-bochum.de

[1] See Chapter 6.

H. Helmchen, N. Sartorius (eds.), *Ethics in Psychiatry*, International Library
of Ethics, Law, and the New Medicine 45, DOI 10.1007/978-90-481-8721-8_7,
© Springer Science+Business Media B.V. 2010

legal authorisation since the 1970s at medical schools and state medical associations. Their members, (researching) physicians,[2] pharmacologists and pharmacists, legal experts, biometricians, etc. vote their approval or disapproval on ethical issues regarding clinical trials on human beings in medical research.

These bodies working in the field of research ethics should be distinguished from clinical ethics committees that advise on ethical issues in the day-to-day clinical care of patients. Such clinical ethics committees (CEC) first came into being in the 1990s in Europe. Their members consist of physicians, nursing personnel, other health care professionals, pastoral counsellors, psychologists or social workers and frequently also a hospital administrator as well as a patient representative. Clinical ethics committees usually do not have any legal status and are established by hospitals on a voluntary basis – often in connection with certification processes within the context of quality assurance measures (see Table 7.1).

Different values and attitudes in Western societies (value pluralism), the ever-growing possibilities of modern medicine (medical progress) and the increased impact of economic factors are essential to the development of professional ethics consultation structures in clinical medicine. Internationally, different forms of ethics consultation have evolved. The clinical ethics committee (CEC) – also known as ethics forum, ethics council or ethics panel – is the form most often realised. Clinical ethics committees provide forums for ethical issues in the day-to-day care of patients in hospitals, nursing homes and homes for the elderly, in care for persons with disabilities, both in partly-inpatient settings and outpatient settings. Besides case-specific ethics consultation for health professionals, patients and family members, the CECs develop ethical guidelines and offer opportunities for training. To date, only a few studies dealing with clinical ethics committees and clinical ethics consultation in psychiatry can be found in the reference literature (PubMed database search. keywords: "Psychiatry" [Mesh = Medical Subject Heading] AND "Ethics Committees, Clinical" [Mesh] OR "Ethics Consultation" [Mesh], as of:

Table 7.1 Comparison: research ethics committee and clinical ethics committee

	Research ethics committee	Clinical ethics committee
Established	1970s	1990s
Tasks	Medical research	Clinical care
Institution	Medical school	Hospital
	State medical association	Nursing home
Members	Researching physicians, lawyers, biometricians	Physicians, nursing personnel, pastoral counsellors, management
Legal basis	German drug law	No legal basis
	German act on medical devices	Certification procedures
	Professional code of conduct	

---

[2] For brevity and readability, the masculine form is used here to refer to mixed groups of men and women and is meant to be inclusive for both genders.

4. Sept. 2008). More studies explore the role of the consultation/liaison psychiatrist with respect to ethical issues in somatic medicine. For that reason, this paper refers mainly to general literature on clinical ethics committees and clinical ethics consultation as well as to the few papers with a special focus on psychiatry (Ford 2008). Furthermore, the author draws on his practical knowledge from more than 10 years of experience in clinical ethics as a psychiatrist and medical ethicist.

Since the 1990s clinical ethics committees have been established in many European countries. In Germany CECs are found particularly in church-related hospitals (both Catholic and Protestant denominations), which comprise about one-third of all German hospitals. The key initiative for their founding came from the two Christian hospital associations, which in 1997 called on their hospitals to found CECs patterned after the American model (Deutscher Evangelischer Krankenhausverband e.V., Katholischer Krankenhausverband Deutschlands e.V. 1997). A survey of all approximately 800 church-related hospitals in 2000 showed that 30 hospitals had established a CEC (Simon 2001). In a Germany-wide survey of all circa 2,300 German hospitals in 2005, 483 hospitals responded (response rate: 22%). Of those who answered the survey, 312 hospitals reported some form of clinical ethics consultation or were in the process of establishing it ($n = 77$). At these hospitals there are 149 clinical ethics committees, 38 ethics forums or ethics councils, 15 ethics consultation services and 33 other forms of clinical ethics consultation (Dörries and Hespe-Jungesblut 2007).

Clinical ethics committees arose out of decentralised initiatives in various patient care hospitals; often they were established without any exchange of information and professional advice. This is one reason why they have different names, concepts and forms of doing their work. The stricter certification procedures for the quality assurance of health care institutions, which often reward the presence of clinical ethics consultation structures as quality-increasing factors, are also conducive to the establishment of CECs. It is only in very recent years that also university hospitals have intensified their efforts to establish CECs and clinical ethics consultation services. According to a current survey of experts, the 36 German university hospitals have a total of 26 clinical ethics committees or ethics consultation services. Eleven university hospitals or medical schools have established positions for academically qualified ethicists (Vollmann 2008a) (see Table 7.2).

Table 7.2 Clinical ethics committees – development in Germany

Year	Institutionalisation
1997	Recommendation of church-related hospitals
1999	First report based on experiences
2000	30 clinical ethics committees
2003	59 clinical ethics committees
2004	70–100 clinical ethics committees
2005	312 clinical ethics consultation structures, of them 149 clinical ethics committees
2007	26 clinical ethics consultation structures at university hospitals ($n = 36$)

Comparable developments have taken place in other European countries, such as Great Britain, the Scandinavian countries and the Benelux countries (Hurst et al. 2007a, b). Since the past few years, CECs and other structures of clinical ethics consultation are also being established in a number of eastern and middle European countries and in Turkey (Kovacs and Strech 2008, Schildmann et al. 2010).

7.2 Clinical Ethics Committees in Psychiatry

The same development described here in general hospitals also applies in key areas to mental hospitals and other health care structures for people with mental disorders (Molewijk et al. 2008a, b, Zacchia and Tremblay 2006). In a survey of all Swiss hospitals by the Swiss Academy of Medical Sciences, 44% of the responding mental hospitals reported that they provided ethics consultation. This corresponded to the frequency of clinical ethics consultation services in general hospitals (44%) (Salathe et al. 2008). Of the responding mental hospitals in a survey of all German hospitals, 38% reported having a clinical ethics committee or another form of clinical ethics consultation (Dörries and Hespe-Jungesblut 2007) (Andrea Dörries, personal communication 2008). In Great Britain local clinical ethics committees are organised in a nation-wide "clinical ethics network". The networks organise regular meetings and education sessions. It further provides contact information for all UK clinical ethics committees, up to date and reliable information on ethical issues that commonly present to clinical ethics committees or arise in clinical practice and runs a internet platform with links to national policy and guidance on ethical issues relating to clinical practice (www.ethics-network.org.uk). However, relatively few CECs in UK mental hospitals were currently reported (Jacinta Tan, personal communication 2008).

Besides the incidence of CECs and ethics consultation services offered by institutions, it is decisive for the clinical practice whether and how often clinical ethics consultation is requested in mental hospitals or in psychiatric wards in general hospitals. A British child and adolescent psychiatrist reports as member of a CEC:

> My own experience as a member of a psychiatric clinical ethics committee is that it's remarkably difficult to get clinicians to bring cases, although there must be lots and lots of ethical dilemmas daily. My own guess is that mental health professionals are so used to dealing with dilemmas they are probably less willing or likely to want to bring the 'everyday' dilemmas (Jacinta Tan 2008, personal communication).

The author can confirm this experience from his professional work in the field of clinical ethics in Germany. In clinical practice, there is an obvious discrepancy between the numerous areas of ethical problems encountered in psychiatry (Helmchen and Vollmann 2001, Austin et al. 2008, Vollmann 2009) and the small number of requests for clinical ethics consultations. However, in interpreting these facts one must take into account that the gravity of the ethical problem is not the only reason for requesting a clinical ethics consultation. Rather, the small number of requests may be due to the competence of the participants in the communication and professional handling of the problem particularly in the field of mental health. This problem-solving competence is exceptionally well developed in the field of psychiatry compared to many somatic fields of medicine. Mentally ill patients are often "difficult" patients, so that in the patient–psychiatrist relationship it is not seldom the case that there are differences of opinion and emotional tensions which the psychiatrist is familiar with and which he generally knows how to deal with professionally (Engel 1992a, b, Hayes 1986, Sexson and Sexson 2008, Steinberg 1997). Here he is aided by his training in psychiatry and psychotherapy and his work in a multi-professional therapeutic team. The pronounced team culture and the cooperative attitude in the field of psychiatry which results from working with nursing personnel, psychologists, social workers, ergotherapists etc. leads to a high competence in communication and in resolving conflicts. These pronounced psychosocial competences and their structural implementation in the day-to-day hospital routine of psychiatric institutions is an important factor for dealing with ethical conflicts within the team, without need for a clinical ethics consultation.

7.3 Clinical Ethics Consultation and Psychiatric Consultation Services

Although professional communication and dealing with interpersonal conflicts is a competence characteristic of both clinical psychiatry and clinical ethics consultation, the two professional fields need to be distinguished (Bauer and Vollmann 2004, Lederberg 1997). In clinical practice the consultant psychiatrist is sometimes called on in medical-ethical conflict situations, because from him as representative of the biopsychosocial model – in contrast to the often purely biological model of the somatic disciplines – a holistic approach to the patient is expected, also qualifying him to solve medical-ethical problems (Youngner 1997). Furthermore, for procedural reasons in the case of ethical conflict situations such as discontinuing treatment at the end of life or in the case of refusing an urgent therapy it is often necessary to consult a psychiatrist, in order to exclude mental disorders or to assess the ability to give consent. Despite obvious overlaps, a lack of differentiation between ethical and psychiatric issues is associated with the risk of masking and unnecessarily psychiatrising a primarily ethical problem, (Perl and Shelp 1982) or of mistaking the presence of a treatable psychiatric disorder for an ethical conflict situation (Leeman 1995).

What the two disciplines have in common is that psychiatric consultation services and clinical ethics consultation have developed as response to medicine as a field that is becoming increasingly technical and more specialised. Both complement the modern, primarily science and technology oriented approach to medicine with psychosocial and ethical dimensions (Youngner 1997). Interdisciplinarity requires competence in communication and mediation from both clinical ethicists and psychiatrists. In the majority of cases there is no simple way to respond to the consultation requests originating from the wards. Rather, the consultant psychiatrists often face the expectations of a larger group of people besides the patient – people who have, in part, diverging interests (family members, physicians, nursing personnel, social workers). What is helpful for both professional groups is training in conflict solution strategies. Here both the clinical ethicist and the consultant psychiatrist are obligated first and foremost to the patient.

Besides these aspects which they have in common, the different professional roles and tasks must be taken into account: Due to his medical specialty a psychiatrist has diagnostic, therapeutic and prognostic competence and bears personal responsibility as a physician. On the other hand, the task of a clinical ethicist is the identification and analysis of ethical issues and conflicts which arise in relation to patient health care (Youngner 1997). For this he needs professional training in medical ethics and its application in concrete situations. Due to the normative character of an ethical conflict situation, responsibility for the decision to be made can nevertheless not be transferred to the clinical ethicist but rather remains with those responsible for the treatment. What the clinical ethicist can do in ethical conflict situations is to provide ethical expertise and to contribute to the decision-making process in the role of a "catalyser" (Bauer and Vollmann 2004).

7.4 Structures

In practice there are many different forms of institutionalising clinical-ethical structures in hospitals and nursing homes, whereas each organisation must find its own structure and style of working in order to make the best possible contribution with respect to its particular local situation. For that reason there is no standard model for a clinical ethics committee. Therefore the information provided in the following section should be considered ideal-typical.

The composition of a clinical ethics committee should reflect the diversity of the different work areas in an institution; at the same time, in order to function, it may not be too large. In practice a size of 7–20 members has proven to work well. The members should reflect the broadest spectrum possible with respect to fields of work and professional groups, but they should not see themselves as lobbyists of a particular body or group. Besides doctors and nurses, the CEC should include members of other health professions in the hospital such as medical-technical assistants, physiotherapists, speech therapists, psychologists and social workers, pastoral counsellors, hospital administrators, etc. To complement the expertise of the CEC, external members (e.g. legal experts, medical ethicists, citizen or patient representatives) are often appointed. In addition, different levels of hierarchy and job profiles should be taken into account within the individual professional groups. The members of the clinical ethics committee are appointed by the hospital management. A chairman and, if necessary, a vice-chairman are to be elected from the circle of appointed members. Professional working committees formulate and adopt rules of procedures and statutes and have regular, fixed meeting dates (usually 6–12 times a year). The hospital management must support the work of the CEC, e.g. through corresponding working time regulations, training opportunities, material resources etc. The independence and credibility of the body must be ensured through the composition of the personnel and the organisational position within the institution. With the aid of an extensive network and broad acceptance, it can be effectively communicated that the purpose of implementing ethics structures is not to delegate ethical responsibility to a "specialist body" but to raise ethical awareness and to foster the competence of all staff members.

The structure of clinical ethical committees:

- 7–20 members
- appointment by the hospital administration
- 2–3 year term of service
- election of a chairman and, if necessary, a vice-chairman
- rules of procedure/statutes
- fixed monthly meetings
- working time regulations for members
- advanced training for members
- material resources

7.5 Implementation

In practice, clinical ethics committees are often founded on the initiative of the hospital management in the context of certification processes because structures of clinical ethics are given a positive assessment in different certification procedures. This initiative "from the top" (top-down model) has the advantage that structural decisions are made by individuals in responsible positions and can be implemented in a goal-oriented manner. In other cases the initiative primarily comes from dedicated staff members, who "from below" see a need for a regular exchange regarding ethical problems in their everyday work (bottom-up model).

This approach has the advantage that where the ethical decision making occurs there are already activities and competence in place which are of central significance for the acceptance and vitality of a clinical ethics committee. Both aspects must be considered to be successful in founding a clinical ethics committee: Without the support of the hospital management no CEC will be able to function successfully in the hospital, without the commitment of the staff members and their identification with the goals of the committee, the CEC will not play a significant role in the day-to-day- work of the hospital. Inadequate or unclearly articulated support in these areas is the main reason for delayed or failed implementation of CECs in German hospitals.

To understand the goal and structure of ethics committees, it is helpful to distinguish between the organisational level and the case-related level which converge in the clinical ethics committee. On the case-related level the focus is on the treatment and care of the patient. Here it is a matter of concrete decisions and ethical conflicts in the individual case, which must be dealt with by physicians, nurses and other health care professionals during the day-to-day work on the ward. When ethics are discussed in the hospital, this case-related level generally predominates in the group of health professionals. By providing ethical consultation with respect to individual cases, a CEC can make an important contribution to support staff members and to improve patient care. The organisational level, on the other hand, has the whole institution in view. Here it is primarily a matter of structures, organisation and personnel development, and economic and legal issues. On the organisational level, the hospital administration, the management, the legal and economic professions are the bodies which view the hospital as a whole, keeping *corporate identity* and quality assurance in mind and who have to secure the position of the hospital in a rapidly changing hospital market. Decisions on the organisational level in the hospital often have an impact on individual patients' treatment options. For that reason they are ethically relevant and must be taken sufficiently into account with respect to the complex structure and interaction in an institution (Vollmann 2007). Recent publications from the Netherlands illustrate the complexity of implementing clinical ethics consultation in psychiatric hospitals and their evaluation (Molewijk et al. 2008a, b).

That is why it is so essential to establish the goal and tasks when founding a CEC. It has become accepted to speak of *clinical* ethics committees, when the task of the committee work clearly focuses on immediate clinical problems (Zentrale Ethikkommission 2006). These include the fields of ethics consultation, development of guidelines and the training and advanced training of staff members, which are listed below. In practice, ethical issues often overlap with communication problems and conflicts between participants. Despite this, the focus should be placed on clinical-ethical issues, in order to retain a central theme for the scope and competence of the committee. A CEC cannot do the work of a staff council, an equal opportunity commissioner or of a psychosocial consultant. Ethical conflicts in the field of labour law, mobbing, tension between the managing director and the staff council, aspects of superordinate development of the company and the management and questions of fair distribution on the macro level do not belong to the primary tasks of a clinical ethics committee. For these issues, organisation ethics committees can provide ethical consultation for the management and those responsible in the institution, especially with respect to the development of the company.

Besides the CECs at individual hospitals which were already mentioned, in many countries a new development in ethics consultation structures is becoming apparent in larger hospital associations. These are being formed increasingly due to the economically motivated fusion of formerly independent hospitals – sometimes far apart geographically – to hospital associations and holding companies. In practice, three levels of institutionalisation can be found among them: (1) On the level of the superordinate, supporting organisation a superordinate ethics committee or ethics forum deals with ethical principles, corporate identity and opportunities for advanced training, (2) on the level of the local hospital there is a clinical ethics committee, and (3) on request, ethics consultation services are available or mobile clinical ethics consultation is provided (see Table 7.3). For the decentralised establishment of ethics work in individual wards and functional units, local ethics circles can be founded

Table 7.3   Three levels of clinical ethics institutionalisation in hospital associations

Body	Level of influence
Superordinate ethics committee/ethics forum	Hospital association
Clinical ethics committee	(individual) Hospital/location
Flexible ethics consultation on request	Ward

that are moderated by the mobile clinical ethics committee. In this way the inhibition threshold for an acute, case-relevant ethics consultation can be lowered and at the same time, e.g. by discussing past complex and difficult cases, the continuing ethical training of the staff members can be promoted.

7.6   Duties and Responsibilities

The duties and responsibilities of a clinical ethics committee include (1) clinical ethics consultation in the individual case, (2) development of guidelines as well as (3) the continuing and advanced training of staff members in clinical ethics. In practice, the work of the CEC is frequently constricted to clinical ethics consultation, and too little value is placed on the other two areas. While it is true that clinical ethics consultation in the specific case provides direct and practice-relevant support, only very few individual cases in the hospital will be reached with clinical ethics consultation alone. Apart from consultation in individual cases it is important to also strive to develop guidelines and organise training workshops and seminars with which – albeit indirectly – more patients and staff members can be reached. That is why over the long term a CEC should cover all three task areas, which will have to be built up and developed step by step. Here the emerging synergy effects can be utilised and a sustainable process of transformation of the "culture of a hospital" can be achieved. Let us now look at the three task areas in detail.

7.6.1   Clinical-Ethical Consultation for Individual Cases

The concrete work forms and definitions of tasks of a CEC vary depending on the structure and demand in the different hospitals. In practice, case-specific ethics consultation for difficult clinical decisions is often predominant. The effectiveness of clinical ethics consultation has not only been demonstrated in numerous practice reports, but also through controlled, randomised studies (Schneiderman et al. 2003, overview: Schildmann and Vollmann 2009).

Models of clinical ethics consultation

- Clinical ethics committee
- Working group of the CEC
- Clinical ethics consultant

A clinical ethics consultation can be conducted by the ethics committee, a working group of the committee or by a professional clinical ethics consultant. This last-mentioned model is offered in Germany by few hospitals, mainly university hospitals, which employ full-time clinical ethicists on a regular basis. The professional medical ethicist provides his specialised ethical expertise in the ethics consultation (Gerdes and Richter 1999, Reiter-Theil 2005). Here the ward team requesting an ethics expert receives direct support as soon as possible. The discussions of cases moderated by a clinical ethicist on the ward can have the effect of advanced training for the staff members and support a further implementation of clinical ethical structures in the hospital.

In the original form of ethics consultation by a clinical ethics committee, the treatment team members bring a conflictive ethical decision situation before the clinical ethics committee of their hospital. After describing the case and following open discussion, the clinical ethics committee formulates an ethical opinion on the case. This individual case consultation carried out by the ethics committee allows the integration and consideration of the many divergent moral perspectives of the CEC members from the different professional groups. In practice, however, this form of consultation has been shown to be disadvantageous since it often cannot be conducted on the ward at short notice, due to the size of the committee. In clinical practice, treatment decisions often have to be made at short notice and cannot be delayed until the next scheduled meeting of the clinical ethics committee. Furthermore, the clinical ethics committee meets in a room separate from the ward, so that only few team members can participate in the CEC meeting. There is also a symbolic difference whether the clinical ethics consultants come to the ward or whether the team of staff members from the ward must come before the committee. In this model physicians often have the feeling that they have to justify their decision or their actions as physicians before an ethics court or a tribunal (tribunal misconception) (Dörries 2003). This can lead to the false impression of a "know-it-all" ethics tribunal which has led in practice to many misunderstandings. Every form of ethics consultation, even by a clinical ethics committee, only takes place at the request of those involved. Clinical ethics consultation should provide support in moral conflicts in the individual case, whereas the professional groups always retain their responsibility and their freedom to make decisions. For that reason the vote of a clinical ethics committee is only the result of a consultation, but never an order or directive to act in a certain way. Ethical responsibility and the authority to make decisions cannot be delegated to third parties by the attending physicians, nor can they be delegated to an advisory body.

Clinical ethics consultation by a fixed working group of the CEC has proven to be a good solution. The clinical ethics committee selects about five people out of its members to form a fixed, interdisciplinary working group for a predetermined period of time. Its task: To carry out prompt, on-site ethics consultation in the hospital on behalf of the clinical ethics committee. This has the advantage that a small and constant number of ethics consultants, who are at the same time members of the CEC, can be deployed continually and flexibly. In this approach, the formal vote of a committee is not in the foreground, but rather the discussion and decision-making process, which is accomplished by the health professionals themselves working in the hospital with the assistance of moderators in ethics consultation. Here a member of the mobile ethics consultation team takes over the moderation of the discussion of the case, while other members co-moderate or write down the case protocol. Within this process of moral deliberation it is important to focus on the ethical aspects of the concrete case. Furthermore all participants regardless their profession or hierarchy have the same rights to speak up and contribute to the discussion, which is often a big challenge of the moderator of the consultation process (Neitzke 2003, Steinkamp and Gordijn 2005, Vollmann and Weidtmann 2003, Wernstedt and Vollmann 2005). The results of the ethics consultation should be documented in written form. Clinical ethics consultation cannot offer any patent solutions as "ethical body" or take over the responsibility and decision-making competence from the treatment team. All participants in ethics consultation are obligated to medical confidentiality. Because during an ethics consultation information related to the patient is discussed, the patient should, if possible, be informed about the ethics consultation to secure his consent (Zentrale Ethikkommission 2006).

In practice, a new offering of ethics consultation services requires a rather extensive start-up period in order to actively publicise the service, its way of working and the participants. Even with a positive development, individual case consultations are only carried out for a very small percentage of the patients (Fig. 7.1).

Fig. 7.1 Development of the numbers of clinical ethics consultations after the foundation of a clinical ethics committee at a university hospital (cumulative) (Vollmann 2008b)

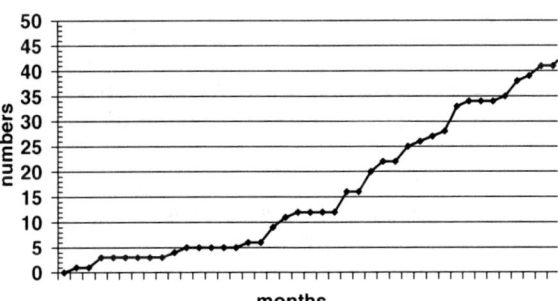

7.6.2 Development of Guidelines

Hospital-internal guidelines can be developed for frequently occurring problems and for superordinate ethical issues in the hospital. Generally these are proposed and written by a special group of the CEC, to which external experts can also belong who are not members. The guidelines are practice-oriented and tailored to the specific conditions of the institution. They thus concretise the legal provisions and guidelines e.g. of the German Medical Association and the respective science associations. Apart from the guideline text, a form of documentation should be stipulated for the practical work of the committee (e.g. checklists, medical record documentation sheets). Common subjects of guidelines are e.g. forgoing cardiac-pulmonary resuscitation (CPR), discontinuing therapy on the intensive ward, dying in the hospital, dealing with patient directives and PEG tubes, or particularly in psychiatry dealing with suicidal patients. Professional competence and continual inclusion of the participating professional groups and authorities are important prerequisites for successful guidelines development and implementation. All guidelines must be approved by the ethics committee and put into force by the board of directors of the hospital. From then on they are valid as guidelines for all professional groups in the hospital and all participants can refer to them. The ultimate goal of developing guidelines is to improve patient care.

7.6.3 Training and Advanced Training Workshops and Seminars

Training and advanced training events on ethical topics have an often underestimated significance for the clinical-ethical structures in the hospital. Since in the education and training of the health professions ethics often only plays a marginal role, there is a great need for advanced training opportunities. This applies not only to the members of the CEC itself, who need professional training, but to all staff members of the hospital. Therefore CEC themselves are involved in many European countries in education and training of health care professionals in clinical ethics. Often they organise and sponsor different forms of educational sessions in hospitals: e.g. through regular events such as ethics days, ethics cafés, scheduled lectures and discussions as well as workshops for the staff, an important contribution can be made to making staff members more aware of ethical issues and to providing further training. Apart from that, ethical topics can be integrated into the existing training structures of the individual professional groups in the hospital, such as on-the-job training for physicians or for nursing personnel. Especially in the fields of psychiatry and psychotherapy it seems advisable to integrate ethical knowledge and competences into the advanced training modules for physicians and other health care professionals in mental health. The same is true for other health professions in the field of psychiatry. Members of the CEC can themselves serve as disseminators of information about ethical issues in the hospital. In doing so, they can make the work of the CEC more well-known and increase its credibility and acceptance. They can make it clear that ethics is an issue for all staff members which should not be simply delegated to a committee. In particular, retrospective case discussions on the ward are suitable for practice- and need-oriented further training in ethics, e.g. in psychiatry each suicidal event should be discussed. In these case discussions, experience can be gained both in case moderation and in analysis of ethical problems, which is beneficial to both the advisers and those who are being advised.

7.6.3.1 Training for Specific Professional Groups

For physicians, topics concerning medical ethics can be integrated into the hospital's advanced training modules for physicians. Ward meetings and shift changes of the treatment team on-duty can offer a more informal opportunity for training.

Specialised advanced training in patient care provides the opportunity to present and discuss ethical topics within the scope of existing training structures. In all training modules, the exclusive use of the lecture format in presenting ethical topics should be avoided. Adequate time for questions, discussion and work in small groups should be planned.

7.6.3.2 Ethics Café/Ethics Salon

At regular intervals, e.g. every 2 months, an *ethics café* takes place in the hospital, lasting about one and a half hours in the afternoon. It is an informal get-together for staff members, who can inform themselves, or discuss ethical issues that arise in their day-to-day work. Over coffee and tea a previously announced topic will be presented in a short talk and then discussed. The discussion is moderated by CEC members, and the participants are given much time to express their concerns and opinions in the discussion. At the same time the participants can become more aware of the structures of an ethical discussion and can practice these structures.

7.6.3.3 Ethics Day

With the organisation of an Ethics Day the CEC annually presents the work of the ethics committee and of the working groups to the people within the hospital. This full-day or half-day event begins with a word of welcome from the director of the hospital and the chairman of the CEC. Apart from informative talks with discussions it can also include short role-plays and scenic representations of ethical problems which arise. Sufficient time is allowed for questions and discussions, and the long coffee break is important to exchange ideas informally and for the staff members to become acquainted with each other.

7.6.3.4 Hospital Newspaper

The hospital magazine or newspaper is a good way to announce events as well as to present different topics such as patient directives or to present a case analysis under a heading "Medical Ethics". This heading can also be written as an interactive case presentation "How would you decide?", giving the reader the opportunity to comment on an anonymised case report.

7.6.3.5 "Ethics Ring Folder"

Ring folders can be distributed to the different wards of the hospital, which aside from information about the CEC and ethics consultation contain ethical guidelines and data sheets of the respective hospital, relevant guidelines of the medical associations, samples of patient consent forms, etc. The ethics ring folder should be continually updated and the individual sheets contained in it should be chosen for the particular ward or medical department. To improve access for the whole region, a model ring folder can be posted on the Internet or Intranet website of the hospital.

7.6.3.6 External Training Courses

For physicians, introductory courses in "Clinical Medical Ethics" can be offered in cooperation, which have proven to be successful both as an individual event and as a series. For local or regional external training courses and events for all professional groups there is the option of cooperating with other psychiatric institutions of the region (Vollmann 2007).

7.7 Professionalisation

In recent years transregional training seminars and workshops have been further developed and professionalised: For example in Germany, to promote the further training and quality assurance of the work of clinical ethics committees, the working group "Ethics Consultation in the Hospital" of the Academy for Ethics in Medicine has developed a curriculum in which the goal and contents of further training for health professionals in clinical ethics consultation were formulated. Dividing this curriculum in a basic course and in theme-and-method-oriented advance courses was suggested (Simon et al. 2005). The contents of the basic course include: fundamentals of ethics and legal foundation, organisation and organisation ethics, forms, tasks and methods of individual case ethics consultation. Topics for advanced courses in the field of psychiatry could include e.g. patient self-determination as well as education and consent of the patient, dealing with patient directives in psychiatry, forced treatments, therapy conflicts with chronically ill mental patients and dement patients. Practice-relevant topics for advanced courses in methods are the moderation of ethics case discussions, the organisation of training and advanced training modules and the development of ethical guidelines.

In Germany the "Qualification Programme Hannover" is oriented on this curriculum, which has been carried out since 2003 by the cooperation partners Academy for Ethics in Medicine, the Centre for Health Ethics Hannover, Hannover Medical School and Ruhr-University Bochum (Dörries et al. 2008). To date more than 500 physicians, nurses, hospital chaplains and other health care professions from Germany and neighbouring countries have received further training (Dörries et al. 2005, 2008), which makes this training programme the largest in Europe. Furthermore, in cooperation with the Academy for Ethics in Medicine a distance-learning course "Consultant for Ethics in the Public Health Sector" was offered by the Centre of Communication, Information and Education of the Nuremberg Hospital Centre (www.cekib.de). In 2005 an Internet platform was set up to create a Germany-wide network of clinical ethics committees "Ethics Consultation in the Hospital. Internet Portal for Clinical Ethics Committees, Consultation and Liaison Services" and to facilitate an exchange of ideas and experiences (www.ethikkomitee.de). In Great Britain a "UK Clinical Ethics Network" (www.ethics-network.org.uk) was established. Clinical ethics consultants from

13 European countries have joined together in the "European Clinical Ethics Network (ECEN)".

A position paper of the Central Ethics Commission with the German Federal Board of Physicians on "Ethics Consultation in Clinical Medicine" asserts the great need for information and development in this area and demands a further professionalisation of the field. The Commission welcomes the development of ethics consultation in clinical medicine and encourages all hospitals to establish such services. Besides information on the development of clinical ethics committees, the structures of ethics consultation, the founding process and the specific tasks, the position paper also addresses problems and different approaches to solutions of these problems. It discusses the issues of integrity and credibility, forms and professionalisation of ethics consultation and the voluntariness of the consultation, time requirements, effectiveness and legal aspects, such as confidentiality and patient consent (Zentrale Ethikkommission 2006).

7.8 Outlook

In summary, the development of clinical ethics consultation in many European countries in the last 10 years has been characterised by an increase and growing differentiation and professionalisation of different forms of clinical ethics consultation. The appointment of clinical ethicists who work on a regular basis as part of the hospital staff, the provision of structured advanced training courses, curricula and recommendations have all contributed to a professionalisation of the field. Although much progress has been made over the last years in making the field of clinical ethics consultation more professional, one has to point out that in most European countries only a very limited numbers of professionally trained clinical ethicists exist. Therefore most of the every day work of clinical ethics consultation is provided by doctors, nurses, chaplains and other health professions beside their main duties in the hospitals. Although a growing number of these CEC members has received some training in clinical ethics and moderation skills beside their main professional obligations we have not reached a level of professionalism in clinical ethics consultation which we expect in all other fields of clinical medicine. To improve this situation more priority and resources within the heatl care sector must be invested into the field of clinical ethics.

To a great extent, this development has also taken place in mental hospitals. But in contrast to somatic medicine, there are differences due to the disease pictures, professional requirements and institutional patient care structures, which were discussed in this article. Only few studies exist on clinical ethics committees and/or on clinical ethics consultation in the field of psychiatry. Here is a great need for research.

References

Austin WJ, Kagan L, Rankel M et al (2008) The balancing act: psychiatrists' experience of moral distress. Med Health Care Philos 11(1):89–97

Bauer A, Vollmann J (2004) Ethische Fragen in der Konsiliar- und Liaisonpsychiatrie. In: Arolt V, Diefenbacher A (eds) Psychiatrie in der klinischen Medizin. Steinkopff, Darmstadt

Deutscher Evangelischer Krankenhausverband e.V., Katholischer Krankenhausverband Deutschlands e.V (1997) Ethik-Komitee im Krankenhaus. Selbstverlag, Berlin Freiburg

Dörries A (2003) Mixed feelings: physicians' concerns about clinical ethics committees in Germany. HEC Forum 15(3):245–257

Dörries A, Hespe-Jungesblut K (2007) Die Implementierung Klinischer Ethikberatung in Deutschland. Ethik Med 19(2):148–156

Dörries A, Neitzke G, Simon A et al (eds) (2008) Klinische Ethikberatung. Ein Praxisbuch. Kohlhammer, Stuttgart

Dörries A, Simon A, Neitzke G et al (2005) Ethikberatung im Krankenhaus. Qualifizierungsprogramm Hannover. Ethik Med 4:327–331

Engel CC Jr. (1992a) Psychiatrists and the general hospital ethics committee. Gen Hosp Psychiatry 14(1):29–35

Engel CC Jr (1992b) Exploring the role of the ethics committee psychiatrist. HEC Forum 4(6): 360–371

Ford PJ (2008) Special section on clinical neuroethics consultation: introduction. HEC Forum 20(4):311–314

Gerdes B, Richter G (1999) Ethik-Konsultationsdienst nach dem Konzept von J. C. Fletcher an der University of Virginia, Charlottesville, USA. Ethik Med 11:249–261

Hayes JR (1986) Consultation-liaison psychiatry and clinical ethics: a model for consultation and teaching. Gen Hosp Psychiatry 8(6):415–418

Helmchen H, Vollmann J (2001) Ethical Questions in Psychiatry. In: Henn FA, Sartorius N, Helmchen H, Lauter H (eds) Part 2: general psychiatry, vol 1. Springer, Berlin, Heidelberg, New York, NY, pp 315–349

Hurst SA, Forde R, Reiter-Theil S, Slowther AM, Perrier A, Pegoraro R, Danis M (2007b) Physicians' views on resource availability and equity in four European health care systems. BMC Health Serv Res 7:137, Aug 31

Hurst SA, Reiter-Theil S, Perrier A, Forde R, Slowther AM, Pegoraro R, Danis M (2007a) Physicians' access to ethics support services in four European coutries. Health Care Anal 15(4):321–335, Dec

Kovacs L, Strech D (2008) BMBF-Klausurwoche: clinical Ehtics Consultation: theories & methods – implementation – evaluation. Ethik Med 20(2):148–150

Lederberg MS (1997) Making a situational diagnosis. Psychiatrists at the interface of psychiatry and ethics in the consultation-liaison setting. Psychosomatics 38(4):327–338

Leeman CP (1995) Ethics consultation masking psychiatric issues in medicine. Arch Intern Med 155(16):1715–1717

Molewijk B, Verkerk M, Milius H et al (2008b) Implementing moral case deliberation in a psychiatric hospital: process and outcome. Med Health Care Philos 11(1):43–56

Molewijk B, van Zadelhoff E, Lendemeijer B et al (2008a) Implementing moral case deliberation in Dutch health care; improving moral competency of professionals and the quality of care. Bioethica Forum 1(1):57–66

Neitzke G (2003) Ethik im Krankenhaus: Funktion und Aufgaben eines Klinischen Ethikkomitees. Arztebl Baden Wurtt 4:175–178

Perl M, Shelp EE (1982) Sounding board. Psychiatric consultation masking moral dilemmas in medicine. N Engl J Med 307(10):618–621

Reiter-Theil S (2005) Klinische Ethikkonsultationen – eine methodische Orientierung zur ethischen Beratung am Krankenbett. Schweizer Ärztezeitung 86:346–352

Salathe M, Amstad H, Jünger M et al (2008) Institutionalisierung der Ethikberatung an Akutspitälern, psychiatrischen Kliniken, Pflegeheimen und Einrichtungen der Rehabilitation der Schweiz: Zweite Umfrage der Schweizerischen Akademie der Medizinischen Wissenschaften. Bioethica Forum 1(1):8–14

Schildmann J, Gordon J, Vollmann J (eds) (2010) Clinical ethics consultation: theories-methods-implementation-evaluation. Springer, Heidelberg (in print)

Schildmann J, Vollmann J (2009) Effekte klinischer Ethikberatung: Eine systematische Übersichtsarbeit zur Methodik quantitativer Evaluationsstudien. In: Vollmann J, Schildmann J, Simon A (eds) Klinische Ethik. Aktuelle Entwicklungen in Theorie und Praxis. Campus, Stuttgart

Schneiderman LJ, Gilmer T, Teetzel HD et al (2003) Effect of ethics consultations on nonbeneficial life-sustaining treatments in the intensive care setting: a randomized controlled trial. JAMA 290(9):1166–1172

Sexson SB, Sexson WR (2008) The role of the child and adolescent psychiatrist on health care institutional ethics committees. Child Adolesc Psychiatr Clin N Am 17(1):209–224

Simon A (2001) Ethics committees in Germany: an empirical survey of christian hospitals. HEC Forum 13:225–231

Simon A, May A, Neitzke G (2005) Curriculum "Ethikberatung im Krankenhaus". Ethik Med 17:322–326

Steinberg MD (1997) Psychiatry and bioethics. An exploration of the relationship. Psychosomatics 38(4):313–320

Steinkamp N, Gordijn B (2005) Ethik in der Klinik und Pflegeeinrichtungen. Ein Arbeitsbuch. Luchterhand, Neuwiedt

Vollmann J (2007) Klinische Ethikkomitees und Klinische Ethikberatung im Krankenhaus. Ein Praxisleitfaden über Strukturen, Aufgaben, Modelle und Implementierungsschritte. Medizinethische Materialien Heft Nr. 164. 9. edition. Bochum

Vollmann J (2008a) Ethikberatung an deutschen Universitätskliniken. Empirische Ergebnisse und aktuelle Entwicklungen. In: Groß D, May A, Simon A (eds) Beiträge zur klinische Ethikberatung an Universitätskliniken. LIT, Münster

Vollmann J (2008b) Klinik: Aufgaben und Kriterien für Klinische Ethikkomitees. Bundesgesundheitsbl – Gesundheitsforscl – Gesundheitsschutz 51:865–871

Vollmann J (2009) Ethische Probleme in der Psychiatrie. In: Berger M (ed) Erkrankungen. Klinik und Therapie, 3. Urban & Fischer, München

Vollmann J, Weidtmann A (2003) Das Klinische Ethikkomitee des Erlanger Universitätsklinikums. Institutionalisierung – Arbeitsweise – Perspektiven. Ethik Med 15:229–238

Wernstedt T, Vollmann J (2005) Das Erlanger Klinische Ethikkomittee. Organisationsethik an einem deutschen Universitätsklinikum. Ethik Med 17:44–45

Youngner SJ (1997) Consultation-liaison psychiatry and clinical ethics. Historical parallels and diversions. Psychosomatics 38(4):309–312

Zacchia C, Tremblay J (2006) Clinical ethics in psychiatry: the experience at Douglas Hospital. Sante Ment Que 31(1):95–105

Zentrale Ethikkommission (2006) Ethikberatung in der Medizin. Stellungnahme der Zentralen Kommission zur Wahrung ethischer Grundsätze in der Medizin und ihren Grenzgebieten (Zentrale Ethikkommission) bei der Bundesärztekammer zur Ethikberatung in der klinischen Medizin. Dtsch Arztebl 103(24):1703–1707

Part II
Principles of Ethics in Psychiatry

Chapter 8
Ethical Principles in Psychiatry: The Declarations of Hawaii and Madrid

Otto W. Steenfeldt-Foss

The medical profession has not only a formal, professional contract with society and the individual patient, but also a psychological, moral contract where the expectations of the doctor as representing the healing art, dedicated to the need of our suffering fellow human beings, is our overall goal and existential basis for our role. Our continuous striving to secure quality of care development, on the basis of human rights and medical ethics, is an implicit reflection of our dedication to this moral contract.

(Steenfeldt-Foss 2005)

Contents

Abbreviations

DNA   Desoxiribonucleid Acid
DRG   Diagnose-Related Groups
ENMESH   European Network for Mental Health Service Evaluation

O.W. Steenfeldt-Foss (✉)
University Health Services of Oslo, Oslo N-0260, Norway
e-mail: ottost@extern.uio.no

H. Helmchen, N. Sartorius (eds.), *Ethics in Psychiatry*, International Library
of Ethics, Law, and the New Medicine 45, DOI 10.1007/978-90-481-8721-8_8,
© Springer Science+Business Media B.V. 2010

NGO   Non-Governmental Organisation
WHO   World Health Organisation
WMA   World Medical Association
WPA   World Psychiatric Association

8.1 Ethical Principles: Declarations, Codes, and Law

8.1.1 Normative Principles in Ethics and Law

The profound changes of social and cultural structures of societies, due to tremen-
dous economic, scientific and technical developments not the least those of
medicine, challenge the normative context of the traditional physician–patient-
relationship. Society responded by a vast variety of normative regulations, from
very general and basic to more or less specific ones, from laws with legally binding
power over morally binding codes to guidelines and recommendations. Common
grounds as well as differences 'between legislation and bioethics', particularly tensions between human rights and ethical principles such as autonomy or dignity,
have been dealt with in the preceding chapter with the example of the European
Convention of Human Rights and Biomedicine. This Convention has been developed by a large body of lawyers, physicians and other experts for the whole range
of medicine and has legally binding power in all countries which have ratified and
signed this document. In contrast the following chapter describes ethical standards
specifically developed by and for psychiatrists in the Declarations of Hawaii and
Madrid.

8.1.2 Ethical Principles and the Socio-Cultural Context

Although there are traditional norms for the acting of physicians since the ancient
times they had been adapted to changes of the social and cultural environment in
which they are embedded: the Hippocratic Oath, a mixture of ethical principles
and etiquette of the medical profession, had been modernized by the Declaration
of Geneva of 1948–1994 (WMA). Much more specifically the upcoming visibility
of the UN Declaration of Human Rights (1994) and human rights activists challenged traditional psychiatric customs and mental health legislation. Since the 1970s
some national psychiatric associations developed ethical standards (Helmchen and
Okasha 2000).[1]

  Slowly the attitude changed towards more openness for the rights of the individual, even the mentally ill. Furthermore, the increasing awareness of the mechanisms
of devaluation, stigmatization, coercion against the mentally ill which led to the

---

[1] In 1995 perhaps 15 national psychiatric associations had special codes for psychiatrists; e.g.
American Psychiatric Association (1973).

terrible peak of perverted psychiatric acting in the totalitarian transformation of society in national socialistic Germany (Lifton 1986) made psychiatrists at least in western countries sensitive for evident abuse of psychiatric concepts, methods and services.[2] One important example was the abuse of psychiatry for political reasons in some countries, particularly for involuntary admission of political dissidents which had been declared mentally ill. Against this background the World Psychiatric Association (WPA) adopted at its World Congress 1977 in Hawaii the first international declaration of ethical principles for the acting of psychiatrists the primary aim of which was 'to encourage the psychiatrist in conflicts of loyalty in contemporary societies and to help him in conflicts of decision-making' (Helmchen and Okasha 2000). The declaration was developed by the Swedish psychiatrist, philosopher and first professor of medical ethics in Sweden Clarence Blomquist which he prepared during a stay at the Hastings Center (Ottosson 2000) where at the same time Tom Beauchamp and James Childress worked for their 'Principles of Biomedical Ethics'. This book, particularly its four ethical principles of (i) respect for autonomy of the patient, (ii) beneficience, (iii) non-maleficence, and (iv) justice became very influential after its first publication in 1978 (Beauchamp and Childress 1978[3]).

8.2 The Declaration of Hawaii

The vision of this declaration is illustrative in the following introduction:

> Ever since the dawn of culture ethics has been an essential part of the healing art. Conflicting loyalties for physicians in contemporary society, the delicate nature of the therapist-patient relationship, and the possibility of abuses of psychiatric concepts, knowledge and

---

[2] A major step was the publication of the Nuremberg Codex in 1947. It was developed for the Nuremberg trial against some of those national socialist physicians who participated in cruel medical experiments with prisoners of concentration camps and who were involved in killing patients with mental disorders. This codex was a major step since it made the principle of informed and voluntary consent for all participants of medical research mandatory, and because it brought this principle to public awareness by the worldwide publicity of the Nuremberg trial. It was a major step also insofar as prior instructions did not prevent these inhumane experiments not the least because they were almost unknown to the physicians in Germany: already in 1900 the Prussian Ministry of Culture instructed the directors of university hospitals that all research patients must consent voluntarily, and in 1931 the German Ministry of the Interior strengthened and differentiated this rule by a very clear decree. The broad and intense progress of medical science after World War II brought hope also for severely ill people and provoked the demand for research to improve their treatment. With the development of consequentialistic ethics in therapy research it also became acceptable with patients who were incompetent to give informed consent under the condition that the research intervention is expected to benefit the participating patient and his/her consent was substituted by that of a legal guardian. This was codified for the first time by the World Medical Association in 1964 with the Helsinki Declaration, and later on also in national laws.

[3] See footnote 1.

technology in actions contrary to the laws of humanity, all make high ethical standards more necessary than ever for those practicing the art and science of psychiatry. As a practitioner of medicine and a member of society, the psychiatrist has to consider the ethical implications specific to psychiatry as well as the ethical demands on all physicians and the societal duties of every man and woman. A keen conscience and personal judgement is essential for ethical behaviour. Nevertheless, to clarify the profession's ethical implications and to guide individual psychiatrists and help form their consciences, written rules are needed.

The declaration explicates 'the ethical principles of respect for autonomy and of beneficience: by formulating the components of informed consent: by calling to mind the obligation of confidentiality; by stating rules for forensic evaluation and compulsory interventions; by demanding the possibility of independent proof of compulsory measures; and by obliging the psychiatrist not to misuse his professional possibilities and particularly to abstain from any compulsory intervention in the absence of a mental disorder. Even the accentuation of these rules constituted evidence of the current need for the rejection of political misuse of psychiatry' (Helmchen and Okasha 2000).

8.3 Steps for a Revision of the Declaration of Hawaii

'New challenges during the following 2 decades led to further responses, of the WPA and other international bodies, to safeguard the human rights of the mentally ill and to improve their treatment and care in order "to reflect the impact of changing social attitudes and new medical developments on the psychiatric profession". Major documents of this kind are: (i) the Declaration of Hawaii/II, adopted by the WPA in 1983 in Vienna, with only minor changes, mainly attenuating some binding formulations and a slightly more paternalistic perspective; (ii) a charter on the rights of mental patients, adopted by the WPA in 1989 in Athens, a condensed catalogue of statements and viewpoints of WPA bodies concerning ethical problems; (iii) the UN resolution 46/119 on "The Ten Principles for the Protection of Persons with Mental Illness and the Improvement of Mental Health Care", adopted by the 1991 General Assembly and including an annex with corresponding "Principles for Policy on Mental Health" with a major implication for just resource allocation for mental health care' (Helmchen and Okasha 2000). The ten principles are: (1) Promotion of mental health and prevention of mental disorders; (2) Access to basic mental health care; (3) Mental health assessments in accordance with internationally accepted principles; (4) Provision of the least restrictive type of mental health care; (5) Self-determination; (6) Right to be assisted in the exercise of self-determination; (7) Availability of review procedure; (8) Automatic periodical review mechanism; (9) Qualified decision maker; and (10) Respect of the rule of law.

These needs and new ethical dilemmas of the 1990s led the WPA to develop further the Declaration of Hawaii on the duties of psychiatrists resulting in the Declaration of Madrid, adopted by the WPA in 1996 in Madrid.

8.4 The Declaration of Madrid

This declaration underlines the responsibility of the psychiatrist to follow accepted scientific and ethical principles, built on continuous research and education, and includes also securing a just allocation of health resources, stresses the importance of treating every patient with respect for individual dignity, integrity and autonomy, stresses the special aspects of forensic psychiatric functions, and underlines the patient's rights to be fully informed of the nature of the condition of the proposed diagnostic and therapeutic procedures, including possible alternatives and the right to choose between the available methods.

It further regulates the conditions related to compulsory treatment, warns especially against the political abuse of psychiatry in the absence of psychiatric illness, stresses the importance of confidentiality, and describes the procedure related to patient participation in research projects and the importance of informed consent. In the final paragraph, the responsibility of the psychiatrist to stop all therapeutic teaching or research programs that may evolve contrary to the principle of this declaration is underlined (WPA 1996).

The Madrid Declaration is not static but regularly updated and amended at the WPA General Assemblies, latest in Yokohama 2002 and Cairo 2005. Upcoming new challenges require the need for developing further guidelines related to specific situations. Examples are the role of psychiatrists related to Euthanasia, Torture, Death Penalty, Genetic Research and Counselling etc. The latest amendments cover Ethics of Psychotherapy in Medicine, Conflict of Interests in Relationship with Industry, Protection of the Rights and dual Responsibilities of psychiatrists etc (http://www.wpanet.org/content/madrid-ethic-engish.shtml).

8.5 Society's Reaction

User's comments on the Declaration of Madrid can be summarized using the following key words: autonomy, integrity and dignity. We will now examine them, making reference to the Declaration's items.

Item 1 is concerned with the promotion of mental health and prevention of mental disorders according to accepted scientific standards. Giving psychiatry the responsibility of securing a just allocation of health services is lifting priority-setting in health care to an ethical dimension. Hitherto, mental disorders have not been sufficiently included in normative analysis of quality assessments and nationally and internationally approved lists of priority. The commentaries are reflecting and highlighting the importance of a broad-spectred service apparatus.

The strongest reaction is however related to the term 'therapeutic interventions that are least restrictive to the freedom of the patient'. This formulation is built on principle 4 of the UN General Assembly Resolution 46/119 of 1991 related to the definition of a mental case. The principle includes the following components to be considered in the selection of the least restrictive alternatives: the disorder involved;

the available treatment; the person's level of autonomy; the person's acceptance and cooperation; the potential that harm is caused to self and others.

There is a fundamental difference between somatic and psychiatric diseases. Mental disorders represent to a much higher extent an unstable situation with exacerbations and remissions (this is valid also for other medical chronic diseases such as some carcinomas or metabolic diseases or neurological diseases such as sclerosis multiplex). However, the "fundamental difference" is of course related to the societal breakdown condition with functional disability that is to some extent identical to the disease picture itself. In the more severe psychiatric conditions as manifest psychoses, according to the ICD-10, one of the main symptoms is loss of insight into reality and inability to make proper evaluation and judgement of own situation.

But according to UN principle 5 on self-determination, consent is required before any type of interference can occur. Interference includes bodily and mental integrity (e.g. diagnostic procedure, medical treatment such as use of drugs, electroconvulsive therapy and irreversible surgery) and liberty (e.g. mandatory commitment to hospital).

In case the person is unable to consent, which happens occasionally but not systematically, there should be a surrogate decision maker to decide on the patient's behalf. Special precautions should be taken to make sure that mental health care providers do not systematically consider mental patients unable to make their own decisions and exercise self-determination with regard to all components of integrity and liberty because the patient was found to be unable with regard to one.

Item 2 states that psychiatrists have a responsibility to keep updated about scientific developments so as to secure the patients the most proper and modern treatment. In many parts of the world, this can be experienced as pure phraseology, since the development of social and health services in general is poor and psychiatry especially has a tendency to be given a low priority. Nevertheless, the responsibility put on the individual psychiatrist to keep updated, and giving this responsibility an ethical dimension as related to allocation of resources, is of value.

The importance of the therapist–patient relationship is highlighted in *item 3*. One of the main complaints of the users of mental health services is a growing retreat from talking with patients. Shorter working time combined with increasing case loads, increasing patient circulation, increasing emphasis on technologic procedures, reduces the available time for talking with patients at length and over time. The organization of the services has put psychiatrists more and more in the role as diagnosticians and medication prescribers to patients treated psychosocially by other professional groups. Uncritical use of neuroleptic drugs and minor tranquilizers instead of proper psychosocial and psychotherapeutic procedures are an increasing danger in services lacking human professional resources, thus threatening the patient–doctor relationship. Securing this relationship is also critical for obtaining relevant information to help the patient coming to a rational decision about their own choices as to different treatment options. Compliance is a necessity for providing the alliance giving the patient a feeling of being a real partner.

The individual patient's dignity and legal rights are underlined in *item 4*. According to international mental health acts and the ten basic principles drawn from the UN General Assembly Resolution, the criteria for compulsory treatment are stated. Regretfully, psychiatric diseases sometimes imply the need for compulsory admission and treatment provided in the best interest of the patient. This inequity between treatment for somatic and psychiatric disorders is illustrated in the dilemma between patient autonomy and societal control – a theme for intensive debate between human rights activists and the medico-legal professions.

The importance of mental health assessments being provided in accordance with internationally accepted medical principles is covered in *item 5*. Refraining from reference to nonclinical criteria such as political, economic, social, racial and religious criteria is a central issue.

The importance of confidentiality in the patient–doctor relationship is discussed in *item 6*. In a time when electronic data processing and communication is increasing, securing confidentiality is more urgent than ever and should be monitored.

Item 7 underlines the importance of psychiatric research being conducted according to the ethical canons of science. The fundamental distinction must be recognised between medical research in which the aim is essentially diagnostic and/or therapeutic, and research where the essential object is purely scientific and without direct value to the persons objected to the research, according to the World Medical Association's Declaration of Helsinki 1964, revised 1975. In any research on human beings, each potential participant should be adequately informed of the aims, methods, anticipated benefits and potential hazards of the study and the discomfort it may entail. When obtaining informed consent, the doctor should be particularly cautious if the individual is in a dependent relationship to him or her or may consent under duress. In that case the informed consent should be obtained by a doctor who is not engaged in the investigation and who is independent of this official relationship. This is of special importance related to' psychiatric patients.

Finally, the comments are related to the importance of user involvement in all stages of planning, running and evaluating services. This concern is implicitly covered in item 1 in the Declaration of Madrid. The planned parallel development of alternatives to hospital into a comprehensive treatment chain with continuity of care, has hitherto not occurred at the same speed as the reduction of hospital beds. More systematic studies are needed to evaluate the effect of different elements, as well as development of more reliable methods in evaluation. Main themes are need assessment, quality of life, family/caregiver burden and satisfaction with services as stated by the European Network for Mental Health Service Evaluation (ENMESH) in 1994. The active involvement of users, including their families, is of central importance to secure services that are meeting real needs of the population to serve. In this connection also, the proposal regarding the establishment of an Ombudsman institution is relevant.

8.6 Reflections of a Clinician

In the last decades we have been moving towards the postindustrial society. The mental health services are finding themselves in a turmoil where the original values and priorities are being threatened by new norms and values of a materialistic type. Health services all over the world are in rapid transformation, being influenced by new economic steering mechanisms, changing the priorities in a way that is not always professionally and scientifically based. Although there is an international agreement on putting a higher priority to the development of services for the underprivileged like the long-term mentally ill, the mentally retarded, and the elderly, the reality is different. We have frames of reference for further improving and developing services that should secure equality for all patient groups. Hitherto, this equality is mainly verbal, and the 'decibel method' is in most countries more effective. Although health political slogans continuously state the importance of treatment and care standards based on genuine respect for the individual patient's dignity and integrity, experience demonstrates the opposite. In the time of crisis patterned not only by economic, but also of ethical confusion, it is even more important to have a clear ideal frame of reference for our clinical and public health activities. The Declarations of Hawaii, Madrid and their continuous development should be used as such frame of reference.

One should, however, be warned against the growing tendency of being so enchanted by fine plans and resolutions that we are loosing track of clinical reality. Political statements have an inborn danger of attaining a value of their own, requiring an importance greater than those for whom the plans were supposed, namely the patients. The ongoing technological development in medicine and psychiatry, is creating new opportunities, but at the same time fostering a danger for further technological fixation with inbuilt contempt for more abstract humanistic values.

Our time is patterned by 'The Dictatorship of Relativism' – nothing is either absolutely right or wrong. This social disintegrative process is counteracting our functions as therapists, aiming at giving the patient insight based on honesty and truth. Our psychiatric activity must also in the future, and to an increasing degree be based on solid clinical and scientific insight and knowledge with an overall loyalty to the professional and humanistic declarations and conventions steering our functions as doctors (Steenfeldt-Foss 2008).

8.7 Perspectives

Future strategies for securing ethical standards for psychiatry should be:

1. An updating of the educational content with broader emphasis on behavioural sciences, ethics, and the special elements which pattern the doctor–patient relationship. The teaching of ethics should include medical aspects of human rights and the special obligation of doctors, integrated in the regular clinical training and as life-long medical learning. The central role of the teachers in model learning for the candidates must be underlined (see Chapter 6).

2. A continuous revision and clarification of the professional ethical codex, i.e. for psychiatrists the Declaration of Madrid including the international conventions and declarations. Special responsibility rests with the World Psychiatric Association in the context of developments in the World Health Organization, the United Nations, and the World Medical Association as well as the national psychiatric associations throughout the world (see Chapter 5).
3. Increasing awareness of psychiatric professionals as well as of health administrators on the psychiatric ethical codes, i.e. the Declaration of Madrid and its amendments. Constant revision of national and international health acts to secure patient rights and prevent the political abuse of psychiatry.
4. Provisions must be taken to secure the special role of the psychiatrist compared with other medical specialists as a caretaker of the humanistic ideals, harmoniously integrated into a broad based biologic, psychological and social theory and method.
5. Epidemiologic studies and need assessments as well as quality assessment procedures must be developed with a clear intention of securing the individual patient's needs, eventually as opposed to society (see Chapter 26).
6. Research models for analysis of ethical conflicts in priority setting in public health services should be developed, leading eventually to the establishment of special protective legislation for services to weaker groups (see Chapters 3 and 12).

References

American Psychiatric Association (1973) The principles of medical ethics with annotations especially applicable to psychiatry. Am J Psychiatry 130:1056–1064

Beauchamp TL, Childress JF (1978) Principles of biomedical ethics. Oxford University Press, New York, NY

Helmchen H, Okasha A (2000) From the Hawaii declaration to the declaration of Madrid. Acta Psychiatr Scand 101:20–23

Lifton JR (1986) The Nazi doctors. Medical killing and the psychology of genocide. Basic Books Inc., New York, NY

Ottosson JO (2000) The declaration of Hawaii and Clarence Blomquist. Acta Psychiatr Scand 101:16–19

Steenfeldt-Foss OW (2005) Patient- and human rights and biopsychosocial development in psychiatry. Cairo: Jean Delay prize lecture, WPA XIII World Congress of Psychiatry, Kairo

Steenfeldt-Foss OW (2008) The planning and management of mental health services – ideal and realities. International Psychiatrist Lecture. American Psychiatric Association Annual Convention, Washington, DC

UN Declaration of Human Rights (1994) Human rights, a complication of international instruments, vols. I and II. United Nations, New York and Geneva

UN General Assembly Resolution 46/119 (1991) United Nations principles for the protection of persons with mental illness and the improvement of mental health care, December 17. UN General Assembly, New York

World Medical Association (1948) Declaration of Geneva, International code of medical ethics, October 1949, IIIrd General Assembly, London

World Psychiatric Association (1996). The Declaration of Madrid on Ethical Standards for Psychiatric Practice, Madrid, Aug 25

Chapter 9
Informed Consent in Psychiatric Practice

Hanfried Helmchen

The patient who is armed with information, who wants to ask questions, should be seen as an asset in the process of care and not an impediment to it.

(Donaldson, cited by Maclean 2009)

Contents

Abbreviations

SGB   Sozialgesetzbuch Germany
WMA   World Medical Association
WPA   World Psychiatric Association

H. Helmchen (✉)
Department Psychiatry & Psychotherapy, CBF, Charitè – University Medicine Berlin,
Berlin D-14050, Germany
e-mail: hanfried.helmchen@charite.de

H. Helmchen, N. Sartorius (eds.), *Ethics in Psychiatry*, International Library
of Ethics, Law, and the New Medicine 45, DOI 10.1007/978-90-481-8721-8_9,
© Springer Science+Business Media B.V. 2010

9.1 Introduction

The right of self-determination is a fairly young cultural development. It arose during the past 300 years mainly in the European and North American culture determined by the ideas of the humanistic enlightenment of the seventeenth and eighteenth century (see Chapter 5). Today it is established as a fundamental human right not only by international declarations such as the Universal UN Declaration of Human Rights (1948), the WMA Declaration of Helsinki (1964) and its amendments (last in Seoul 2008) with regard to patients' participation in medical research, or the WPA Declarations of Hawaii (1978) and Madrid (1996) and amendments (last in Cairo 2005) with regard to ethical challenges in psychiatry, but also legally by international law such as the International UN Covenant on Civil and Political Rights (1966) and specifically the European Convention on Human Rights and Biomedicine (1997)[1] (see Chapter 5) as well as by national law either as part of constitutions, e.g. in Germany the Grundgesetz, or at least in special laws, e.g. in Germany the Drug Law and particularly in social law (SGB XI: insurance of care; SGB IX: rehabilitation and participation of handicapped people). The consequence of this right is in medicine that every medical intervention can be done only with the consent of the patient. In order to exercise this right by consent the patient must be informed before giving his consent. However, informing the patient is not only a legal obligation, but also an ethical demand in order to build up trust in the patient. It is an essential part of the patient–physician relationship.

This chapter deals with questions such as why, when, what about, how, and by whom the patient will be informed as far as this information is related to the patient's decision making. This involves the question for the patient's capacity to consent, i.e. to understand the necessary information for a valid consent and to decide competently with this knowledge. Obviously this is particularly relevant in mentally ill patients because mental disorders might reduce this capacity more frequently and often less clearly than in other medical conditions. Since informed consent is a complex issue (Koch et al. 1996) as well as at the core of the patient's care, some aspects will be dealt with in separate chapters such as the assessment of the capacity (ability, competence) to consent (see Chapter 14), the consequences of an impairment or even a lack of this capacity for both the legal substitution of this capacity (see Chapter 10), and for research with such particularly vulnerable patients (see Chapter 25).

9.2 Components of Informed Consent in Practice

Validity of a consent requires that the patient understands the intervention-related medical information, understands its significance and consequences, and can value its meaning for himself. Furthermore, the patient must be able to build a stable

---

[1] Which up to now has not become national law for all members of the EU, e.g. not in the United Kingdom and in Germany (see Chapter 6)

judgement and to express it. Thereby the existence of competence to consent depends upon various abilities. Its assessment is determined by the weight that will be attached to these individual prerequisites. But because there are no standard rules for weighing these items, generally accepted criteria for assessment of the capacity to consent are missing. Also there do not exist generally accepted standardised procedures how to assess these blurred criteria (see Chapter 14): there is not only a variance of criteria but also a variance of assessment. However, after all it can be said that the threshold of assuming competence to consent depends upon the difficulty and significance of the necessary decision. The consent to an urgently needed intervention without serious risks makes fewer demands on the capacity to consent than a decision of application or omission of a life-saving treatment in a terminal stage of a fatal disease. The definition of such differences of thresholds for the assessment of the competence to consent is ethically important because thereby the most possible respect for the self-determination of the patient is joined with a far reaching protection of the patient against irrational decisions, e.g. distorted by depression.

9.3 The Patient–Psychiatrist Relationship as the Context of Informing the Patient

Respecting seriously the patients' autonomy and his right of self-determination is more than only to convey to him the basic facts of what will (or better: should) be done. However, in practice and particularly under time pressure informing the patient is often seen as more of a legal requirement than a task to empower the patient for self-determined decisions as well as more a single act than a process. Furthermore, the process of informing the patient is recognised too seldom as a chance to gain the trust of the patient; but it should be part of and embedded in the development of the psychiatrist–patient relationship. This is especially valid in psychiatry since the majority of mental disorders has a longer time perspective and to build up trust between the patient and the therapist is often more difficult than in other medical disciplines.

Hence, the psychiatrist should have the ability to observe precisely and to listen with concentration in order to find out thoroughly the complaints of the patient and their embedding both in the patient's present life situation and biography as well as his concepts of life and value system. The demand of many patients to be understood often remains unsatisfied as inquiries show that patients most frequently lament over their experience that the physician has not enough time, and thereby cannot listen attentively and does not really understand them. The needed attitude against these complaints can be called a communicative-dialogical one (Epstein et al. 2004, Gask and Usherwood 2002, Price and Leaver 2002).

The patient–psychiatrist relationship is based on communication. Apparently it depends not only on the patient but also on the behaviour of the psychiatrist, particularly his ability to produce an atmosphere of openness and trust by listening and recognising and asking the right questions in order to clarify the relevant facts as well as the implied values. In this context the psychiatrist should convey his knowledge and experience to the patient in a way that he can understand and use this knowledge for a reasonable decision with regard to the intervention proposed by the psychiatrist. The talk should enable the patient to participate actively and to ask questions. It requires quietness and time. Empathetic understanding, emotional support and readiness to help characterise the "therapeutic basic behaviour". This behaviour and also the ability to find the adequate level of verbalising are essential elements of communication.

The dialogical process proceeds on the part of the patient in an interplay of hesitating, concealing, opening and telling, on that of the psychiatrist in exploring the history of the suffering and biography as a togetherness and succession of listening, expressing empathy and checking back, sometimes trying to understand by verbalising the patient's experiences. This procedure may build up a relationship of trust. It presents the psychiatrist the chance to help the patient's understanding of his illness and to involve him in measures that are in his health interest; in doing this he has to take into account the risk of crossing the border between motivation and persuasion or even manipulation (see Chapter 20). A lasting working alliance often can be established only if by this procedure a common model of explanation and acting will be found that will reconcile the subjective experience of the patient with an understandable scientific concept and makes possible a congruence of both worlds of experience. Thereby the psychiatrist has to balance his empathetic engagement with the patient on the one hand and his control of its distortions by counter-transference and culture-bound behaviour on the other hand: "How personal can the physician–patient relationship become and how depersonalized must it remain?" (Katz 1986, p. xix).

In this context specific questions will be answered.

9.4 Procedures: Informing the Patient Why? What About? How? When? And by Whom?

9.4.1 Why?

Consent of the patient as a justification for medical interventions has had a long legal tradition since the nineteenth century. However, only during the second half of the past century the legal construct of informed consent has been elaborated, not the least in the courts against physicians who as a rule did not question the (tacit) acceptance or non-refusal of a proposed intervention that was in the best interest of the patient. Moreover, a mentally ill patient who has no clear cognitive or behavioural disturbances might experience doubt expressed in his or her ability to consent as discrimination and an impairment of trust in the psychiatrist. Generally the competence to consent was only called into question in practice if a patient refused an urgent or life-saving intervention without a recognisable or understandable reason. However, today the consent generally will be asked for, sometimes more in a formal way and without proof of its validity. But the validity of consent must be the more certain the more potential risks are associated with the intervention, i.e. the competence to consent must be assessed with regard to the subject in question because it is not a global but a relative one; the patient may be competent for one subject but not for another one. Furthermore, since this capacity is not a time-invariant quality its validity is established only for the date of assessment. Therefore, it should be documented very well. This could be valuable in cases in which patients later on complain of not having been informed; however, decisive is the competence here and now but not the recollection of the information. With regard to the validity of consent it must be based on all information that will be relevant for an adequate decision of the patient.

9.4.2 What About?

Information should be given upon (i) the illness, its probable course with as well as without treatment and the reason for the proposed intervention, (ii) the aim, procedure, and probable duration of the intervention, (iii) expected wanted effects and possible unwanted side-effects including inconveniences and risks, (iv) alternatives to the proposed measures, and (v) consequences of refusal (Helmchen 1996).

The patient should also become aware of the fact that information on the prognosis of the mental illness and on the benefits and risks of the intervention are probability statements loaded with uncertainties, furthermore, that medical knowledge is limited, and that sometimes several options of treatment may be given. Therefore, the psychiatrist will choose with regard to the individual situation of the patient a treatment that most probably promises success. At least in long-term treatments the patient should be urged to consider that, in view of such uncertainties, his consent and some responsibility for his own health is important.

9.4.3 How?

Information should be given verbally and individually. The increasingly demanded written form of informed consent gives the patient–psychiatrist relationship the formal character of a contract as is usually made between persons who are unknown to each other. This may provoke perplexity, uncertainty or anxiety and contradict a trustful relationship between the patient and the psychiatrist.

Information should be given with empathy. For instance, patients with a beginning dementia should be prepared clearly and cautiously on the possibility that their competence to consent could be impaired: "Although many patients would like to know the truth, the rights of those who do not want to know should also be respected. Therefore, the diagnosis of dementia should not be routinely disclosed, but just as in other disorders, health care professionals should seek to understand their patients' preferences and act appropriately according to their choice" (Marzanski 2000). If such a patient does not refuse to be informed his attention should be called to the possibility that it will be appropriate to document his living will in time in order to enforce his wishes for carrying out omission-defined treatments in case he may become incompetent (see Chapter 10).

Information should be given with sensitivity in view of specific vulnerabilities according to the cultural, intellectual and emotional conditions of the patient. This is particularly valid in countries with migrants and minority groups from different cultural backgrounds. With regard to the different resilience of patients some information could be given later or even withheld, justified by the so-called "therapeutic privilege" in order not to impair the welfare of the patient. Or withholding information about rare side-effects is allowed in some European countries but not in others.

9.4.4 When?

At best informing the patient is a step-by-step process starting with some basic facts in an order according to the immediacy of needed decisions and the resilience of the patient. The patient must have the opportunity to check back his understanding. Even in emergency situations the psychiatrist should try to inform the patient about what and why an intervention must and will be done.

9.4.5 By Whom?

As has been said, the informed consent process is part of the development of the patient–psychiatrist relationship; thus it is clear that informing the patient is an essential task of the patient's treating psychiatrist. If the psychiatrist is not responsive the patient may ask other members of the team, nurses or social workers. This is the more important the more the patient himself will seek information from relatives, friends, other patients, and not the least from unknown others and lay people, i.e. increasingly from the media and the internet, chatrooms, blogs etc. Thereby he runs the risk of collecting information that may be wrong or does not fit his special case. Perhaps this may stabilise misconceptions on his disorder or the treatment and may increase the risk of non-adherence to treatment. Therefore, it is the task of the psychiatrist, not the beginner but the experienced one, to engage himself in the process of informing the patient.

9.5 Informed Consent and Ethical Conflicts

Respect for the autonomy of the patient is expressed by appreciating informed consent as a basic right. However, in psychiatry two major obstacles have to be mentioned.

First, to realise the ideally described elements of informing the patient in day-to-day reality is often only possible in a more or less restricted mode. Limits are given on the part of the patient by a low level of cognitive ability or disorganised or incoherent thinking, or by the non-responsive behaviour of a suspicious or doubtful or unwilling patient; on the part of the psychiatrist by time pressure, counter-transference or inadequately trained ability to communicate with the patient. However, if the psychiatrist can build up a trustful relationship with the patient, this may lead the patient to abandon his self-determination. If such trust means an accepted dependency, the question is raised whether trust is compatible with the concept of the patient as a self-determined partner of the psychiatrist (Welie and Welie 2001).

> The proponents of informed consent and patient-self-determination have insufficiently appreciated that trusting oneself and others to become aware of the certainties and uncertainties that surround the practice of medicine, and to integrate them with one's hopes, fears and realistic expectations, are inordinately difficult tasks.... The opponents of informed consent, on the other hand, have insufficiently appreciated that disclosure and consent do not abolish trust (Katz 1986, p. xv f).

Second, the principle of respect for the self-determination of the patient may come into conflict with the old ethical principle to act in the best interest and welfare of the patient. There is still a "great difficulty letting patients make choices that doctors feel are not in their best interest" (Fargot-Largeault 1996). Apart from obviously immediate risks in emergency cases, in other cases of a patient at risk to harm himself or others the psychiatrist has to answer the question how intense the actual risk must be to induce a legal action that authorises him to overcome the refusal of a mentally ill patient's consent to any psychiatric intervention. Furthermore, he should be clear about the aim of the intervention, i.e. only protection against harm by non-voluntary admission or treatment of the mental disorder that causes the risk and impairs the autonomy of the patient. In some national laws the reasons and criteria for involuntary treatment are identical with those of involuntary admission (McClelland 1996), in others they do not include the involuntary treatment of an involuntarily admitted mentally ill. For instance, in Switzerland the procedure requires two steps of decisions by a judge: first, a decision on involuntary admission, and second, a decision on involuntary treatment. In Germany it was only after a clarifying judgement of the Federal Constitutional Court in 2006 that the involuntary treatment of involuntarily admitted severely mentally ill patients became legally justified.

Such differences and uncertainties gave and give rise to controversies between liberal positions that see the law as too restrictive and families or caretakers of people with severe mental disorders who may view some aspects of them as too permissive. But even below the threshold of legally authorised overriding of the refusal of a patient's consent, psychiatrists experience the tensions between respecting the will of the patient and acting according to their obligation to care for the patient. The responsible psychiatrist cannot leave the patient alone with his right of self-determination. His obligation to care may justify ethically the demand for a "mild paternalism". However, he also has to be aware that in this fortunately rarely regulated area sometimes the borderline between an explicit consent or assent of the patient and beyond down to persuasion and even coercion may be blurred (see Chapter 20).

9.6 Conclusion

Informed consent should be viewed not only as a legal requirement but also as a chance to build up a trustful patient–psychiatrist relationship. It is an essential part of this relationship.

References

Epstein RM, Alper BS, Quill TE (2004) Communicating evidence for participatory decision making. JAMA 291:2359–2366

Fargot-Largeault A (1996) National report: France. In: Koch HG, Reiter-Theil S, Helmchen H (eds) Informed consent in psychiatry. European perspectives of ethics, law and clinical practice. Nomos, Baden-Baden, p 78

Gask L, Usherwood T (2002) ABC of psychological medicine. The consultation. BMJ 324(7353):1567–1569

Helmchen H (1996) Common European standards and differences, problems and recommendations. In: Koch HG, Reiter-Theil S, Helmchen H (eds) Informed consent in psychiatry. European perspectives of ethics, law and clinical practice. Nomos, Baden-Baden, pp 383–409

Katz J (1986) The silent World of doctor and patient. The Free Press, Macmillan, New York, NY

Koch HG, Reiter-Theil S, Helmchen H (eds) (1996) Informed consent in psychiatry. Nomos, Baden-Baden

Maclean A (2009) Autonomy, informed consent and medical law. Cambridge University Press, New York, NY

Marzanski M (2000) Would you like to know what is wrong with you? On telling the truth to patients with dementia. J Med Ethics 26(2):108–113

McClelland R (1996) The basic problem of informed consent. In: Koch HG, Reiter-Theil S, Helmchen H (eds) Informed consent in psychiatry. European perspectives of ethics, law and clinical practice. Nomos, Baden-Baden, pp 268–277

Price J, Leaver L (2002) ABC of psychological medicine: beginning treatment. BMJ 325:33–35

Welie JVM, Welie SPK (2001) Is incompetence the exception or the rule? Med Health Care Philos 4:125–126

Chapter 10
Advance Directives: Balancing Patient's Self-Determination with Professional Paternalism

Hans-Martin Sass and Arnd T. May

Contents

10.1 The Case

Walter Jens, famous Professor of Rhetoric at the University of Tuebingen, in 1995 together with his friend Hans Kueng, Professor of Theology, published a book 'Dying in Dignity. Pleading for Self-Responsibility', in which both strongly argued in favor of individual self-determination, including respecting autonomous decisions for euthanasia in cases of severe pain or dementia. In 2006, his wife wife Inge and Walter executed individual advance directives.

> The model form allowed to indicate when 'all medical interventions have to be stopped which will prevent me from dying' and mentions the following three situations 'when
> (a) I am in an irreversible final stage of an incurable fatal disease or in the process of dying,
> (b) when I am mentally so confused that I don't know any more who I am, where I am, and

---

H.-M. Sass (✉)
Kennedy Institute of Ethics, Georgetown University, Washington, 20057 DC, USA;
Institute of Philosophy, Ruhr Universität, Bochum 44780, Germany
e-mail: sass@ethikzentrum.de

H. Helmchen, N. Sartorius (eds.), *Ethics in Psychiatry*, International Library
of Ethics, Law, and the New Medicine 45, DOI 10.1007/978-90-481-8721-8_10,
© Springer Science+Business Media B.V. 2010

do not recognize family and friends any more, (c) I am for a prolonged period in a coma and there is no or only minimal hope to regain conscience' (Kielstein and Sass 2007, p. 13).

Shortly after executing the advance directive Walter very rapidly developed dementia; Inge became his legal guardian entrusted with honoring his last wishes as documented in his Living Will. Most of the time Walter has no pain, in rare moments he seems to be clear, most of the time he is extreme confused, does not know who he is, where he is, does not recognize family and friends'. Mrs. Jens being very well aware of the trust and guardianship she owes Walter asks herself:

> How do I, who has committed herself to be his guardian, deal with my husband's wishes to die and his requests to assist in dying? How binding are his words 'I want to die', 'I cannot bear it anymore', 'no, no, no!', how binding his appeal 'Help me, I want to be dead!'? – I have at this moment – at this time of writing – no answer to the question (Jens and Kueng 2009, p. 205).

In well developed systems of law, competent adults have the right to decide for themselves in health care matters. They also may appoint a legal guardian for future situations of incompetence or express their preferences, wishes, directives in advance for those situations, i.e. writing so called Living Wills. Professor Jens and his wife have exercised their right to self-determination in executing advance directives naming each other as guardians. They have done what self-responsible people are supposed to do: to protect themselves in future scenarios of incapacity and to direct or at least assist their doctors and caretakers in making decisions in their 'best interest'.

There are different personal, professional and legal cultures in accepting or respecting individual self-determination. There is also a more or less stringent supervision, when the individual capacity for 'self-determination' is questionable or potentially harmful to the person. Courts also may appoint legal representatives to determine the welfare of persons incompetent to make their own decisions; but the term 'welfare' quite often is very loosely used such as in section 1901 of the German Civil Code and in other national legislations. In regard to Professor Jens' situation his wife Inge says:

> For me the problems start there, when the patient is a person who has no terrible pain and handicaps to suffer from, thus when he is in the colloquial sense "physically healthy" (Jens and Kueng 2009, p. 205).

She defines a moral, a medical, and a legal issue of uncertainty in decision making. In other family or partnership cultures or on other continents her question might be phrased differently, but it always will fall within the tensions (a) of paternalism versus autonomy, (b) of balancing the 'old person' who expressed wishes and values versus the 'new person' who is in a new situation, he or she might or might not have been able to envision concretely enough, and (c) of different possible hermeneutic models of how to interpret previously voiced or written wishes, values, directives. In the care of psychiatric or otherwise mentally challenged patients, there will be many additional conflicts between the 'good of the patient', as defined paternalistically, and the 'will of the patient', as defined by the patient or as interpreted by others based on written and/or oral statements (Henderson et al. 2008, Sass 2003).

10.2 Changing Values and Wishes: The 'Old' and the 'New' Person

Mrs. Jens is the guardian of the 'new Person' her husband was, but she signed the guardian contract with the 'old person', the Professor Jens, who had his very distinct values, wishes and expectations regarding life, quality of life, and a life worth living. There is an extended debate whether or not previous directives should be honored or whether the 'presumed actual will' of the patient at the time of intervention should guide intervention decisions. The normative conflict can best be illustrated by the German legal situation where all medical interventions have to based on the 'patient's will' and in the absence of a patient's will based on the 'presumed will of the patient'; an advance directive is not understood as representing the will of the patient, rather only as an instrument to assess her or his 'presumed will' at the actual moment. Supreme Court decisions in many countries, including Germany and Austria (Marschner 2000, May 2001) hold that previous directives are binding, if the person giving those directives was competent at the time of making those decisions, and that so-called objective criteria of futility may only be used as a default position in the clear absence of the expression of individual preference. The German Chamber of Physicians recommends to take the 'presumed wish' of the patient into account when the patient herself or himself is unable to communicate; they recommend to recognize a written advance directive or previous oral statement as an important information for assuming what the presumed actual will would be. The German Chamber of Physicians does not make it binding to base medical decisions exclusively on advance directive documents if those are present (Kielstein and Sass 2007, Sass et al. 1998). Quite a number of controversial positions surrounding the concept 'presumed actual will' persist. A possible conflict between the previous and the actual person is real and not only theoretical (May 2001, pp. 199–201, Sass 2001, 2003). Practical reasoning and common sense in interpreting expectations, angst, wishes, hopes and directives of patients in individual situations seems to be the prudent way to solve individual cases and conflicts (Davis 2002, Quante 1999, Kielstein and Sass 2007).

(a) As long as competent persons have not changed oral or written statements and directives those statements should be taken as their true position and others should act accordingly in respect for persons; it is the right and obligation of competent persons who change their views and preferences to let others know, and if they do not do so they carry the risk of being misunderstood and mistreated.

(b) If someone falls in a state of wishlessness such as deep and prolonged coma, this person will have no new experience on which a change of values and wishes could be based and therefore should be treated according to wishes and values expressed previously.

(c) Patients suffering from chronic and progressive illnesses and persons suddenly confronted with physical disabilities will or will not adapt to new and quite different parameters of quality of life. Clinicians are very aware that many chronically patients and also those suddenly in a situation, which they previously would not heave thought to be worth living in, indeed, do adjust to new challenges over and over again. This phenomenon of human adaptability has been widely described in the literature. As long as these patients are competent, they have ample opportunity to accept or reject treatment; for those patients the use of advance directives is not indicated, and they, if they choose so, may adjust their previously stated preferences according to their new experiences and visions for life.

(d) Severely demented patients, not knowing who they are, where they are, and unable to recognize friends and loved ones, who while fully competent have executed advance directives refusing or requesting certain interventions in given situations, should be honored as the persons they were when they made those decisions, which they then felt would be the most appropriate expression of their visions and values. This does not exclude, to honor those actual present preferences which are based on the person's adaptability to new situations and changing understandings and experiences of quality of life issues as long as the 'grande vision' of the advance value-and-wish directive is kept as a overall guiding force. As Professor Jens is already in a state of severe dementia, are the 'do-gooder' actually 'trouble maker' in not respecting his original values and wishes? (Savalescu 1997).

(e) Hard cases, however, seem to be those, where (a) patients oscillate between different stages of confusion or (b) request forms of comfort care which would contradict their previous wishes or directives. These are situations full of ambiguity in making the most appropriate benevolent paternalistic clinical decision and often the care for the 'good' as presently expressed by the patient should be honoured over previous statements (Lauter and Helmchen 2006, Quante 1999, Savulescu and Dickenson 1998, Spellecy 2003, Srebnik and Kim 2006); other recommend to use a least invasive treatment and care (Olsen 1998, Sass 2001). Research, development and review of validity and improvement of advance care planning for mental goes on (Atkinson et al. 2003, Henderson et al. 2008, Sass 2003). Thus, the improvement and introduction of better models of advance care documents is one of the most promising instruments to reduce coercive crisis intervention (Henderson et al. 2009, Swanson et al. 2008). National standards and legislations are different (Loetjoenen 2006, Swanson et al. 2006a).

10.3 Models of Advance Care Documents and Directives

Contemporary debates advance care documents for mental health patients are not centered around 'whether or not', but on 'how much' decision making capacity psychiatric patients should be given, balancing the autonomy principle with the harm – to others and themselves – principle (Lynn et al. 1999, Marschner 2000, Olson 1998,

Sabat 1998, Sehgal et al. 1996, Srebnik and Kim 2006, Voelzke 1998). We may differentiate between three groups of patients for which the legal guardianship situation and their human and civil right to respect and to honor their wishes, values and directives are different:

10.3.1 Episodically Compromised Patients

There are psychiatric diseases such as affective and schizophrenic disorders with an episodic course, where patients are only compromised some of the time, not all the time, where there are acute and non-acute stages. In non-acute situations most patients are quite capable to understand, communicate logically, judge prudently and give directives in advance for future acute situations, similar to those they had experienced before in the course of their illness. It is indispensable and, if not done, it would be a violation of professional ethics and medical standard to assist patients in establishing as detailed as possible in advance the forms and modes of care they prefer in the case of acute illness. Ulysses arrangements or contracts (Behandlungsvereinbarungen) e.g. for epileptic patients can be quite extensive; some might have half a dozen pages and contain detailed clauses on physical or pharmacological restraints, names of persons requested and names of those rejected for bedside vigil in case of four-point or other physical restraint (Brock 1993, Dresser 1982, Olsen 1998, Voelzke 1998, Borbe et al. 2009). Ulysses contracts are care-and-protection contracts in advance (Vereinbarungen), discussed, negotiated, and signed in advance by caregivers and those patients who are familiar with those situations and who have the right, the competence and the duty to make those decisions and not burden their caregivers with heteronomously choosing between intervention options in the absence of the patient's wish and will. Given the temporary character of compromised competency and the patient's experience in such situations, it is a civil and medical right of psychiatric patients to have Ulysses arrangements, whenever those can be established.

10.3.2 Permanently Compromised Patients

There are psychiatric illnesses, where patients are compromised all the time and where their wishes are either a threat to themselves or others or cannot be established coherently at all. Four models of paternalistically identifying the 'good for the patient' where the 'will of the patient' is in fact, or is presumed to be, self-destructive and dangerous, have been recommended: a community based standard, a provider based standard, a dispositional-preference standard (Sass 1991, Srebnik and Kim 2006, Swanson et al. 2006b, Wilder et al. 2007).

10.3.2.1 Community Based Standard

A 'community standard' is based on a 'local patient community medical directive' originated from shared preferences and values within a small community as compared to another community nationally and internationally. Patients have morally significant preferences and attitudes; these preferences are informed and build on the basis of their respective family and culture. The benefits of community – based - directives over expert - consensus include the indirect recognition of the autonomy of the constituents, the experience of patient's preferences. Community - based preference lists for good clinical practice reflect preferences, attitudes, and expectations of a particular community; ethics consultation groups or various forms can be very helpful to review such community standards, also in the light of human and civil rights of patients (Lynn et al. 1999, May 2001).

10.3.2.2 Provider Based Standard

Within larger communities of diverse preferences, a 'provider - based standard' might be more efficient and justifiable. In communities of culturally diverse populations, various approaches compete to define the best clinical care and support a dignified process of dying, according the provider's corporate ethics as chosen by the patient (Hare 1981, Szasz 2007). The provider defined standard of good clinical practice and dying in dignity serves a multicultural community only when a multitude of options are offered. In regard to complex psychiatric illnesses characterized by severity, iatrogenic influence, undeterminable competence of the patient, the provider-based standard and peer review of caregivers including nursing specialists might be better for determining and reviewing the good of the patient.

10.3.2.3 Dispositional-Preference Standard

Savulescu and Dickenson have proposed a 'dispositional-preference standard', allowing psychiatric patients to express present preferences relevant for care and treatment, including the refusal of medically beneficial intervention. 'The imperative to respect a person's present preferences in a liberal society is great. According to the dispositional analysis, it is not necessary for clinicians to show that a mentally ill person was competent, informed, and acting freely at the time he expressed his preference in an advance directive, but only that the person has a present dispositional preference similar to that expressed in the advance directive, and he is free, competent, and fully informed at some point during the time he had the preference. We distinguish this dispositional analysis of advance directives from an analysis in terms of substituted judgment' (Savulescu and Dickenson 1998, p. 224). To prove the human-right and civil-right point of respecting autonomy, they 'suggest that the mentally ill should be able not only to consent to treatment in advance, but also to refuse it, and indeed to refuse life-saving medical treatment or psychiatric treatment when they are a risk to themselves' (Savulescu and Dickenson 1998, p. 226). The dispositional preference standard is a new and yet not fully proven way to import into the theory and ethics of psychiatry again the century-old debate on how to best respect and to honor human dignity and individual choice, when self-determination might be confused or compromised and harmful the person herself or himself.

10.3.2.4 Precedent Autonomy

More recent models of advance health care planning for patients with mental disorders use traditional methods of balancing respect for self-determination with professional paternalism (Davis 2002, Quante 1999, Sass 1991, 2003).

10.3.2.5 Ulysses Contracts

Ulysses, according to Greek mythology, asked his comrades to tie him tightly to the mast – but to close his ears and eyes – when his ship would approach the Sirens, so that he could not follow their sweet whispers. Upon approaching the sirens, Ulysses was temporarily insane, struggling to free himself and follow their call; after the period of great danger was over, Ulysses again took control of his ship and his crew. In psychiatry the Ulysses model (Dresser 1982) is used to sign a contract with patients who previously and routinely had disturbing psychiatric attacks, and who want with the help of their physicians and caretakers determine aspects of their treatment during the next attack (Gremmen et al. 2008, Borbe et al. 2009, Dresser 1982).

10.3.2.6 Proxy Model

Finally, there is a fourth form of paternalistic decision making. It is the 'proxy model' widely used, where legally accepted, in designating a friend or family member as legal representative to make medical decisions by requesting, accepting or refusing treatment. Given the complex nature of decisions to be made, proxies or family members might not be well informed or act under distress or conflict of interest (Kielstein and Sass 2007). But we see in the case of Walter and Inge Jens how difficult it is to determine the patient's best actual interest. In all four situations, the basic philosophical and cultural issue is, how far the state or professional or societal paternalism may decide 'what is best for the patient'. All models therefore are within the realm of more or less 'soft paternalism' within different societal, cultural, and professional cultures.

10.4 Narratives and Other Methods to Establish Value-and-Wish Profiles

There are different methods of communication to prepare in advance for medical and mental uncertainties of unknown character, for guardianship and for advance care documents. We can differentiate between three models: the Ulysses directive, the pro-active negotiation procedure, the narrative information model. All three models are similar as far as they intend to establish a value anamnesis or value history, additional to the medical history, in order to guide future person-oriented individualized treatment and care.

10.4.1 Ulysses Contracts

See above Section 10.3.2.5; in general, all advance care documents and guardianship documents, not just those for psychiatric patients, are more or less based on the Ulysses contract model.

10.4.2 Negotiations

The content of determining and directing in advance future actions of intervention and care needs to be negotiated and defined in communication-in-trust and then executed by cooperation-in-trust between the various stakeholders. Experienced authors put emphasis on the negotiation skills and the details of the content to be covered, also on the latitude of final arrangements (Dietz and Poerksen 1998, Borbe et al. 2009, Voelzke 1998). In this regard, 'experienced' psychiatric patients are in a more educated positions than Professor Jens. They know what to expect in a specific situation; Professor Jens could not have known the specific details of 'when I am mentally so confused that I don't know any more who I am, where I am, and do not recognize family and friends any more'. Negotiations do not need necessary have to lead to formal and signed contracts; they help caretakers to understand the value-wish-fear profile of their patients better and to provide a more individualized care and treatment.

10.4.3 Narratives

Stories have played an important role in the history of philosophy and ethics (Kielstein and Sass 1993, Kielstein and Sass 1997). Plato and Aesop used a story rather than theoretical argumentation to identify ethical issues and to formulate moral judgments, principles, and virtues. According to Paul Ricoeur 'the interpretation of a text is a proposal of a world in which I would like to live and develop my possibilities' and that there is 'a correlation between the telling of a story and the temporal character of human experience' (Ricoeur 1988, p. 121). Margaret Walker has described the role of narratives in medical ethics consultation; she understands moral as 'points in continuing histories of attempted mutual adjustments and understandings among people' as compared to bioethical principles resembling 'law-like directives of high generality, like those giving autonomy, sanctity of life, or beneficence absolute or relative priority' (Walker et al. 1993, p. 35). Physician-patient interaction is based on communicating in the hermeneutics of the patient's narrative of her or his stories surrounding his well-being, well-feeling, pain, scare, distress, anxiety, quest for help. Communication-in-trust and cooperation-in-trust defines the success in predictive and preventive medicine and in acute intervention. The methodological approach to use narratives methods for the preparation and validation of advance medical care documents for competent patients or future patients builds on the hermeneutical insight into the superior role of narratives for dealing with specific topics in human communication and on the tradition and experience of medical practice. Over the last decade the use of narrative methods to assess competent patient's values and wishes and to establish preferences for medical and nursing care in advance has been well established. As Kielstein describes, the narrative method uses stories to confront competent prospective patients or patients with typical future situations of surrogate decision making, they might encounter. Carefully selected and tested stories of dementia, multimorbidity, prolonged coma, severe stroke, and suicide can be employed in order to invoke patient reactions and enable them to picture themselves in the situation and determine their preferences of treatment. Those narratives confront the healthy and competent person with 'real' situations or scenarios and allow for differentiated responses, for actual self-understanding and self-value-anamnesis and for future self-determination and for voicing in advance wishes and values for medical treatment care. Experience tells, that not more than four or five well chosen stories are necessary to cover the most basic clinical situations where the will of the patient or the presumed will of the patient has to play an essential role in guiding medical decisions, to provide, to modify, to withhold or to terminate certain medical interventions (Kielstein and Sass 1993, 1997). For specific diseases, in particular degenerative chronic diseases, where patients and physicians should communicate about the probable future course of degeneration, specific narratives could be developed, in particular for assessing attitudes of adaptability, loss of zest, and increase in frustration or depression; the same is true for intellectually or emotionally compromised persons. In particular, when hospice care is preferred or when withholding of clinically available life-prolonging technology is contemplated, decision must always be based on the individual's value-and-wish profile, thus integrating value status, clinical status, blood status, and other information into a truly differential diagnosis and prognosis.

Using narrative methods in the care of the mentally disturbed and demented patient will provide similar benefits and will prepare and assist caregivers in making good and patient-oriented choices in the care and treatment of those patients. Nursing care and medical care of patients with intellectual disabilities needs to be individualized care the same way, competent patients are given individualized care. As Meininger points out, narrative methods can and will change attitudes of nurses and physicians and result in better care:

'A new approach to personalized care and support should therefore be based upon a narrative and hermeneutical model of interpretation and support as opposed to an empirical model of diagnosis and treatment': 'Life stories, and the knowledge they convey, can be used for several purposes. The purpose may depend on the occasion on which the story is told, written, read, or listened to. Persons with intellectual disabilities themselves should be seen as prime authors of their own life stories. However, in the construction of their written life stories, family, friends, care workers or researchers may play an important part. Sometimes they serve as assistants or ghost-writers; more often they are readers and interpreters' (Meininger 2000, p. 2).

Well chosen narratives containing issues of health and disease, pain and suffering, shame and dependency, death and dying seem to be an excellent way to gain access to problems which in societies primarily build on principles of success and power, self-determination and self-rewarding, good life and quality of life have been pushed out of the prevailing culture, put under taboo and into those institutions we have created precisely for the purpose of collectively and individually not being bothered with: hospitals, psychiatric institutions, geriatric wards and old age homes. Mordacci and Sobel have pointed out:

> The possibility for us to tell a story about our illness rests on the availability of communicable meanings that can give an acceptable sense to our frailty, to which we can give out assent. We cannot be storytellers of our illness if our culture is deprived of any useful contents for that purpose; this is the tragedy of an exclusively technological culture (Mordacci and Sobel 1998, p. 36).

As the basic narrative approach cannot be employed to some demented patients, a special form of a narrative value history, called value diary, has been developed by Grundstein-Amado. Through a series of repeated interviews and statements of previously recorded interviews with patients, the caregiver records a 'Value Diary' and establishes a systematic index to the patient's preferences, wishes and values (Grundstein-Amado 1992). For multimorbid dialysis patients Kielstein has utilizesd a model based on an 'Individual Care Group' (Betreuungsgruppe) of two or three nurses and physicians, supported by friends or family members of the incompetent patient (Kielstein and Sass 1993, 1997). Separately, they write private reports on their communication with the patient and on the patient's preferences in treatment and his or her pronouncement of values or wishes regarding quality of life. They then compare notes and device a long term care plan, which routinely is reviewed. Clearly, the Grundstein-Amado and the Kielstein approach are paternalistic methods to gain access to value-and-wish information and directives, which otherwise would be expressed autonomously by the patient with or without communication and advice by a health care expert. But their approaches seem to be the most adequate ones to indirectly base nursing and clinical decisions on the patient's life narrative and value profile, even though only through the veil of the patient's symptoms and shortcomings. While developed for persons suffering from various degrees of dementia, such a communication-narrative approach could be used for different groups of psychiatric patients in different and disorder-specific ways.

10.5 Balancing Paternalism and Self-Determination

The methods and models described for optimizing and individualizing treatment of psychiatric patients are all based on a principles of 'soft paternalism' and 'good-for-the-patient'. The dispositional-preference model gives priority to the recognition of personal liberty and free choice, even though the free will of patients suffering from psychiatric disorders might be questioned and the choice might be self-destructive. The peer-group standard and the community-based standard also orient themselves on the principle of the 'patient's good' as they stress a professional or neighbourly obligation to prevent harm and to provide support, care and human solidarity. These models are paternalistic as far as they are based on a philosophical methodological decision regarding the best-interest standard for 'the other': a patient with psychiatric disorders (ADRA 1994, Hare 1981, Lynn et al. 1999, Vollmann 2001; May 2001).

Under a least-restrictive standard 'only intervention directed at the patient's harmful behavior is allowed; intervention is never justified because the provider prefers goals other than those expressed by the patient'; such an approach might lead to undertreatment and can be interpreted as unethical refusal of professional. A most-therapeutic standard, on the other hand 'promises a broader range of treatment to more people who are in need of treatment, but there is a risk that the heart of liberty, the right to self-definition, will be lost by those individuals' (Olsen 1998, p. 245). Theory and practice of psychiatry and the active involvement of patients in treatment decisions and making advance directives and having their present preferences been honored will always have to oscillate between these two poles. Academic, social, cultural, and professional debates, also the rewriting of rules and regulations, might never come to a final solution, as the positions taken are classical options in how to deal with fellow humans, suffering from mental disorders. As far as we see, neither position has empirically proven to be superior yet, nor has any one of these models be given overwhelming cultural or political support over the others. For ethicists, this is an uncomfortable situation in the transfer of reasoning and values into the complex every-day work of psychiatric wards and old-age homes. Therefore the most prudent common-sense consequence would be to understand the different models as theoretical default models, to be used as guiding visions only, not prescriptions or instruments to be followed strictly. While most psychiatrists seem to be open to a new partnership model with their patients, given them input into treatment decisions and care options based on patients' values and wishes, Thomas Szasz in a recent discussion on the legal right of patients to have their advance directives honored by medical staff strongly expressed a different view: 'the perceived moral-psychiatric need to prevent harm to self and others precludes the use of advance directives in psychiatry' (Szasz 2007, p. 900). Savalescu on the other hand stresses the quest and respect for self-determination, even in the light of potentially wrong and even totally mistaken decisions, in calling paternalistic 'good-doers' alike Szasz 'trouble makers' (Savulescu 1997). Whatever model is preferred in guiding geriatric or psychiatric care and intervention decision making, narrative methods seem to be a superior choice for individualized care, or in the words of Meininger:

A new approach to personalized care and support should therefore be based upon a narrative and hermeneutical model of interpretation and support as opposed to an empiricist model of diagnosis and treatment. Consistent with the latter model is an interventionist character of professional practice. Conversely, processes of (re)constructing and interpreting stories demand a professional commitment in which "presence" and "engagement", as distinct from "intervention" and "distancing" are central characteristics (Meininger 2000, p. 3).

References

ADRA [Alzheimer's Disease and Related Disorders Association Inc] (1994) Guidelines for the treatment of patients with advanced dementia. ADRDA, Chicago

Atkinson JM, Garner HC, Stuart S et al (2003) The development of potential models of advance directives in mental health care. J Ment Health 12:575–584

Borbe R, Jaeger S, Steinert T (2009) Behandlungsvereinbarung in der Psychiatrie. Joint Crisis Plans in Psychiatry. Psychiat Prax 36:7–15

Brock D (1993) A proposal for the use of advance directives in the treatment of incompetent mentally ill persons. Bioethics 7:247–256

Davis J (2002) The concept of precedent autonomy. Bioethics 16:114–133

Dietz A, Poerksen N (1998) Verhandeln als Leitlinie psychiatrischen Handelns. In: A Dietz, N Poerksen, W Voelzke (Hrsg) Behandlungsvereinbarungen. Psychiatrie, Bonn, 9–15

Dresser R (1982) Ulysses and the psychiatrists. Civil Lib Law Rev 16:833–835

Gremmen I, Widdershoven G, Beekman ZR, Sevenhuijsen S (2008) Ulysses arrangements in psychiatriy. A matter of good care? J Med Ethics 34:77–80

Grundstein-Amado R (1992) Value inquiry: a method of eliciting advance health care directives. Humane Med 8:31–39

Hare R (1981) The philosophical basis of psychiatric ethics. In: Bloch S, Chodoff P (eds) Psychiatric ethics. Oxford University Press, New York, NY, pp 31–43

Henderson C, Flood C, Leese M et al (2009) Views of patients and providers on joint crisis plans: single blind randomized trial. Soc Psychiatry Psychiatr Epidemiol 44:369–376

Henderson C, Swanson JW, Szmukler G, Thornicroft G, Zinkler M (2008) A typology of advance statements in mental health care. Psychiatr Serv 59:63–71

Jens W, Kueng H (2009) Menschenwuerdig sterben. Ein Plaedoyer fuer Selbstverantwortung. Piper Erweiterte und aktualisierte Ausgabe, Muenchen

Kielstein R, Sass HM (1993) Using stories to assess values and establish medical directives. Kennedy Inst Ethics J 3:303–325

Kielstein R, Sass HM (1997) Die Wertanamnese: ein narrativer Ansatz zur Erstellung und Interpretation von Betreuungsverfügungen. Wiener Medizinische Wochenschrift 147(6): 125–129

Kielstein R, Sass HM (2007) Die persoenliche Patientenverfuegung. Ein Arbeitsbuch zur Vorbereitung mit Bausteinen und Modellen. Muenster: Lit, 5. Aufl.

Lauter H, Helmchen H (2006) Vorausverfügter Behandlungsverzicht bei Verlust der Selbstbestimmbarkeit infolge persistierender Hirnerkrankung. Nervenarzt 77(9):1031–1039

Loetjoenen S (2006) Medical research on patients with dementia. The role of advance directives in European legal instruments. Eur J Health Law 13(3):235–261

Lynn J, Teno J, Dresser R, Brock D, Lindemann H, Lindemann N, Kielstein R, Fukuchi Y, Lu D, Itakura H (1999) Dementia and advance-care planning. Perspectives from three countries on ethics and epidemiology. J Clin Ethics 10:273–285

Marschner R (2000) Verbindlichkeit und notwendiger Inhalt von Vorsorgevollmachten und Patientenverfugungen in de Psychiatrie. Recht und Psychiatrie 18(4):161–163

May A (2001) Autonomie und Fremdbestimmung bei medizinischen Entscheidungen für Nichteinwilligungsfähige. Lit, Münster

Meininger HP (2000) Life stories: narratives and professional practice. Ethics Intellect Disabil 5(2):1–3

Mordacci R, Sobel R (1998) Health, a comprehensive concept. Hastings Cent Rep 28(1):34–37

Olsen DP (1998) Toward an ethical standard of coerced mental health treatment: least restrictive or most therapeutic? J Clin Ethics 9:235–246

Quante M (1999) Precedent autonomy and personal identity. Kennedy Inst Ethics J 9:365–381

Ricoeur P (1988) Time and narrative. University of Chicago Press, Chicago, IL

Sabat SR (1998) Voices of Alzheimer's disease sufferers: a call for treatment based on personhood. J Clin Ethics 9:35–48

Sass HM (1991) Differentialethik und Psychiatrie. In: Poeldinger W, Wagner W (eds) Ethik in der Psychiatrie. Springer, Berlin, 95–118

Sass HM (2001) Balancing autonomy with paternalism for psychiatric patients. Formos J Med Humanit 2(1/2):18–28

Sass HM (2003) Advance directives of psychiatric patients? Balancing paternalism and autonomy. Wiener Medizinische Wochenschrift 17/18:380–384

Sass HM, Veatch RM, Kimura R (eds) (1998) Advance directives and surrogate decision making in health care. Johns Hopkins University Press, Baltimore, MD

Savulescu J (1997) The trouble with do-gooders. The example of suicide. J Med Ethics 23:108–115

Savulescu J, Dickenson D (1998) The time frame of preferences, dispositions, and the validity of Advance Directives for the mentally ill. Philos Psychiatr Psychol 5(3):225–246

Sehgal AR, Weisheit C, Miura Y, Butzlaff M, Kielstein R, Taguchi Y (1996) Advance directives and withdrawl of dialysis in the United States, Germany, and Japan. JAMA 276:1652–1656

Spellecy R (2003) Reviving ulysses contract. Kennedy Inst Ethics J 13:373–392

Srebnik DS, Kim SY (2006) Competency for creating, use, and revocation of psychiatric advance directives. J Am Psychiatry Law 34:501–510

Swanson JW, Swartz MS, Elbogen EB et al (2008) Psychiatric advance directives and reduction of coercive crisis intervention. J Ment Health 17:255–267

Swanson J, Swartz M, Ferron J, Elbogen E, van Dorn R (2006a) Psychiatric advance directives among public health consumers in five US cities: prevalence, demand, and correlates. J Am Acad Psychiatry Law 34:43–57

Swanson JW, Van McCrary S, Swartz MS, Elbogen EB et al (2006b) Superseding psychiatric advance directives. Ethical and legal considerations. J Am Psychiatry Law 34:385–294

Szasz T (2007) You don't in psychiatry. Br Med J 35:900

Voelzke W (1998) Sinn und Zweck, Chancen der Behandlungsvereinbarung. In: A Dietz, N Poerksen, W Voelzke (eds) Behandlungsvereinbarungen. Psychiatrie, Bonn, 16–28

Vollmann J (2001) Advance directives in patients with Alzheimer's disease. Med Health Care Philos 4:161–167

Walker, Margaret Urban (1993) Keeping moral space open. Hastings Cent Rep 23(2):33–40

Wilder CM, Elbogen EB, Swartz MS, Van Dorn RA (2007) Effects of patient's reasons for refusing treatment on implementing psychiatric Advance Directives. Psychiatr Serv 58(10):1348–1350

Chapter 11
Confidentiality

Roy McClelland

Contents

Abbreviations

ECHR   Council of Europe's Convention for the Protection of Human Rights
                and Fundamental Freedoms (ETS no 005, 1950 as amended)
ECtHR   European Court of Human Rights
ETS   European Treaty Series

R. McClelland (✉)
School of Medicine, Dentistry & Biomedical Sciences, Queen's University, Belfast, UK
e-mail: r.j.mcClelland@qub.ac.uk

H. Helmchen, N. Sartorius (eds.), *Ethics in Psychiatry*, International Library
of Ethics, Law, and the New Medicine 45, DOI 10.1007/978-90-481-8721-8_11,
© Springer Science+Business Media B.V. 2010

EU   European Union
EuroSOCAP   European Standards in Privacy and Confidentiality in Healthcare
UK   United Kingdom

11.1 Introduction

All patients have the right to privacy and the reasonable expectation that the
confidentiality of their personal information will be rigorously maintained by all
healthcare professionals.[1] Each patient's right to privacy and the professional's
duty of confidentiality apply regardless of the form (for example, electronic,
photographic, biological) in which the information is held or communicated.

11.2 Ethical and Legal Foundation

The protection of privacy and confidentiality is both an ethical obligation and a
legal obligation. These are two different kinds of obligations, although normally
what they require will be the same in a particular situation. They are not absolute
obligations and must often be considered in the light of other obligations. Healthcare
professionals have an ethical obligation to be aware of the nature and extent of
their legal obligations. Health professional organisations should work to ensure that
such legal obligations are in keeping with the ethical obligations of their profession.
While psychiatrists have an obligation to obey the law, doing so does not guarantee
that they have behaved ethically.

11.2.1 Principles

There are core principles of medical confidentiality and privacy which find expression in both ethical and legal norms. These are not simply rules, but rather guides
for all healthcare professionals in decision-making and aids for the promotion of
ethical conduct in particular situations.

   The importance of maintaining confidentiality in the practice of healthcare has
been recognised continuously over the two and a half millennia since the composition of the Hippocratic Oath. It remains a core value to this day and in modern
Europe finds expression in three key principles of healthcare confidentiality.

---

[1]This chapter on confidentiality is grounded on the European Standards on Confidentiality and
Privacy in Healthcare (2006) developed through the work of the EuroSOCAP Project (QRLT-2002-
00771). The Standards were written following detailed consideration of the needs of vulnerable
patients – particularly children and young people, older people, homeless people, people with
mental health problems, prisoners, people with an intellectual disability, and people who lack
decision-making capacity.

- Individuals have a fundamental right to the privacy and confidentiality of their health information.
- Individuals have a right to control access to and disclosure of their own health information by giving, withholding or withdrawing consent.
- For any disclosure of confidential information healthcare professionals should have regard to its necessity, proportionality and attendant risks.

These principles find application in specific ways with different patient groups.

11.2.2 The Ethical Basis of Privacy and Confidentiality

11.2.2.1 Ethical Justifications

Privacy refers here to the general interest in control of one's private sphere broadly conceived. The right to privacy, the right to respect for private life, is a well-established right in the European tradition. This right guarantees the protection of the person against the intervention or interference of public authorities in the private sphere and it embraces, but is not restricted to, the protection of personal information.

A major source of the requirement of confidentiality in psychiatric practice is the fact that the relationship between the psychiatrist and the patient is, or should be, one of 'fidelity' or 'trust'. Within the relationship there exists a tacit understanding on the part of the patient that confidential information will not be further used or disclosed without the awareness and consent of the patient. The patient has, thus, a reasonable expectation that information shared will not be further shared with anyone else.

A different, though related, reason for not using or disclosing personal information is that the patient may not want it to be used or disclosed. Just as the patient has a right to self-determination in various other healthcare matters, it is the patient's decision as to who should have access to personal healthcare information and how it should be used.

The confidential nature of the relationship between psychiatrist and patient and respect for the patient's autonomy constitute prima facie reasons for protection of personal information. Taken together they strengthen the case for the non-use or non-disclosure of private information about a patient. There are also other justifications. For example, one reason for respecting confidences in healthcare is that doing so enables patients to disclose sensitive information that the psychiatrist needs to carry out treatment. Without an assurance that confidentiality will be maintained, patients might be less willing to disclose information, resulting in negative effects for their health, for public health and for healthcare practice.

The patient's right to self-determination in matters of information sharing are also be justified on other grounds. These include the view that the patient is in the best position to understand and therefore protect his or her own interests, and that there is an intrinsic value in people deciding about and taking responsibility for their own lives. In ethical theory, possible justifications can be in terms of the consequences of actions or rules, or can be in terms of duty. While their reasons differ, both of these approaches are united in their commitment to a confidentiality requirement.

11.2.2.2 Ethical Boundaries to Confidentiality

None of the ethical arguments stated above lead to the conclusion that the duty of confidentiality is absolute. The confidentiality requirement exists within a wider social context in which psychiatrists have other duties, which may conflict with their duty of confidentiality. In particular, psychiatrists may have other ethical duties to disclose confidential information, without consent, if serious and imminent dangers are present for third parties and where the psychiatrist judges that the disclosure of that information is likely to reduce or eliminate the danger. In assessing such risks and whether they outweigh the duty of confidentiality both the probability of the harm and its magnitude need to be considered. In situations where both the probability and seriousness of harm to a third party are high, the moral duty to disclose to prevent harm is greater.

11.2.3 The Legal Basis of Privacy and Confidentiality

11.2.3.1 Privacy and Confidentiality in Law

The relationship between psychiatrists and their patients carries with it legal obligations of confidentiality as well as ethical ones.[2] For the Member States of the European Union (EU), the disclosure and use of personal information about health are regulated by laws on privacy, confidentiality and data protection.

International Norms on Privacy

At an international level the protection of privacy, including healthcare privacy, is required by the following general instruments:

- Universal Declaration of Human Rights (1948)
- International Covenant on Civil and Political Rights (1966)
- Universal Declaration on Bioethics and Human Rights (2005)

European Norms on Privacy and Confidentiality

Within the broader legal context, confidentiality in a professional relationship (such as psychiatrist and patient), is part of privacy, and already protected by the general right to privacy. Added protection stems from the fact that confidentiality imposes an obligation on the person who obtained information in confidence not to disclose this information.

---

[2]See also Chapters 5 and 8.

Directive 95/46/EC, the 'data protection directive', refers to certain data being 'processed by a health professional subject under national law or rules established by national competent bodies to the obligation of professional secrecy' (Article 8 (3)). Recital 9 of Directive 2001/20/EC on clinical trials refers to 'the rules of confidentiality'.

Charter of Fundamental Rights of the European Union (2000/C 364/01). Two articles of the Charter emphasize the importance of the protection of privacy: Article 7 states: 'Everyone has the right to respect for his or her private and family life, home and communications.' Article 8 states: '1. Everyone has the right to the protection of personal data concerning him or her. 2. Such data must be processed fairly for specified purposes and on the basis of the consent of the person concerned or some other legitimate basis laid down by law. Everyone has the right of access to data which has been collected concerning him or her, and the right to have it rectified. 3. Compliance with these rules shall be subject to control by an independent authority.'

Council of Europe's Convention for the Protection of Human Rights and Fundamental Freedoms (ECHR) (ETS no 005, 1950 as amended). The ECHR is an international treaty which is binding on all those countries that have ratified it, which includes all EU Member States. Article 8 (1) of the Convention states 'Everyone has the right to respect for his private and family life, his home and his correspondence'. The case law of the European Court of Human Rights (ECtHR) makes clear that the essential object of Article 8 is to protect the individual against arbitrary interference by the public authorities. There are in addition obligations on States to take positive steps to ensure that the right is respected, not merely to avoid measures which interfere with the right. The ECtHR has held: 'Respecting the confidentiality of health data is a vital principle in the legal systems of all the Contracting Parties to the Convention. It is crucial not only to respect the sense of privacy of a patient but also to preserve his or her confidence in the medical profession and in the health services in general. Without such protection, those in need of medical assistance may be deterred from revealing such information of a personal and intimate nature as may be necessary in order to receive appropriate treatment, and, even, from seeking such assistance, thereby endangering their own health and, in the case of transmissible diseases, that of the community' (Z v Finland 1997; MS v Sweden, 1997).

Council of Europe 'Convention for the Protection of Individuals with regard to automatic processing of personal data' (No. 108) (1981). While decisions of the ECtHR show that the ECHR does not grant an absolute right to personal data confidentiality (see below), the protection granted to confidentiality is extended in the Council of Europe 'Convention for the Protection of Individuals with regard to automatic processing of personal data' (No. 108). This Convention was the first international legally binding text on data confidentiality. It applies to all 'automated personal data files and

automatic processing of personal data in the public and private sectors'
(Article 3), as long as the data relates to an 'identified or identifiable individ-
ual' (Article 2), whatever their nationality or place of residence (see also the
Chapter 26).

*Council of Europe 'Convention for the Protection of Human Rights and Dignity
of the Human Being with Regard to the Application of Biology and Medicine:
Convention on Human Rights and Biomedicine' (No. 164) (1997).* The
Convention on Human Rights and Biomedicine expands on many of the
rights contained in the European Convention on Human Rights and elabo-
rates how they apply in the field of medicine. Unlike the ECHR which applies
to all EU Member States, the Convention on Human Rights and Biomedicine
has not been signed or ratified by many States, including most of the larger
States. In spite of it not applying directly to many EU States, it is nevertheless significant in that it has been drawn upon by the European Court of
Human Rights in making judgments involving States who are not parties to
this Convention.

11.2.3.2 Legal Boundaries to Privacy and Confidentiality

Whilst patient privacy and confidentiality find legal protection at the national, the
European and the international level, this protection is not absolute. The right to
privacy of Article 8 (1) of the ECHR is limited by Article 8 (2) which states,

> There shall be no interference by a public authority with the exercise of this right except
> such as is in accordance with the law and is necessary in a democratic society in the inter-
> ests of national security, public safety or the economic well-being of the country, for the
> prevention of disorder or crime, for the protection of health or morals, or for the protection
> of the rights and freedoms of others.

Therefore to be compatible with the ECHR, any interference with the right to privacy must meet certain conditions. It must be 'in accordance with the law', which
means that any interference must have some basis in national law, and the law must
be precise enough so that people can reasonably understand its requirements and
consequences. It must be 'necessary in a democratic society', which means that the
interference must also both correspond to a 'pressing social need' and be 'proportionate to the legitimate aim pursued'. Such 'legitimate aims' are exhaustively listed
in Article 8(2).

Within the EU, Article 8 of Directive 95/46/EC 'on the protection of individuals
with regard to the processing of personal data and on the free movement of such
data' deals with the processing of special categories of data and in particular with
data concerning health. Member States must prohibit the processing of those special
categories of data, except in the situations (a) where the data subject has given his
or her explicit consent; (b) where the processing is necessary to protect the vital
interests of the data subject or of another person; and (c) where the data subject is
physically or legally incapable of giving consent. Paragraphs 3 and 4 provide for
other exceptions:

§3 '(...) where processing of the data is required for the purposes of preventative medicine, medical diagnosis, the provision of care or treatment or the management of healthcare services, and where those data are processed by a health professional subject under national law or rules established by national competent bodies to the obligation of professional secrecy or by another person also subject to an equivalent obligation of secrecy'.

§4 'Subject to the provision of suitable safeguards, Member States may, for reasons of substantial public interest, lay down exemptions (...) either by national law or by decision of the supervisory authority'.

Directive 95/46/EC is thus broadly in keeping with other international and European norms in this area.

11.2.3.3 Country Specific Legislation and Their Commonalities

In many Member States of the EU, country specific human rights legislation often incorporates the ECHR and thereby underpins all other legislation concerned with privacy.[3] Country specific legislation giving effect to EC Directive 95/46/EC sets standards of information processing.

With the individual Member States, laws on privacy and confidentiality are enshrined in statutes, civil and criminal codes or, in jurisdictions such as the United Kingdom (UK), in the common law. In most Member States confidentiality and privacy are protected by statutory laws. For example, while the French Constitution does not expressly protect the right to privacy the Constitutional Court has confirmed that privacy is a constitutional principle. Again in Germany the Federal Constitutional Court has argued that Articles 1 (1) and 2 (1) of the Constitution grant every individual an inviolate sphere of private life. Unlike in many European countries, UK law does not recognise a general criminal offence of breach of professional secrecy. In the UK statutory duties of confidentiality are limited to special circumstances such as abortion or venereal disease.

Despite the variety of legal provisions, the overall direction across the Member States is a strong protection of confidentiality in healthcare. The following points summarise some shared European legal principles on confidentiality:

(a) there is a prima facie obligation to maintain confidentiality when information has been imparted to a professional within a confidentiality relationship;
(b) this obligation to maintain confidentiality can be discharged when the subject of the confidence affords appropriate consent to the disclosure of the information; and
(c) in providing a justification for the non-consensual disclosure of confidential information healthcare professionals should have particular regard to issues such as:

---

[3] See also Chapter 8.

(i) the necessity of any particular disclosure;

(ii) the proportionality of any particular disclosure;

(iii) the risks attendant upon any particular disclosure; and

(iv) the existence of identifiable risks of serious harm to identifiable third parties arising from non-disclosure.

11.2.4 Vulnerability

The nature of vulnerability

Vulnerability refers to a circumstance in which a person finds himself or herself particularly susceptible to injury or harm. All patients in healthcare are vulnerable to misuses or abuses of their private information by healthcare professionals, healthcare providers, and by healthcare researchers. The circumstances of some persons (for example, children, the homeless or those with disabilities) can create particularly challenging situations for ethical and legal conduct.

The vulnerability of patients is a significant factor warranting ethical consideration and the perspective of those who are vulnerable should be at the centre of considerations for decision-making about the protection, use or disclosure of their confidential information.

Article 8 of the Universal Declaration on Bioethics and Human Rights (2005) states: 'In applying and advancing scientific knowledge, medical practice and associated technologies, human vulnerability should be taken into account. Individuals and groups of special vulnerability should be protected and the personal integrity of such individuals respected.' While the provisions necessary for good practice in information sharing are fairly straightforward in many clinical situations, the presence of specific vulnerabilities, either because of the patient's condition or their situation, poses significant challenges for healthcare providers and healthcare professionals.

There are many overlapping sources of vulnerability, but a key one in information sharing is patient vulnerability because of a lack of decision-making capacity (see also Chapters 9 and Chapter 14). The consequences of this vulnerability for the effective protection of patient rights are not adequately addressed through the protections of the ECHR.

Patients who lack capacity can also be harmed in their basic rights through a failure to identify such patients correctly. Effective measures must be in place at all levels of an institution to ensure that patients lacking decision-making capacity are correctly identified and that they receive the added protection and empowerment they need.

While a legal determination of a lack of decision-making capacity with respect to the use and disclosure of their information is a valid reason for added protection, not all vulnerable patients lack that capacity. Although legally competent to make decisions, many patients remain vulnerable to undue influence and/or exploitation through an inability to assert their own interests and rights.

All patients, including the vulnerable, must be treated with respect. In particular, their exercise of their right to decide about the use and disclosure of their confidential information should be facilitated.

11.2.5 Balanced Decision-Making

Within the framework of fundamental rights of the patient, there is a need for balanced ethical decision-making about the protection, use and disclosure of confidential information. In healthcare decision-making, privacy and confidentiality, although important values in their own right, often have to be balanced against other values. The value placed on privacy and confidentiality can vary between patients and for the same patient in different contexts. Good decision-making will take full account of the values and fundamental beliefs of the patient concerned.

Balanced decision-making about the use and disclosure of confidential patient information in day-to-day practice may require difficult judgements and these judgements need to be supported by a clear framework of ethical and legal obligations. However, there are limits to the extent to which regulations alone can provide for balanced decision-making. Balanced decisions, and the judgements on which these are based, also depend on good *process* in applying the general guidance defined by ethical and legal regulation in individual cases.

11.3 Good Practice in the Protection, Uses and Disclosure of Patient Identifiable Information

11.3.1 Protection, Use and Disclosure of Patient Information – General Considerations

In principle, patient information is confidential and should not be disclosed without adequate justification. Nevertheless in many instances disclosure of confidential information, or its use, are desirable or necessary.

First, patient information might need to be shared with members of the multidisciplinary healthcare team for that patient's healthcare needs, or it might be needed for auditing purposes, in order to improve the patient's care. Second, in some situations, the disclosure or use of confidential patient information might be important for purposes that are related to healthcare, but not to the care of the particular patient, for example, where patient information is used for healthcare research. Third, it is possible that confidential patient information held by a healthcare professional may have important uses outside the healthcare context, for example where a health care professional has information about the dangerousness of the patient to the public.

These three kinds of situations to some extent require different considerations when deciding according to what criteria disclosure can be justified and are dealt with separately under Sections 11.3.2, 11.3.3 and 11.3.4 below. Some considerations, however, are common to all situations and these are outlined below.

11.3.1.1 Patient Consent

The justification for disclosure should normally be consent.[4] Where the patient is competent, only the patient can give consent to disclosure. Consent is a means by which the competent patient can exercise control over the dissemination of confidential patient information. Valid consent requires that the patient has been informed as to what information it is intended to disclose, and for which purposes disclosure is proposed. Consent also presupposes choice, which means that the patient who is asked to consent must have the possibility to refuse or withdraw such consent.

If the competent patient refuses to consent to disclosure, the information cannot be disclosed, unless, exceptionally, a justification other than consent exists (see Section 11.3.4). The psychiatrist should discuss with the patient why he/she thinks that disclosure is in the patient's best interests. However, it can never be justified to disclose information in the best interests of the competent patient who refuses to consent to disclosure, as it is the competent patient, not the psychiatrist who should decide what the competent patient's best interests are.

11.3.1.2 Circumstances Where a Patient Is Unable to Consent

There are circumstances where a patient is unable to consent to the use or disclosure of their confidential information and in such circumstances, special considerations apply.

Incapacity. The precise definition of incapacity, how it is to be determined, and the status of any legal representative, who would have the right to give proxy consent to uses and disclosures on behalf of an incompetent patient, depend on country-specific law (see also Chapter 10). The control a legal representative exercises over the patient's information is usually more limited than that exercised by the patient him/herself while competent, as legal representatives have to act in the patient's best interests. Where a psychiatrist thinks that disclosure would be in the best interests of a patient unable to consent, he/she should raise this with the patient's legal representative. If the consent of the legal representative is withheld, the healthcare professional might involve the court to settle the dispute.

Emergency Situations. In emergency situations it may be impossible to keep a patient and/or their legal representative properly informed and to gain their consent. In such situations, uses or disclosures may be made, but only the minimum

---

[4]See also Chapter 9.

necessary information should be used or disclosed to deal with the emergency situation.

11.3.1.3 Disclosure to Protect Interests that Override the Patient's Right to Confidentiality

Exceptionally, it might be justified to disclose confidential patient information where disclosure is necessary to protect interests that override the patient's right to confidentiality. This is dealt with in more specific terms in Section 11.3.4.2, but as a general principle, it is important to remember that the interests of the competent patient cannot justify disclosure against the patient's wishes. Thus, where the patient is competent, disclosure without consent can only be justified if it is exceptionally necessary to protect the overriding rights of *others*, or there are overriding legally protected public interests. With regard to the incompetent patient, disclosure might also be justified to protect overriding interests of the incompetent patient, for example where disclosure is necessary to protect the incompetent patient from sexual abuse.

11.3.1.4 Disclosure After a Patient's Death

The confidential nature of a patient's healthcare information and the psychiatrist's obligation to respect that confidentiality are not changed by the death of that patient. However, just as in life, the right to privacy and the duty to maintain patient confidentiality after their death are not absolute, but are subject to ethical and legal limitations.

The death of a patient never in itself permits disclosure, but it does represent a changed situation for balanced decision-making. After the death of a patient it will be more common that the balanced ethical decision will favour disclosure, as the possible harm to which the dead patient is subject is considerably reduced. The death of the patient does not automatically favour disclosure and an ethical balance must still be struck by the healthcare professional. Disclosures after death remain subject to the ethical considerations governing any disclosure, such as whether disclosure serves a legally protected public interest and that any disclosure should be as minimal as possible.

A competent patient can give or withhold consent to disclosure before their death and such wishes should be respected as they would in other circumstances. In particular, where a competent patient has made an explicit request before his or her death that their confidence be maintained following requests from family members or carers for disclosure, then that request should normally be respected.

11.3.1.5 Patient Access to Their Healthcare Records

Patients have a right, both ethical and legal (EC Directive 95/46/EC on data protection), to know what information is held in relation to them. There may be exceptional

circumstances where access to specific information poses a significant threat of life
or serious harm to the patient or a third party which may justify the withholding of
information.

11.3.2 Protection, Use and Disclosure of Patient Information for Their Healthcare

11.3.2.1 Keeping Patients Informed

That patients must be kept informed about the possible uses and disclosures of
their information is a binding legal obligation across the EU. Keeping patients fully
informed is also essential for maintaining the relationship of confidentiality. Better
communication with patients (and/or their legal representative) will also improve the
partnership between patients and professionals enhancing the quality and experience
of their care.

Modern health services often involve sharing information between healthcare
professionals to provide optimal care and treatment. Patients may be unaware of
what information is held about them, the purposes for which the information is used
or the people with whom such information may need to be shared to provide their
care. Patients and/or their legal representative must be made aware that information
given may be recorded and shared to provide the patient with care. It may also
be used to support clinical audits and other work to monitor the quality of care
provided. Patients and/or their legal representative, also need to be aware of the
choices they have for the use and disclosure of the information shared in confidence
with their psychiatrist.

It is an ethical and legal requirement that patients are both kept informed of
all circumstances in which they can give or withhold consent to the use of their
information and given information necessary for that consent.

11.3.2.2 Consent to the Use and Disclosure of Patient Information

As with any other intervention in healthcare, patient consent occupies a pivotal role
in legitimising the uses and disclosures of patient information. Patients and/or their
legal representative must be informed of what information sharing is necessary for
their healthcare. Provided they are informed in this way, explicit consent is not necessary, implied consent is sufficient for the ethical sharing of patient information for
their healthcare.

11.3.2.3 Clinical Audit

Patient identifiable information will often be required for purposes which aim to
support or assure the quality of patient care, for example clinical audit. Processes of
clinical audit are an essential part of healthcare provision for which personal health
information may need to be used. Patients in general (and the wider public) have a

clear interest in the health services being subject to effective audit. Audits are part of the primary uses of patient information. Patients and/or their legal representative must be aware of such uses.

From an ethical perspective, a wide range of activities by health service staff providing that care or treatment may be covered under the heading of audit. Clinical audit which makes use of confidential patient information is usually carried out within the health service by staff directly involved in that patient's care. Implied consent is sufficient.

Provider institutions must ensure that patient express consent (or that of their legal representative) is obtained for processes of clinical audit by staff not involved in the care of that patient or where it is proposed to make information available outside the health provider institution.

11.3.2.4 Disclosure to a Patient's Carers

All people employed by or working in organisations providing healthcare should be under an obligation of confidentiality.

Families and other persons who are caring for a patient have an understandable desire or need for information about a patient's healthcare problems and management. Such knowledge may benefit both the patient and the carer by, for example, creating a better understanding of the patient's illness, or by promoting more appropriate responses to the patient and their needs. However, the fact that such information sharing may be beneficial does not diminish the duty of confidentiality owed to the patient by the psychiatrist. In situations of ongoing need for care and support, the potential benefits of information sharing with their informal carers should be discussed with the patient and/or their legal representative.

11.3.2.5 Multidisciplinary and Inter-agency Working

It is good practice that when a healthcare professional legitimately discloses information in a multidisciplinary team or in inter-agency working, that such disclosure takes place on a clear basis of agreed protocols for information sharing.

Multidisciplinary work. Psychiatrists as part of their work will have contact with other professionals and other agencies delivering aspects of care. Healthcare professionals may have different criteria and thresholds for the disclosure of confidential information, for example in relation to public safety. It is essential that each healthcare professional familiarise him or herself with such differences and moderate disclosures accordingly.

Inter-agency work. It is common practice in many areas of healthcare provision to involve outside agencies in providing services for patients. This inevitably involves discussions about patients at various points in their treatment. Issues about sharing information may arise in the context of verbal or written reports, or attendance at case conferences.

11.3.2.6 Dual Roles and Obligations

Psychiatrists may work in situations where they may have dual roles with dual and conflicting responsibilities and obligations.[5] This includes work in prisons and for court liaison schemes where there are duties to both the patient in their care and to the authority. Such dual roles and obligations may cause conflict about the confidentiality of patient information. For example, a prisoner or defendant may have consulted the psychiatrists and divulged information that they do not wish an outside agency to know, while in their current role the healthcare professional may be obligated to disclose that information.

11.3.3 Protection, Use and Disclosure of Patient Information for Healthcare Purposes Not Directly Related to Their Healthcare

11.3.3.1 Secondary Uses

Many uses of confidential healthcare information not directly related to the healthcare of the patient are legitimate for limited and specified healthcare purposes provided certain criteria are met. Secondary uses of confidential patient information are uses in healthcare which do not contribute directly to or support the healthcare that a patient receives. Such uses are increasingly required for evidence based practice and a rational approach to service planning, management and commissioning. The following are some examples of secondary uses:

- planning of services;
- payment for services;
- management of services;
- contracting of services;
- risk management;
- patient safety;
- investigating complaints;
- auditing accounts and performance;
- local and national inquiries;
- teaching;
- research.

11.3.3.2 Keeping Patients Informed About Secondary Uses

One cannot assume that patients seeking healthcare, or their legal representative, are aware of or content for patient information to be used in these ways. Under the Data

---

[5] See also Chapter 22.

Protection Directive patients must be informed about such secondary uses and their purposes, and have a right to object to the use or sharing of confidential information that identifies them. All health service organisations must have policies for informing patients and/or their legal representative of the protections, uses and disclosures of their information for secondary purposes. Patients and/or their legal representative must also be informed of the categories of people and organisations to which information may need to be passed for health services to function. Patients and/or their legal representative should be told how information will be used before they are asked to provide it and should be given an opportunity to discuss any aspects. It should be made clear to patients and/or their legal representative that they may object to specific secondary healthcare uses of identifiable patient information and that their objection will be respected.

11.3.3.3 Consent for Secondary Use or Disclosure of Confidential Patient Information

Express consent from the patient or their legal representative should wherever possible be obtained before any proposed secondary uses of patient personal information. Where there is agreement to disclosure, only the minimum necessary patient identifiable information should be used for each legitimate healthcare purpose.

While many uses (administrative, management, public health monitoring) of confidential patient information are essential to the provision of healthcare, research to develop healthcare is not itself healthcare. For research uses express consent should be sought and refusals respected (see also Chapter 25).

Possible justifications for not seeking consent for a secondary use of patient information are that it is impracticable or impossible because of particular circumstances.[6] These grounds for not seeking consent for a secondary use of patient information can be considered as follows:

(a) It might be *impracticable* to obtain consent for the use of patient information for a secondary use (for example a public health study) where the patient information had been obtained some time previously. A possible ground for justifying not obtaining consent would be disproportionate effort (for example obtaining consent for a large sample of patients on whom the information had been obtained many years earlier. However, in some countries stronger limitations on disclosing confidential patient information for such purpose may exist, e.g. by the Data Protection Law in Germany which states that without consent confidential patient information can be used only on a high level of anonymity).

(b) It might be *impossible* where the confidential information was obtained with consent for a particular secondary use, and the potential for use for another purpose is now being considered. If the data has subsequently been irretrievably unlinked from those who initially gave consent, then although they have a moral

---

[6]This may be valid especially in epidemiologic research: see Chapter 26.

interest in its further use, gaining their consent to the second use is impossible. Likewise, previously gathered information may have a value beyond the death of the individual. It is sometimes impossible to gain consent for the secondary use of information from a patient who lacks decision-making capacity but in such circumstances there are additional protections which must be observed (See Section 11.3.1.2).

Independent data protection officers or Ethics Committees should be involved whenever judgments of impracticability or impossibility are given as grounds for secondary uses of confidential information without receiving consent. It is also appropriate for such prior independent checking to occur whenever there is any claim of exemption from the duty to provide information to patients and/or their legal representative about uses or disclosures.

11.3.3.4 Maintaining Information in a Form Which Protects the Identity of the Patient

Personal information should wherever possible be maintained in a form that protects the identity of the patient from disclosure to unauthorised persons.

11.3.4 Obligations and Justifications for the Disclosure of Patient Identifiable Information for Purposes Not Related to Their Healthcare

In some situations, psychiatrists might be under a legal obligation to disclose information, or disclosure might be legally justified. Where a legal obligation to disclose exists, non-disclosure might have legal consequences. A legal justification of disclosure, on the other hand, means that while the healthcare professional does not have to disclose confidential information, disclosure might under certain circumstances be regarded as legally acceptable. Every disclosure must also be ethically acceptable. In democratic societies laws and ethical codes in relation to patient privacy should be aligned. However psychiatrists may be working within political regimes in which laws, in particular, laws relating to individual citizen privacy, do not accord to international codes of ethics. In such situations psychiatrists should seek to follow agreed international ethical codes.

11.3.4.1 Legal Obligations to Disclose

In a number of European countries there are legal regulations governing the disclosure of confidential information that require the duty of confidentiality to be overridden, for example notification requirements with regard to certain communicable diseases. Where there is a legal obligation a healthcare professional is required to disclose the relevant information to the appropriate authorities. The psychiatrist must bear in mind that failure to do so may lead to legal sanctions. However, given that every disclosure is an interference with the patient's right to privacy, disclosure should not be made uncritically and should be kept to the absolute minimum.

In some European countries courts and other authorities that have a legal right of access to certain confidential information have powers to order the disclosure of documents before and during proceedings. They can also order the production of that material to an applicant and to their legal and professional advisers. Also during court proceedings a judge may order that medical records be disclosed. A psychiatrist of a defendant can also be compelled to answer questions about what the defendant has said to her/him, as well as providing details of the patient's medical history and condition. The psychiatrist must do his or her best to ensure that every argument that can properly be put against disclosure is put before the court. Any disclosure must be limited to what is strictly relevant to the court proceedings.

11.3.4.2 Justifications to Disclose

Disclosure of confidential information to third parties outside the health services may be justifiable in order to protect overriding interests of third parties or a legally protected public interest. However, every decision to disclose confidential patient information outside the healthcare services violates the patient's right to privacy, and is in breach of the healthcare professional's obligation of confidentiality. The disclosure will only be justified in exceptional circumstances, that is, if the disclosure serves an interest that in the particular circumstances outweighs the patient's right to privacy. Potential outweighing interests could be the protection of the rights and freedoms of others, national security, public safety, the economic well-being of the country, the prevention of disorder or crime, or the protection of health or morals (as suggested by Article 8 (2) of the ECHR).

In all of these cases, there is no obligation to disclose, but whether or not disclosure can be justified rather depends on balancing the interests that are in conflict in each case. It needs to be borne in mind that every instance of disclosure leads to a certain violation of the patient's right to privacy, while the benefits of disclosure will often be less certain. While a balancing of the patient's right to privacy against other rights and interests is always difficult, it is usually more easily performed where the conflict is with rights of identifiable third parties, than where there is a conflict with a more diffuse public interest such as national security or public health. It is not sufficient that it might be more convenient for the protection of such interests that information is disclosed, but the test is instead one of strict necessity in the specific circumstances of each case.

Disclosure to protect the best interests of the patient. Disclosure to protect the interests of the competent patient against his/her wishes can never be justified, as on balance, the right of the patient to decide autonomously what is in his/her interests always prevails.

Where the patient is incompetent, disclosure can be justified to protect the best interests of that patient. Whether disclosure is justified in the individual case depends on a careful weighing of the patient's interest in having the confidentiality of his/her information maintained and the interests that are at risk without disclosure.

Good practice for justified disclosures. When a decision has been reached that disclosure is justified in a particular situation, there are requirements for how that disclosure should best be made. In all instances where judgement is involved psychiatrists should discuss the case with colleagues without revealing the identity of the patient and if necessary seek legal advice. Most situations where decisions to disclose have been reached require good communication with and support for patients whose confidentiality is to be breached. Once a decision has been made an explanation of the reasons for sharing information should be given to the patient or their legal representative. The exception to this is where informing the patient of the disclosure in advance that the disclosure will be made would prevent achieving the justified aim of the disclosure.

11.3.5 Security of Patient Information

The quality and integrity of patient information, information protection and the controls required to ensure that patient information sharing is secure, confidential and responsive to patient preferences are inextricably linked. A coherent institutional framework for information governance is required. Within such a framework the principal means of enhancing the security of personal information are restriction of access and the maintenance of information in a form which protects the identity of the patient.

11.4 Conclusion

While each patient's healthcare information is protected under both ethical and legal obligations of confidentiality, there are a variety of situations where the use and disclosure of personal information may occur for legitimate purposes. For practical purposes it is helpful to consider:

- the purpose of any planned use or disclosure of confidential healthcare information; and
- the criteria which must be satisfied to allow such use or disclosure.

In general, any use or disclosure of confidential healthcare information without consent, (for example, to the appropriate authorities at State level for health monitoring) should clearly serve one of the purposes specified in international human rights law as being a legitimate limitation on the right to privacy. Such disclosures must also meet the criteria of being proportionate to the legitimate aim of the disclosure, in accordance with (domestic) law, and taking place within the highest levels of data protection and data security.

References

Charter of Fundamental Rights of the European Union (2000/C 364/01) http://www.europarl.europa.eu/charter/default_en.htm

Council of Europe 'Convention for the Protection of Human Rights and Dignity of the Human Being with Regard to the Application of Biology and Medicine: Convention on Human Rights and Biomedicine' (No. 164) (1997) http://conventions.coe.int/treaty/EN/Treaties/Html/164.htm

Council of Europe 'Convention for the Protection of Individuals with regard to automatic processing of personal data' (No. 108) (1981) http://conventions.coe.int/treaty/EN/Treaties/Html/108.htm

Council of Europe's Convention for the Protection of Human Rights and Fundamental Freedoms (ECHR) (ETS no 005, 1950 as amended) http://conventions.coe.int/Treaty/en/Treaties/Html/005.htm

Directive 95/46/EC of the European Parliament and of the Council of 24 October 1995 on the protection of individuals with regard to the processing of personal data and on the free movement of such data http://eurlex.europa.eu/LexUriServ/LexUriServ.do?uri=CELEX:31995L0046:EN:NOT

International Covenant on Civil and Political Rights (1966) http://www.unhchr.ch/html/menu3/b/a_ccpr.htm

Universal Declaration of Human Rights (1948) http://www.un.org/Overview/rights.html

Universal Declaration on Bioethics and Human Rights (2005) http://portal.unesco.org/shs/en/ev.php-URL_ID=1883&URL_DO=DO_TOPIC&URL_SECTION=201.html

Chapter 12
Justice in Access to and Distribution of Resources in Psychiatry and Mental Health Care

Ruud ter Meulen

Contents

Abbreviations

CVZ   College voor Zorgverzekeringen (Dutch Health Insurance Council)
DSM   Diagnostic and Statistic Manual, American Psychiatric Association
EbM   Evidence-Based Medicine
GDP   Gross Domestic Product
NHS   National Health Service
NICE   National Institute for Health and Clinical Excellence
OECD   Organisation for Economic Co-operation and Development
QALYs   Quality Adjusted Life Years
RCT   Randomised Controlled Trial
SSRI   Selective Serotonin Reuptake Inhibitor

R. ter Meulen (✉)
Centre for Ethics in Medicine, University of Bristol, Hampton House, Cotham Hill, Bristol BS 6 6AU, UK
e-mail: r.termeulen@bristol.ac.uk

H. Helmchen, N. Sartorius (eds.), *Ethics in Psychiatry*, International Library of Ethics, Law, and the New Medicine 45, DOI 10.1007/978-90-481-8721-8_12,
© Springer Science+Business Media B.V. 2010

12.1 Introduction

Since the end of the 1980s, health care systems in the developed world have been experiencing increasing difficulties to guarantee universal access to a comprehensive package of health services. Rising costs of care, due to increasing demand and the introduction of expensive technologies, have led to a limitation of the supply of care: economic assessment of services, priority-setting, delisting, rationing by way of guidelines and needs assessment, waiting lists and co-payments are all trying to limit both the supply and demand of care services. The scarcity of resources is particularly noticed in mental health care. In The Netherlands there are waiting lists for mental health care services, particularly for outpatient care. In the United Kingdom, mental health patients are experiencing increasing difficulties to access services as these are becoming concentrated in larger treatment centres at a far distance from local communities. Quality of care may decline because of lack of resources to fund staff or adequate housing.

The chapter gives an overview of the general ethical principles of justice to health care, and the application of these principles to resource allocation in relation with mental health care. The chapter starts with an introduction to the concept of necessary care and the moral importance of healthcare as argued by Norman Daniels. According to Daniels, health care and mental health care restores normal species functioning and the normal opportunity range of individuals. The supply of adequate mental health care is morally important as mental illness can have a severely negative impact on the functioning of individuals in society and the access to the range of opportunities that normally would have been open to them. The chapter continues with the concept of solidarity that plays an important role in the moral foundation of health care systems in Europe. Solidarity is a strong moral motive in our society to finance a basic package of services for individuals in our society, particularly for individuals who are not able to take care of themselves, like people with mental health problems. The chapter discusses the concept of need as a criterion for distribution of health care resources. However, this concept is often seen as too subjective. For example, it is difficult to distinguish needs from wishes or preferences as can be illustrated by the use of psychopharmaceutical drugs like Prozac for enhancement purposes. Because the subjectivity of the concept of need, there is nowadays a growing interest in the concept of burden of disease which is linked to the concept of need, but gives more concrete criteria for the allocation of resources. A different way to allocate scarce healthcare resources is the use of effectiveness and cost-effectiveness as a criterion for prioritizing health care services. Such an approach uses the outcome of treatments as a criterion for distributing healthcare services at a societal or clinical level. This approach is popular as it seems to be more 'objective' than need for distributing health care resources. Evidence-based medicine, which tries to base clinical and policy decisions on the outcomes of randomised controlled trials, is seen as the 'value-free' approach to deal with the problem of allocating scarce resources. However, the chapter argues that effectiveness is also a normative concept and the study of effectiveness of treatments in randomised controlled trials is also based on various normative assumptions. Moreover, evidence based approaches to distribution of scarce health care resources may affect vulnerable groups in our society and could violate the principle of solidarity. Approaches to the distribution of scarce resources should never be based on evidence about the effectiveness of treatments only, but should also include an assessment of the needs of patients, or, as has been argued recently, a measurement of the burden of disease.

12.2 Necessary Care

According to some people scarcity of resources in healthcare can easily be resolved: why don't we spend just more on healthcare? In countries where healthcare is financed by insurance premiums, like Germany and The Netherlands, one could raise the premiums of healthcare insurance. In that case every individual would be able to receive the service and treatment that is deemed necessary. The consequence would be that the share of healthcare in the Gross Domestic Product (GDP) would rise significantly. In the countries of the OECD, the average share of the GDP spent on healthcare was 8.9% in 2006. In Europe, the largest share of the GDP spent on healthcare is for Switzerland with 11.3%, followed by France with 11.1% and Germany with 10.6%. The Netherlands spent 9.3% of its GDP on health care and the United Kingdom 8.4% (www.oecd.org). Why shouldn't these European countries raise this percentage to 12% of the GDP or even more? An argument against increased allocation of resources to healthcare is that such a policy probably would not lead to an improvement in health. For example, in the United States the share of healthcare in the GDP is about 15.3% in 2006 (www.oecd.org). However, in spite of the large amount of money spent on healthcare in the US the health status of the population, as measured in terms of life expectancy, falls behind countries like the United Kingdom and Japan where the share of healthcare in the GDP is significantly lower than the one in the United States. More money to healthcare does not automatically mean better health. On the contrary, increased spending on health care could negatively affect conditions of (health related) living, as money will be withdrawn from other services in a society, like education, environmental protection, housing and living conditions, and job creation. A deterioration of conditions of living, work and education is likely to have a negative impact on people's health thereby increasing the pressure on the health care system. This is particularly true for mental health care, where adequate housing, social support and other living conditions are important to prevent mental illness or to prevent recurrence or worsening of the mental health of individuals.

The answer to the question, how much money should be spent on healthcare is subject to ethical and political debate. An important concept in this debate is *necessary care*. Necessary care can be considered as the minimum level of care that society should guarantee to each individual. This minimum should be the starting point for decision making about the allocation of resources to the various healthcare services including mental health care. What this minimum should contain depends on our view on the *moral* importance of health care: why should we as a society provide health care services, including mental health services, to all members of a society, and what services should have priority? The American philosopher Norman Daniels presents a theory on the moral importance of healthcare in his book *Just Health Care* (1985), and more recently in *Just Health* (2008). According to Daniels each member of a society has the right to a fair share in the total of opportunities that are open to individuals in the particular society. This is what Daniels calls the *normal opportunity range*. This normal opportunity range which depends upon individual talents and abilities is seriously limited in the case of disease and disability. Disease and disability affect what Daniels calls *normal species functioning* which is a condition for the enjoyment by an individual of the opportunities that are open to him or her in a specific society. The moral importance of healthcare is that it restores normal species functioning and in doing so, enables individuals to take their share in the societal opportunities to which they have a right to on the basis of their personal abilities and talents. Basic healthcare is not just a commodity to be distributed according to ability to pay, but must be regarded as a societal obligation. The government should guarantee that each individual has equal access to healthcare independent of income, age or gender. Each member of society has the right to a basic package of healthcare services to enable individuals to participate in society. An important condition of the basic package is that it is adequate, that means that it enables people to fulfil their potential and to pursue their life plans.

An important question is whether mental health care falls within the range of our obligations to preserve or restore fair equality of opportunity. A difficulty in Daniels theory is raised by his use of the concept of normal species functioning. This concept is based on the naturalistic concept of health and disease of Christopher Boorse (1975, 1977). According to Boorse health can be conceptualised according to the functions that are typical of the individuals of the species they belong to. In a medical-biological context, it is important to define the purpose of these functions. These purposes can be derived from the 'natural design' of the organism. The functions of examples of a species can be empirically determined, without conferring any value to these functions and the purposes they are contributing to. A statistical analysis of these observations will produce a 'species design' of these organisms, which can be specified for age and sex. Biological functions for the human species are essentially contributing to survival and reproduction of individual examples. Boorse concludes that normal functioning or 'health' means then normal functional ability. According to Boorse this is a 'value-free' definition of health: a human being is healthy if his body functions according to its 'design'.

Boorse's definition of health is heavily criticised as it considers health and disease purely within a biological framework. Many 'healthy' functions of the human body have nothing to do with survival or reproduction, while some kinds of illness can hardly be seen as interference with 'normal species functioning'. By reducing health and disease to just biological phenomena, Boorse's concept prioritises medical-biological treatments, to the expense of treatments that take sociological or cultural aspects into account. If this concept would be used for setting priorities in mental health care, it could lead to a dominance for biologically oriented psychiatry to the expense of psychotherapeutic treatments.

Though Daniels is basing his theory of justice in health on a narrow, naturalistic concept of health and disease, he does not exclude mental health care from the basic package of health care services. In a chapter written with Sabin and Daniels (1999) he argues that mental health care is not different from general medical care in contributing to fair equality of opportunity. According to Sabin and Daniels, 'mental health conditions cause substantial impairments of species-typical capacity, equal to or greater than many chronic medical diseases about which there is no controversy regarding insurance coverage' (Sabin and Daniels 1999) However, it is not clear to which extent Daniels' approach favours a biological approach as compared to psychotherapy when setting priorities within mental health care.

In the developed world, mental health care is usually seen as an important part of the basic health care package. In the United Kingdom for example, the Department of Health argued in 1994 that people who are most disabled by mental illness should be awarded the highest priority (Jenkins 1994). Under this policy the National Health Service (NHS) established three targets for mental health resource allocation: 'to improve significantly the health and social functioning of mentally ill people', 'to reduce the overall suicide rate by at least 15% by the year 2000' and 'to reduce the suicide rate of severely mentally ill people by at least 33% by the year 2000' (Sabin and Daniels 1999). Moreover, the NHS allocated resources to general practitioners to give priority to more common but less severe mental health conditions they regularly encounter in practice (idem). In 2004 the British Government stated that people with mental illnesses are considered to be among the socially excluded and that they should be recognised in government policies. This could be provision of better care but also policies to improve pathways to work, provision of welfare benefits and reducing discrimination. However, despite clear advances in treatment and provision of services the standards for mental health services fall below those in the rest of the NHS (Ikkos et al. 2006). Funding for mental health services remains low, poorly distributed while national service frameworks have promised increased spending on the health service as a whole and mental health services in particular, these policies have not resulted in a substantial improvement of mental health care particularly at the local level (idem).

Though the concept 'necessary care' is often used in the debates on resource allocation it is far from clear what the boundaries are of this concept. For example, the Dutch Committee Choices in Health Care argued in its report Choices in Health Care (1991) that the basic health care package should be prioritised on the basis of four criteria: necessary care, effectiveness, efficiency and individual (financial) responsibility. According to the Committee, necessary care is care that enables individuals to function in society according to the values of that society. Important values are equality and protection of vulnerable people, like people with dementia, learning difficulties and people with severe psychiatric disorders. However, the Dutch government did not agree with the Committee's recommendations: though effectiveness and efficiency (see below) were seen as important criteria for distributing health care resources, 'necessary care' was seen as 'too vague' for making such decisions.

Necessary care has got a new interpretation by linking this concept to the concept 'burden of disease' or 'disease burden' (see below) which represents a more measurable approach to 'necessary care'. According to the Dutch Health Insurance Council (College voor Zorgverzekeringen CVZ) a health service is more necessary when the disease that this service is supposed to treat or dealt with has a bigger impact on the life of the individuals as for example measured by acute threat to the life of the individual or by quality of life scores in case of chronic conditions (CVZ 2003). The Dutch Health Council (Gezondheidsraad) also introduced this concept in its report on the content and extent of the basic package (Gezondheidsraad 2003). In the report of the Health Council, individual disease burden is defined as 'reduced quality of life or life span as a result of disease or some other somatic or mental health problem in cases where no health care service would be utilized' (Gezondheidsraad 2003). According to the Health Council, the burden of disease should be combined with measures of cost-effectiveness (see below) when defining a basic package of health care. Cost-effectiveness denotes the effectiveness of a treatment in reducing the burden of disease also in relation with the costs in terms of financial resources. This approach should include mental health services as 'necessary care', as the burden of disease can be very high for people with mental health problems, while there are also effective treatments available for many of these conditions. However, one should be aware that some mental conditions need long-term care without resulting in complete recovery. This should not be a reason to give such long-term services a lower priority or to exclude them from the basic package because of a narrow definition of effectiveness.

12.3 Solidarity

In European countries, equal access to necessary care is based on the ethical principle of *solidarity*. In general, the idea of solidarity is associated with mutual respect, personal support and commitment to a common cause. In European welfare states, the basic understanding of solidarity is that everyone is assumed to make a fair financial contribution to a collectively organised insurance system that guarantees equal access to health and social care for all members of society. This equally applies to other systems of social protection, which are operating in Europeans welfare states, such as social insurance systems covering the financial risks of unemployment and work related illness and disability, as well as old age insurance systems and pension schemes (ter Meulen et al. 2001). In this context, solidarity is an *instrumental* value insofar as it tries to promote equal access to healthcare for all citizens, independent of income or health risk. This means that individuals with a high risk of a disease like older people, people with chronic diseases or people who are suffering from mental health problems should have equal access to healthcare compared to people with a lower risk to such conditions.

However, in the area of health care and social policy solidarity has acquired a particular meaning that goes beyond solely transferring income or in kind benefits to protect the vulnerable and needy in society. In the domain of health and social care, solidarity is first and foremost understood as a moral value and social attitude regarding those in need of support. Solidarity with vulnerable groups in modern societies, in particular the chronically ill, the handicapped, the political refugees and the frail elderly, is taken as an expression of personal concern and responsibility by the care giver, no matter whether she or he is a professional care-worker, a relative or a friend (ter Meulen et al. 2001). Solidarity in this sense has an *intrinsic* value: it means to stand for and protect the other not out of self-interest but because the other needs this protection.

In its report *Choices in Health* Care (1991), the Dutch Committee Choices in Healthcare argued that necessary care should include the care for individuals who are limited by their disease or disability. The Committee asked for special protection by our society for individuals who due to their health problems cannot function independently any more. The Committee based its arguments on the value of equality and protection of all human life as well the principle of *solidarity* with people who are in need. Care for people who, for whatever reason, cannot take care of themselves should be part of the basic healthcare package. This can be care for people in nursing homes, care for people with learning disabilities but also people with psychiatric illnesses.

While solidarity in the past resulted in homogeneous care services without any involvement of the care-receivers, modern 'reflexive' solidarity sees care receivers as active participants who need our support and solidarity, but who also want to be included as critical citizens with their own views how solidarity should be practiced (Houtepen and ter Meulen 2000). In mental health care 'reflexive solidarity' means an engagement to get people with mental health problems involved in the debate on the distribution and delivery of care services. Mental health care receivers should not just be seen as passive bearers of rights or mere carriers of disease or handicaps, but as citizens with specific views on their own care. 'Reflexive solidarity' is strongly linked to the concept of citizenship which wants to *include* care receivers in our society by empowering them and giving them a voice in the debate on the delivery and content of care services.

According to Ikkos et al. (2006) fairness and justice in mental health care should come together under the umbrella of citizenship. Following the argument about social inclusion, it is important that rights of citizenship are given to people with mental illnesses and these are distributed fairly and allow access to justice. This means in psychiatry, amongst others, that psychiatrists should allow people to participate more actively in the decision-making regarding their own care.

12.4 Allocation on the Basis of Need

The ethical debate about allocation of healthcare resources is basically dealing with the question: what is necessary care? What kind of care do we owe to our fellow human beings? How much of its resources should society spend on meeting this obligation? However, it is not just a matter of the amount of money (or percentage of GDP) we think our society should spend on health care. We also face the question on the basis of which criteria this package should be defined. Should it be defined on the basis of *the need* of medical care, or on the basis of the *effectiveness* of medical treatments in relation to the costs, i.e. *efficiency*? In other words should we define basic care on the basis of the most urgent needs of individuals or on the basis of care that gives 'the greatest benefit for the greatest possible number'?

Although need is an important criterion for allocating healthcare resources there is always the problem how to rank order needs or prioritise these services on the basis of need. Mental healthcare and other care services and long term care services often have to compete with acute care services the need for which is often higher prioritised than long term care or mental health care. The example of mental health care in the NHS (see above) speaks for itself. Mental illness is often seen as less important or less worthy of funding than physical illnesses which can be treated effectively by medicine, by way of surgery or medical drugs (see Chapter 2.1.1). According to Sabin and Daniels there is a tendency within mental healthcare to prioritise 'biologically-based brain disorders' (schizophrenia, schizo-affective disorder, bipolar disorder, major depression, obsessive compulsive disorder and panic disorder) as part of a tendency to get mental healthcare at the same funding level as physical illness. This may be an effective political strategy, as Sabin and Daniels argue, but it is hard to see why a certain ideological view should give priority to one disorder over another. Moreover, less serious mental disorders could be equally based on biological processes while some extreme disorders may be partly the result of experiences and environmental circumstances. In the end, the debate on what mental health care services should provide reflects different views about the concept of mental health and mental disease, and of the political struggle within mental healthcare about ideology of mental disease and its treatment. When mental illness is seen as primarily caused by a biological process (for example resulting in lack of serotonin), there will be more emphasis in drug based treatment than on psychotherapy (though cognitive therapy is often accepted as complimentary to drug treatment in the case of depression). A psychological interpretation of mental illness will be more inclined to treat mental health problems by way of psychotherapy, like psychoanalytic therapy or counselling.

Another difficulty is the specificity of needs. What some people call an objective medical need is just a preference or a wish for somebody else. How can we distinguish between 'real' needs and mere 'wishes' or preferences? If a so-called need for treatment is seen as based on a subjective preference, society is in no way obliged to meet such a need, certainly not when it can only be met by expensive technology. An example in medicine is the treatment with growth hormone for children who are lagging behind in the development of their body. Is such a treatment meeting an objective medical need or is the need for height a subjective desire induced by modern medical research? (Parens 1998; see Chapter 30).

In mental health, the distinction between health and disease, and the correspondent distinction between needs and preferences, is often difficult to make. An example is the treatment of depression with selective serotonin reuptake inhibitors (SSRIs) like Prozac. Most people using SSRIs meet DSM IV criteria for some psychiatric disorder, although not necessarily major depression: dysthymia, social phobia, premenstrual dysphoric disorder and various anxiety and eating disorders respond well to SSRIs (Farah and Wolpe 2004). Antidepressants are most efficacious in preventing recurrence of depressive disorders in people who have such recurrent depression. While the importance of drug treatments in psychiatry is widely recognized, there is an ethical discussion on the use of SSRIs and other mood treatments on people who do not have a recognized psychiatric disorder (even though the boundaries are hard to define). Originally developed for the treatment of major depression and other affective and anxiety disorders where disability, morbidity and distress indicate the presence of mental illness, such mood-elevating drugs presently are also, and increasingly, prescribed for people whose problems are not recognised mental illnesses, whose disability and morbidity are either not documented or severe enough and where brain function may not be abnormal (President's Council on Bioethics 2003). One of the concerns is the progressive extension of the boundaries of psychiatry and psychopharmacology, leading to the further medicalisation of 'normal' emotional and social problems into medical 'needs' (Berghmans et al. 2010). Should society fund the use of SSRIs for individuals who have only mild symptoms of depression and who may be better off with psychotherapy or counselling?

Some people are using SSRIs who in fact have no recognised illness at all. The use of Prozac or other anti-depressants makes them feel 'better than well' (Kramer 1993, Elliott 2003). They are energized, more alert, more able to cope with the world and to understand themselves and their problems (Elliott 1999). The so-called 'enhancement of mood' for non-clinical purposes raises ethical and societal concerns, including the issue whether society should fund the use of such drugs for such purposes (see Chapter 30). The problem is that there is no simple discontinuity between the characteristic mood of patients with diagnosable mood disorders and the range of moods found in the general population. A clear distinction between health and disease, or preferences and needs is difficult to make (Butler 1999).

The concept of need may get a more concrete quantifiable basis when it is linked to the concept of 'burden of disease' (see above). Burden of disease can be measured by various quantitative and qualitative parameters, like acute threat to life in case of acute diseases and 'quality of life' in the case of burdening chronic conditions. These parameters will not only express the burden of a specific disease or handicap, but also need for treatment. When using this concept to prioritize treatments for specific conditions, enhancement would score low as it is not responding to acute life threatening diseases or to chronic conditions with poor quality of life. Enhancement would get a low priority as it does not relieve any burden of disease.

12.5 Outcome-Oriented Approaches

The example of the use of SSRIs shows the difficulty to distinguish 'real' needs from individual preferences, while the definition of need also depends on changing conceptions in our society and culture. According to some authors it is better not to base allocation decisions on the basis of need, but on the *outcomes* of treatment or services on the physical and mental wellbeing of individuals (Butler 1999). On the basis of such measurements various treatments can be evaluated according to their effectiveness in relation to costs.

While costs of medical treatments can be expressed in quantitative figures such an approach is much more difficult for the outcome of treatments. In the past success of a treatment was measured on the basis of the extent of curing of illness and the prevention of untimely death (life years gained). Such a measure says nothing about the impact of the treatment on the physical and mental wellbeing of the patient. Some very advanced medical treatments can stop the progression of an illness without full recovery. After or during treatment, the patient may be confronted with various disabilities and limitations. An example is renal dialysis, a technology which has several advantages but also various disadvantages like dependency on the dialysis machine, dietary prescriptions and medical complications. Moreover, the use of outcome measures to allocate resources can be disadvantageous for older people (Harris 1985). A heart transplant will add more years to life for a younger patient than for an older patient. For this reason, the outcome of medical treatments is not just based on the number of life years gained by treatment but also on the *quality of life* during and after treatment that means the physical and psycho-social situation of the patient and the personal assessment of his or her situation by the patient. The outcome of medical treatment is nowadays measured by way of QALYs (*Quality Adjusted Life Years*). A QALY means the number of life years gained multiplied with a factor for quality of life (between 0 meaning death to 1 – high quality). Twenty QALYs can mean 40 years of life extension with moderate to poor (0.5) quality of life.

The introduction of quality of life as a measurement of the effects of medical treatment has been criticised. An important problem is, for example, how to define quality of life. Quality of life depends on subjective preferences and variations. This is particularly true in the field of mental healthcare (see below). A second point of criticism is that not all individuals are better off with correction by quality of life. In the end, the most important measure is the number of life years gained which is lower for older than for younger people.

Information about the effects of treatments and services in relation to the cost are important within *outcome-oriented* or *consequentialist* approaches. These are approaches in which the moral weight of an act or a decision is based on the consequences of this action or decision. The most important consequentialist approach is *utilitarianism*. According to utilitarianism an act is morally right in case it contributes to the 'greatest possible happiness of the greatest possible number' (utility principle). In relation to the allocation of resources in healthcare utilitarianism supports an allocation of resources that results in the greatest possible health of the greatest possible number of individuals. The outcome of this approach will be calculated on the basis of gained life years, possibly with a correction for quality of life (QALYs). Allocation of resources that promote the health of a large part of the population ('breadth') have priority over investments that promote the health of a smaller group ('depth').

The counterpart of utilitarianism is *egalitarianism*. While the utilitarian model tries to achieve for the greatest possible health for the greatest possible number, egalitarianism tries to realise the greatest possible *equality* in health. This means that patients with the worst condition should have priority. Investments in 'depth' should have priority within an egalitarian approach. A person with a serious mental or physical condition has a higher priority than the treatment of a large group of patients with a less burdensome condition. An egalitarian approach to the allocation of resources will focus on the health of a small group with rare conditions that need intensive treatment. However, for some people with serious health conditions or serious disabilities, treatments or health services can result in improvement of quality of life, but not in the removal of their complaints and illnesses. This is particularly true for a patient with chronic mental illness and for patients with other chronic conditions or with learning disabilities.

An important problem within an egalitarian approach is: 'who is the worst off?'. The person with Down's Syndrome, the patient with schizophrenia or the patient who just had a heart attack and possibly will not survive a second attack? At the level of society such a weighing is very difficult to make. An egalitarian approach is more successful on the clinical level that is the micro-level, where one does not compare groups of patients or service or treatment categories, but one weighing the urgency of treatment between patients within the same category (for example patients on a waiting list). But also in that case it is not always easy to determine which patient is the worst off.

A conclusion might be that utilitarian approaches are more effective and more acceptable on the macro-level when allocating resources to various sectors of the public health system. Egalitarian approaches will be more helpful and acceptable on the micro-level when making decisions about the urgency of treatment for specific individuals. However, a utilitarian approach at the macro-level should always be balanced by an egalitarian approach, in order to prevent that the needs of people with severe, but rare health problems are ignored. Similarly an egalitarian approach at the clinical level should be balanced by a utilitarian approach in order to prevent a reduction in effectiveness of the specific health service.

12.6 Cost-Effectiveness in Mental Health Care

A consequentialist approach to the allocation of resources will look at the effectiveness of treatments as the basis for funding decisions. Usually effectiveness of a treatment is compared to its costs in order to determine which treatments have the best outcome for the lowest costs. In other words, which treatments give the 'best value for money'? Cost-effectiveness measures are usually applied within health service categories, like for example mental health care: which treatment for mental health problems gives better results as compared to other competing approaches. They are also used to compare various health services and treatments in terms of effectiveness in terms of Life Year of QALY gained by the various treatments in relation to their costs. Sometimes an amount of money is put forward as the limit for one Life Year or QALY gained. For example a treatment should not be funded when the cost per Life Year gained rise above a specific amount, for example 50,000€.

However, the use of cost-effectiveness studies to allocate resources within mental health care is highly contested (Reis et al. 2005). An important issue is the definition of what counts as effectiveness in relation to the goals of treatment (see Chapter 15).

The history of mental healthcare and psychiatry has seen the emergence of different schools of thought, each of which advocates the 'single best way' to explain and treat mental disorders. Today, this divergence in the goals and methods of mental healthcare still exists. One can point at the struggle between the biological and the psychological schools in the treatment of depression. In biologically oriented psychiatry (and some behavioural therapy), therapy of depression focuses on the control and reduction of specific symptoms. According to biological psychiatry, this objective can best be achieved with psychopharmacological means. By contrast, in a psychodynamic approach, the objective of therapy lies beyond mere symptom control. The goal of psychodynamic therapy is generally for the patient to gain insight in the structure of his or her personality. This is considered to be a means to gain reduction of and control over symptoms as well as to lessen the vulnerability of patients in the long run (Berghmans et al. 2004; see Chapter 18).

Biologically oriented psychiatrists will generally use scores of symptom scales, e.g. Hamilton scores as a criterion for treatment effect and success, while for psychoanalytically oriented psychiatrists experienced insight of the patient into unconscious conflicts will be a more relevant criterion for effective treatment. A socially oriented psychiatrist by comparison will put more emphasis on the change of social circumstances as a means to help the patient. As a result, effectiveness is more difficult to define or to make operational in a psychodynamic or socially oriented therapy than in the case of symptom control by antidepressant drugs, which can be made operational and measured by standardized instruments. As cost-effectiveness studies have a strong bias towards the use of such 'objective measures' (i.e. Hamilton scores), therapeutic strategies with other goals than mere symptom control are considered to be less cost-effective (Berghmans et al. 2004). If the psychiatrist and/or the patient consider the gain of insight or the change of social circumstances to be an important goal of treatment above symptom control, then they may be discriminated against in case only Hamilton scores are used as effect measure in cost-effectiveness analyses.

This example illustrates how studies of the cost-effectiveness of treatments may include normative assumptions about the definition of effectiveness. Psychiatric treatments are considered effective to the extent in which they succeed in measurable control and reduction of specific symptoms. Insight in the structure of a personality or changes in social circumstances are less easy to measure and disappear as acknowledged goals of psychiatric treatment. Still, patient specific factors, like personal values, social class and life style determine for a large part the effectiveness of treatment. Also the level of knowledge about himself and the causes of his problems will influence the decision which therapy will be applied and which goals should be achieved. To avoid narrowing down the effectiveness of psychiatric treatment to mere symptom reduction, one should ask in every individual case what is the best goal for the patient and what is the best method to reach this goal (Berghmans et al. 2004).

12.7 Evidence-Based Medicine

One way to cope with scarcity of resources in mental health care is to have more specific diagnoses and to limit the duration of treatment by introducing guidelines based on scientific studies on the effects and costs of treatments. Instead of 'opinion-based' medicine, decisions in health care and mental health care, should be based on 'evidence', particularly evidence on the effectiveness of treatments as determined in scientific studies. The moving idea behind Evidence-based Medicine (EbM) is that doctors should only make use of therapies that have been proved to be effective, instead of using unproven therapies that may be ineffective or even harmful to the patient. According to its founding father, David Sackett, EbM helps doctors to make the best decisions for patients, by offering them the best available evidence according to their experience with regard to the individual patient (Sackett 1996). The best evidence is evidence that is produced in randomised controlled clinical trials (RCT), where the association between a specific intervention and its outcomes is researched within very strictly controlled conditions. While Sackett and his colleagues wanted to restrict EbM to the context of individual patient care, other advocates of EbM have extended the range of EbM to the area of health care systems and policy-making (ter Meulen and Dickenson 2002). Evidence-based medicine is seen not only as an important means to improve the quality of medical care, but also as an instrument to control costs. In view of the scarcity of health care resources, decisions on the allocation of care will have to be made more explicitly and should be made more transparent and accountable.

Health policy makers frequently require that new, promising interventions can only be allowed or reimbursed after proven effectiveness within the context of randomized controlled clinical trials. In the United Kingdom for example, the NHS Executive orientates its priority setting by means of an evidence-based medicine agenda, both in research and development. The National Institute for Health and Clinical Excellence (NICE) coordinates the efforts to provide evidence on the effectiveness of (new) treatments in relation to costs and gives advice to the NHS about the funding of treatments on the basis of the (predominantly RCT based) research.

While there is an increasing body of evidence on the effectiveness of psychiatric and psychotherapeutic methods, there are significant ethical concerns about the use of RCTs as the main method to produce evidence in mental health care, e.g. for a large range of treatments such as multimedications no RCTs exist. Moreover, there is a concern that funding and allocation decisions based on RCTs will result in unfair priorities between the various treatment options which may not be in the interest of the patient. According to Reis et al. (2005), many are concerned that the positivist-reductionist approach of the traditional RCT may lead to a systematic bias against the evaluation of the outcomes of discussion therapy. The rise of EbM as a powerful tool for the evaluation of the outcomes of pharmacological treatments seems to support the dissemination of the same criteria for the evaluation of other therapeutic interventions. Reis et al. raise the question whether the 'hierarchy of evidence' of EbM with RCTs at its top is transferable to psychotherapy where there are specific demands for valid studies. Many authors, mainly from psychodynamic schools of psychotherapy, have argued that by their very nature, psychiatric patients and psychotherapy do not lend themselves to the methods of EbM (Reis et al. 2005, p. 68). Evidence based research in psychotherapy is not able to 'measure' inter-personal and contextual factors which have a thorough influence on the therapeutic outcome. RCTs are natural-scientific experiments which are designed to find highly specific relationships between supposed causes and describable effects. According to Hennigsen and Rudolf (2000) such experiments are very adequate to verify the special effectiveness of a single substance in pharmacology. However, they wonder whether similar experiments can be conducted to assess the impact of the therapeutic relationship between psychotherapist and patient (Hennigsen and Rudolf 2000).

Psychotherapies can and should be evaluated like somatic therapies and one can define certain outcome criteria like maturing of personality, patient satisfaction or other symptom reducing goals. However, such measures are often not regarded as 'hard' evidence. Moreover, there are many methodological problems to conduct randomised trials into the effectiveness of psychotherapies. For example the use of control groups or placebo controls is difficult when researching the effectiveness of psychotherapy. What is for example the definition of 'usual care' for a control group? How to randomise patients between control and experimental group? How to deal with the impossibility of blinding the study population and the impact of the lack of blinding on the control group? How to single out the effect of the therapy versus control or placebo when there are many variables that may have influenced the outcome of the study? Because of these methodological problems, particularly the difficulty of establishing a contrast between control and experimental study groups, there have not been many randomised controlled trials into the effects of psychotherapy. Similar problems are encountered in efforts to conduct to research the effects of chronic disease management by way of randomised controlled trials (Van Vliet Vlieland 2002). One solution could be to use different methods for outcome evaluation, like observational studies. However, such studies have a lower status in the hierarchy of evidence which sees the randomised controlled trial as the most reliable kind of evidence.

In spite of the many methodological problems, psychotherapeutic medicine is facing an ever increasing pressure from third-party payers to be based on scientific evidence. This phenomenon is not specific for psychotherapy but part of intense efforts throughout all medical fields to contain the constantly rising costs in the health care sector. Third parties are gaining more power to control the delivery of care services and will in fact determine which kinds of treatment and how many will be reimbursed. In continental health care systems, for example, the introduction of market approaches in the health care system will result in a shift of power towards insurance companies. Because those companies bear greater financial risks, they could force providers, including hospitals and institutions for mental health care, to make use of the most cost-effective treatments, such as drug-based therapies instead of psychotherapy or social care services for which there is less 'hard' evidence available.

12.8 Conclusion

Health care systems are confronted with an increasing scarcity of resources and the problem how to allocate health care services in a fair way while maintaining universal access to care. This situation affects the availability and quality of health care services, particularly mental health care. In many countries, funding of mental health care is lagging behind acute care services, which are more 'appealing' and are seen as more 'successful' than chronic care (including mental health care). This situation is unfair and unjust, as the needs of mental health care patients are widely acknowledged as the most important in our health care systems. While need is an important concept to distribute health care services, it is also seen as subjective and difficult to distinguish from preference or wishes of individuals. However, recent reports (Gezondheidsraad 2003, CVZ 2003) argue that allocation policies should take into account the 'burden of disease' which can be seen as a quantitative and qualitative measurement of the need of individuals for a specific treatment. However, when used in the context of allocation of resources 'burden of disease' should be combined with measures of cost-effectiveness, as for example proven within randomised controlled trials (Gezondheidsraad 2003). For mental health care this would mean that funding for mental health services should have high priority because of the severe burden of mental health problems. It would also mean that within mental health care the interests of individuals with severe and chronic mental problems should be adequately dealt with. A specific problem for mental health, and particularly for chronic mental illness, is that treatments are often seen as not 'effective', meaning that treatment does not result in complete control or disappearance of symptoms. However, long-term care and support are important for individuals with chronic mental health problems. Services offering such long-term support should be funded by the health care system even if the outcome of such support is not measurable by effectiveness studies as proven by randomised controlled trials. Though effectiveness is an important measure when allocating healthcare resources, a health care system, and particularly a mental health service, should always express our solidarity and compassion with its most vulnerable individuals even though the outcome of these services will be difficult to measure.

References

Berghmans RLP, Berg M, van de Burg M, ter Meulen RHJ (2004) Ethical issues of cost-effectiveness analysis and guideline setting in mental health care. J Med Ethics 30:146–150

Berghmans RLP, ter Meulen RHJ, Malizia A, Vos R (2010) Feeling better. Scientific, ethical and social issues in mood enhancement. In: Savulescu J, ter Meulen RHJ, Kahane G (eds) Enhancing human capacities. Blackwell, London (forthcoming)

Boorse C (1975) On the distinction between disease and illness. Philos Public Aff 5:49–68

Boorse C (1977) Health as a theoretical concept. Philos Sci 44:542–573

Butler J (1999) The ethics of health care rationing. Principles and practices. Cassell, London

Committee Choices in Health Care (1991) Choices in health care. Ministry of Health, Welfare and Sports, Rijswijk

CVZ (2003) Ziektelast toegepast op het geneesmiddelen pakket. CVZ College voor Zorgverzekeringen, Diemen

Daniels N (1985) Just health care. Cambridge University Press, Cambridge

Daniels N (2008) Just health. Meeting health needs fairly. Cambridge University Press, Cambridge

Elliott C (1999) A philosophical disease: bioethics, culture and identity. Routledge, New York, London

Elliott C (2003) Better than well: American medicine meets the American dream. Norton, New York, NY

Farah MJ, Wolpe PR (2004) New neuroscience technologies and their ethical implications. Hastings Cent Rep 34(3):35–45. May–June

Gezondheidsraad (2003) Countouren van het basis pakket [Countours of the basic health care benefit package]. Gezondheidsraad, The Hague

Harris J (1985) The value of life. Routledge, London

Hennigsen P, Rudolf G (2000) Zur Bedeutung der Evidence-based Medicine fuer die Psychotherapeutische Medizin. Psychother Psychosom Med Psychol 50:366–375

Houtepen R, ter Meulen RHJ (2000) The expectations of solidarity: matters of justice, responsibility and identity in the reconstruction of the health care system. Health Care Anal 8:355–379

Ikkos G, Boardman J, Zigmond T (2006) Talking liberties: John Rawls's theory of justice and psychiatric practice. Adv Psychiatr Treat 12:202–213

Jenkins R (1994) The health of the nation: recent government policy and legislation. *Psychiatr Bull* 18:324–327

Kramer P (1993) Listening to Prozac. Viking, New York, NY

ter Meulen RHJ, Arts W, Muffels R (eds) (2001) Solidarity in health and social care in Europe. Kluwer Academic Publishers, Dordrecht

ter Meulen RHJ, Dickenson DL (2002) Into the hidden world behind evidence based medicine. Health Care Anal 10:231–241

OECD (2008) Health data 2008. www.oecd.org

Parens E (ed) (1998) Enhancing human traits: ethical and social implications. Georgetown University Press, Washington, DC

President's Council on Bioethics (2003) Beyond therapy. Biotechnology and the pursuit of happiness. Regan Books, New York, NY

Reis A, Lenk Ch, Reis-Streussnig C, Biller-Andorno N (2005) Evidence-based medicine in mental health: toward better and fairer treatment? In: ter Meulen RHJ, Lenk Ch, Biller-Andorno N, Lie R (eds) Ethical issues of evidence-based medicine. Springer, Berlin, pp 65–75

Sabin JE, Daniels N (1999) Ethical issues in mental health resource allocation. In: Bloch S, Chodoff P, Green S (eds) Psychiatry ethics. Oxford University Press, Oxford

Sackett DL (1996) Evidence-based medicine: what it is and what it isn't. Br Med J 312:71–72

Van Vliet Vlieland Th (2002) Managing chronic disease: evidence-based medicine or patient centred medicine? Health Care Anal 10:289–298

Part III
The Applications of the Ethical Principles in Psychiatric Practice and Research

Chapter 13
Ethics of Diagnosis and Classification in Psychiatry

Juan José López-Ibor, Maria-Inès López-Ibor, and Hanfried Helmchen

Contents

Abbreviations

DRG   Diasgnosis Related Groups
DSM-IV   Diagnostic and Statistic Manual (of Mental Disorders), 4th Revision
          (American Psychiatric Association)
ICD   International Classification of Diseases (World Health Organisation)

13.1 Introductory Remarks About Diagnosis and Classification

Present medical action is increasingly ruled by evidence-based and consequently formalised and standardised procedures. This is counterbalanced by an increasing emphasis on value-based, qualitative, individual aspects of the patient–physician relationship. The topic will be elaborated from this latter perspective.

J.J. López-Ibor (✉)
Institute of Psychiatry and Mental Health, San Carlos Hospital, Complutense University, 28035 Madrid, Spain; CIBERSAM (Spanish Network for Research on Mental Health), Complutense University, 28035 Madrid, Spain
e-mail: jli@lopez-ibor.com

H. Helmchen, N. Sartorius (eds.), *Ethics in Psychiatry*, International Library of Ethics, Law, and the New Medicine 45, DOI 10.1007/978-90-481-8721-8_13, © Springer Science+Business Media B.V. 2010

Classification reflects the taxonomically ordered area of phenomena. Like every reflection it has two limitations: it reduces the diversity and complexity of the phenomena and views them from the viewpoint of the taxonomist. It is thus an artefact, even though a necessary one, as a human instrument to comprehend the world. Two questions are especially important: (1) how *useful*, i.e. how valid and reliable, is this instrument for its intended purpose? (2) what *consequences* does its usage have?

This is also true about psychiatric classification. It orders mental illnesses, which are described as psychopathological characteristic configurations (syndromes) and are expanded by dimensions such as course (age of illness, episodic, recurring, chronic-progressive) or 'etiology' (disposing, causing, and establishing conditions) to (multiaxial) nosological diagnoses (Helmchen 1983). Diagnoses are short instrumental formulas (Kendell 1975). In *practice* they serve the understanding among psychiatrists and the approach to specific knowledge about the particular illness; in *research* they serve the formation of (more or less) homogenous classes and thus the limitation of the research objects; in *health politics* they serve the assessment of frequencies of morbidity groups and thereby the planning of care facilities and now also cost calculation (Helmchen 1978/1981, 2006, Gethmann et al. 2004). The operationalised diagnoses of ICD and DSM have made worldwide understanding among psychiatrists more reliable, but their relevance for psychiatric therapy must be considerably improved. These questions of *usefulness* (Helmchen 1994) seem ethically less relevant than questions of the consequences of its usage.

Ethically relevant *consequences* of psychiatric classification and diagnosing result from the fact that neither the process of diagnosing and classifying nor the individual classification systems are free from valuation. The diagnosis involves separating an observed psychopathological phenomenon from its context, describing it in words, and evaluating it in relationship to an ordering system. To be sure, there are, according to criteria, operationalised and thus objectivising standards for the scientific form of this evaluation, but at the same time moral valuations, usually unreflected, are included in the observation of a psychopathological phenomenon separated from its context, regardless of whether observed by the patient or the examiner, as well as its description, e.g. those with stigmatising effects, which are transported, among other paths, by language. These include subjective appraisal of the individual as well as socially accepted values of society (Wakefield 2005, Wakefield et al. 2002) in their individual regional and cultural specificity, which is also dependent on social class (Sartorius 2005). These appraising subjective and normative influences are not only effective on the level of the identification of the phenomenon (1), but also on the level of the diagnosis (2), and also on the level of nosological classification (3), especially in the choice of the ordering criteria. This socially and culturally related and thus implicitly evaluating context dependency of shaping, processing, describing, and ordering the illness characteristics is for the relationship to the individual patient and thus the individualisation of the treatment, e.g. for the treatment adherence, of considerable, if not decisive significance (First 2005).

This will be discussed by first illustrating ethically relevant evaluating influences on the above-mentioned three levels with several examples and then by clarifying, through their relationship to the illness concept, what this means for the individual patient who comes to the psychiatrist because of disturbing psychopathological phenomena and who is diagnosed within a psychiatric classification system by the psychiatrist.

13.2 Description of Psychopathological Phenomena

The pragmatic but imprecise *everyday language* of the patient is opposed to the increasing precision of medical language. In the relationship of both languages of course the tension between the individual closeness to life and standardised precept knowledge becomes obvious. This scientific influence on the language with the shrinking of the semantic halos is probably embedded in a general development of the century, characterized by the change from quality to quantity. The words of the medical language as those of a *scientific language* are, indeed, relatively unequivocal and thus necessary for communication among physicians, but are also often too simple to describe adequately the complexity and diversity of specific phenomena of reality. But unequivocal does not mean without value. One can consider the implicit difference in nosological category, for example, between the description of a special mood as 'high spirits' (related to mania) or as 'euphoric' in the sense of a quiet cheerfulness (related to an organic state). The high spirits of the mood will first but not exclusively be judged on the change in the individual basic mood, because the (regionally of course differing) average mood of the individual regional or social group is included in the evaluation. In the same sense but ethically much more relevant are terms with negative implications and stigmatising effects. Therefore, terms such as 'hysteric' or 'psychopathic' are left due to its moralizing meaning.

13.3 Formulating Psychiatric Diagnoses

The formulation of *psychiatric diagnoses* can also include moral evaluations (Fulford et al. 2005). For example, it took a long time to recognize and accept alcohol dependence as a psychiatric illness, because the concepts used in describing it, such as drunkenness and drinking addiction included and socially still includes strong moral implications. Or homosexuality, which involved a risk of psychiatric morbidity at a time when this socially and thus morally outlawed sexual orientation was kept secret, is now no longer a taboo and has been taken out of the psychiatric diagnosis catalogue. The value criteria of psychiatric judgments are especially evident in the determination of guilt incapacity, which is especially pertinent in discussions today. It corresponds to socially generally accepted values to judge and punish a perpetrator according to moral categories such as responsibility, freedom of will, and guilt, unless, because of psychic illness, he could not act otherwise. The risk of the forensically involved psychiatrist to include moral values in judging the psychiatric reasons for the limitation or loss of guilt capacity is always great if he

does not restrict himself to the medical and scientific identification of a psychic illness and its particularity, but also judges its consequences for responsible action or
even takes on the task of the judge to establish guilt incapacity – and thus to enter
explicitly the area of moral categories (see Chapter 22).

13.4 Value Conceptions in Psychiatric Classifications

At the same time *psychiatric classifications* contain value concepts and moral evaluations; indeed, they are now, even if seldom, stated explicitly. On the other hand
the border between medical-scientific concepts, as established in nosological systems and moral value conceptions remains generally overlooked. Even the latest
psychiatric classification, DSM-IV, contains social value judgments in spite of the
pre-eminence of evidence and thus fact basing; for example the concept of disorder
used in the psychiatric classification can have many meanings and is thus more liable
for moral valuation than the otherwise customary concept of illness. How far must
a disorder depart from which order in a negative and deficient sense in order to gain
the status of an illness? This question contains three components (order, extent, deficiency), whose determination is not possible without valuation. The establishment
of deficiency seems most liable to moral evaluation if one thinks only about our
normative pictures of sexual roles and sexual orientation or personality structures
(Sadler 2005). Value judgments are expressed explicitly if, for example, criterion
A for behaviour disorders covers 'behaviour that injures the basic rights of others
or important age-appropriate norms or rules', or if criterion B for schizophrenia is
related to 'social or professional dysfunction', and thus not exclusively to a change
as a fact but, with the 'dys', to a change for the worse and is thus a value judgment
(Fulford et al. 2005). Further examples are found in the commentaries to a recent
study by Fulford et al. (2005).

The influence of a surely not valueless spirit of the day becomes clear with the
appearance and disappearance of diagnoses, for example neurasthenia. In the western world at the time already meaningless, it had great importance in the China of
the Cultural Revolution, perhaps as a possibility of allowing psychic disorders to
be socially acceptable as somatic illnesses (Lee 1998). Culturally specific implications also influence the diagnostic association, as shown by an example from
Los Angeles: with an identical symptoms profile Asian-Americans were diagnosed
with neurasthenia, but Caucasian Americans were diagnosed with chronic fatigue
syndrome (CFS) (Lin et al. 1996).

13.5 Culturally Specific Diagnoses

The recognition of evaluating semantics in the psychopathological language and the
psychiatric diagnostics and classification has led to the recommendation (Fulford
et al. 2005), and even to the demand (King 2005) to make these valuations part of

a patient-centred psychiatric diagnostic procedure. The predecessors of this are culturally specific diagnoses such as, e.g. Latah (with Malays) or Imudo (with Ainos). They evolve when, as is (still) usually the case in psychiatry, the diagnosis of an illness is symptom-oriented and the individual symptoms are interculturally different in their emphasis. But a closer analysis shows that the 'symptoms are not so different and only their culturally specific halo makes them attractive; even when they are specific for a particular culture, they can easily be interpreted as a penis invagination, as, e.g. Latah; indeed they represent a way of experiencing widespread feelings as fear (e.g., the "susto" in Latin America, which is a typical panic illness)' (López-Ibor 2003). It is especially criticised that 'cultural adaptations challenge the principles of universality of science and ethics' and 'contain the risk of giving in to social or cultural pressure' (López-Ibor 2003). This pressure becomes evident when, with the global spread of the international classification of disease, the ICD, and especially the national American DSM, in some areas of the world is criticised as cultural imperialism (Baker 1997).

But in the end it is all about, on the one hand, the *diagnosis of mental illness*, the scientifically based establishment of an illness, and ordering it, with the help of known phenomena, into a class within a nosological system and thus to gain access to the standard knowledge about the illness and its treatment, and, on the other hand, the *establishment of the individuality* of the individual patient, in order to fit the standard knowledge optimally to the constitution and situation of the individual patent in his social and cultural context and to transform this into a successful treatment. It is of course difficult to speak of a diagnosis in connection with the establishment of individuality, because individual uniqueness (to date anyhow and in principle to a certain extent) generally keeps itself from a regular classification.

13.6 The (Psychiatric) Illness Concept

The examples describe the ethically relevant problems of an *illness concept* that moves like a Janus head between medical and social definitions of illness (Helmchen, 2006). Just the relationship to the social dimension of illness opens its diagnosis and classification to the influence of value judgments. Consideration of this was last forced on psychiatry by the antipsychiatric movement in the 1970s (Laing 1960, Szasz 1960, Scheff 1966, Kisker 1979) – and this will become necessary once again under pressure of making medicine more economic, since the borders of illness as opposed to non-illness are being pulled together more narrowly. For example, this is the case with the introduction of the DRG system or the thinking about solidarity financing of only a basic insurance, which means only clearly evident illnesses, in order to limit demands of patients for help with illnesses (Helmchen 2003). A physician can most clearly sense the social components in the determination as an illness of emotional disturbances that lie between obvious illness and obvious good health, when he, in front of a patient complaining about chronic fatigue or the loss of emotional tension or about diffuse problems 'beneath the ribs' ('hypo chondrium'), has to decide whether this is merely an unpleasant condition that sometimes happens in life that can be compensated by a healthy person himself, or whether this is indeed an emotional illness, perhaps depression, perhaps in disguised form – and with a decision for this latter diagnosis of illness he establishes a demand for support from the solidarity community (Helmchen 2006). In any case the physician should not overlook the development of a (later) obvious illness and not sufficiently treated illness residue (Helmchen 2003), nor should he limit the self-help potential of the patient by an only insufficiently based illness diagnosis. Rather he should strengthen the patient's own responsible *usage* of this potential, especially with persons with light disorders that objectively do not hinder them. Häfner pointed this out with his question, 'What to do with illnesses that do not exist?' (Häfner 1997).

13.7 The Consequences of Psychiatric Diagnosis and Classification for the Individual Person

Establishing an illness diagnosis for a psychiatrically conspicuous person and thus placing him into a category of a psychiatric classification system makes possible, as discussed above, the access to special knowledge about the illness phenomena constituting this category. But it can also be misunderstood as a label if the physician only concentrates on the illness and ignores the diversity and complexity of the individuality of the patient. Thus in the past this accusation was made when, for example, a forensic judgment about driving capability or enforced custody was based only on the psychiatric diagnosis and only thereafter on the individual psychopathological findings. The diagnosis opens the view toward the standard knowledge of the special illness and is also the prerequisite of evaluating a disturbed behaviour as based on illness, but only the understanding of the individuality of the patient enables his appropriate treatment (First 2005). The task of the psychiatrist is thus to include the entire patient in his view and only then to determine a possible illness.

For this he must recognize the illness character of inner causes, such as delusion or a depressive loss of interest, and their proportion of the – perhaps still healthy – inner experiences, motivations, and drives of the individual patient. This is more definitely possible with a strong emphasis of qualitatively new experiences, such as hallucinations or delusion, especially in acute condition, than with only quantitatively abnormally determined changes in feelings or thoughts that healthy persons also experience. These latter are often more difficult to recognize, because their borders to the normal, especially with only a mild emphasis, are not sharp; threshold measurements have been established in order to meet this problem. This limitation, necessary for practical reasons, contains something arbitrary. This arbitrary and thus of course also evaluating moment in the definition of 'already' ill on the border to being 'still' normal creates an ethical problem, when the consequences of the definition are considered. This problem exists in the fact that the medical model of the illness gives the patient protection on the one hand: medical insurance takes over the expenses, and the physician assumes the responsibility for his treatment and care; his abnormal behaviour is excused. At the same time it can take away the patient's own feeling of responsibility for his strange behaviour and for efforts to overcome this and can stigmatise him as mentally ill, often in the negative connotation of incurable, disturbing, or failing. Then the self determination of the mentally ill person, also of the formerly ill person, is limited not only by the inner causes of the illness, the 'not-able-to-do-otherwise', but also by the consequences of the role of patient. Finzen (2000) spoke of a 'second illness'. If the border is drawn far into the normal from clear illness, from the complete picture of an illness type, then still present self determination, self responsibility, and social freedom can be lost. However, if the border is drawn too narrowly, even to giving up the illness model, the patient and his family will be given a responsibility that he cannot carry. Psychiatric experience teaches too that even within an illness concept, regardless of its breadth, self determination is limited to an individual extent and can hardly ever be completely and permanently be abolished. Thus the responsibility of the psychiatrist includes recognizing correctly the kind and extent of the limitation of self identity and self-determination or else the still present autonomy of each patient. But in this he moves between Scylla and Charybdis; if he draws the border too broadly, he will work against the self-determination of the patient; if he draws it too narrowly, he will offend against the ethical maxim of the physician to do everything for the good of the patient and to keep harm from him. In any case the wrong limitation can harm the patient. In the first case mentioned, a regressive passive development or hospitalism can take place; in the latter case an illness progression and a self or external burden or even danger can occur.

13.8 Illness and Ill Being

The delineation between illness and not ill being of the patient is sometimes difficult, especially with the frequently underlying mental illnesses, whose emphasis is not sufficient to cross over the threshold set by convention to a diagnosis in a psychiatric classification system (Helmchen 2001), but which nonetheless lead to suffering pressure and increased usage of medical help. This is also valid for the division between degenerative, chronically progressive illnesses such as old-age dementia that slowly reach the level of clinical symptoms and normal changes of aging and old persons (Helmchen and Lauter 2001). But especially this division can be very difficult with a long-lasting, chronic course of illness. The reaction of the patient to the experience of changes caused by illness and processing them sometimes develops after a longer period to a thick layer of experience including the usually countless experiences with doctors and many other persons in medical facilities and in which a self-developed illness concept comes to the fore for the patient, supported today increasingly with the help of the media all the way to the internet and complete with superficially read and misunderstood specialized expressions. These can cover up the core of the illness so much that they can hardly be penetrated. But they determine the patient's feeling of ill *being* in the sense that he feels like an expert for his own illness.

The far-reaching technical possibilities today, including standardised procedures for the diagnosis of an illness, might give a young physician the feeling that a real dialogue with the patient is unnecessary, and to diagnose some acute illnesses this is neither possible nor necessary. However, with longer-lasting illnesses or even with those that have become chronic, ongoing dialogues cannot be waived, if the physician wants not only to establish the diagnosis of an illness but also to understand the individually emphasised and, within a certain context, ill *being* of the patient, as a prerequisite of an appropriate treatment, to understand, and especially to support the patient in overcoming his ill being.

References

Baker R (1997) Transkulturelle Medizinethik und Menschenrechte. In: Tröhler U, Reiter-Theil S (eds) Ethik und Medizin 1947–1997. Wallstein, Göttingen, pp 433–459

Finzen A (2000) Psychose und Stigma. Psychiatrie Verlag, Bonn, zit. n. Fritze J et al (2005) Entwurf des Antidiskriminierungsgesetzes. Nervenarzt 76(5):653

First MB (2005) Keeping an eye on clinical utility. World Psychiatry 4:87–88

Fulford KWM, Broome M, Stanghellini G, Thornton T (2005) Looking with both eyes open: fact and value in psychiatric diagnosis. World Psychiatry 4:78–86

Gethmann CF, Gerok W, Helmchen H, Henke KD, Mittelstraß J, Schmidt-Assmann E, Stock G, Taupitz J, Thiele F (2004) Gesundheit nach Maß? Eine transdisziplinäre Studie zu den Grundlagen eines dauerhaften Gesundheitssystems. Akademie, Berlin, S 345

Häfner H (1997) Was tun mit Krankheiten, die keine sind? Munch Med Wochenschr 139:158–160

Helmchen H (1978/1981) Functions and consequences of psychiatric diagnoses. In: Agassi, J(ed) Psychiatric diagnoses. Proceedings of an International Interdisciplinary Interschool Symposium, Bielefeld University 1978. Balaban International Science Services, Philadelphia, PA 1981, pp 99–105

Helmchen H (1983) Multiaxial classification in psychiatry. Compr Psychiatry 24:20–24

Helmchen H (1994) Relevanz der Diagnostik in der psychiatrischen Therapie. In: Dilling H, Schulte-Markwort E, Freyberger H (eds) Von der ICD-9 zur ICD-10. Neue Ansätze der Diagnostik psychischer Störungen in der Psychiatrie, Psychosomatik und Kinder- und Jugendpsychiatrie. Huber, Bern, pp 89–100

Helmchen H (2001) Unterschwellige psychische Störungen. Nervenarzt 72:181–189

Helmchen H (2003) Krankheitsbegriff und Anspruch auf medizinische Leistungen. Nervenarzt 74:395–397

Helmchen H (2006) Ökonomische Determinanten ärztlichen Handelns. In: Schneider F (Hrsg) Entwicklungen der Psychiatrie. Springer, Berlin, Heidelberg, New York, NY, pp S93–S105

Helmchen H, Lauter H (2001) Diagnostic problems in geriatric psychiatry. In: Henn FA, Sartorius N, Helmchen H, Lauter H (eds) Contemporary psychiatry, vol 2. Psychiatry in special situations. Springer, Berlin, Heidelberg, New York, NY, pp 117–127

Kendell RE (1975) The role of diagnosis in psychiatry. Blackwell, Oxford

King C (2005) Coloring our eyes. World Psychiatry 4:95

Kisker KP (1979) Antipsychiatrie. In: Kisker KP, Meyer JE, Müller C, Strömgren E (Hrsg) Psychiatrie der Gegenwart, 2. Aufl., Bd. I/1 Grundlagen und Methoden der Psychiatrie. Springer, Heidelberg, pp 811–825

Laing RD (1960) The divided self. Tavistock, London

Lee S (1998) Estranged bodies, simulated harmony, and misplaced cultures: neurasthenia in contemporary chinese society. Psychosom Med 60:448–457

Lin K, Cheung F, Zheng Y, Weiss M, Nakasaki G, Ren Y (1996) A cross cultural study of neurasthenia and CFS in LA. Tenth World Congress of Psychiatry, Madrid, 23–28 August, p 184

Lopez-Ibor JJ Jr (2003) Cultural adaptations of current psychiatric classifications: are they the solution? Psychopathology 36(3):114–119

Sadler JZ (2005) Bug-eyed and breathless: emerging crises involving values. World Psychiatry 4:87

Sartorius N (2005) Recognizing that values matter. World Psychiatry 4:90

Scheff TJ (1966) Being mentally ill: a sociological theory. Aldine, Chicago, IL

Szasz TS (1960) The myth of mental illness. Am Psychol 15:113–118

Wakefield JC (2005) On winking at the facts, and losing one's Hare: value pluralism and the harmful dysfunction analysis. World Psychiatry 4:88–89

Wakefield JC, Pottick KJ, Kirk SA (2002) Should the DSM-IV diagnostic criteria for conduct disorder consider social context? Am J Psychiatry 159:380–386

Chapter 14
Competence Assessment

Lienhard Maeck and Gabriela Stoppe

Contents

Abbreviations

ACE   Aid to Capacity Evaluation
AD   Alzheimer's Disease
BPSD   Behavioural and Psychological Symptoms of Dementia
HCAT   Hopkins Competency Assessment Test

L. Maeck (✉)
Universitäre Psychiatrische Kliniken, CH-4025 Basel, Switzerland
e-mail: lienhard.maeck@upkbs.ch

H. Helmchen, N. Sartorius (eds.), *Ethics in Psychiatry*, International Library
of Ethics, Law, and the New Medicine 45, DOI 10.1007/978-90-481-8721-8_14,
© Springer Science+Business Media B.V. 2010

MacCAT-T	MacArthur Competence Assessment Tool for Treatment
MacCAT-CR	MacArthur Competence Assessment Tool for Clinical Research
MCI	Mild Cognitive Impairment
MMSE	Mini Mental State Examination
PD	Parkinson's Disease
SICIATRI	Structured Interview for Competency Incompetency Assessment Testing and Ranking Inventory
UBACC	University of California, San Diego Brief Assessment of Capacity to Consent

14.1 Definition of Competence

The terms competence, competency, capacity, ability or performance are frequently used synonymously throughout the literature. This might reflect the heterogeneity of contexts in which they are used, for example in the legal or the rehabilitation context. In the following we will use the term competence in a broader sense taking together its medical, social as well as legal meaning.

Many studies refer to a definition of competence given by Appelbaum and Grisso (1988). The person should be able to *understand* a task or situation, *appreciate* the relevance, the emotional impact, the rational requirements or the future consequences of a decision, to *reason* on risks and benefits and weigh the arguments, and at last to *express a choice*. In the legal situation the person should also be able to follow a conversation or trial and promote his or her own interests. This implies that competence varies with regard to the purpose under consideration. It also depends on various aspects of personality, affective regulation and intellectual or cognitive functioning of the person. In addition, somatic factors play a role: visual and hearing function as well as the influence of substances or drugs might influence competence and its assessment.

There are two main approaches to assess competence: the *categorical* and the *functional* operationalisation. The categorical approach, if done by a psychiatrist often regarded as gold standard and widely used, has been shown to have a bad interrater reliability. As to be expected, it depends from education and training and might be increased by structured assessment tools (Marson et al. 2000). The functional approach often includes psychometric instruments and delivers reliable data. However, to define a clear threshold required for the decision whether competence is sufficient or not, is crucial.

The present situation is characterised by a lack of – empirically proven – standards or consensus either on the legal or on the medical domain (Kim et al. 2002). The variations between countries and even within a country are substantial, considering for example the role of the individual and the family. An overview concerning the situation for the demented population has been published recently (Stoppe 2008). Differentiating capable from incapable subjects remains an issue at stake for both, physicians and legal experts.

14.2 Areas of Competence

In the following, we will briefly discuss those areas of patients' competence, which have potential impact on relevant legal or medical decisions. In view of the fact that patients might only be incompetent concerning special domains of competence or at ascertained points in time, a differentiated view on distinct areas of competence is required. In other words, regarding competence, no global and definite decision can be made.

14.2.1 Informed Consent

Except in the case of an emergency, any medical intervention like hospitalisation, diagnostics, treatment, but also participation in research requires the patients' agreement (informed consent).[1] To be capable to agree, the patient has to be informed on the nature and natural course of his or her disease, treatment options and alternatives with any typically associated risks and their possible consequences. The educating physician should be conducted by what he assumes to be in the patients' best interest (Arie 1996). To assure, that informed consent becomes a *valid* informed consent, the doctor has to apply an individually adapted diction and to ascertain, that the patient understands the key points. Of questionable relevance is the fact that patients won't recall the given information only minutes later. In a recent study, for example, 82 healthy volunteers in a clinical trial – 90% with university level education – were given a written information sheet and allowed to ask questions. After having indicated they were ready to give consent, only 12% could name the three study drugs, only 17% could indicate three or more potential risks of the medication, and the maximum number of risks remembered was 6 out of 23 (Fortun et al. 2008).

The 'sliding scale strategy' is frequently considered as a pragmatic approach to assess capacity to consent. According to this, the standard of decisional capacity should depend on the (potential) harm or benefit that may arise due to a specific intervention. With increasing risks or decreasing benefits, more capacity is required for the patient to be considered competent to consent to a treatment, and vice versa.

If a patient is incapable of giving informed consent, this can be done by a guardian, usually being installed by court or a judge. Competencies of a guardianship may be limited to distinct areas like finances, the right to determine the place of residence, or medical treatment. They are usually regulated in the national civil codes (see Chapter 10).

For research purposes, the World Medical Association (WMA) gives authoritative guidelines to be applied in almost every country regarding ethical principles for medical research involving human subjects. Among other things it states that 'in medical research involving *competent* human subjects, each potential subject must be adequately informed of the aims, methods, sources of funding,

---

[1] See Chapter 9.

any possible conflicts of interest, institutional affiliations of the researcher, the anticipated benefits and potential risks of the study and the discomfort it may entail, and any other relevant aspects of the study' (WMA 2008). Ethical problems arise in non-competent patients (see Chapter 25), like severely demented or acutely psychotic individuals, especially, when taking part in the study is not accompanied by any foreseeable (clinical) benefit. Like a transeuropean survey on the acquaintance of approval from medical research committees pointed out, researchers', 'group-specific' and incapacitated individuals' interests frequently don't seem to be adequately balanced yet (Helmchen and Lauter 1995, Olde Rikkert et al. 2005). During recent years, a shift from normatively oriented discussions to more empirically based investigations seems to have developed (Helmchen 2008).

The question of *involuntary psychiatric treatment and hospitalisation* is subject to controversial ethical debates in the field of mental health (see Chapter 20). From a legal point of view, both assume one out of two of the following conditions: (1) a subject suffering from severe mental disorder making it impossible to competently decide on treatment itself or express the treatment wishes, or (2) a subject being in danger of harming itself or others due to mental disorder. The determination of incompetent and competent individuals is essential, as competent individuals only exceptionally may be subject to compulsory treatment. The Board of German Judges, for example, acknowledges the 'right of being ill' and states, that many decisions on the necessity of mandatory measures take place in a grey zone (Board of German Judges 2000). On the subject of involuntary treatment, there are two ethical considerations justifying compulsory measures under certain conditions. Being part of a teleological ethical approach, the *utilitarism* proposes to act in favour of the greatest possible benefit. According to this theory, the moral virtue of a given action is determined by the benefit resulting from its consequences and might allow considerable compulsion. As a 'middle principle', the *principle of beneficence* is partially dependent for its content on how the concepts of 'the good' and 'goodness' are defined. Beneficence assumes that doctors have to care for those who are incapable of caring for themselves. Involuntary treatment is considered as necessary to ideally recover the incompetent patients' autonomy. Supporting these ideas, a recent study on the long-term outcome of involuntary admissions demonstrated that at least 40% of the affected clientele considered their original admission justified 1 year later (Priebe et al. 2009). However, beneficence is not seen as an ethical rule; for this it can't tell by itself what actions constitute doing good and avoiding bad. Objectors of arbitrary measures refer to the *liberitarism*, which stands for the right of existence without any external control. According to this perspective, the prevention of harm to others is the only condition justifying the exercise of mandatory measures (Milone 2001). In contrast to ethical considerations supporting involuntary measures, here, the question of being capable or not is of subordinate impact. The *principle of non-maleficence* (*primum non nocere*) may be regarded as a superordinate maxim according to which, in a given situation, it may be better to do nothing than to risk causing more harm than good. Superficially, this concept has a root in the Hippocratic Oath including the promise to 'abstain from doing harm'.

However, since harm and benefit might differ when comparing the immediate and later future, the ethical challenge remains for the individual case and therapy; it does not only apply to treatment of mental disorders but also e.g. to some cancer therapies.

14.2.2 Advance Directives

Advance directives, frequently referred to as *living wills*, deal with the kind and amount of medical care an individual wants in case of severe disease or incapacity of expressing its wishes.[2] They represent legal documents and anticipate decisions about end-of-life care. Typical issues to be addressed encompass the use of dialysis and breathing machines, tube feeding, organ or tissue donation, terminal and irreversible brain damage, dementing illness, and reanimation. A document naming a (health) care proxy is called *durable power of attorney*; the proxy is enabled to make health relevant decisions in case the authorising person is unable to do so. Without any advance directives, treatment decisions may rely on decision makers, such as courts, judges, physicians, who might not have the required knowledge for acting in accordance to the patients' wishes. New technology might help to find better solutions in the future. In a study, a computer- and population-based treatment indicator that predicts which treatment a given patient would prefer based on the treatment preferences of a peer group has been proposed to predict patient preferences more accurate as patient-appointed surrogates (Varma and Wendler 2007). On the other hand, there is evidence that living wills are frequently poorly understood by patients and may not be given in a state of optimal information concerning e.g. intensive care or nursing home admission. The usefulness of these documents in the clinical context is limited (Manthous 2003).

To give valid advance directives, it is essential, that the concerning person has a basic understanding of a living wills' nature (see above). For this, any abnormal state of mind, which may have an impact on this understanding and the wills' potential consequences, has to be excluded. In this context it could be demonstrated, for example, that depression may change the preferences for life sustaining therapies (Blank et al. 2001, Ganzini et al. 1994, Rosenblatt and Block 2001). As depression has a high prevalence and remains frequently unrecognised, many living wills might be of questionable validity (Charney et al. 2003). The person should also realise the fundamental problem that it is basically impossible to have an overview on any clinical situation, which may arise in future. Last but not least the likelihood of a living will remaining in force in a situation the individual has not anticipated is low, when the will stands against the patients' *presumed interest*. When expressing a living will, it is recommended to consult a physician, who might serve as a witness to confirm competence for the time the will was shaped.

---

[2] See Chapter 10.

14.2.3 Testamentary Competence

The elderly are in possession of considerable asset values. In Germany, for example, during the years 2000 and 2010 assets of an estimated total of 2.25 trillion € will have been transmitted to the subsequent generations. For this, testamentary competency deserves thorough attention. It generally refers to an individuals' capacity to make a valid will. Its legal requirement is usually present in the national civil codes. To prove his capacity, traditionally, the testator has to be able to (1) know that he or she is going to make a will, i.e., to have a basic understanding of a wills' nature, (2) to have knowledge about the legal heirs, (3) to be aware of the extent of property to make decisions on, and (4) to know how the property is going to be divided (Blau 1998). Regarding a will-making process there are usually two distinct situations possibly requiring a specialists' statement on the willmakers' testamentary capacity. One is before the solicitor is taking instructions from the willmaker (testator is still alive); ideally, the solicitor would therefore ensure the testators' testamentary competency at the time the will is recorded. However, there is no operationalisation available for the legal expert how to assess testamentary competence (Stoppe and Lichtenwimmer 2005). The more common second situation is after the willmakers' death and prior the execution of a will; here, an expert will have to retrospectively determine whether or not the testator, at the time of making the will, was suffering from mental disorder or had a medical condition, which might have affected the validity of a will. To determine, whether a testator lacks competence for the task of making a will, a semi-structured interview might be useful. In addition, the expert assessor should be aware of hints for undue influence (see later).

Many other legal transactions like *marriage or divorce* require a person's capacity to contract which is defined in the corresponding civil codes. In some legal systems, incapacity to contract may be partial, i.e. related to a distinct legal transaction (e.g. partial incapacity to contract might apply in an otherwise healthy person, that is subject to serious delusional convictions related to his or her potential marriage partner).

14.2.4 Driving Competence

In the industrialised world, the ability of driving a car is considered as a matter of course and as indispensable for social mobility. Any driving related constrictions may result in a considerable (subjective) loss in quality of life. Nevertheless, driving requires driving ability. The latter may be narrowed by age-related decline in driving performance, by somatic or mental disorders as well as by medication and the influence of alcohol or illegal drugs. Deficits regarding distinct requirements may be compensated by an adapted driving style in case the driver disposes of a sufficient amount of self-criticism and insight into disease. Likewise, both may be reduced under certain conditions, like for example personality disorder, psychosis or dementia. If the correspondence to reality or the behaviour control is decreased due to severe mental illness, driving ability is not given (Römer and Dittmann 2007).

Although many countries introduced obligatory health checks in the elderly, old-aged drivers play a subordinate role in causing road accidents; in fact they are at greater risk of being victims in traffic (Cerelli 1989). Accordingly, the European Council Directive 91/439/EEC constituting the rationale for national criteria in view of medical fitness supposed to be required for driving, is not age-related and primarily focuses on somatic disorders (White and O'Neill 2000).

14.3 Factors Affecting Competence

A persons' capacity to give informed consent, to make advance directives, a valid will, or to drive a motor vehicle deserves a complex set of comprehension, encoding, information processing, decision making, and communication abilities. These premises are subject to an explicit neurobiological basis which might be affected by physical and mental clinical conditions (Marson and Harrell 1999). It is important to state, however, that factors like old age, physical disease, psychosis or failing memory do not account for a lack of competence by themselves. Competence has to be considered as a state rather than a trait variable, as lucid intervals even may arise in irreversible types of dementia. However, these lucid intervals seem to be overestimated in legal practice. When in doubt, for example, whether a will might be valid or not, it has to be focused on the point in time when the will was expressed. If a person is subjected to multiple impairing conditions (e.g. mild dementia and loss of hearing), the resulting extent of potential incompetence will not necessarily increase in an 'additive' way. Only those restrictions with impact on a distinct competence issue will have to be considered (e.g. loss of hearing usually will not impair a person's ability to make a living will; a coincident dementia, however, might be reason for testamentary capacity assessment).

14.3.1 Physical Factors

Physical factors mainly concern brain dysfunction as well as the influence of certain medication on the persons' ability to reason clearly. The probably most important physical condition affecting competence is dementia. It is characterised by a non-specific progressive decline of cognitive function beyond normal aging, typically affecting memory, attention, language, and executive functions like problem solving. Depending upon the aetiology, dementia syndromes can be more or less reversible (less than 10%) or irreversible (more than 90%). However, disease modification is possible yet. Whereas primary dementia like Alzheimer's disease (AD) origins from the brain itself, other (non-cerebral) physical disease or exogenous influences like drugs and their side effects may result in secondary dementia. In most cases, dementia syndromes go along with behavioural and psychological symptoms (BPSD) which add to the risk of competence impairment (Stoppe 2008).

Patients suffering from AD have been demonstrated to have an impaired performance on the measures of decisional capacity, particularly with respect of understanding disclosed information (Palmer et al. 2009). Remarkably, with respect to participation in research, around 70% of patients eligible for randomised controlled trials in mild to moderately severe AD were actually incompetent or marginally competent (Pucci et al. 2001). These results were recently replicated in another setting with regard to legal criteria (Warner et al. 2008). The perception of information as a basis for reasoning and coming to a decision seems to be impaired also in patients with mild cognitive impairment (MCI). Patients with the amnestic type of MCI showed a progressive decline in the ability to understand consent information over a 3-years period, which accelerated after conversion to AD (Okonkwo et al. 2008). In view of AD patients' capacity to drive, study data are inconsistent. Trobe et al. (1996) could demonstrate that 49% of 143 investigated AD patients had stopped driving before diagnosis was made. Their mean remaining driving time after diagnosis was 1.6 ± 1.3 years, so that the problem of demented people driving may be overestimated.

Longitudinal studies in patients with idiopathic Parkinsons' disease (PD) show that up to 75% may develop secondary dementia. Even more, a study on the decision-making capacity of PD patients unravelled decision-making deficits in the non-demented but cognitively impaired PD subgroup. It could be demonstrated, that executive functions and – to a lower extent – memory and orientation were key predictors of performance (Dymek et al. 2001). Like MCI and AD, this impairment increases, when PD patient develop overt dementia (Martin et al. 2008).

Structural brain lesions may also result from stroke, subacute ischemic lesions, primary cancer or metastasis, infectious abscesses as well as autoimmune processes (e.g. multiple sclerosis). Especially when affecting frontal system structures, deficits regarding executive functioning are likely, whereas temporomesial lesions are prone to memory impairment. Moreover, some metabolic states like hypoglycaemia, hypoperfusion, or hypoxia can lead to a *functional* brain disruption.

Although related literature is rare, decision making capacity in patients with epilepsy has been shown to depend on frequency, type, predictability and consequences of seizures, on the type and effectiveness of protective medication as well as on additional disabilities. Older adults with chronic partial epilepsy have been demonstrated to perform significantly below controls with regard to evidencing choice, appreciation and understanding, all of them being assessed by the Capacity to Consent to Treatment Instrument.[3] The patients' performance with regard to understanding is related to the number of antiepileptic drugs, the duration of epilepsy and the age of seizure onset (Bambara et al. 2007).

Although the association between medical illness and the capacity to consent has not been systematically studied, there is some evidence that medical disorders like diabetes, hypertension, respiratory and cardiac diseases, renal failure, neoplasia, and HIV come along with an impairment of executive functions. Deficits in patients with diabetes or hypertension tend to persist when covarying for depression and alcohol use pointing to an inherent effect of the medical condition (Schillerstrom et al. 2005). With respect to driving capacity, a permanent diastolic blood pressure

---

[3] See Section 14.5.

of more than 130 mm Hg augments the risk of retina and cerebral bleeding as well as myocardial failure. Those patients are therefore not eligible to drive a vehicle (Römer and Dittmann 2007).

Myopia, hyperopia, amblyopia, retinal degeneration, albinism, cataract, and glaucoma are frequent causes of visual impairment, whereas presbyacusis, ear canal obstruction, abnormalities of the middle ear, and otosclerosis will result in hearing impairment. Although not directly interfering with the capacity to reason, both conditions might prevent a person from understanding relevant information; the latter, however, is indispensable to come to a valid decision. Regarding the capacity to drive, it is plausible, that aural and visual senses have to meet minimal standards. In Switzerland, for example, a best corrected vision of at least 0.6 regarding the better eye as well as 0.1 with respect to the worse eye is mandatory; moreover, a field of sight of 140° horizontally is required and double images have to be excluded. Deafness combined with monophthalmia is a criterion for exclusion of roadworthiness.

There are several groups of drugs which can cause memory disturbances and other cognitive impairment. Even over-the-counter medication, e.g. some antihistamines, may make a negative impact on cerebral functioning. Anticholinergic side effects can reduce a persons' ability to pay attention and to keep something in mind. They have to be considered when treating with tricyclic antidepressants (e.g. amitriptyline, trimipramine, clomipramine), some antiparkinson agents (e.g. benzatropine and biperidine), antitussives (e.g. codeine), first generation antihistamines (e.g. diphenhydramine), antispasmodics (hyoscine and oxybutynine), antinauseants (e.g. scopolamine), and some antiarrhythmics (e.g. disopyramide). Moreover, benzodiazepines, opioids, and alcohol affect arousal, attention, and episodic memory; chronic alcohol abuse may result in an amnestic syndrome (Korsakows' syndrome) with prominent deficiencies of episodic memory, executive functioning and judgement skills. Depending on the dosage, traffic accident hazard is 1.5- to 6.5-fold elevated when driving under the influence of benzodiazepines. This effect is even stronger in elderly patients; however, the risk varies considerably interindividually (Römer and Dittmann 2007). By means of blocking frontostriatal dopaminergic neurons, especially first generation antipsychotics may have a negative impact on attention and executive functioning (O'Luanaigh and Lawlor 2008).

14.3.2 Mental Factors

A study addressing the proportion of psychiatric in-patients who lack capacity to make treatment decisions unravelled that 43.8% lacked treatment-related decisional capacity when employing the MacArthur Competence Tool for Treatment.[4] Mania, psychosis, poor insight, delusions, black and minority ethnic group were associated with mental incapacity (Cairns et al. 2005). It is important, however, to recognize, that many mental disorders do not impact a persons' capacity to validly consent or

---

[4] See Section 14.5.

to express a will in general (e.g. somatic delusions, which do not affect a testators' capacity to know, that he or she is going to make a will, to have knowledge about his or hers legal heirs, to be aware of the extent of property, and to know how the property is going to be divided). Even sufferers from severe mental illnesses, like for example schizophrenia, not necessarily show (irreversible) impairment. Judging the competence to consent requires considering the mental condition of the specific patient at a distinct point in time and with regard to a specific area of consent, for example antipsychotic drug treatment.

Regarding their ability to consent for treatment, as a group, patients with dementia show significantly more often impaired performance than those with schizophrenia; the latter are more impaired than depressed patients (Vollmann et al. 2003). Patients with schizophrenia and severe depression, however, demonstrate executive limitations, which are similar to those suffering from dementia (Hanes et al. 1996); especially schizophrenic patients tend to have worse understanding of disclosed material, appreciating, and reasoning when compared to healthy subjects. In many cases, a reduced capacity can be compensated by an educational intervention as part of the informed consent process (Carpenter et al. 2000). Outpatients with major depression, however, have been shown to perform well on capacity measures and to maintain their level of performance over time. Regarding consent to research, they displayed no correlation between performance and severity of depressive symptoms (Appelbaum et al. 1999). However, treatment decisions and advance directives might be influenced by depression related pessimism and hopelessness (see above).

To decide whether decision-making capacity is existent in psychotic or manic episodes of bipolar affective disorder, the extent of insight, i.e. the ability to recognize one's own mental illness, might be helpful. According to a recent study applying the MacArthur Competence Assessment Tool for Treatment,[5] in such patients, insight was demonstrated to be a reliable indicator of their capacity status. In non-psychotic patients, however, it was depressed mood, which discriminated capacity status best. Interestingly, cognitive performance did not predict capacity status between psychotic patients (Owen et al. 2008). With respect to driving competence, acute and severe psychosis leading to an altered recognition of reality and/or to an impaired behaviour control gives rise to driving incapacity. After remission, an observation period for 1 year is recommended until re-granting a driving permit; it should include regular medication (Römer and Dittmann 2007).

14.4 Problem of Undue Influence

The problem of undue influence is especially relevant for testamentary capacity. Although being nearly undebated, it is also of importance in the areas of informed consent to diagnosis, therapy or research.

---

[5] See Section 14.5.

Undue influence is a legal construct, and is defined by the courts according to the respective national jurisdiction. The clinicians' role is to advise the court about a person's vulnerability to undue influence, but the determination that undue influence has actually occurred is a decision that remains with the court. If undue influence is found by the court to have been present at the time instructions for a will were given or when it was executed, it has the effect of rendering a will invalid. Undue influence is given, when coercion is proven. However, from a clinical perspective, the concept of 'subversion of will' is a more useful term that allows for influence to be defined relative to the vulnerability of the testator. Therefore, in a cognitively or emotionally vulnerable individual a less 'coercive' influence could still be determined to be 'undue' (Peisah et al. 2008, Shulman et al. 2007). This broadening of the concept of 'undue influence' also allows discussing the effects of stigma, ageism and altruistic attitudes (Sartorius and Schulze 2005, Tornstam 2005; see Chapter 2). It is plausible that stereotypes held by the expert assessor, the societies, the courts and even the patients themselves, influence the decisions for e.g. life sustaining therapies or the expression of a will.

In order to help the individual to make a valid will, many law systems recommend to give a will or to appoint an attorney before a quasi judicial officer or even a court. The presumption of capacity afforded by this process is said to render wills 'practically immune' to later or post-mortem challenge on grounds of incapacity or undue influence. However, there is nearly no guideline or training for assessment of capacity available for these legal persons/notaries (see above, Stoppe and Lichtenwimmer 2005). This might be one reason why the courts – at least in many European countries – rely on experts' opinions.

Nevertheless, some rules are available, e.g. to apply (Shoet 2001 cited according to Peisah et al. 2008):

- the test of independence – the determination of the extent to which the testator was physically and, particularly, mentally independent
- the test of aid – the determination of the extent to which the beneficiary aided the testator
- the test of relationships – the determination of the nature and strength of relationships between the testator and persons other than the beneficiary
- the test of circumstances under which the will was executed

For the expert assessor as well as the court it is important to be sensible for risk factors or 'red flags'. From a person-oriented point of view, their presence should cause the assessing psychiatric professional to reason about the concerning persons' vulnerability. It is up to a court, however, to assess whether undue influence is actually present or not (situation-oriented point of view).

14.4.1 Social Environment of the Testator

Relationships that create opportunity for undue influence are diverse and occur in many social situations. Situations of adverse influence usually occur between a mentally ill person and e.g. a cohabiting family member such as an adult child, a helpful neighbour or friend or a formal or informal carer.

Certain circumstances may predispose to undue influence. Among these are sequestration and isolation of the impaired person such that outside contact is inhibited, a family conflict or the physical and/or psychological dependency on a carer. Changes in documents such as wills may be made in a desperate attempt to garner care, support or comfort at a time when the impaired person feels increasingly vulnerable or threatened (Shulman et al. 2007).

14.4.2 Factors of Potential Influence

The physical, psychological and mental factors which have been explained to affect competence also increase the vulnerability of a testator regarding undue influence. Individuals with substance abuse may be susceptible to undue influence by virtue of the effects of dependency or substance-induced neurotoxicity and accompanying cognitive as well as emotional changes. For example, with regards to alcohol abuse, the compounded effects of self or other-enforced isolation, a dependent relationship between friends or relatives who supply alcohol to the person who craves it, and cognitive changes such as retrograde amnesia and impaired planning and judgement, may all render vulnerability to undue influence. As already pointed out, a range of mental disorders such as delirium, dementia, chronic schizophrenia, paranoid and mood disorders might affect the capacity for driving, give informed consent or to formulate a will. In a similar way, they may predispose a person to undue influence, e.g. when a testator suffers paranoid ideation, his or hers suspiciousness of potential beneficiaries may be fuelled by an 'influencer' (Peisah et al. 2006, 2008). Being incompetent by some means or other, however, is neither *sine qua non* nor a sufficient condition to assume a person being predisposed to undue influence.

Cognitive disorders such as intellectual disability (i.e., developmental disability or mental retardation) and dementia, may render a person vulnerable to the influence of others (Shulman et al. 2007) suggested a threshold concept. The severity of cognitive impairment affects vulnerability of a person to undue influence. As the disease progresses to more advanced dementia, the person would be more susceptible to undue influence and even subtle influences might be considered undue. Also the pattern of progression is of importance with a person affected by frontal lobe deficits being earlier impaired than a typical AD patient.

14.4.3 Context of Capacity Assessment or Making the Will

Circumstances surrounding the capacity assessment or the making of the will may be considered to arouse suspicion that undue influence might have occurred or may be occurring. Examples for this are if a person of influence instigates an assessment or a change to a will (e.g. finds the lawyer) or a radical change in the pattern of distribution of the assets occurred recently.

14.5 Structured Assessments of Competence

Throughout the literature, measurement of decision-making abilities and decisional competence is highly heterogeneous (Kim et al. 2002). Research on decisional capacity has become boosted by the availability of structured assessments allowing the reliable measurement of abilities related to decision making. The use of operationalised standards has been shown to classify substantially more patients impaired than by clinical assessment (67.7% vs. 48.4% of patients with dementia, 20.0% vs. 2.9% of patients with depression, 53.5% vs. 18.4% of patients with schizophrenia) (Vollmann et al. 2003). In view of this discrepancies the question arises which is the legitimate method to ascertain decisional capacity. The answer to this question, however, highly depends from the ethical assumptions under which structured assessments of competence have been constructed. The below mentioned MacArthur Competence Assessment Tool for Treatment, for example, is based on the ethical *principle of autonomy*, derived from Kantian ethics. An important way of respecting patients' autonomy is to recognize their right to self-determination. In the medical context this implies the right to decide what is and is not done to their body. In other words, any therapeutic intervention basically requires the patient's informed consent; the latter, however, presupposes capacity to consent (understanding of disorder and treatment; reasoning; appreciation of disorder and treatment). Alternative ethical groundings of capacity to consent are currently discussed, in part deriving from narrative, hermeneutic, or care ethics (Berghmans et al. 2004).

To allow for best valid results, any assessment should be repeated over a period of time. A reassessment of competence should take place in case of a marked change with respect to the mental state (e.g. when recovering from severe depression) or the impairing physical condition. Competence should also be reassessed when the subject to be decided on changes or when the patient retracts on a former decision without being able to give any rational motives.

The *University of California, San Diego Brief Assessment of Capacity to Consent (UBACC)* is a relatively new instrument to administer decision-making capacity in view of participating in research (Jeste et al. 2007). It is a 10-item scale with questions focusing on the understanding and appreciation of the information given in a research protocol. The UBACC has been shown to have good internal consistency, interrater reliability, concurrent validity, high sensitivity, and acceptable specificity. It typically takes about 5 min to administer and does not make great demands on the assessor; however, it primarily has to be regarded as a screening instrument for larger populations.

The *MacArthur Competence Assessment Tool for Treatment (MacCAT-T)* has been established as a 'gold standard' questionnaire for the evaluation of capacity to consent to treatment (Grisso et al. 1997). Based on the same principles, the *MacArthur Competence Assessment Tool for Clinical Research (MacCAT-CR)* (Appelbaum and Grisso 2001) is a questionnaire to evaluate decision-making capacity. By means of 21 items, each scored on a 3-point scale from 0 to 2 (inadequate – partial – adequate), performance with respect to the four legal standards of competence is measured. Typically, this semi-structured interview can be administered in 15–20 min and may be adapted to the attendants of any given research project. Subjects' responses may be quantified allowing for comparisons across subjects and any given subgroup of subjects. For this, the MacCAT-CR has been proposed to be employed for conducting research on the characteristics of subject populations and for assessing the effectiveness of interventions designed to increase subjects' capacities. As a result of the contextual nature of the MacCAT-CR, no cut-scores for the categorical determinations of capacity or incapacity exist. It is therefore not astonishing, that the usually employed practice only to use the understanding subscale as a basis for including a patient into a study may lead to unfunded conclusions (Dunn et al. 2007). The MacCAT-CR has been demonstrated to have excellent content validity as well as good inter-interviewer and interscorer reliabilities. Other questionnaires derived from the MacCAT format have been developed (measures for assessing a persons' capacity to complete a health care proxy, to complete a psychiatric advance directive, and to make decisions about continuing to drive).

The *Aid to Capacity Evaluation* (*ACE*) has been developed as a semi-structured decisional aid prompting inquiries into seven relevant areas (ability to understand the medical problem, the proposed treatment, the alternatives to proposed treatment, the option of refusing treatment, to appreciate the reasonably foreseeable consequences of accepting and refusing treatment, and to make a decision that is not substantially based on hallucinations, delusions, or cognitive signs of depression) (Etchells et al. 1999). For each area, the tool suggests questions as well as guidelines to rate patients' answers; it also delivers scoring examples. In each area the investigator gives a rating of yes, unsure, or no and chooses one of four overall assessments (definitely incapable, probably incapable, probably capable, or definitely capable). It could be shown, that ACE scores agreed closely with expert assessments of patient capacity to consent to treatment.

Taking on average 20 min to complete, the *Structured Interview for Competency Incompetency Assessment Testing and Ranking Inventory* (*SICIATRI*) (Kitamura and Kitamura 1993) has been developed to cover the three dimensions understanding, appreciation, and decision making in order to test for the capacity of patients to give informed consent. It is a 12-item structured interview covering on several aspects of a persons' capacity to consent. By use of a ranking inventory for competency, the latter is categorised in one out five competence levels based on the cognitive performance regarding the 12 items. The SICIATRI has reached a relatively high inter-rater reliability (more than 50% of the items had kappa $>= 0.60$) and useable validity (sensitivity $= 0.83$, specifity $= 0.67$ when compared to the global judgement of competency rating by the attending physician) (Tomoda et al. 1997).

The *Hopkins Competency Assessment Test* (*HCAT*) (Janofsky et al. 1992), which takes about 10 min to be administered, is a brief instrument to evaluate a persons' ability to give informed consent or advance directives on basis of its capability to understand. The patient is requested to read a four paragraph text about the process of giving informed consent and advanced directives. Simultaneously, the essay, which is available on three different language levels, is read aloud by the investigator, who may also be a non-clinician. Afterwards the person to be assessed is asked 10 questions referring to the text and may obtain one point for every right answer (score range: 0–10 points). When comparing test results with those derived from a detailed forensic psychiatric interview, Janofsky et al. found a cut-off score of less than 4 points to decide for incapacity (sensitivity and specificity 100%). Remaining unclear what was meant by competency as judged by clinical staff performing the psychiatric interviews, however, the HCAT has been criticised for its insufficient operationalisation of the term competence (Grisso 2001).

Although not measuring decision-making capacity itself, the *Mini Mental State Examination (MMSE)* may add a time and cost efficient value to the capacity evaluation process in elderly persons (Kim et al. 2002). A cut-off <= 23 had been shown to identify decisional impairment with a sensitivity of 53–65% and a specifity of 80% in a nursing home population (Fitten et al. 1990). In higher functioning elderly, a sensitivity of 100% and a specifity of 97% were reached (Krynski et al. 1994). Nevertheless, Warner et al. found the MMSE not suitable to predict capacity to consent in a study with mild to moderate AD patients (2008).

So far, to our knowledge, there is no structured assessment for evaluating testamentary capacity meeting scientific criteria. As a rough guideline, however, the Australian Medical Association has proposed to apply the following 'semi-structured' approach in order to evaluate a persons' testamentary capability (AMA Victoria 2009). The testator should be able

- to explain the effect of a will and to ask whether there is knowledge about what would happen to his or hers property if no will was made
- to give a general estimate of the property and its value
- to describe the reasoning behind his or hers decision to include or exclude potential heirs
- state, whether he or her has knowledge that a will revokes all previous wills

Any answers should be documented and it is recommended that the interview should take place more than one time. For the evaluation of general financial competency in AD patients, Bassett proposed a 5-item financial competence questionnaire with a cut-off >= 3 right answers. The obtained score significantly correlated with patients' medical decision-making competency (Bassett 1999).

14.6 Conclusion

The assessment of competence in persons suffering from psychical and mental disorders means an increasing challenge for the medical and legal systems. The strengthening of individual and patient rights in most European legislations makes the persons' condition even more important. In addition, even in the elderly generation the relative number of single households increases and family proxies are not 'automatically' available. Due to their absence, a persons' potential incapacity remains 'uncompensated' and ideally should be administered even more resolute. At last, migration and its cultural interface is a still marginalised topic in psychiatry as well as in the special area of competence assessment.

Competence assessments should differ with regard to their purpose. Main areas are informed consent, advance directives and living wills, testaments and the licensing for driving or weapons. Competence may be affected by many conditions either alone or in combination. For example a patient with dementia might also suffer from vision impairment and comorbid depression. Treatment, especially drug treatment, must also be considered. Undue influence may affect the results of e.g. testamentary decisions much more in the vulnerable psychiatric population.

The determination of competence in a given patient at a distinct point in time and with regard to a specific domain still is a medical task. The assessing physician, e.g. psychiatrist, should not rely on his or hers clinical experience alone (low interrater reliability!) but rather employ one or more structured tools for the assessment of competence. However, although the latter clearly help the assessment process, they also cannot be relied on alone. Ideally, an assessment should take place each time a decision is made. No diagnosis or treatment allows the assumption of global lack of competence per se.

Finally, the competence of experts for assessing competence is not sufficiently defined yet. This applies to the medical, as well as to the psychological and legal experts. There is also a need for more consensus – on basic requirements and training.

References

AMA Victoria (Australian Medical Association) (2009) http://www.amavic.com.au/page/Media/Whats_New/Assessing_testamentary_capacity/. Accessed 18 Jan 2009

Appelbaum PS, Grisso T (1988) Assessing patients' capacities to consent to treatment. N Engl J Med 319:1635–1638

Appelbaum PS, Grisso T (2001) MacCAT-CR: MacArthur competence assessment tool for clinical research. Professional Resource Press, Sarasota

Appelbaum PS, Grisso T, Frank E et al (1999) Competence of depressed patients for consent to research. Am J Psychiatry 156:1380–1384

Arie T (1996) Caring for older people: some legal aspects of mental capacity. BMJ 313:156–158

Bambara JK, Griffith HR, Martin RC et al (2007) Medical decision-making abilities in older adults with chronic partial epilepsy. Epilepsy Behav 10:63–68

Bassett SS (1999) Attention: neuropsychological predictor of competency in Alzheimer's disease. J Geriatr Psychiatry Neurol 12:200–205

Berghmans R, Dickenson D, Ter Meulen R (2004) Mental capacity: in search of alternative perspectives. Health Care Anal 12:251–263

Blank K, Robinson J, Doherty E et al (2001) Life-sustaining treatment and assisted death choices in depressed older patients. J Am Geriatr Soc 49:153–161

Blau TH (1998) The psychologist as expert witness. Wiley, New York, NY

Board of German Judges (2000) Stellungnahme des Deutschen Richterbundes zum White Paper über den Schutz der Menschenrechte und der Würde von Menschen, die an einer Geistesstörung leiden [...]. http://www.drb.de/cms/index.php?id=400. Accessed 10 Jan 2009

Cairns R, Maddock C, Buchanan A (2005) Prevalence and predictors of mental incapacity in psychiatric in-patients. Br J Psychiatry 187:379–385

Carpenter WT, Gold JM, Lathi AC et al (2000) Decisional capacity for informed consent in schizophrenia research. Arch Gen Psychiatry 57:533–538

Cerelli E (1989) Older drivers: the age factor in traffic safety. DOTHS 807 402, NHTSA Technical Report, Feb 1989, Washington, DC

Charney DS, Reynolds CF, Lewis L et al (2003) Depression and bipolar support alliance consensus statement on the unmet needs in diagnosis and treatment of mood disorders in late life. Arch Gen Psychiatry 60:664–672

Dunn LB, Palmer BW, Appelbaum PS (2007) Prevalence and correlates of adequate performance on a measure of abilities related to decisional capacity: differences among three standards for the MacCAT-CR in patients with schizophrenia. Schizophr Res 89:110–118

Dymek MP, Atchison P, Harrell L et al (2001) Competency to consent to medical treatment in cognitively impaired patients with Parkinson's disease. Neurology 56:17–24

Etchells E, Darzins P, Silberfeld M (1999) Assessment of patient capacity to consent to treatment. JGIM 14:27–34

Fitten LJ, Lusky R, Hamann C (1990) Assessing treatment decision-making capacity in elderly nursing home residents. J Am Geriatr Soc 38:1097–1104

Fortun P, West J, Chalkley L et al (2008) Recall of informed consent information by healthy volunteers in clinical trials. QJM 101:625–629

Ganzini L, Lee MA, Heintz RT et al (1994) The effect of depression treatment on elderly patients' preferences for life-sustaining medical therapy. Am J Psychiatry 151:1631–1636

Grisso T (2001) Evaluating competencies: forensic assessments and instruments, 2nd edn. Springer, New York, NY

Grisso T, Appelbaum PS, Hill-Fotouhi C (1997) The MacCAT-T: a clinical tool to assess patients' capacities to make treatment decisions. Psychiatr Serv 48:1415–1419

Hanes KR, Andrewes DG, Pantelis C (1996) Subcortical dysfunction in schizophrenia: a comparison with Parkinson's disease and Huntington's disease. Schizophr Res 19:121–128

Helmchen H (2008) Ethical questions in clinical research with the mentally ill. Nervenarzt 79:1036–1050

Helmchen H, Lauter H (1995) Dürfen Ärzte mit Demenzkranken forschen? Forschungsbedarf und Einwilligungsproblematik bei Demenz. Thieme, Stuttgart

Janofsky JS, McCarthy RJ, Folstein MF (1992) The Hopkins competency assessment test: a brief method for evaluating patients' capacity to give informed consent. Hosp Community Psychiatry 43:132–136

Jeste DV, Palmer BW, Appelbaum PS et al (2007) A new brief instrument for assessing decisional capacity for clinical research. Arch Gen Psychiatry 64:966–974

Kim SYH, Karlawish JHT, Caine ED (2002) Current state of research on decision-making competence of cognitively impaired elderly persons. Am J Geriatr Psychiatry 10:151–165

Kitamura T, Kitamura F (1993) Structured interview for competency/incompetency assessment testing and ranking inventory (SICIATRI). Department of Sociocultural Environmental Research, National Institute of Mental Health, NCNP, Ichikawa

Krynski MD, Tymchuk AJ, Ouslander JG (1994) How informed can consent be? New light on comprehension among elderly people making decisions about enteral tube feeding. Gerontologist 34:36–43

Manthous CA (2003) Are living wills useful? In search of a new paradigm. Conn Med 67:283–290

Marson DC, Earnst KS, Jamil F et al (2000) Consistency of physicians' legal standard and personal judgements of competency in patients with Alzheimer's disease. J Am Geriatr Soc 48:911–918

Marson D, Harrell L (1999) Executive dysfunction and loss of capacity to consent to medical treatment in patients with Alzheimer's disease. Semin Clin Neuropsychiatry 4:41–49

Martin RC, Okonkwo OC, Hill J et al (2008) Medical decision-making capacity in cognitively impaired Parkinson's disease patients without dementia. Mov Disord 23:1867–1874

Milone RD (2001) Involuntary hospitalisation. In: Ethics Primer of the American Psychiatric Association, American Psychiatric Press, Arlington, VA

Okonkwo OC, Griffith HR, Copeland JN et al (2008) Medical decision-making capacity in mild cognitive impairment: a 3-year longitudinal study. Neurology 71:1474–1480

Olde Rikkert MGM, Lauque S, Frölich L et al (2005) The practice of obtaining approval from medical research ethics committees: a comparison within 12 European countries for a descriptive study on acetylcholinesterase inhibitors in Alzheimer's dementia. Eur J Neurol 12:212–217

O'Luanaigh C, Lawlor B (2008) Drugs that affect competence. In: Stoppe G (ed) Competence assessment in dementia. Springer, Wien, pp 41–50

Owen GS, David AS, Richardson G et al (2008) Mental capacity, diagnosis and insight in psychiatric in-patients: a cross-sectional study. Psychol Med 22:1–10

Palmer BW, Dunn LB, Appelbaum PS et al (2009) Assessment of capacity to consent to research among older persons with schizophrenia, Alzheimer disease, or diabetes mellitus. Arch Gen Psychiatry 62:726–733

Peisah C, Brodaty H, Quadrio C (2006) Family conflict in dementia: prodigal sons & black sheep. Int J Geriatr Psychiatry 21:485–492

Peisah C, Finkel S, Shulman K et al, for the International Psychogeriatric Association Task Force on Wills and Undue Influence (2008) The wills of older people: risk factors for undue influence (review). Int Psychogeriatr 21(1):7–15. Dec 1:1–9

Priebe S, Katsakou C, Amos T et al (2009) Patients' views and readmissions 1 year after involuntary hospitalisation. Br J Psychiatry 194:49–54

Pucci E, Belardinelli N, Borsetti G et al (2001) Information and competency for consent to pharmacologic clinical trials in Alzheimer disease: an empirical analysis in patients and family caregivers. Alzheimer Dis Assoc Disord 15:146–154

Römer KD, Dittmann V (2007) Fahren im Alter. Die Psychiatrie 4:50–56

Rosenblatt L, Block SD (2001) Depression, decision making, and the cessation of life-sustaining treatment. West J Med 175:320–325

Sartorius N, Schulze H, Global Programme of the World Psychiatric Association (2005) Reducing the stigma of mental illness: a report from a global programme of the World Psychiatric Association. Cambridge University Press, Cambridge

Schillerstrom JE, Horton MS, Royall DR (2005) The impact of medical illness on executive function. Psychosomatics 46:508–516

Shulman K, Cohen CA, Kirsh FC et al (2007) Assessment of testamentary capacity and vulnerability to undue influence. Am J Psychiatry 164:722–727

Stoppe G (ed) (2008) Competence assessment in dementia. Springer, Wien

Stoppe G, Lichtenwimmer A (2005) Die Feststellung der Geschäfts- und Testierfähigkeit beim alten Menschen durch den Notar – ein interdisziplinärer Vorschlag. D NotZ 11:806–813

Tomoda A, Yasumiya R, Sumiyama T et al (1997) Validity and reliability of structured interview for competency incompetency assessment testing and ranking inventory. J Clin Psychol 53:443–450

Tornstam L (2005) Gerotranscendence: a developmental theory of positive aging. Springer, Wien

Trobe JD, Waller PF, Cook-Flannagan CA et al (1996) Crashes and violations among drivers with Alzheimer disease. Arch Neurol 53:411–416

Varma S, Wendler D (2007) Medical decision making for patients without surrogates. Arch Intern Med 167:1711–1715

Vollmann J, Bauer A, Danker-Hopfe H et al (2003) Competence of mentally ill patients: a comparative empirical study. Psychol Med 33:1463–1471

Warner J, McCarney R, Griffin M et al (2008) Participation in research: rates and correlates of capacity to give informed consent. J Med Ethics 34:167–170

White S, O'Neill D (2000) Health and relicensing policies for older drivers in the European Union. Gerontology 46:146–152

World Medical Association (WMA) (2008) Declaration of Helsinki. http://www.wma.net/e/policy/pdf/17c.pdf. Accessed 4 Jan 2009

Chapter 15
General Overview of Ethical Issues in Psychiatric Treatment

Hanfried Helmchen

Contents

Abbreviations

APA	American Psychiatric Association
BfArM	Bundesinstitut für Arzneimittel und Medizinprodukte, Germany
DSM-IV	Diagnostic and Statistical Manual of the APA, 4th revision
EMEA	European Medicines Agency, Europe
FDA	Food and Drug Administration, USA
ICD-10	International Classification of Diseases of the WHO, 10th revision
ICH-GCP-Guideline	European Guideline for Good Clinical Practice, Europe
IQWIG	Institut für Qualitätssicherung und Wirtschaftlichkeit, Germany
NICE	National Institute for Health and Clinical Excellence, United Kingdom
TM	Therapeutic Misconception
WHO	World Health Organisation

H. Helmchen (✉)
Department of Psychiatry and Psychotherapy, CBF, Charitè – University Medicine Berlin,
D-14050 Berlin, Germany
e-mail: hanfried.helmchen@charite.de

H. Helmchen, N. Sartorius (eds.), *Ethics in Psychiatry*, International Library
of Ethics, Law, and the New Medicine 45, DOI 10.1007/978-90-481-8721-8_15,
© Springer Science+Business Media B.V. 2010

15.1 First Step: Establishing and Discussing the Benefit-Risk Ratio with the Patient *Before* Treatment

The psychiatrist has to evaluate the intended intervention according to the above-mentioned ethical principles. If he is convinced that the intervention corresponds to them, he must then obtain the patient's informed consent. *Respect for autonomy* requires improving the patient's understanding of the disorder (its causes, symptoms, course, and outcome) and of the benefits and risks of both the recommended and alternative therapeutic interventions as well as the elucidation of their meaning for the patient's personal values, especially his expectations towards the therapy with the consequence of an informed consent.

Taking the patient seriously means learning about his preferences and sharing the necessary decisions about interventions as far as possible with him (Hamann et al. 2006). This implies the obligation of the psychiatrist to make sure that the patient understands the information or, in case of doubt, to assess the patient's competence to consent (see Chapter 14). In case of an impaired capacity the legal requirements according to the national law must be utilized as a substitute for the consent by authorised persons. The valid assessment of competence to consent is ethically relevant, because an incorrect estimation either leads to an invalid consent and leaves the responsibility for decisions with an incompetent patient or else discriminates against a competent patient.

Most therapeutic interventions do not fulfil the therapeutic ideal (Ehrlich's 'therapia magna sterilisans'): to achieve immediate and complete elimination of the disorder in all patients with this disorder for which the therapy is indicated (specific effectiveness) with no unwanted side effects (safety). Usually some benefit can be obtained only at the cost of some more or less serious unwanted side effects. Therefore, the relationship between beneficence and non-maleficence must be evaluated and this *risk-benefit-estimation* should be imparted to the patient – but only in case the physician comes to the conclusion that this estimation ethically justifies the intervention, e.g. to expect stronger or faster effectiveness than that of to other existing treatments and/or lesser or only minimal risks and burdens for the patient.

15.2 Second Step: Assessment of Benefits and Risks *During* Treatment

After having obtained the informed consent of the patient and having started the intervention, a second step is treatment control. This means assessment of both efficacy and unwanted effects during the course of the intervention. In general, the relationship of therapeutic *effectiveness* (or efficacy) to its costs, both medically in terms of side effects and risks and today particularly economically in terms of financial burdens, is labelled as *efficiency*. The assessment of these different outcome aspects of therapeutic interventions has some implications (Helmchen 2001b).

In general the assessment of effectiveness implies presuppositions with regard to more or less complex theoretical constructs such as illness or needs or therapeutic aims (Winterer and Herrmann 1996). A behavioral model of mental illness will give reason for aiming psychotherapeutically at symptom relief whereas an intervention based on a psychodynamic model may be oriented towards a maturing development of personality, i.e. treatment follows either an adaptational or an emancipatory model of psychotherapy. A sociogenic model of mental illness may lead rather to social interventions with the aim of acceptance of the disorder or of the disordered subject by both the individual him/herself and society, whereas the medical model more likely aims at symptom relief by drug treatment. Furthermore, the aims of therapy may be different from the perspectives of the patient (e.g. well-being), of the therapist (e.g. welfare and best interest of the patient; maturation of personality or protecting the patient against himself), and of the insurance company (e.g. fitness for work) (Kottje-Birnbacher and Birnbacher 1998). Last but not least, these constructs depend on the contemporary context.

Hence, *assessment of effectiveness* of a therapeutic intervention means evaluating its ability to achieve these construct-related aims. This will be done either in general by research in order to establish the strength and specificity of the relatedness of the intervention to its aim, i.e. to find an evidence-based indication, or, in the individual case, by judging whether the indicated intervention actually did achieve this patient's aim. The strength of the specific relatedness of an intervention to a defined aim may be a continuous variable, but will be categorized by cut-offs to be usable as an indicator of therapeutic effectiveness, e.g. the reduction of a symptom score below a defined threshold. Here, again, different standards are in use: evidence of efficacy may already be satisfied by a symptom reduction of 50%, whereas patients will admit only a complete relief of symptoms to be therapeutic effectiveness. More difficult will be the assessment of more complex outcome criteria such as for example the improvement of quality of life. Therefore, universally accepted aims for therapeutic interventions do not exist, and 'there are no universally accepted standards of efficacy and safety' (Kane and Borenstein 1997).

This seems to be even more valid for the *assessment of efficiency* because this has to deal with much more complexity: a divorce as a side effect of strengthening the autonomy by psychotherapy may be acceptable for one patient but not for the other (notwithstanding the problems of informed consent with regard to such a potential side effect); tardive dyskinesia as a potentially irreversible side effect of a neuroleptic long-term medication urgently needed for suppression of agonizing schizophrenic symptoms will be accepted by one patient but not by another. Furthermore, although from the medical point of view in the latter case the application of another drug with the same antipsychotic effectiveness but without the risk of this unwanted effect is undoubtedly indicated, to be efficient with regard to a limited budget the doctor is nevertheless urged to decide in favor of a less expensive drug; if he prescribes the needed best but expensive drug, he perhaps cannot prescribe drugs urgently needed by other patients. However, with regard to subsequent burdens to society, it could be much more efficient to choose a cost-intensive treatment with fewer unwanted effects on adherence to treatment even for help-seeking people with low or subthreshold morbidity (morbid states below the threshold of officially used operationalised diagnostic systems such as the Diagnostic and Statistical Manual of the American Psychiatric Association, 4th revision (DSM-IV) or International Classification of Diseases of WHO, 10th revision (ICD-10) than to withhold therapeutic interventions from them (Helmchen 2001a).

Such ethical considerations are valid for the *application* of standard therapies in everyday practice as well as for the *development* of new therapeutic interventions through research. The methods of assessment and their ethical status may be more elaborated in drug treatment, but nevertheless they are also valid for other biological, e.g. neurostimulative treatments, or psychotherapeutic or social interventions, e.g. preventive or coercive measures or the deinstitutionalization of the chronically mentally ill. Compared with drug treatment also the awareness of risks and ethical implications of the latter mentioned interventions have been delayed, and the methods of assessment may be more complicated:

> There is nowadays a fairly general acceptance of the view that new drugs should not be introduced without being tested. There is no similar consensus in respect of social treatments, most of which become firmly adopted before they have been thoroughly examined. The harm that may come from the application of misguided social theories, or the misapplication of sensible theories, is at least as great as that which can follow the prescription of a harmful drug or an unnecessary course of psychotherapy. In fact, it can be much greater, since harmful social practices can become institutionalized into the structure of a complete psychiatric service. The 'custodial era' in psychiatry, although it was not as black as it has sometimes been painted, nevertheless illustrates how the practices inherent in the concept of the 'total institution' can become generally and uncritically adopted, even though many of them were quite unnecessary and demonstrably harmful (Wing 1981, p. 278).

However, the following chapters demonstrate the current state of ethical awareness and the assessment of benefits, risks and burdens of these therapeutic interventions. Although there are considerable difficulties to transfer the methodology of clinical trials with new drugs, particularly the gold standard of the randomized controlled clinical trial, to other types of therapeutic interventions, the basic ethical considerations will be the same for all types of interventions. However, due to particular characteristics of each therapeutic intervention, its specific ethical implications will be elucidated.

15.3 Development of Therapeutic Interventions Through Research

In therapeutic research the basic ethical problem on the one hand is the exposition of the participating subject to a potentially ineffective intervention or to unknown risks. However, without such research on the other hand, the risk is the exposition of the patient to the uncontrolled risks of an intervention, the effectiveness and safety of which are not really known. This has been called the *ethical paradox of therapeutic research* (Helmchen and Müller-Oerlinghausen 1975): it may be unethical to subject a person to an experimental intervention, e.g. a controlled clinical trial, and it may likewise be unethical to introduce an unproven intervention into the market, because in practice the application of an ineffective intervention will not only produce unnecessary costs for the patient (or the health insurance company) but, much worse, may keep the sick person from receiving an effective therapy. Thus, the scientific evaluation of therapeutic effectiveness is a moral as well as a legal imperative (Wing 1981).

The need and the risks of therapeutic research have produced highly specified *regulations* with different binding character (recommendations, guidelines, and laws).[1] Among other things they define standards of efficacy and safety as well as of participation of fully informed and voluntarily consenting subjects in such therapy research.

They oblige the researcher to minimize the risks and inconveniences for patients (Emanuel et al. 2000), particularly by

- *observing the principle of equipoise* as a basic ethical presupposition of clinical testing of a new intervention. 'This principle states that the patient should be enrolled in a randomized, controlled clinical trial only if there is substantial uncertainty ("equal bet") about which of the treatments would benefit a patient most' (Djulbegovic 2001). Otherwise a randomization would be unethical because of an unequal chance of benefit for health in the compared test group and the control group (Kienle 2008). 'Acknowledging equipoise... is the best mechanism available for choosing an adequate control group. When the principle of equipoise is applied, patients do not lose out prospectively and are not required to sacrifice themselves for others.' (Djulbegovic et al. 2000)[2]
- *formulating the research question* as specifically as possible as the basis of a biometric calculation of the needed sample size in order to avoid the inclusion of more persons than needed and, thus, to worry patients unnecessarily, as well as to avoid the inclusion of fewer persons than needed for a valid result and thus to burden all participants for nothing;
- *choosing adequate selection criteria* in order to minimize the drop-out rate and to achieve at least some generalisability of the results. Otherwise the results may be invalid or not applicable to other patients than to those of the study

---

[1] Major guidelines are the Helsinki Declaration of the World Medical Association from 1964 and its revisions, particularly in Tokyo 1975, Edinburgh 2000, Washington 2002, and Seoul in 2008, the European Guideline for Good Clinical Practice (ICH-GCP-Guideline E6) in 1996/Directive 2001/20/EC on clinical trials, which became part of national laws in some European countries, e.g. in Germany by the 12th amendment of the Drug Law in 2004, guidelines of the drug licensing authorities, particularly the US Food and Drug Administration (FDA), the European Medicines Agency (EMEA), or national authorities such as the German Bundesinstitut für Arzneimittel und Medizinprodukte (BfArM) or the Schweizerisches Heilmittelinstitut Swissmedic. In addition, the standards of national institutes for quality assessment influence the clinical testing of drugs, e.g. the National Institute for Health and Clinical Excellence (NICE) in the United Kingdom, or the Institut für Qualitätssicherung und Wirtschaftlichkeit (IQWIG) in Germany.

[2] This principle seems to be more precise than the hitherto dominating general risk-benefit estimation; however, with regard to uncertainty as the core component of this principle, it should be clarified for each defined study: who is uncertain and what about? Some limitations of this principle with regard to its justification as well as to its application are discussed in detail in Boos et al. (2008).

sample, thus causing sample patients to be burdened unnecessarily. Often therapy resistant patients will be excluded from clinical trials; however, the need to overcome therapy resistance is much more urgent than, for example, the testing of a further antidepressant drug of an assumed known therapeutic mechanism for a questionable quantitative improvement of existing standard therapies;

- *not withholding effective therapies* from patients, e.g. by use of placebo controls or waiting lists in psychotherapy research. However, in order to control the effect of subjective influences and spontaneous remissions on the evaluation of effectiveness, such procedures may be unavoidable. Thus, e.g. in the case of depression a placebo controlled trial may be ethically justifiable only for mild depressions in which the effectiveness of available antidepressant drugs is questionable (Paykel et al. 1988), or in therapy resistant depressions. Otherwise there remains no alternative but to test the experimental intervention against a standard intervention.

Furthermore, the clinical researcher has to decide which statistical method will be the best to burden only the least needed number of patients.

Examples: in order to avoid both the beta error and a large sample size, one should test for falsification of the zero hypothesis, i.e. that the investigational intervention is more effective than the standard. However, with regard to the principle of equipoise it would be more adequate to test twotailed for both positive or negative falsification of the zero hypothesis – at the expense of a small size of the sample.[3]

15.4 Therapy and Research: The Therapeutic Misconception

Apart from various different definitions of research (Helmchen 2002), it is important to recognize that the borderline between research and care may be blurred, and that a categorical decision implies values: a decision for defining an actual procedure as research increases the level of protection for participating patients.

Examples: (i) assessment of laboratory values in a multimorbid and multimedicated demented patient in order to understand an unexpected deterioration as a possible unwanted drug interaction is an act of care with potential individual benefit. However, systematic sampling of such laboratory values in a group of such patients can be viewed as research with only group-specific benefit. (ii) In another case, above what number of deviations from the standard therapy, e.g. of a specified

---

[3] With regard to clinical drug trials it can be said that in many countries regulatory agencies do not demand evidence of superiority of the investigational drug; therefore, an investigator can choose for himself the criteria of superiority, e.g. more than 15% of effectiveness. In comparison to test for non-inferiority often seems to be only in the interest of the pharmaceutical industry but not in the interest of the individual patient, and thereby could be unethical – at least if probands are not informed that the effectiveness of the investigational drug is expected to be not less than 5 (or 10?)% of the standard; sometimes the lower threshold (delta) can be found as more than 5% for lesser effectivity (Garattini and Bertelé 2007).

off-label use in individual patients ('Heilversuch') does this individually oriented experimental therapeutic intervention become a research trial?

Clinically more important is the patient's misconception of research as care, i.e. 'to confuse the design and conduct of research with personalized medical care' (Miller and Joffe 2006). This situation was labelled 25 years ago by the term 'therapeutic misconception' (TM) (Appelbaum et al. 1982). Recently this concept has been discussed controversially.

It was suggested that the term TM supports the 'assumption that clinical trial participation disadvantages research participants as compared with receiving standard medical care' (Miller and Joffe 2006) as well as the reproach that some of its newer interpretations 'exaggerate the distinction between research and treatment' (Kimmelman 2007). But such statements contradict the principle of equipoise, and were clearly repudiated by the inventors of the term, who stated 'Our concerns about TM's impact on informed consent do not derive from the belief that research subjects have poorer outcomes than persons receiving ordinary clinical care. Rather, we believe that subjects with TM cannot give an adequate informed consent to research participation, which harms their dignitary interests and their abilities to make meaningful decisions.... In the absence of empirical studies on the steps required to dispel TM and the impact of such procedures on subject recruitment, it is premature to surrender to the belief that TM must be widely tolerated in clinical research' (Appelbaum and Lidz 2008, Lidz 2006). An investigation by the latter authors resulted in the conclusion that 'subjects often sign consents to participate in clinical trials with only the most modest appreciation of the risks and disadvantages of participation' (Lidz et al. 2004).

15.5 Conclusion

The following chapters will show how these general ethical considerations can or cannot be applied to or must be modified with regard to the development and application of therapeutic interventions in psychiatry and what the ethical consequences are.

References

Appelbaum PS, Lidz CW (2008) Re-evaluating the therapeutic misconception: response to Miller and Joffe. Kennedy Inst Ethics J 16:367–373
Appelbaum PS, Roth LH, Lidz CW (1982) The therapeutic misconception: informed consent in psychiatric research. Int J Law Psychiatry 5:319–329
Beauchamp TL, Childress JF (1994) Principles of biomedical ethics (Last edition 2008). Oxford University Press, New York, NY
Boos J, Merkel R, Raspe J, Schöne-Seifert B (Hrsg) (2008) Nutzen und Schaden aus klinischer Forschung am Menschen. Abwägung, Equipoise und normative Grundlagen. Deutscher Ärzteverlag, Köln Djulbegovic B (2001) Placebo-Controlled Trials. Ann Intern Med 135(1):62–63
Djulbegovic B (2001) Placebo-controlled trials. Ann Intern Med 135(1):62–63

Djulbegovic B, Lacevic M, Cantor A, Fields KK, Bennett CL, Adams JR, Kuderer NM, Lyman GH (2000) The uncertainty principle and industry-sponsored research. Lancet 356:635–638

Emanuel EJ, Wendler D, Grady C (2000) What makes clinical research ethical? J Am Med Assoc 283(20):2701–2711

Garattini S, Bertelé V (2007) Non-inferiority trials are unethical because they disregard patients' interests. Lancet 370:1875–1877

Hamann J, Langer B, Winkler V, Busch R, Cohen R, Leucht S, Kissling W (2006) Shared decision making for in-patients with schizophrenia. Acta Psychiatr Scand 114:265–273

Helmchen H (2001a) Subthreshold psychiatric disorders. Dir Psychiatry 21:125–138

Helmchen H (2001b) Assessment and application of therapeutic effectiveness, ethical implications of. In: Smelser NJ, Baltes PB (eds) International encyclopedia of the social and behavioral sciences. Elsevier, Amsterdam

Helmchen H (2002) Biomedizinische Forschung mit einwilligungsunfähigen Erwachsenen. In: Taupitz J (ed) Das Menschenrechtsübereinkommen zur Biomedizin des Europarates – taugliches Vorbild für eine weltweit geltende Regelung? Springer, Berlin, Heidelberg, New York, NY, pp 83–115

Helmchen H, Müller-Oerlinghausen B (1975) The inherent paradox of clinical trials in psychiatry. J Med Ethics 1:168–173

Kane JM, Borenstein M (1997) The use of placebo controls in psychiatric research. In: Shamoo AE (ed) Ethics in Neurobiological research with human subjects. The Baltimore conference on ethics. Gordon & Breach Publishers, Amsterdam

Kienle GS (2008) Evidenzbasierte Medizin und ärztliche Therapiefreiheit. Vom Durchschnitt zum Individuum. DÄ 105:1161–1163

Kimmelman J (2007) The therapeutic misconception at 25: treatment, research, and confusion. Hastings Cent Rep 37:36–42

Kottje-Birnbacher L, Birnbacher D (1998) Ethische Aspekte bei der Setzung von Therapiezielen. In: Ambühl H, Strauß B (eds) Therapieziele. Hogrefe, Göttingen

Lidz CW (2006) The therapeutic misconception and our models of competency and informed consent. Behav Sci Law 24:535–546

Lidz CW, Appelbaum PS, Grisso T, Renaud M (2004) Therapeutic misconception and the appreciation of risks in clinical trials. Int J Law Psychiatry 58:1689–1697

Miller FG, Joffe S (2006) Evaluating the therapeutic misconception. Kennedy Inst Ethics J 16: 353–366

Paykel ES, Hollyman JA, Freeling P (1988) Predictors of therapeutic benefit from amitriptyline in mild depression: a general practice placebo-controlled trial. J Affect Disord 14:83–95

Wing J (1981) Ethics and psychiatric research. In: Bloch S, Chodoff P (eds) Psychiatric ethics. Oxford University Press, Oxford

Winterer G, Herrmann WM (1996) Effect and efficacy – on the function of models in controlled phase III trials and the need for prospective pharmacoepidemiological studies. Pharmacopsychiatry 29:135–141

Chapter 16
Prevention and Early Treatment

Joachim Klosterkötter and Frauke Schultze-Lutter

Contents

Abbreviations

AD   Alzheimer's Disease
APA   American Psychiatric Association
ARR   Absolute Risk Reduction
DSM   Diagnostic and Statistical Manual of Mental Disorders
ED   Early Detection
EDI   Early Detection and Intervention
EI   Early Intervention
FTLD   Frontotemporal Lobal Degeneration
HD   Huntington's Disease
ICD   International Classification of Diseases
NNT   Number Needed to Treat
RRR   Relative Risk Reduction
WHO   World Health Organization
YLD   Years Lived with Disability

F. Schultze-Lutter (✉)
Research Department, University Hospital of Child and Adolescent Psychiatry Bern (UPD),
CH-3000 Bern 60, Bern, Switzerland
e-mail: frauke.schultze-lutter@kjp.unibe.ch; frauke.schultze-lutter@web.de

H. Helmchen, N. Sartorius (eds.), *Ethics in Psychiatry*, International Library
of Ethics, Law, and the New Medicine 45, DOI 10.1007/978-90-481-8721-8_16,
© Springer Science+Business Media B.V. 2010

16.1 The Prevention Paradigm

The psychiatrist sees too many end states and deals professionally with too few of the pre-psychotic (Sullivan 1927, p. 135).

This early observation has not only been true for psychosis but for most neuropsychiatric disorders. Psychiatrists have traditionally focussed treatment of chronic states and, for nearly a century since George S. Huntington (1850–1916), Emil Kraepelin (1856–1926) and Alois Alzheimer (1864–1915), had little interest in the early states prior to the full-blown disorder. And even in today's 'age of preventive medicine' (Post 2001, p. 107), a significant proportion of psychiatrists still gives top priority to the care of the most severe, chronic cases (David 2004, Machleidt and Brüggemann 2008). They argue against preventive efforts, e.g., in psychosis, as 'a waste of valuable resources' (David 2004, p. 111); focussing prevention would violate fundamental medical principles such as the paramount of continuity of care and of treating the sickest first as well as the avoidance of doing undue harm, i.e., nonmaleficence (David 2004).

This traditional precedence of acute and chronic states started to change in the wake of growing preventive efforts in somatic medicine and of studies demonstrating the immense burden that neuropsychiatric disorders continue to have on affected persons, their families and society (WHO 2003): According to the World Health Organization's (WHO) Global Burden of Disease study 2001, 33% of the years lived with disability (YLD) are due to neuropsychiatric disorders, which account for four of the six leading causes of YLD, i.e., depression, alcohol-use disorders, schizophrenia and bipolar disorder. Taken together, neuropsychiatric conditions account for 13% of disability adjusted life years (WHO 2003), and severe mental disorders like psychosis, bipolar disorder and dementias still tend to take a chronic or even lethal course in a significant portion of cases. In addition, Alzheimer's disease (AD), depression and schizophrenia present high prevalence-cost ratios even if compared to other chronic conditions such as asthma, cancer, osteoporosis and hypertension (Berto et al. 2000). These facts are further aggravated by the expected increase of incidence rates of neuropsychiatric disorders due to the ageing of the population, worsening social problems and civil unrest (WHO 2003), e.g., the current number of 25 million people worldwide suffering from dementia is predicted to double every 20 years and to exceed 81 million by 2040 (Ferri et al. 2005). This growing burden of neuropsychiatric disorders along with the hope to learn more about their aetiology by studying the early states (e.g., McGlashan 2001) could introduce a potential conflict into the estimation of individual and common good in the decision about the ethics of preventive measures or studies. It might also lead to increasing public pressure to give aspects of common good priority over the individual good (Helmchen 2005c), e.g., in order to reduce health care expenses and individual insurance contributions.

Neuropsychiatric disorders differ greatly in major characteristics such as age of onset, aetiological models, validity and mode of risk assessment and availability and safety of (preventive) treatments. Therefore, ethical considerations of an early detection and intervention (EDI) in the premorbid and/or prodromal state will selectively put main emphasis on different aspects of the four major ethical medical principles, *patient's autonomy, nonmaleficence, beneficence* and *fairness* (Brüggemann 2007), in accordance with the disorder's characteristics and the related preventive approach (see Table 16.1).

Other than in autosomally dominant inherited neuropsychiatric diseases with a distinct clinical entity of known aetiology like Huntington's disease (HD; Meincke et al. 2003), most severe psychiatric illnesses have multiple interacting genetic-biological and environmental causes of as yet insufficiently understood aetiological mechanisms (Helmchen 2003). For this lack of risk and aetiological factors of sufficient validity and specificity, current approaches to the prevention of mental disorders, and especially of severe ones like psychoses, bipolar disorder, AD or dementias of the frontotemporal lobal degeneration (FTLD) type, mainly follow an indicated approach (Mrazek and Haggerty 1994). Whereas universal and selective preventions aim at the unselected general population and certain risk populations, respectively, an *indicated prevention* targets persons with already detectable signs and symptoms who do not (yet) fulfil diagnostic criteria of the full-blown disorder (e.g., Barrera et al. 2007, Chertkow et al. 2008, Conus et al. 2008, Correll et al. 2007, Dickerson et al. 2007, Diehl-Schmid et al. 2007, Förstl et al. 2009, Hauser et al. 2007, Machleidt and Brüggemann 2008, McGlashan and Johannessen 1996, McGorry 1998, Post 2001). Thus, its main focus is on persons who already show mental health impairments according to the WHO's definition of health as 'a state of complete physical, mental and social well-being and not merely the absence of disease or infirmity' (WHO 2001, p. 1). Thereby, 'concepts of *mental health* include subjective well-being, perceived self-efficacy, autonomy, competence, intergenerational dependence and recognition of the ability to realize one's intellectual and emotional potential. It has also been defined as a state of *well-being* whereby individuals recognize their abilities, are able to cope with the normal stresses of life, work productively and fruitfully, and make a contribution to their communities.' (WHO 2003, p. 7).

Mental as well as physical health, however, might not be immediately obvious to those living under this fortunate condition, and presence of health might in fact not be recognized at all times but simply be experienced as a general feeling of well-being, of having a future and, especially in the first decades of life, of living eternally (Gadamer 1993). By the introduction of prevention, and especially of universal or selective approaches targeting non-symptomatic populations, this unaware, untroubled way of living under the impression of nonpalpable health might be transformed to a knowing, conscious way with awareness of possible diseases (Machleidt and Brüggemann 2008). When risk assessments lead to a positive result, knowledge of one's risk status might force on existential questions about life's finiteness and sufferings. Although such self-reflections can have positive effects, they can also be devastating and even result in suicidal crisis – particularly if the risk or prediction of a lethal or severely disabling illness for that no cure exists is imparted without sufficient supportive measures, e.g., in case of a positive genetic test for HD or of an increased risk for developing AD (Machleidt and Brüggemann 2008). On the other hand, in case of disorders for that efficient treatments though not necessarily

Table 16.1 Ethical principles in preventive approaches and characteristics of illness: exemplarily comparing schizophrenia,[a] Alzheimer's and Huntington's disease

	Schizophrenia/psychoses (Machleidt and Brüggemann 2008, Post 2001)	Alzheimer's disease (Post 2001)	Huntington's disease (Meincke et al. 2003, Taylor 2004)
Age of onset	Late adolescence to early adulthood	Old age[b]	Middle to late adulthood
Course and outcome	Recovery possible but frequently severe outcome (chronification and disability)	No recovery, progressive and irreversible course with fatal outcome	No recovery, progressive and irreversible course with fatal outcome
Availability/safety of treatment	Pharmacological and psychological treatments but no cure available, risk of considerable adverse side-effects in neuroleptic treatment; benign neuroprotective early interventions studied[c]	No treatment (as yet) available; some benign substances like antioxidants studied, current antidementive drugs only temporarily ameliorate but do not treat symptoms, studies of their value in mild cognitive impairment are inconclusive[d]	No treatment available
Aetiological model	Interaction of multiple genetic/biological and environmental factors	Emphasis on multiple genetic factors, several susceptibility genes identified	Autosomally dominant inherited neurodegenerative disorder
Preventive approach	Indicated (early 'prodromal' symptoms, different symptomatic criteria proposed and studied)	Indicated (subthreshold, i.e., mild cognitive impairment, genotyping considered as adjunct diagnostic test)	Selective (genotyping)
Target population	Help-seeking persons	Help-seeking persons	Clarity-seeking persons with a positive family history of Huntington's disease

Table 16.1 (continued)

	Schizophrenia/psychoses (Machleidt and Brüggemann 2008, Post 2001)	Alzheimer's disease (Post 2001)	Huntington's disease (Meincke et al. 2003, Taylor 2004)
Validity of early detection	Depending on criteria and follow-up period, correct predictions between 16% within 2[e] and 79% within 10 years[f]	Correct prediction in 41% (62% including cognitive decline)[g] to 84% within 5 years[h], at an annual rate of approximately 10–20%[i]	Correct prediction in more than 99%, independent of follow-up
	Frequent unnecessary exposure to positive result	About 20% unnecessary exposure to positive result	Hardly any unnecessary exposure to positive result
Main reasons for testing	Diagnostic clarification of symptoms already present	Diagnostic clarification of symptoms already present	Certainty rather than uncertainty about own risk status
Autonomy/competence	Patients generally fully capable to consent, can involve minors	Patients with frequently impaired capacity to consent	Healthy persons; minors and psychiatric patients with currently impaired capacity to consent generally excluded
	Minors have to consent or at least not express dissent in addition to the consent of their parents		Possible conflict between person's autonomy and doctor's paternalism
	Primacy of autonomy does not allow for outreach teams in competent older adolescents refusing service utilisation and not falling under the mental health act[j] (different national regulations)		Possible conflict between persons' autonomies, if test result conflicts with another's 'right not to want to know'

Table 16.1 (continued)

	Schizophrenia/psychoses (Machleidt and Brüggemann 2008, Post 2001)	Alzheimer's disease (Post 2001)	Huntington's disease (Meincke et al. 2003, Taylor 2004)
Nonmaleficence/beneficence	Possible therapeutic benefit, but also risk of unnecessary benefit from treatment and, even more, of unnecessary suffering from adverse side-effects	No direct therapeutic benefit[k]; some indirect therapeutic benefit from treatment of risk factors such as somatic or psychiatric co-morbidities	Problem of assessing free will vs. pressure of others in decision to get tested
			No therapeutic benefit
			Preference of knowing over uncertainty and wondering
	Negative result can significantly reduce fears (esp. in those with genetic risk)	Negative result can significantly reduce fears and 'Alzheimer phobia'[l] (esp. in those with genetic risk and/or first cognitive impairments)	Negative result can significantly reduce fears, but also evoke feelings alike 'survivor's guilt'
	Need-for-treatment in majority of at-risk persons beyond meeting prodromal criteria		
	Positive result enables preparing for the (potential) disorder (reduction of duration of untreated psychosis, improvement of outcome) and understanding of present symptoms	Positive result enables preparing for the disorder, e.g., choice of a future legal representative/proxy, advance directives	Positive result can be traumatic and evoke severe social and mental problems including suicidality, but enables preparing for the disorder

Table 16.1 (continued)

	Schizophrenia/psychoses (Machleidt and Brüggemann 2008, Post 2001)	Alzheimer's disease (Post 2001)	Huntington's disease (Meincke et al. 2003, Taylor 2004)
Fairness/discrimination	Privacy of risk probability vs. obligation to inform, e.g., in applying for insurances	Avoidance of discrimination and/or interpersonal conflicts due to misinterpretation of early symptoms[m]	Privacy of genetic test results vs. obligation to inform, e.g., in applying for insurances (dependent on health care system) Discrimination in various areas including family planning, occupation
Stigmatisation	Early and potentially unnecessary stigmatisation by risk 'label' vs. avoidance of subsequent stigmatisation by symptoms of illness and/or higher treatment intensity No early stigmatisation of risk samples reported	Avoidance of stigmatisation by early symptoms	Early stigmatisation (e.g., self image, family, friends, colleagues) vs. stigmatisation by occurrence symptoms of illness
Suggested solutions	Detailed psychoeducation about risk status balancing external and internal evidence Improvement of prediction (with both good sensitivity and specificity)	Assessment of competence for decision-making	Obligatory pre- and post-test counselling, psychotherapy and psychoeducation of persons (and their significant others) including assessment of competence for decision-making in cases of doubt

Table 16.1 (continued)

Schizophrenia/psychoses (Machleidt and Brüggemann 2008, Post 2001)	Alzheimer's disease (Post 2001)	Huntington's disease (Meincke et al. 2003, Taylor 2004)
Shared well-informed decision-making about treatment options/symptom monitoring		
Development/use of benign treatments, esp. in the lower-risk states		
Accompanying anti-stigma-campaigns to reduce general stigma, empowerment against self-stigmatisation		

If not indicated otherwise, statements relate to references given in the headline.

[a] Research on prevention of bipolar disorder is still in its infancy, yet its ethical considerations will be much along the lines of those for schizophrenia due to similar basic characteristics (Conus et al. 2008, Correll et al. 2007, Hauser et al. 2007).

[b] Other dementias such as those of the FTLD type might develop earlier, in late adulthood (Diehl-Schmid et al. 2007).

[c] Berger et al. (2007), Amminger et al. (2008).

[d] Chertkow et al. (2008), Förstl et al. (2009).

[e] Yung et al. (2008).

[f] Schultze-Lutter et al. (2007), Schultze-Lutter and Ruhrmann (2008).

[g] Dickerson et al. (2007).

[h] Morris et al. (2001).

[i] Förstl et al. (2009), Grundman et al. (2004).

[j] Brüggemann (2007).

[k] Levey et al. (2006).

[l] Förstl et al. (2009, p. 39).

[m] Reported for FTLD dementias (Diehl-Schmid et al. 2007).

a cure exist, e.g., psychosis or bipolar disorder, an indicated prevention might enable an alleviation of current troublesome symptoms, forestall negative consequences of these early disturbances such as psychosocial deficits and actually prevent development of symptoms into the full-blown clinical picture of the disorder or at least improve outcome and recovery (McGlashan and Johannessen 1996, McGorry 1998).

To be successful in preventing the target disorder, i.e., in reducing its incidence (Mrazek and Haggerty 1994), preventive approaches will have to reach as many at-risk persons as possible and ideally all. This aim might conflict with a person's autonomy and fundamental 'right not to want to know'. The decision against wanting to know, however, can only be regarded as a sufficiently informed one when enough knowledge about the costs and benefits of an early detection (ED) and related early intervention (EI) has been acquired (Machleidt and Brüggemann 2008). Therefore, by also including adequate details on potential costs associated with preventive measures, awareness and information campaigns that are an integral part of efficient an EDI (e.g., Joa et al. 2008, Kaduszkiewicz et al. 2008) should provide a base for an informed decision about whether or not an individual wants to participate in an ED or risk assessment, in particular if not suffering from current symptoms and not seeking help or if no EI options are available in case of a positive test result. The fact, however, that an informed decision against an ED can only be made by a potential at-risk person after thorough information about the illness, the expected course of the untreated illness and the treatment options creates an ethical paradox of 'the obligation to know and the right not to know'.[1]

16.2 Cost-Benefit Estimation of Early Risk Assessment

A special challenge in preventive approaches in psychiatry is that, in the majority of neuropsychiatric disorders, ED in symptomatic and help-seeking patients deals with risk *probabilities* and, other than in presymptomatic genetic testing for HD, not with predictive *certainties* (e.g., Post 2001, Taylor 2004). This is aggravated by the fact that, again in contrast to the distinct clinical and aetiological entity of HD, even diagnosable neuropsychiatric disorders such as schizophrenia do not represent distinct disease entities (Candilis 2003) but, for lack of aetiological knowledge, consensus categories compiled by experts in the field and predominantly defined by the DSM-IV (APA; American Psychiatric Association 1994) and ICD-10 (WHO 1991). Thus, the very targets of prevention might be subject to change when knowledge progresses (Candilis 2003). Furthermore, and

---

[1] Sass (1992, S. 13): 'Pflicht zum Wissen und [. . .] Recht auf Nichtwissen' (cited according to Strech 2005).

HD is again holding a special position of dichotomy in this, neuropsychiatric disorders are mainly regarded as representing one end of a continuum with unclear or even arbitrary borders between 'normality' and 'illness' (e.g., Chertkow et al. 2008, Corcoran et al. 2005, van Os et al. 2009). Against this background and the heterogeneous clinical picture of psychiatric disorders and their early prodromal states, an EDI was long thought to be impossible even if prodromal states of considerable length and severity have been acknowledged as a characteristic of the disorder.[2]

An as yet unsolved problem arising from dealing with risk probabilities is that they depend on the studied population, the applied risk criterion and on the importance given to the various prospective accuracy measures in guiding potential therapeutic actions (e.g., Boyko 1994, McNeil et al. 1975, Sackett 1991, 1992, Strech 2005). As a solution to the different 'rules of thumb' (Boyko 1994, p. 175) in selecting the appropriate test or criterion, the positive and negative likelihood ratios, which weigh the major classic accuracy measures of sensitivity and specificity against each other and are regarded as independent of the prevalence of the disorder or outcome in the study sample, had been proposed along with guidelines for their evaluation (Sackett et al. 1985). Another approach to address the problem of different risk probabilities, which is a well-established and widespread risk modelling procedure in somatic medicine such as oncology or pneumology (Briggs et al. 2008, MRC-BTWP 1990, Sperduto et al. 2008) but has just recently been introduced in psychiatry (Ruhrmann et al. 2010), is the use of prognostic indices for a multivariate clinical staging of risk. Such an approach might not only allow a more precise assessment of current risk but also a tracing of risk improvements or aggravations by repeated measures, thus minimizing risks of worrying patients disproportionately over time.

When relating to cost-benefit estimation associated with risk probabilities such as percentages of false-negative and false-negative predictions, the quality of the consequences associated with each kind of false prediction have to be weighted against each other: When the expected outcome is severe or even fatal, it might outweigh the negative consequences from false-positive predictions if these are decidedly less severe, whereas the same negative consequences might not be tolerable in a disorder generally leading to only little distress and constraint (Candilis 2003, Strech 2005). Consequently, ethical considerations of the ED of a given disorder depend not only on the precision of risk assessment but also on the severity of the disorder in question, the severity of negative consequences related to false predictions and the effectiveness, efficacy and safety of available preventive interventions (see Table 16.1).

---

[2] In ICD-10 (WHO 1991) and DSM-IV (APA; American Psychiatric Association 1994), a prodrome is explicitly acknowledged for the majority of first-episode schizophrenia patients and is considered as part of both the time and the disability criterion, but, for its assumed unspecific nature, not as a diagnostic criterion per se.

Besides these aspects of the *external evidence* of preventive measures, another factor of significant impact on the actual management of risk probabilities in everyday practice is the *internal evidence*, i.e., the degree of experience that the individual professional has with the ED of the disorder (Strech 2005). Developing adequate internal evidence in preventive psychiatry requires the possibility to follow up patients for a longer time. Consequently, developing an internal evidence corresponding to research results is often hindered in mental disorders that tend to start during adolescence but will develop into the full-blown disorder predominately in early adulthood such as schizophrenias and other psychoses. Due to the rather strict separation between child and adolescent and adult psychiatry in most health care systems, developing an adequate internal evidence for these disorders will be hindered in child and adolescent psychiatrists by not evidencing the majority of conversions into the full-blown disorder and in adult psychiatrists by missing early signs typically showing during adolescence.

Another problem that might result in conflicting internal and external evidence is that, particularly in early states of mental illness, symptoms, and even 'predictive' symptoms, are often subtle, thus, escaping the attention if not explicitly asked for (e.g., Schultze-Lutter 2009). As a result of this, clinicians will develop the internal evidence that these symptoms were rarely found and not fit for an ED, even if research gives evidence to the opposite.

Thus, already informing a patient about the costs and benefits associated with an ED of risk probabilities confronts doctor and patient with several problems in everyday practice including weighing internal against external evidence (Candilis 2003, Strech 2005). Thoroughly informed consent to any risk assessment, however, is essentially indispensable (Corcoran et al. 2005).

A heated debate about ethics of ED has been related to psychoses (Bentall and Morrison 2002, Brüggemann 2007, Machleidt and Brüggemann 2008, McGlashan 2001, McGorry 1998, Rosen 2000, Schultze-Lutter et al. 2008). In addition to the insufficient validity and precision of risk assessments (Table 16.1), critics of an ED of psychosis frequently argue with two possible hazards: (i) of positive test results becoming self-fulfilling prophecies (Larsen et al. 2001) and/or (ii), according to labelling theories of psychiatric stigma (Penn and Martin 1998), of exposing the individual to early and unnecessary *stigmatisation* that might violate patient's *right to dignity* (Brüggemann 2007, Corcoran et al. 2005, Larsen et al. 2001, Machleidt and Brüggemann 2008). Yet, symptoms and illness related behaviour such as anergia, perceived strangeness, social skills deficits and positive symptoms were shown to be related to an increased social distance, a proxy measure of stigma, and were discussed as having at the very least an additive effect with labels on stigma (Gaebel et al. 2006, Penn et al. 2000). ED when accompanied by sufficiently effective treatments and psychoeducation (see below) might thus help to prevent stigmatization and discrimination by others by preventing the occurrence of alienating, more severe symptoms and odd, disorganized behaviours. Moreover, an ED could help to avoid self-stigmatization by increasing empowerment early (Rüsch et al. 2005).

In addition, a risk assessment and related psychoeducation about options of actions can give patients and their family a perspective as well as feelings of being prepared and of control over observed changes, which would otherwise remain inexplicable and lead to perplexity and strain (McGlashan et al. 2007). Since individuals with (developing) neuropsychiatric disorders are frequently highly sensitive to stress and react to stressors with a deterioration of their mental state (e.g., Myin-Germeys et al. 2001), additional stress introduced by such puzzlement and perplexity might further destabilise the patient and increase symptoms (McGlashan et al. 2007). Putting a name and a model to these changes can therefore offer significant relief to patients and their families. And the more mental problems are causing turmoil and the more the patient is already distressed, the more the idea of being at risk will be 'just a footnote' (Corcoran et al. 2005, p. 177). A negative test result, on the other hand, might help to considerably alleviate fears of getting 'mad' or demented (Förstl et al. 2009, Strech 2005).

Further supporting a positive view on prevention, the ED concept was reported to be well received by the families of patients with a full-blown psychotic disorder who would have expected a range of positive consequences from an earlier 'diagnosis' for themselves and their affected relative and who would have rarely feared an earlier stigmatisation as a result of it (Lauber and Rössler 2003, Lauber et al. 2001). A positive rather than negative effect of an ED (and EI) is also anticipated by the WHO (2004) who argued that a successful prevention may offer a chance to positively change the negative public perception of mental disorders as being unpredictable and untreatable underlying public stigmatization and discrimination of those suffering from mental disorders. Accordingly, the ability to relate treatment options to risk assessments and to offer a chance to alter the predicted or likely course and current symptoms seems to be important in avoiding possible stigma related to an early diagnosis or in even improving the very image of mental disorders on a larger, general scale.

Despite subject to ethical debates (see below), offering treatment, though without guarantee of success, is possible in the ED of psychoses (and bipolar disorders). This treatment perspective is one major difference to the ED of AD or HD that can be predicted with much greater accuracy than psychoses or even certainty, but do not (as yet) result in offering direct preventive or significantly course-altering treatment (Table 16.1; Meincke et al. 2003, Post 2001). In case of AD, for lack of treatment that delays progression by more than 3–11 months even in preclinical or mild cases (e.g., Levey et al. 2006, Post 2001), a preventive approach targeting risk factors by controlling somatic and especially vascular morbidities, nutrition, life style and environmental toxins was suggested as a means to a significant incidence reduction (Förstl et al. 2009, Jansson 2005, Patterson et al. 2008).

Another concern in the ED of psychosis that is related to concerns about the predictive validity and accuracy of some currently used criteria originates from epidemiological studies reporting that about one person in ten would experience psychotic symptoms like delusions and hallucinations with the majority of them not seeking treatment (e.g., Bak et al. 2005, Bentall and Morrison 2002, Hanssen et al. 2005, Rössler et al. 2007, Verdoux and Cougnard 2006, Verdoux and van Os 2002, Verdoux et al. 1998). The authors (ibid.) drew three main conclusions from these results: (i) they questioned the pathognomonic and pathologic significance of symptoms currently employed in prodromal criteria, in particular of attenuated and transient psychotic symptoms for their seemingly wide distribution in the general population, (ii) based on the obvious continuum between subjects from the general population and those with frank psychosis, they argued that the introduction of a boundary between (prodromal) psychosis and normality was a decidedly arbitrary process, and (iii) they pointed out that many persons seem to lead happy and productive lives despite being 'psychotic'. Consequently, psychosis might not be dreadful for a significant number of those affected implicating that an application of treatment with potentially severe side-effects to avoid serious illness would not be justified (ibid.). Yet, lay-administered fully-structured interviews or self-report questionnaires might not be valid measures for clinically relevant psychotic symptoms and result in an overestimation of their prevalence (Ochoa et al. 2008). The exact prevalence of at-risk criteria in non-clinical populations would therefore have to be explored before broadening the current scope of EDI from help-seeking persons to a less selected or even the general population could be considered (Machleidt and Brüggemann 2008).

Albeit not resulting in direct preventive actions, an ED of AD in symptomatic patients is less subject to ethical concerns than ED of psychoses (Post 2001). Nonetheless, a widespread reluctance of GPs to recognise dementia at an early state and to communicate positive findings to the patients and their families has been reported (e.g., Kaduszkiewicz et al. 2008, Vernooij-Dassen et al. 2005). While under-diagnosing in particular might be related to the perceived stigma in aging and dementia, accompanied by a physician's sense of helplessness for lack of treatment (Vernooij-Dassen et al. 2005), it forestalls timely preparation for the full-blown disorder that will ultimately lead to incompetence and loss of self-identity (Hogan et al. 2008, Post 2001). Hence, delay in diagnosis or communication of risk can violate a patient's autonomy as well as the principle of nonmaleficence by missing the chance to express his/her wishes for future decisions and to appoint a proxy of his/her choice as well as by missing the change to diagnose and treat somatic risk factors for AD in hope for a positive effect on progression (Förstl et al. 2009). It might also lead to strained interpersonal relationships due to misinterpreting early symptoms for lack of knowledge about the risk status and the nature of AD and other dementias itself (Diehl-Schmid et al. 2007, Hogan et al. 2008).

Presymptomatic testing for HD is often regarded a model in the context of genetic predictive testing and, for the lack of treatment and the certainty of lethal outcome, puts special emphasis on ethical considerations of its ED (see also Chapter 27). Although only a fraction of persons with a known family history of HD decides for genetic testing, an ED was mainly approached, because knowledge of the own HD genetic status had been preferred over wondering and uncertainty and offered relief from these (Codori and Brandt 1994). At-risk persons regarded testing as a significant life decision with important implications for themselves and their significant others; and besides their own, the perceived value of and need for the test information of significant others was given major consideration in appraising test options along with the degree to which such information could be tolerated and managed by all those affected by it (Taylor 2004). This high psychosocial vulnerability and potential pressure of others who want to know could pose a threat to patient's autonomy that might be aggravated by today's health policy imperatives towards individual responsibility and self-governance (Meincke et al. 2003, Taylor 2004). Moreover, in case of an ED of HD, one person's autonomy and 'right to know' can directly conflict with another person's autonomy and 'right not to want to know', e.g., if, only one monozygotic twin decides for testing or if, in families with an affected grandparent, only the grandchildren but not the siblings of the stricken search clarity (Meincke et al. 2003).

Whereas testing negative for HD will mainly offer great relief, it could also lead to negative emotional reactions such as survivor's guilt, especially in multiple affected families (Meincke et al. 2003). A positive result, however, can be traumatic to the extent of severe depression and suicidality, and doctor's paternalism may outweigh a person's autonomy if there are indications that the person might be prone to react in such a way. A positive test result can also lead to discrimination in multiple settings including personal relationships, work or insurances,[3] and a person should be very aware of these costs before deciding for testing (Meincke et al. 2003).

Autonomy and related concerns about pressurisation by others and patient's competence as well as nonmaleficence have thus been the main principles in guiding the general agreement on not exposing minors to pre-symptomatic genetic testing of adult-onset untreatable conditions (Meincke et al. 2003, Richards 2008). To further minimize psychological costs of testing for HD (Codori and Brandt 1994, Meincke et al. 2003, Taylor 2004), a thorough pre- and post-test counselling, psychotherapy and psychoeducation of risk persons (and their significant others) including assessment of competence for decision-making in cases of doubt, yet limiting medical paternalism to prevail patients' autonomy in deciding for or against testing is an imperative today (Meincke et al. 2003). Lacking treatment options, however, a positive testing for HD, does not result in any preventive efforts, but, similar to the ED of AD (see above), has its major benefit in enabling patients and their significant other to get prepared for and adjust early to future challenges (Table 16.1) – a process that should be assisted by appropriate professional guidance to maximize beneficence (Hogan et al. 2008, Meincke et al. 2003, Post 2001).

16.3 Cost-Benefit Estimation of Early Intervention

Preventive approaches targeting specific neuropsychiatric disorders have to be discriminated from general or group-specific, i.e., *primary universal or selective strategies* to improve mental health, e.g., by providing family support or stress

---

[3] Concerns about confidentiality to protect persons identified as being at-risk of a serious (neuropsychiatric) disorder from insurance discrimination are highly dependent in their relevance from the national health insurance system and a particular concern in countries with mainly private insurances such as the US (Corcoran et al. 2005).

reducing programmes (e.g., Bayer et al. 2007, LaMontagne et al. 2007). These primary preventive strategies are generally considered to have no adverse effects and, consequently, to not violate the principle of nonmaleficence even if administered to persons at no risk for mental disorder at all. Universal and selective preventive interventions have rarely been related to specific multicausal psychiatric disorders such as psychoses due to the lack of specificity and the multitude of known risk factors, each of little predictive value in itself (Machleidt and Brüggemann 2008, Mäki et al. 2005). Yet, even for psychosis, it had been argued that, at least theoretically, a primary universal prevention might be possible in future (McGrath 2003, Suvisaari et al. 1999, Veijola et al. 2000) and 'that trying to prevent schizophrenic disorders could be helpful, paradoxically, even if we could not reduce even one case of schizophrenia. As a matter of fact, risk factors for schizophrenia are not specific for this disease, but are relevant for a broad range of mental disorders. Even if we do not prevent schizophrenic disorders with primary, secondary or tertiary measures, our beginning to counteract well established risk factors for mental health (and promote protective factors) is an outstanding enterprise to promote health in the general population and reduce a broad spectrum of mental disorders.' (Grispini 2003, p. 26).

A more targeted *universal approach* to detect persons at-risk of a serious mental disorder, e.g., psychoses, by *screening the general population* is currently not regarded as feasible, because most single neuropsychiatric disorders fail to fulfil all of the following requirements: (1) sufficiently high prevalence and severity of the disorder, (2) importance as expressed by the community, (3) controllability, i.e., availability of an effective preventive treatment, and (4) good financial cost-benefit ratio of the service (Rosen 2000). The main approach to prevention in psychiatry is therefore an *indicated strategy* targeting symptomatic and help-seeking patients meeting current (research) risk criteria of the respective beginning disorder.

In dealing with risk probabilities (see above), cost-benefit estimations of preventive interventions are strongly dependent on the accuracy of outcome prediction and the *safety* and *efficacy/efficiency* of the available treatment options. They also have to be considered against the background of the severity of the full-blown illness and the preventive approach, i.e., whether treatment is offered to presymptomatic or symptomatic and help-seeking persons (see Table 16.1). When targeting already symptomatic 'at-risk' *patients*, the cost-benefit estimation of early treatment has to additionally include aspects of the *pathologic significance of current mental problems* and the *need* or *right for treatment*.

Beyond all consideration of costs and benefits of treatment or, in case of at-risk states, of 'wait and see', a prerequisite to respect patient's autonomy is to share these with patients and or their proxies[4] in a way that enables a comprehensively informed decision (Machleidt and Brüggemann 2008). Yet, individual benefits and costs of preventive treatment can only be roughly estimated, because, just as the individual

---

[4] A patient's competence to consent might already be compromised by the evolving neuropsychiatric disorder or s/he might still be a minor (see related discussion in Section 16.4).

risk, the individual treatment effect is only a probability. In relating this probability to the patient, information about the benefit of participating in preventive measures often refer to the relative risk reduction (RRR), an estimate of the *relative* benefit of participating (Strech 2005). Yet, especially if the outcome is rare, RRR might be misleading in estimating the individual benefit of participation, i.e., the *absolute* risk reduction (ARR), which takes into consideration the base rate of the event to be prevented, or the number needed to treat (NNT) to prevent this event[5] (Strech 2005). It has therefore been argued that an adequate education about risks or rather risk probabilities and the expected benefit of preventive programs should relate rather to ARR and/or NNT than to RRR (Strech 2005). Since preventive treatments based on probabilities are caught in an intrinsic dilemma between the principles of nonmaleficence and beneficence, ensuring patient's autonomy and informed consent by the most careful and detailed communication of costs and benefits is most important (Machleidt and Brüggemann 2008). In addition, less easily misleading, clearly formulated cost-benefit information might also prevent a potential decline of public confidence in modern preventive medicine and, consequently, in prevention service use by avoiding raising expectations that can not (yet) be fulfilled (Strech 2005).

Another general drawback in evaluating costs and benefits of preventive treatments is the general lack of universal agreement on standards of efficacy and safety (see also Chapter 15). This is worsen by the fact that preventive treatment based on risk probabilities faces additional imponderability in estimating outcome of treatment and, in relation to this, beneficence: on the one hand, if the full-blown disorder does not develop, this might either be due to an effective treatment or to an incorrect overestimated baseline risk status; on the other hand, if the disorder develops, this might either be due to an ineffective treatment, to its too late start or to an underestimated high baseline risk status (Corcoran et al. 2005). The latter lines of argument have been criticized as the 'irrefutable' hypothesis of preventive research: 'if it is not working, it means enough has not been done, or, one did not intervene early enough' (David 2004, p. 109).

The following will exemplify the discussion about costs and benefits of preventive treatment within an indicated approach by the prevention of psychosis (e.g., Bentall and Morrison 2002, Brüggemann 2007, Candilis 2003, Corcoran et al. 2005, French and Morrison 2004, Grispini 2003, Klosterkötter et al. 2008, Larsen et al. 2001, Machleidt and Brüggemann 2008, McGlashan 2001, McGorry 1998, Post 2001, Rosen 2000, Schultze-Lutter et al. 2008, Yung et al. 1998). In the matter of the pathologic significance and need for treatment, the majority of at-risk patients according to current at-risk criteria (Schultze-Lutter et al. 2007, Yung et al. 1998)

---

[5] For example, participation in breast cancer screening will lead to a 25% RRR of death (i.e., to a reduction from four to three cases out of 1,000 women of the general population over the age of 35), while the ARR is only 0.1% and, consequently, the NNT to prevent one case of death is 1,000 (Strech 2005). For comparison at a 5%-rate of false-positive predictions, the 'number needed to harm' (Strech 2005, p. 106) would only be 50, i.e., 50 of 1,000 women would unnecessarily be confronted with a positive test result and the related potential negative consequences.

were repeatedly shown to suffer from at least one mental disorder, mainly depression or anxiety disorders (e.g., Rosen et al. 2006, Simon et al. 2007). But even patients below the threshold of current diagnostic criteria felt seriously troubled by their rather persistent and often progressive mental changes, experienced a reduction in quality of life and psychosocial functioning that was comparable if not worse than those of first-episode psychosis patients (e.g., Addington et al. 2008, Bechdolf et al. 2005, Ruhrmann et al. 2008) and sought help and treatment for these problems. Thus, in light of the above definition of mental health as 'a state of well-being and not merely the absence of disease' (WHO 2001, p. 1), symptomatic and help-seeking at-risk patients clearly have to be considered 'not healthy' or even ill and already experiencing disablements[6] as well as being entitled to treatment irrespective of their risk of future psychosis (Machleidt and Brüggemann 2008); in line with this, it was shown that they had frequently undergone diverse treatments including psychopharmacotherapy even prior to their risk status assessment (Preda et al. 2002). Consequently, beyond all questions concerning communication of risk status (see above), current cost-benefit discussions of EI in at-risk states of psychosis mainly focus on indication, safety and efficacy issues but less on need for treatment itself (Table 16.1).

Most studied EI approaches so far had been modelled on strategies established in psychosis (i.e., low-dose atypical neuroleptic treatment and/or cognitive or cognitive-behavioural psychotherapy) and exhibited some ability to prevent or at least delay conversion to frank psychosis in help-seeking persons symptomatically at risk of psychosis with NNTs at the end of intervention between 4 and 7 (review in Ruhrmann et al. 2009, Schultze-Lutter and Ruhrmann 2008). Yet, as any treatment, an EI carries the risk of somatic or psychological side-effects including that of early and unnecessary stigmatisation (Corcoran et al. 2005, Larsen et al. 2001). Stigma, however, was not only related to symptoms or behaviour (see above) but also to treatment intensity, with inpatient treatment and use of depot-neuroleptics having the worst reputation (Gaebel et al. 2006). Thus, an efficient EI that remains on the level of ambulatory psychotherapy, but also an out-patient treatment including use of low-dose antipsychotics, might help to prevent some most stigmatising treatment conditions related to acute or chronic states. This line of argument might be even more true, as a stigmatising effect of EDI has never been proven; and experiences with at-risk populations were reassuring to this point suggesting 'that while the shadow of stigma looms large and threatening in theory, it can be managed, if not prevented, in practice' (McGlashan 2001, p. 51).

---

[6] Not meeting the WHO's criteria of 'health' cannot be equalled with being 'ill' (Helmchen 2003, 2005b). Even more than in somatic medicine, in psychiatry, the lack of clear-cut somatic/genetic diagnostic criteria has led to ongoing debates about the concept or comprehension of 'psychiatric illness' and about the very existence of psychiatric diseases (Helmchen 2003). Therein, the traditional, aetiopathologically oriented concept of illness has increasingly been complemented by an intervention oriented concept of disorder in that 'disablement' including the dimensions of *impairment*, *disability* and *handicap* is given more emphasis (Helmchen 2003).

For the substantial number of false-positive predictions by current at-risk criteria, 'both who to treat (with a psychosis-preventive strategy) and how to treat are subjects for debate' (Corcoran et al. 2005, p. 176). Based on the still insufficient predictive accuracy of current at-risk criteria and the frequent erroneous assumption that psychotherapeutic treatment was completely benign and without negative side-effects (e.g., Bühler and Haltenhof 1992; see also Chapter 18) and thus could safely be offered to at-risk persons (e.g., Bentall and Morrison 2002, French and Morrison 2004), pharmacological treatment has been in the focus of discussions on nonmaleficence in the prevention of psychosis. Thereby, it had been argued that psychosis was not a feasible target of potentially harmful preventive measures such as neuroleptic treatment for the lack of well-established aetiological causes and the shortage of a coherent developmental model (Bentall and Morrison 2002, Grispini 2003). In addition, concerns were raised that the prescription of antipsychotic medication might confirm patients' worries about going mad, hence escalating their distressing appraisals and unhelpful coping strategies, which would be predicted by psychological approaches to understanding the transition to psychosis (French and Morrison 2004). This discussion illustrates the difficulties that arise when different models underlie the cost-benefit assessments of an EI (see Chapter 15), in this case a psychological and a medical model.

Efficacy of treatment, however, has been considered only a necessary but not a sufficient condition, since a treatment also has to be effective, i.e., recognized by the patient as beneficial (Strech 2005). As this facet of EI that also includes aspects of patients' satisfaction has not yet been explicitly studied, it should complement efficacy and safety measures in future EI studies on a regular base.

The superior cost-benefit ratio of novel atypical neuroleptics over that of standard neuroleptics had positive impacted on the cost-benefit estimation of using low-dose neuroleptic medication in EI and given impetus to EDI (McGlashan 2001). Still the main points of discussion about an EI with a low-dose atypical neuroleptic are usually around side-effects and use of placebo in intervention trial (see below and Chapter 17). Significant adverse effects can already be associated with low-dose atypical neuroleptic treatment (e.g., Bentall and Morrison 2002, Corcoran et al. 2005). Therefore, an EI has to be strictly controlled and monitored for unwanted effects. To adhere to the principle of nonmaleficence, these should not be tolerated if a patient fails to benefit from treatment; if the EI, however, is efficient in a patient, i.e., reduces current symptoms and increases psychosocial functioning, less severe side-effects such as weight gain or asymptomatic increase in prolactin levels might be acceptable. Should, in future, the efficacy and, subsequently, effectiveness of an EI be proven, a simple monitoring attitude to potentially prodromal persons could be regarded as 'unprotective, if not unethical' (McGlashan 2001, p. 50) by not providing the best care and risking serious deterioration of potentially progressive nature.

Currently, however, EI is predominantly subject to research, and most arguments and perspectives, including that of the early phase offering the opportunity for alternative neuroprotective intervention strategies of little if any side effects (Berger et al. 2007), concern research settings.

16.4 Cost-Benefit Estimation of Prevention Research

In research, and especially in intervention research, the principle of equipoise – a state of genuine uncertainty and/or exigency for undertaking research – and of not withholding effective therapies from participating patients has to be observed (see Chapter 15). And in case of psychotic disorders, prevention research on both ED and EI has been regarded as justified based on its current state of scientific and additional political equipoise (McGlashan 2001). Scientific equipoise is given mainly by the fact that the main rational for EDI, the positive long-term effect of reducing duration of untreated psychosis or illness, is based on correlative but no causal observations and that an active treatment in the prodromal phase has not (yet) been proven to be efficient, despite first indications of it. Political equipoise, on the other hand, is given by the observation that neuroleptic medication is already prescribed to assumed prodromal persons in everyday practice without both ED and EI being based on empirical knowledge (McGlashan 2001). This latter was partially argued against by Corcoran and colleagues (2005) who pointed to the likeliness of this prescribing pattern actually being promoted by the emerging and publicly discussed findings of efficacy of a pharmacological EI and, thus, was posing a circular argument. Yet, they took a positive look on the increasing use of antipsychotics in clinical practice as a possible 'window of opportunity for naturalistic studies and clinical trials' (ibid., p. 177).

A frequent ethical problem of research on neuropsychiatric disorders such as dementias, psychoses and bipolar disorders is that the patient's capacity to give informed consent might be compromised by the illness; the amount of it will vary with individual characteristics, severity and course of the illness and, thus, will have to be assessed with each person individually (Helmchen 2008).[7] Yet when compared to research in later stages, preventive research in psychosis was argued to offer the indirect benefit that participation in EDI studies is usually negotiated when participants are most likely uncompromised in their capacity to comprehend the study's aims and protocol and to provide informed consent (McGlashan 2001). For example, in case of EDI research in psychoses that might involve minors and thus consent of parents or guardians, any limited capacity to give informed consent of the minor was reported to result from immaturity but not the emerging disorder (McGlashan 2001). Still, in risk states for psychosis, insight might be slightly impaired, and help-seeking persons meeting criteria for an imminent risk of psychosis mainly by attenuated psychotic symptoms showed an impairment of insight that was broadly comparable to that of patients with psychosis with a chronic course (Lappin et al. 2007). Thereby, subjective perception of need for treatment was least and subjective perception of the pathological character of symptoms most impaired. Would that seemingly contrarian pattern of a highly perceived need for treatment in addition to a lowly perceived psychopathological character of symptoms prove

---

[7] Although the ability to consent should always be assessed in each patient, reliable guidelines or instruments for the assessment of competence are still wanting (see also Chapter 14).

to be common in at-risk states – at least of psychoses, this might predispose this population to agree to an EI or EI study for mental illness despite not feeling mentally ill.

Though a compromised capacity to consent will usually not pose a problem in presymptomatic EDI studies of HD and only rarely in persons symptomatically at-risk of psychosis, it might well occur in EDI studies of dementia. Though generally not as severe a problem in preventive research as in research of acute or chronic states, it had been shown that even older adults with only mild cognitive impairments already posses a progressive pattern of compromised decision-making capacity from being marginally capable to being incapable to give informed consent according to stringent standards (Okonkwo et al. 2007). The capacity to give informed consent is not a stable feature but might vary with regard to the decision in question; and persons at increased risk of AD can already be impaired in appropriately understanding and evaluating information in order to reach and express an informed decision but, based on their attitude toward research participation and their knowledge of and trust in a certain third person, can still be fully capable to appoint a proxy acting on their behalf in deciding on research participation (Helmchen 2008, Kim and Appelbaum 2006; see also Chapters 9 and 25). It had been shown, however, that the decision for or against participation in research on AD is not simply based on a general attitude toward participating in scientific trials but depends on the various perceived risks and benefits and on the perspective from that the decision should be made, i.e., from a social-medical political point of view, as a decision for oneself or from the perspective of a proxy for a loved-one (Kim et al. 2005). A positive decision was most likely (about 90%) when perceived risks were minimal, i.e., for observational, interview and blood draw studies, and when the decision was made for oneself. It was least likely in perceived high risk studies such as gene transfer studies and when asked to take the perspective of a proxy. The authors concluded that, despite the highly protective attitude displayed when taking a proxy's perspective, there was a great overall support of proxy consent-based research among people at-risk for AD even if the risks and burdens were significant (Kim et al. 2005). Thus, the common practice of combining patient's assent and proxy's consent could generally be considered a good measure to ensure patients' autonomy when decision-making capacity is already compromised.

Although this study of Kim and colleagues (2005) indicated that study participation is decided rather on grounds of risks than on perceived personal benefits, benefit aspects play an important role in research that involves (partially) incompetent patients or minors. Therein, the definition of the study aim, i.e., to what degree an individual or at least group-specific benefit is expected, is even more crucial than in populations with intact decision-making capacity. Based on the anticipated benefit, Helmchen (2008) defined three groups in that research might be acceptable from an ethical perspective: (1) a group that will gain direct potential personal benefit, e.g., from a treatment of current symptoms or from relieving the patient from troubling worries over his/her risk status by a current estimate of it; (2) a group that will experience no instant but future potential personal benefit, e.g., by significantly improving the course and/or outcome but not current symptoms such as a theoretic genetically based prenatal therapy for carriers of the Huntington's gene and (3) a group with no or at least no direct imminent or future benefit for the individual patient but for future members of his group, e.g., participants in a naturalistic study aiming to learn more about the course of the illness. All research, however, has value beyond any benefit for the patient: thus the ratio of personal and group-specific scientific benefit is crucial (Helmchen 2008). But while the exact identification of the involved group might already be difficult in research on full-blown mental disorders, and especially in intervention studies, it is even more complicated in at-risk states in that the very group membership is only a probability and should be given careful consideration in designing a prevention study.

When *minors* are concerned, e.g., in EI studies of psychosis, the international and national ethical requirements[8] defined for otherwise *incompetent patients* such as the Convention on Human Rights and Biomedicine (Europarat 2003) have to be most closely observed and reasoned (Helmchen 2008). Tough often considered together, it had been argued that minors are different from incompetent adults in that they had not been competent before and, consequently, there is no baseline of autonomy from which to project a substituted judgement but only the possibility to revert to the best-interest standards (DeGrazia 2001). Nevertheless, the following requirements generally apply to both incompetent persons and minors (Helmchen 2008): (1) the research cannot be carried out with adult and competent persons alone – for reasons of age and/or state this is certainly true for research in early psychosis and AD but not HD, (2) significant new knowledge about prevention of the full-blown disorder is expected – sufficient knowledge on preventive or even treatment measures is lacking for psychosis, AD and HD, (3) an adequate cost-benefit ratio can be assumed for each participant – this is certainly the case of EI studies for AD and HD that take a lethal course and lack efficient treatment of their acute states, but will have to be decided individually in EI studies of psychosis (currently) involving a great proportion of false-positives, (4) a proxy or parent/guardian was thoroughly informed and has given informed consent, (5) the incompetent patient or minor had signalled assent or at the very least had not given indication of dissent, (6) the ethic committee had approved the study and information for participants/proxies[9] and, (7) for the third group with no direct or indirect personal but group-specific benefit, only minimal risk/burden is expected from participation – e.g., as in case of naturalistic studies to improve ED of psychosis and evaluate effects of different treatment regimes (Corcoran et al. 2005, see above). Although, in Germany, the argument in favour of an only group-specific benefit in the research on minors with no or only minimal anticipated risks and burden from study participation

---

[8] For the dependence of research on moral and social attitudes and national legal regulations, the general framework in that EDI research is conducted varies across countries. A description of this intertwinement of research, societal attitudes and medical laws in the Netherlands is given by Evert van Leeuwen (2001).

[9] In case of EI studies, this should also include information on NNT and/or ARR (see above) and should ensure that a 'therapeutic misconception' (Appelbaum et al. 1982), i.e., the perception of research as personalized care, is avoided.

was accepted by the central ethic committee mainly on grounds of not excluding this group from scientific and medical progress (Zentrale Ethikkommission bei der Bundesärztekammer 2004),[10] it is still controversially discussed (Helmchen 2008).

The above discussion exemplifies the general *ethical paradox of therapeutic research* described by the fact that there is no therapeutic research without exposing participants to potentially inefficient treatments and/or unknown risks, while without such research patients might be exposed to interventions of insufficient efficacy and safety (Helmchen 2005a; see also Chapter 15). Therefore, the scientific proof of therapeutic efficacy is generally regarded a moral imperative to ensure the principles of nonmaleficence and beneficence and patient's autonomy by being able to relate adequate cost-benefit information about a certain treatment. In preventive research, however, ethical considerations including those on placebo-controlled clinical trials (e.g., Helmchen 2005a; see also Chapter 17) are again complicated by the fact that a proportion of participants or even of those just asked to participate in a certain study does in fact not meet the study's entry criterion by being truly at-risk of a certain disorder. Hence, the formulation of the research question, the calculation of the optimal sample size to guarantee a statistically valid result of sufficient power, the choice of the most adequate selection criteria that minimize drop-outs and ensure maximum representativeness of the sample and the a priori implementation of a strategy to handle with treatment resistance (Helmchen 2008) have to be given the outmost care to reach conclusive results while bothering the least number of at-risk persons. Moreover, EI trails based on risk probabilities should include an exit strategy to determine when to discontinue treatment in individuals who do not go on to develop the full-blown disorder (Cornblatt et al. 2001).

In all, it was concluded that the ethical issues in prevention in psychiatry 'need not to be insurmountable. In fact, they are encountered daily in medical practice' (Candilis 2003, p. 78) and are not unique to this field. Furthermore, EDI research was considered an 'ethical obligation' (Candilis 2003, p. 78; Rosen 2000), especially to help-seeking patients who would otherwise receive no help when clearly in need and searching for it (see above) or treatments that are potentially ineffective and harmful.

---

[10] The German central ethics committee (Zentrale Ethikkommission bei der Bundesärztekammer 2004) stated that the use of pharmaceuticals of insufficiently known age-specific efficacy and safety in minors was surely dubious from an ethical point of view, yet that this was rather common practice in their pharmacological treatment due to lack of research in this age-group. While posing additional safety and protection requirements for this group, the committee even voted for studies of mainly group-specific benefit if no alternative to such a trial would exist. The view that, under some conditions, a study posing a minor increase (that does not exceed the degree of risks ordinarily encountered in daily life) over minimal risk while offering no prospect of direct medical benefit is acceptable, is also shared by the US Common Rule (DeGrazia 2001).

16.5 Conclusion

Prevention in psychiatry is more than just a scientific fashion; it is an obligation in the struggle to reduce the burden of mental disorders. Neuropsychiatric disorders, however, form a most heterogeneous group of disorders that, among others, strike first in different age groups, differ in outcome and in availability as well as safety of treatments, have more or less understood different aetiologies and are subject to different degrees of stigmatisation and discrimination. Current ethics of prevention in psychiatry are therefore neither a unitary nor a static concept. Thus, the best way to adhere to the major ethical principles in medicine – autonomy, nonmaleficence, beneficence and fairness – in research and clinical practice will have to be reassessed continuously for each neuropsychiatric disorder against the background of the current state of knowledge as well as the public opinion. In research specifically, great care has to be given to ensuring that the great expected 'common good' will not overcome an individual patient's right to his or her own good and that, especially in patients more vulnerable to misuse, i.e., minors and those already impaired in their decision-making capacity, autonomy is given priority.

References

Addington J, Penn D, Woods SW et al (2008) Social functioning in individuals at clinical high risk for psychosis. Schizophr Res 99:119–124

Amminger GP, Schäfer MR, Papageorgiou K et al (2008) Relationship between reduced erythrocyte membrane fatty acids and transition to psychosis in ultra high risk individuals. Schizophr Res 98:108

APA, American Psychiatric Association (1994) Diagnostic and statistical manual of mental disorders, 4th edn. DSM-IV. APA, Washington, DC

Appelbaum PS, Roth LH, Lidz CW (1982) The therapeutic misperception: informed consent in psychiatric research. Int J Law Psychiatry 5:319–329

Bak M, Myin-Germeys I, Delespaul P et al (2005) Do different psychotic experiences differentially predict need for care in the general population? Compr Psychiatry 46:192–199

Barrera AZ, Torres LD, Muñoz RF (2007) Prevention of depression: the state of the science at the beginning of the 21st century. Int Rev Psychiatry 19:655–670

Bayer JK, Hiscock H, Morton-Allen E et al (2007) Prevention of mental health problems: rationale for a universal approach. Arch Dis Child 92:34–38

Bechdolf A, Pukrop R, Köhn D et al (2005) Subjective quality of life in subjects at risk for a first episode of psychosis: a comparison with first episode schizophrenia patients and healthy controls. Schizophr Res 79:137–143

Bentall RP, Morrison AP (2002) More harm than good: the case against using antipsychotic drugs to prevent severe mental illness. J Ment Health 11:351–356

Berger G, Dell'Olio M, Amminger P et al (2007) Neuroprotection in emerging psychotic disorders. Early Interv Psychiatry 1:114–127

Berto P, D'Ilario D, Ruffo P et al (2000) Depression: cost-of-illness studies in the international literature: a review. J Ment Health Policy Econ 3:3–10

Boyko EJ (1994) Ruling out or ruling in desease with the most sensitive or specific diagnostic test: short cut or wrong turn? Med Decis Making 14:175–179

Briggs A, Spencer M, Wang H et al (2008) Development and validation of a prognostic index for health outcomes in chronic obstructive pulmonary disease. Arch Intern Med 168:71–79

Brüggemann B (2007) Ethische Aspekte der Frühintervention und Akutbehandlung schizophrener Störungen [Ethics of early intervention and acute treatment of schizophrenic disorders]. Ethik Med 19:91–102

Bühler KE, Haltenhof H (1992) Ethische Aspekte der Psychotherapie [Ethical aspects of psychotherapy]. Z Klin Psychol Psychopathol Psychother 40:364–377

Candilis PJ (2003) Early intervention in schizophrenia: three frameworks for guiding ethical inquiry. Psychopharmacology (Berl) 171(1):75–80

Chertkow H, Massoud F, Nasreddine Z et al (2008) Diagnosis and treatment of dementia: 3. Mild cognitive impairment and cognitive impairment without dementia. CMAJ 178: 1273–1285

Codori AM, Brandt J (1994) Psychological costs and benefits of predictive testing for Huntington's disease. Am J Med Genet 54:174–184

Conus P, Ward J, Hallam KT et al (2008) The proximal prodrome to first episode mania – a new target for early intervention. Bipolar Disord 10:555–565

Corcoran C, Malaspina D, Hercher L (2005) Prodromal interventions for schizophrenia vulnerability: the risks of being "at risk". Schizophr Res 73:173–184

Cornblatt BA, Lencz T, Kane JM (2001) Treatment of the schizophrenia prodrome: is it presently ethical? Schizophr Res 51:31–38

Correll CU, Penzner JB, Lencz T et al (2007) Early identification and high-risk strategies for bipolar disorder. Bipolar Disord 9:324–338

David AS (2004) Is early intervention a waste of valuable resources? In: McDonald C, Schultz K, Murray R, Wright P (eds) Schizophrenia: challenging the orthodox. Taylor & Francis, London, New York, NY

DeGrazia D (2001) Ethical issues in early-intervention clinical trials involving minors at risk for schizophrenia. Schizophr Res 51:77–86

Dickerson BC, Sperling RA, Hyman BT et al (2007) Clinical prediction of Alzheimer disease dementia across the spectrum of mild cognitive impairment. Arch Gen Psychiatry 64: 1443–1450

Diehl-Schmid J, Pohl C, Perneczky R et al (2007) Frühsymptome, Überlebenszeit und Todesursachen – Beobachtungen an 115 Patienten mit Demenz auf der Grundlage frontotemporal lobärer Demenz [Initial symptoms, survival and causes of death in 115 patients with frontotemporal lobal degeneration]. Fortschr Neurol Psychiatr 75:708–713

Europarat (2003) Draft additional protocol to the convention on human rights and biomedicine, on biomedical research. Steering Committee on Bioethics, Strasbourg, 23 Aug 2003

Ferri CP, Prince M, Brayne C et al (2005) Global prevalence of dementia: a Delphi consensus study. Lancet 366:2112–2117

Förstl H, Bickel H, Frölich L et al (2009) MCI-plus: leichte cognitive Beeinträchtigung mit rascher Progredienz. Teil I: Prävention und Therapie [MCI-plus: mild cognitive impairment with rapid progression. Part I: prevention and therapy]. Dtsch Med Wochenschr 134:39–44

French P, Morrison AP (2004) Early detection and cognitive therapy for people at high risk of developing psychosis. Wiley, Chichester

Gadamer HG (1993) Über die Verborgenheit der Gesundheit [On the seclusion of health]. Suhrkamp, Frankfurt am Main

Gaebel W, Zäske H, Baumann AE (2006) The relationship between mental illness severity and stigma. Acta Psychiatr Scand 113(Suppl 429):41–45

Grispini A (2003) Opportunities and limits of preventive strategies for schizophrenic disorders. Implications from an epigenetic-developmental model. In: Grispini A (ed) Preventive strategies for schizophrenic disorders. Giovanni Fioriti Editore srl, Rome

Grundman M, Petersen RC, Ferris SH et al (2004) Mild cognitive impairment can be distinguished from Alzheimer disease and normal aging for clinical trials. Arch Neurol 61:59–66

Hanssen M, Bak M, Bijl R et al (2005) The incidence and outcome of subclinical psychotic experiences in the general population. Br J Clin Psychol 44:181–191

Hauser M, Pfenning A, Özgurdal S et al (2007) Early recognition of bipolar disorder. Eur Psychiatry 22:92–98

Helmchen H (2003) Krankheitsbegriff und Anspruch auf medizinische Leistungen [Comprehension of illness and utilization of medical resources]. Nervenarzt 74:395–397

Helmchen H (2005a) Ethische Implikationen plazebokontrollierter Prüfungen von Psychopharmaka [Ethical implications of placebo-controlled clinical trials for psychotropic drugs]. Nervenarzt 76:1319–1329

Helmchen H (2005b) Zum Krankheitsbegriff in der Psychiatrie [The concept of disease in psychiatry]. Nervenarzt 77:271–275

Helmchen H (2005c) Forthcoming ethical issues in biological psychiatry. World J Biol Psychiatry 6(Suppl 2):56–64

Helmchen H (2008) Ethische Erwägungen in der klinischen Forschung mit psychisch Kranken [Ethical questions in clinical research with the mentally ill]. Nervenarzt 79:1036–1050

Hogan DB, Bailey P, Black S et al (2008) Diagnosis and treatment of dementia: 4. Approach to management of mild and moderate dementia. CMAJ 179:787–793

Jansson ET (2005) Alzheimer disease is substantially preventable in the United States – review of risk factors, therapy, and the prospects for an expert software system. Med Hypotheses 64: 960–967

Joa I, Johannessen JO, Auestad B (2008) The key to reducing duration of untreated first psychosis: information campaigns. Schizophr Bull 34:466–472

Kaduszkiewicz H, Röntgen I, Mossakowski K et al (2008) Tabu und Stigma in der Versorgung von Patienten mit Demenz. Kann ein Fortbildungsangebot für Hausärzte und ambulante Pflegedienste zur Destigmatisierung beitragen? [Stigma and taboo in dementia care – does continuing education for GPs and nurses contribute to destigmatisation?]. Z Gerontol Geriatr. Epub 8 Sept 2008. doi:10.1007/s00391-008-0569-0

Kim SY, Appelbaum PS (2006) The capacity to appoint a proxy and the possibility of concurrent proxy directions. Behav Sci Law 24:469–478

Kim SY, Kim HM, McCallum C et al (2005) What do people at risk for Alzheimer disease think about surrogate consent for research? Neurology 65:1395–1401

Klosterkötter J, Schultze-Lutter F, Ruhrmann S (2008) Kraepelin and psychotic prodromal conditions. Eur Arch Psychiatry Clin Neurosci 258(Suppl 2):74–84

LaMontagne AD, Keegel T, Vallance D (2007) Protecting and promoting mental health in the workplace: developing a systems approach to job stress. Health Promot J Austr 18:221–228

Lappin JM, Dazzan P, Morgan K et al (2007) Insight in individuals with an at risk mental state. Schizophr Res 90:238–244

Larsen TK, Friis S, Haahr U et al (2001) Early detection and intervention in first-episode schizophrenia: a critical review. Acta Psychiatr Scand 1003:323–334

Lauber C, Rössler W (2003) Relatives and their attitude to early detection of schizophrenic psychosis. Psychiatr Bull 27:134–136

Lauber C, Schmid-Diebold H, Rössler W (2001) Die Einstellung von Angehörigen psychisch Kranker zu psychiatrischer Forschung, insbesondere zur Früherfassung von schizophrenen Psychosen [Attitudes to psychiatric research, early detection of schizophrenic psychosis, and stigmatization: a survey with relatives of mentally ill patients]. Psychiatr Prax 28:144–146

Levey A, Lah J, Goldstein F et al (2006) Mild cognitive impairment: an opportunity to identify patients at high ridsk for progression to Alzheimer's disease. Clin Ther 28:991–1001

Machleidt W, Brüggemann BR (2008) Sozialpsychiatrische und ethische Überlegungen zur Prävention schizophrener Störungen [Socialpsychiatric and ethical reflections about prevention of schizophrenic disorders]. Fortschr Neurol Psychiatr 76:97–105

Mäki P, Veijola J, Jones PB (2005) Predictors of schizophrenia – a review. Br Med Bull 73 and 74:1–15

McGlashan TH (2001) Psychosis treatment prior to psychosis onset: ethical issues. Schizophr Res 51:47–54

McGlashan TH, Addington J, Cannon TD et al (2007) Recruitment and treatment practices for help-seeking "prodromal" patients. Schizophr Bull 33:715–726

McGlashan TH, Johannessen JO (1996) Early detection and intervention with schizophrenia: rationale. Schizophr Bull 22:201–222

McGorry PD (1998) "A stitch in time". . .the scope for preventive strategies in early psychosis. Eur Arch Psychiatry Clin Neurosci 248:22–31

McGrath J (2003) Prevention of schizophrenia – not an impossible dream. In: Murray RM, Jones PB, Susser E et al (eds) The epidemiology of schizophrenia. Cambridge University Press, Cambridge

McNeil BJ, Keeler E, Adelstein SJ (1975) Primer on certain elements of medical decision making. N Engl J Med 293:211–215

Meincke U, Kosinski Ch, Zerres K et al (2003) Psychiatrische und ethische Aspekte genetischer Diagnostik am Beispiel der Chorea Huntington [Psychiatric and ethical aspects of genetic diagnostics in cases of Huntington's disease]. Nervenarzt 74:413–419

Morris JC, Storandt M, Miller JP et al (2001) Mild cognitive impairment represents early stage Alzheimer disease. Arch Neurol 58:397–405

Mrazek PJ, Haggerty RJ (1994) Reducing risks for mental disorders: frontiers for preventive intervention research. National Academy Press, Washington, DC

MRC-BTWP (1990) Prognostic factors for high-grade malignant glioma – development of a prognostic index – a report of the Medical Research Council Brain-Tumor Working Party. J Neurooncol 9:47–55

Myin-Germeys I, van Os J, Schwartz JE et al (2001) Emotional reactivity to daily life stress in psychosis. Arch Gen Psychiatry 58:1137–1144

Ochoa S, Haro JM, Torres JV et al (2008) What is the relative importance of self reported psychotic symptoms in epidemiological studies? Results from the ESEMeD – Catalonia study. Schizophr Res 102:261–269

Okonkwo O, Griffith HR, Belue K (2007) Medical decision-making capacity in patients with mild cognitive impairment. Neurology 69:1528–1535

Patterson C, Feighner JW, Garcia A et al (2008) Diagnosis and treatment of dementia: 1. Risk assessment and primary prevention of Alzheimer disease. CMAJ 178:548–556

Penn DL, Kohlmaier JR, Corrigan PW (2000) Interpersonal factors contributing to the stigma of schizophrenia: social skills, perceived attractiveness and symptoms. Schizophr Res 45:37–45

Penn DJ, Martin J (1998) The stigma of severe mental illness: some potential solutions to a recalcitrant problem. Psychiatr Q 69:235–247

Post SG (2001) Preventing schizophrenia and Alzheimer disease: comparative ethics. Schizophr Res 51:103–108

Preda A, Miller TJ, Rosen JL et al (2002) Treatment histories of patients with a syndrome putatively prodromal to schizophrenia. Psychiatr Serv 53:342–344

Richards FH (2008) Predictive genetic testing of adolescents for Huntington disease: a question of autonomy and harm. Am J Med Genet A 146A:2443–2444

Rosen A (2000) Ethics of early prevention in schizophrenia. Aust N Z J Psychiatry 34(Suppl):S208–S212

Rosen JL, Miller TJ, D'Andrea JT et al (2006) Comorbid diagnoses in patients meeting criteria for the schizophrenia prodrome. Schizophr Res 85:124–131

Rössler W, Riecher-Rössler A, Angst J et al (2007) Psychotic experiences in the general population: a twenty-year prospective community study. Schizophr Res 92:1–14

Ruhrmann S, Paruch J, Bechdolf A et al (2008) Reduced subjective quality of life in persons at risk for psychosis. Acta Psychiatr Scand 117:357–368

Ruhrmann S, Schultze-Lutter F, Klosterkötter J (2009) Intervention in the at-risk state to prevent transition to psychosis. Curr Opin Psychiatry 22(2):177–183

Ruhrmann S, Schultze-Lutter F, Salokangas RKR et al (2010) Prediction of psychosis in adolescents and young adults – results from a prospective european multicenter study (EPOS). Arch Gen Psychiatry 67(3):241–251

Rüsch N, Angermeyer MC, Corrigan PW (2005) Das Stigma psychischer Erkrankungen: Konzepte, Formen und Folgen [The stigma of mental illness: concepts, forms, and consequences]. Psychiatr Prax 32:221–232

Sackett DI (1991) Clinical reality, binary models, babies and bath water. J Clin Epidemiol 44: 217–219

Sackett DI (1992) A primer on the precision and accuracy of the clinical examination. J Am Med Assoc 267:2638–2644

Sackett DI, Haynes RB, Tugwell P (1985) Clinical epidemiology: a basic science for clinical medicine. Little & Brown, Boston, MA

Sass HM (1992) Informierte Zustimmung als Vorstufe zur Autonomie des Patienten [Informed consent as a first step towards patient's autonomy]. Medicoethical Materials, Broschure 78. Centre for Medical Ethics, Bochum

Schultze-Lutter F (2009) Subjective symptoms of schizophrenia in research and the clinic: the basic symptom concept. Schizophr Bull 35:5–8

Schultze-Lutter F, Klosterkötter J, Picker H et al (2007) Predicting first-episode psychosis by basic symptom criteria. Clin Neuropsychiatry 4:11–22

Schultze-Lutter F, Ruhrmann S (2008) Früherkennung und Frühbehandlung von Psychosen [Early detection and early treatment of psychoses]. Uni-Med, Bremen

Schultze-Lutter F, Ruhrmann S, Klosterkötter J (2008) Early detection and early intervention in psychosis in Western Europe. Clin Neuropsychiatry 5:303–315

Simon AE, Cattapan-Ludewig K, Zmilacher S et al (2007) Cognitive functioning in the schizophrenia prodrome. Schizophr Bull 33:761–771

Sperduto PW, Berkey B, Gaspar LE et al (2008) A new prognostic index and comparison to three other indices for patients with brain metastases: an analysis of 1,960 patients in the RTOG database. Int J Radiat Oncol Biol Phys 70:510–514

Strech D (2005) Der Umgang mit Wahrscheinlichkeiten und das Vertrauen in die Medizin [Dealing with probabilities and confidence in medicine. Ethical and scientific aspects of evidence-based medicine in cancer screening]. Ethik Med 17:103–113

Sullivan SH (1927) The onset of schizophrenia. Am J Psychiatry 151:135–139

Suvisaari JM, Haukka JK, Tanskanen AJ et al (1999) Decline in the incidence of schizophrenia in Finnish cohorts born from 1954 to 1965. Arch Gen Psychiatry 56:733–740

Taylor SD (2004) Predictive genetic test decisions for Huntington's disease: context, appraisal and new moral imperatives. Soc Sci Med 58:137–149

van Leeuwen E (2001) Dutch law and ethics concerning the experimental treatment of early psychosis. Schizophr Res 51:63–67

van Os J, Linscott RJ, Myin-Germeys I et al (2009) A systematic review and meta-analysis of the psychosis continuum: evidence for a psychosis proneness-persistence-impairment model of psychotic disorder. Psychol Med 39:179–195

Veijola J, Jones P, Mäkikyrö T et al (2000–2001) Early association for schizophrenia in the 1966 North Finland general population birth cohort. Int J Ment Health 29:84–90

Verdoux H, Cougnard A (2006) Schizophrenia: who is at risk? Who is a case? Int Clin Psychopharmacol 21(Suppl 2):S17–S19

Verdoux H, Maurice-Tison S, Gay B et al (1998) A survey of delusional ideation in primary-care patients. Psychol Med 28:127–134

Verdoux H, van Os J (2002) Psychotic symptoms in non-clinical populations and the continuum of psychosis. Schizophr Res 54:59–65

Vernooij-Dassen MJ, Moniz-Cook ED, Woods RT et al (2005) Factors affecting timely recognition and diagnosis of dementia across Europe: from awareness to stigma. Int J Geriatr Psychiatry 20:377–386

WHO, World Health Organization (1991) Internationale Klassifikation psychischer Störungen: ICD-10, Kapitel V (F) [Tenth revision of the international classification of diseases, Chapter V (F): Mental and behavioural disorders]. Huber, Bern, Göttingen, Toronto, ON

WHO, World Health Organization (2001) World Health Report 2001. WHO, Geneva

WHO, World Health Organization (2003) Investing in mental health. WHO, Geneva

WHO, World Health Organization (2004) Promoting mental health. WHO, Geneva

Yung AR, Nelson B, Stanford C et al (2008) Validation of "prodromal" criteria to detect individuals at ultra high risk of psychosis: 2 year follow-up. Schizophr Res 105:10–17

Yung AR, Phillips LJ, McGorry PD et al (1998) Prediction of psychosis. A step towards indicated prevention of schizophrenia. Br J Psychiatry 172(Suppl 33):14–20

Zentrale Ethikkommission bei der Bundesärztekammer [Central Ethics Commitee of the Bundesärztekammer] (2004) Stellungnahme "Forschung mit Minderjährigen" [Statement "research in minors"]. Dtsch Ärztebl 101:A1613–A1617

Chapter 17
Ethical Implications of Psychopharmacotherapy

Hanfried Helmchen

Contents

Abbreviations

AGNP	Arbeitsgemeinschaft für Neuopsychopharmaklogie und Pharmakopsychiatrie
DUO	Drug Utilization Observation
EMEA	European Agency for the Evaluation of Medicinal Products
FDA	Food and Drug Administration, USA
GCP	Guideline for Good Clinical Practice
NICE	National Institute for Health and Clinical Excellence
UDR	Unwanted Drug Reactions
WMA	World Medical Association

H. Helmchen (✉)
Department of Psychiatry and Psychotherapy, CBF, Charitè – University Medicine Berlin,
D-14050 Berlin, Germany
e-mail: hanfried.helmchen@charite.de

H. Helmchen, N. Sartorius (eds.), *Ethics in Psychiatry*, International Library
of Ethics, Law, and the New Medicine 45, DOI 10.1007/978-90-481-8721-8_17,
© Springer Science+Business Media B.V. 2010

17.1 Introduction

The introduction of therapeutically effective psychotropic drugs in 1952 had two fundamental consequences: first, against the hitherto almost only available effective psychiatric treatments, mainly the malaria-fever therapy against general paresis and the shock therapies against psychoses, the new drug treatment seemed to be more efficacious and, above all, gentler or less aggressive, and thereby much better accepted by both patients and physicians; second, the new drugs catalysed a broad expansion of the field of brain research, providing a more rational basis for treatments. Higher standards for treatment research were a logical consequence of these developments. as well as the inclusion of ethical standards which had been hitherto largely ignored.

In the beginning of this new era the antipsychotic, antidepressant, and anxiolytic effects of psychotropic drugs were experienced as beneficial and were hailed as a therapeutic break-through. However, this success became overshadowed by side-effects of drug treatment, the knowledge of which increased in the 1960s and 1970s. This proved to be important the more these negative effects on the adherence and compliance of patients and thereby on the outcome of drug treatments became recognised. Accordingly the attitude of psychiatrists changed: from the dominating (and sometimes only) aim to remove the symptoms of the disease/disorder to an attitude of taking the complaints of patients seriously, and to move the patient into a broader perspective: the focus of therapeutic aims shifted from symptom-reduction, i.e. fighting the disease, to quality of life, i.e. helping the patient to overcome his illness. This change of paradigm meant that the psychotropic medication had to be embedded in social and psychotherapeutic measures of care and rehabilitation.

Simultaneously the awareness of ethical issues in therapeutic interventions in psychiatry with psychotropic drugs increased. Both the efficient application of a marketed psychotropic drug to mentally ill persons in *practice* and the assessment of effectiveness and safety of a potentially therapeutic drug through *research* imply ethical problems (Helmchen 2001).[1]

17.2 Practice: Clinical Applications of Established Treatments

The therapeutic application of psychotropic drugs raises three major questions of ethical significance: (i) is the medication necessary, or do other likewise effective treatments exist with fewer side-effects? (ii) when and about what must the patient be informed before starting a proposed drug treatment? (iii) how to involve the

---

[1] Here only paradigmatically a few but major ethical implications of psychopharmacotherapy are dealt with. Related problems are discussed in other chapters: the problem of preventive medication in healthy persons with specific and high risks of becoming ill in the preceding Chapter 16; the question of influences from the drug industry on the independence of psychiatrists judgements in Chapter 4; the non-medical use of psychotropic drugs in Chapter 30; and also in further Chapter 23; or Chapter 20.

patient into balancing benefits and risks of the proposed treatment? These questions may be answered differently in the application of drugs in acute states of mental disorders and in long-term medication.

All psychotropic drugs have side-effects, i.e. effects that do not correspond with the therapeutically wanted effects. Some of them will not be noticed by the patient or are harmless, others such as sedating or sleep-inducing effects of antipsychotics are even occasionally wanted, but others are unwanted, particularly those with objective risks such as agranulocytosis, or tardive dyskinesia, or excessive weight gain, even if they are unnoticed such as hyperlipidemia. The subjectively bothersome side-effects of psychotropic drugs, e.g. dryness of the mouth, are fairly frequent (2/3 of all applications), whereas objectively severe unwanted drug reactions (UDRs) are much less frequent and potentially acutely life-threatening unwanted side-effects are rare (1–2%) (Grohmann et al. 1993, 2004). As many of these side effects can be either prevented or treated, this still makes psychotropic drugs generally safe medications when compared to the whole medical pharmacopeia.

Nevertheless, the prescribing physician must know the whole spectrum of side-effects in order to do justice to the patient's right of self-determination by his/her adequate information as well as to the ethical principle of non-maleficence (nil nocere). To avoid harm of patients psychotropic drugs must be indicated strictly and the course of treatment must be controlled adequately for unwanted drug effects.

Often this is not the case, and psychotropic drugs are often not used conscientiously enough, including any off-label indications. One example was the excessive indication of benzodiazepines and – after recognition of their dependency potential (Linden and Thiels 2001) – their substitution by sedating antipsychotics, particularly in restless inhabitants of nursing homes (Sonntag et al. 2006) without adequate side-effects monitoring, e.g. Parkinsonian symptoms, and a regular review of the indication for these medications.

Before starting a psychotropic medication the patient must be informed about the expected benefits, risks and procedure of the medication. This is necessary not only from a legal point of view for a valid consent but also in order to establish trust with regard to the necessary adherence to treatment. Information should be focussed on immediately and early expected effects. If there are doubts that the patient has understood the information and thereby of the validity of the consent, the capacity to consent should be assessed. In case of an impaired capacity the legal requirements according to the national law must be settled for substituting the consent by authorised persons.

17.2.1 Short-Term Medication

Control of acute states of mental disorders aims primarily at reducing symptoms and at enabling the patient to receive all diagnostic and therapeutic measures besides a medication, e.g. social or psychotherapeutic interventions that will help him to find his way back into his life before his illness. Especially mental states with danger of damage for the patient him/herself or for others may be a clear indication for the application of sedating drugs, mostly benzodiazepines and/or antipsychotics. Sometimes, this may be possible only against the will of the patient. But even in case of a compulsory medication (see Chapters 20 and 21) the psychiatrist should try to explain the need and the effects of the medication. Especially in case of the first application of antipsychotics the patient must be informed about early side-effects such as acute dystonias or dyskinesias of the tongue and mouth, since such unwanted effects may frighten the uninformed patient and thereby deteriorate the situation. Neither ethically nor legally is it permissible to leave the patient without personal assistance in such a compulsory situation. If a prolongation of the compulsory medication is needed beyond the acute emergency situation, then the legal requirements must be met by asking for a decision by an authorised person, i.e. a judge and/or a legal guardian or legal representative. This also may serve the external control of the medical procedure. Occasionally it can be helpful, because a more paternalistic physician may estimate the need for medication more urgently than a physician who sees the patient more as a self-responsible and competent to consent individual, or one who prefers to adopt the sole role of a consulting expert. The physician has to weigh between two risks: on the one hand he can impair the welfare of the patient by recognising the 'natural' will of the patient too late as distorted and thereby taking safeguarding measures too late; on the other hand he may impair the self-determination of the patient by judging the expressions of the patient's will too quickly as distorted and correspondingly establishing a compulsory medication too early.

17.2.2 Long-Term Medication

Manifold reasons exist for a transition from short-term to long-term medication: suppression of disturbing symptoms, stabilisation and maintenance of a remission, preventing relapses or recurrences. The careful psychiatrist will weigh the desired effects against the undesired ones in every medication but especially before starting a long-term medication. He will do this together with the patient as much as possible; and often this is not only possible but wanted by the patient (Hamann et al. 2006). The psychiatrist will 'titrate' the medication with regard to an optimal relation between efficacy and side-effects and take into consideration the preferences of the patient. Thus, there are patients who put up with the risk of tardive dyskinesias in order to alleviate agonizing psychotic experiences; but other patients refuse the medication due to this risk, or even without giving any reasons. This situation becomes ethically difficult if the psychiatrist considers an antipsychotic long-term medication to be necessary in order to avoid frequent and both subjectively and socially serious relapses, but if the patient refuses, i.e. the (perhaps distorted) will of the patient contradicts his/her well-being.

In such cases the psychiatrist should inform the patient about the necessity of a medication protecting him against relapses, but should tell him that he will not force it upon him. At least in an out-patient medication this is almost impossible since pressure, e.g. by attaching the payment of pocket money to the adherence to dates for depot-injections of the medication, is ethically questionable (Appelbaum and Redlich 2006), or a compulsory depot-injection on the basis of a legal guardianship may destroy trust and seems ethically acceptable only in cases in which the patient is unable to consent due to illness, and thereby his welfare must have priority over his distorted will (Roberts and Geppert 2004). Nevertheless, the refusing patient should be assured that the psychiatrist will help him again if without medication he very probably will experience a relapse. In my own experience, such patients have sometimes been able to accept the beginning of medication only after the second or third relapse. However, it could be asked whether such trust-building patience is justified, because it has brought the suffering of relapses to the patient and high hospital costs to the health insurance company (Kelly et al. 2002). If neurotoxic consequences of relapses (Fleischhacker 2009, personal message) are evidence-based, the threshold of accepting capacity to consent (or refuse) treatment must be higher in order to avoid harm. This is also particularly valid in patients who pose a danger to themselves or others if not treated adequately. In such cases it is not only the patient's welfare which is of concern but also the public's safety.

Special ethical problems result from the risk of late side-effects such as weight gain or tardive dyskinesias. The latter name indicates that these partially irreversible and socially aversive motor disturbances appear late, at the earliest after 3–6 months in the course of a treatment with some antipsychotics. Since it is a serious side-effect, the patients must be informed about it. However, due to the delayed manifestation, there is no necessity to inform the patient before beginning a curative first medication with antipsychotics (Helmchen 1992). Such a wrongly timed and thereby unnecessary information could provoke the risk to discourage the patient from the needed medication. Furthermore, this emphasizes the necessity to perceive a long-term medication as a treatment with its own indication: before its beginning the patient must be informed again even about the risk of tardive dyskinesia. An antipsychotic medication is not allowed to be perpetuated unreflectedly; since the transition from a timely limited curative medication to a relapse-preventing or maintenance treatment is a turning point that should give rise to considerations, particularly for a change from a classic antipsychotic drug, e.g. haloperidol, to an atypical one such as clozapine which is free from this risk (Meltzer and McGurk 1999).

Ethical problems arise not only from conflicts between the ethical principles of respecting the autonomy of the patient and his welfare, but also between the principle of individual welfare and the principle of justice, i.e. the common good. This problem will now be analyzed in more detail.

17.2.3 An Example: The Application of Antidementive Drugs

Due to the shortage of resources the application of antidementive drugs with only modest effectiveness will be questioned. This points to a possible ethical conflict between the principles of the individual as opposed to the common welfare.

Symptoms such as anxiety, restlessness, hallucinations, or delusions that may complicate the course of dementia can already be treated successfully. However, this is not as valid for the medical treatment of the progressive course of the neurodegenerative disease itself. It is true that the currently licensed antidementive drugs have an evidence-based but only small and temporary course-delaying effectiveness. Nevertheless, at least a delay of the move from one's own home to a nursing home even for only a few months may be a large gain for some patients. But many people and experts, e.g. those of the National Institute for Health and Clinical Excellence (NICE), consider such benefit compared to the costs as too small in order to justify this treatment – at least not at the expense of the community.

Considerations of individual benefit and the low efficiency of the presently available antidementive drugs also seem to be familiar to practitioners who think that they can treat other patients with more efficiency with the money saved in demented patients. There are hints that physicians see the indication for an antidementive drug treatment in this sense (Brockmann 2002); a decision that may also be coloured by a negative aging stereotype, i.e. a form of discrimination on the grounds of unreflected stigmatisation (see Chapter 2). However, other physicians feel primarily obliged to each of their patients with an age-dependent dementia. In this contrast the former physicians put a general allocational decision into the personal patient–physician-relationship that belongs to the societal macro-level, whereas physicians of the latter decisional behaviour act responsibly only if they give rational reasons for their medical indication for the individual patient. This is of course a rhetorical contrast because in practice physicians may be influenced by or even consider both aspects. However, again the question of stigmatisation against mental illness and mentally ill patients may arise if one compares this discussion with treatments in somatic medicine, e.g. expensive cancer treatments with limited efficacy.

The special relevance of ethical considerations for the application of antidementive drugs lies in the fact that the physician must reflect the ethical implications of his acting all the more the less the – even if evidence-based – benefit or efficiency of the treatment is profiled. On the other hand the argument of economic efficiency that is justified by the principle of justice or solidarity should avert misplacement or abuse of scarce resources. Abuse develops from unfounded or reckless use of medical resources – reckless by risky life-styles, unfounded by a lack of indication or by insufficient evidence.

The question is: can ethical reasons be given to withhold a treatment without an alternative from a demented patient if other patients can be treated more efficiently by the money saved thereby? To withhold means in this case that the obligation to pay for treatment with antidementive drugs will be taken off the health insurance company since it can be assumed that the majority of demented patients cannot pay for such antidementive medication from their own pockets.

17.2.3.1 Ethical Analysis

A first ethical consideration would conclude that an antidementive treatment must not be withheld from demented patients since basic human rights do not allow us to consider one human life and not another. But this argument loses power since life itself and quality of life are rated differently because antidementive drugs may improve the quality of life but cannot save life directly. Accordingly it could be argued from a utilitarian standpoint that in this case the point is 'only' the quality of life but the resources are not enough to allow a high quality of life for all at the same level. Therefore, the question whether it is ethically acceptable to withhold an antidementive medication from a demented patient could be answered in the affirmative only if other medical interventions with the same cost could result in a considerable better quality of life of other patients.

However, if the efficiency of medical interventions will be compared by the criterion of quality of life it has to overcome the difficulty that neither the concept nor the assessment of quality of life is accepted unequivocally, not the least since this concept will be recognised remarkably differently, and moreover it changes during the course of time by adaptational processes. But even if specific and quantifiable positive aspects of quality of life such as e.g. longer duration of independence or later beginning of the need of care would be taken as criteria for comparison, the difficulties remain considerable. This is valid at least if, in other comparable diseases, completely different aspects of quality of life, e.g. freedom from pain, would prevail. Due to these almost impassable practical difficulties to compare the efficiency of treatment outcomes with regard to quality of life, it seems unjustifiable to withhold a therapy with antidementive drugs from a demented patient.

A further question: is it fair to ask demented patients for their solidarity to give up a treatment without an alternative in favour of other patients' benefit, if, at the same time, society accepts that a lot of people behave totally without solidarity by adopting without hesitation health risks such as dangerous sports, overweight, abuse of drugs, alcohol and nicotine, and take for granted the treatment of the resulting damages of health? Principally this question must be answered in the negative, since it is extremely unlikely that all people can be forced to reduce a risky way of life, without regard for the high costs of expedience, by an overall control of behaviour which reduces freedom for all. Nevertheless, health insurance companies demand solidarity for all of their members with the argument of economic efficiency insofar as every saving of a non-efficient intervention is for the benefit of all. In sum it seems unfair to expect a renunciation of a medication without alternative by demented people for the benefit of other ill people; but this solidarity will be required by the argument of economic efficiency. In the case of antidementive drugs it is the only argument, because to pay for it cannot be refused due to no benefit or no need of its application, since the benefit – even if limited – is evidence-based, and the application will be considered necessary by the medical indication in the individual case.

A provocative question shall elucidate this problem of economic efficiency with regard to justice: is the amputation of a gangrenous leg in a nicotine-dependent smoker more efficient than an antidementive medication in a demented patient? In both cases the intervention would be necessary and beneficial, but the prognosis is bad because the basic disease is progressive as in the freshly amputated smoker who rides in his wheelchair to the ward door in order to smoke the next cigarette, or in the demented patient who sometimes has to move into a nursing home. This question will not be put due to the life-threatening character of the diseases; but in that case cost-benefit-analyses ought to supply evidence for a much more favourable relationship between the high cost of the surgical intervention and longer and qualitatively good survival time by the surgical intervention compared with an antidementive medication. Such analyses are not known. They are not even to be expected because the life-threatening argument taboos all other arguments – rightly so in the medical tradition because the physician is obliged primarily to the individual patient. Therefore, a comparison according to exclusive economic criteria is not possible. But the discussion could become more intense with increasing changes of the societal and economical context.

If the antidementive drug will not be paid by the health insurance company, the psychiatrist cannot apply the indicated antidementive medication in the individual case although he remains obliged to do so. This principle is expressed in the so-called 'curative freedom' ('Kurierfreiheit'); it means the application of the scientifically based knowledge in the single case according to the individual aspects of each patient. However, this will not be possible if the patient cannot pay for the treatment by himself. This example shows how important it is to make transparent for the public procedures and reasons for unavoidable decisions.

17.2.4 Conclusion

The currently available antidementive drugs provoke a question with regard to both their only limited efficiency and the known scarcity of resources. It is the question of allocational ethics: may this treatment be withheld from demented patients for the benefit of all of the members of the health insurance company? Reasons have been given for the following answers: (1) in terms of duty (deontologically) it is not *justified* to withhold currently available antidementive drugs from demented patients; however in terms of consequences of actions (from a utilitarian standpoint) it may be justified in principle but not practically with the current status of knowledge and research; (2) withholding an antidementive treatment seems intuitively not *fair* under the current societal conditions, but for reasons of economic efficiency this kind of solidarity is requested; (3) the question of economic efficiency with regard to *justice* cannot be answered at this point in time, because no sufficient data exist for such comparisons of different interventions. But our real world shows that under the pressure of economic efficiency and rationing consequential approaches gain weight against deontological ones. This brings problems to the psychiatrist who traditionally feels him/herself obliged to the individual patient and to give him the best available treatment – even if its efficiency is low. However, the psychiatrist cannot escape this challenge of his professional identity. But by contrasting this discussion with the same questions in other fields of medicine, e.g. in palliative medicine, he brings the topic of stigmatisation of the mentally ill to the forefront.

17.3 Research: Clinical Trials of New Interventions

17.3.1 Need for Research

A considerable part of established 'standard' treatments in psychiatry is not evidence-based, or their evidence is established only on a low level or at least not reflected in everyday practice. This is valid for most combinations of drug treatments such as multimedications, particularly in multimorbid states in the elderly, or with psychopharmacotherapy plus psychotherapy and/or social interventions. However, since these patients are often severely ill, the methodology of clinical trials must be modified, and specific ethical questions may arise such as not to harm by withdrawal of perhaps necessary medications or by the inconveniences of a stringent research methodology. Furthermore, controlled clinical trials are still considered the gold standard to obtain succinct information about a treatment's efficacy and/or safety. In recent years, the field has started to appreciate that such studies also have their limitations. The latter include sample selection, an increasing placebo response rate, high drop-out rates, and not the least they are expensive (Johnston et al. 2006) just to name a few, all of which jeopardize the generalizability of results obtained from such studies. Therefore new strategies are needed to clarify and expand the knowledge regarding some of the key clinical treatment issues. One example is phase IV research, i.e. the systematic and long term assessment of marketed drugs in the field of everyday clinical practice (Linden 1993). This includes so-called efficiency trials, in which the benefit/risk ratio of a given treatment is assessed in a population reflecting clinical practice as closely as possible by evaluating broad outcome measures which have an impact on patients' lives. Another component of such phase IV research are drug utilization observation (DUO) studies (AGNP, Linden et al. 1997). Specific ethical questions are related to the quality and protection of data (see Chapter 26), or to possible influences by industrial sponsors of these expensive studies (see Chapter 4; Helmchen (2004)).

To elucidate the ethical implications of clinical therapy research the example of placebo-controlled clinical trials for psychotropic drugs will be discussed in more detail.

17.3.2 The Example of Placebo-Controlled Clinical Trials

Paragraph 32 of the Declaration of Helsinki, amended by the World Medical Association (WMA) in 2008, reads as follows: 'the use of placebo... is acceptable in studies where no current proven intervention exists' or, if such intervention exists, as an exception 'where for compelling and scientifically sound methodological reasons the use of placebo is necessary to determine the efficacy or safety of an intervention and the patients who receive placebo or no treatment will not be subject to any risk of serious or irreversible harm. Extreme care must be taken to avoid abuse of this option.' (World Medical Association 2008). The introduction of the second statement in 2000 (World Medical Association 2000) with the possibility for placebo controls in patients for whom a standard therapy exists led to heated questions and a 'note of clarification' in 2001 (World Medical Association 2001), which initiated a comprehensive, intensive, and controversial debate between advocates of a 'placebo orthodoxy' versus those of an 'active control orthodoxy' (Emanuel and Miller 2001).

To assess the strength of the arguments it is important: (i) how effective and how tolerated the standard therapy is, i.e. whether it goes convincingly or only marginally beyond the placebo-response-rate, whether it is evidence-based or only empirically established, whether it is effective causally or symptomatically, furthermore (ii) how serious the disease state is, i.e. whether it is life threatening and/or progressive, whether and how much it disables or handicaps the patient functionally or socially, or whether it is of a transient nature, (iii) how harmful the withholding of an effective therapy is, i.e. the probability of the risk of harm or disadvantages, e.g. relapses or suicides. There is a need for intensified ethical analyses and empirical examination as well as of operationalising benefits and risks in the use of placebo.

The essential arguments of this debate, especially about the use of placebo in cases where an effective standard therapy exists, are as follows:

17.3.2.1 Pros

The evidence of efficacy and safety of new drugs must be found on the methodologically safest level, since trials are ethically not justified that, for methodological reasons, are not qualified to permit stringent statements about the efficacy of a new drug (Addington et al. 1997). The placebo-controlled clinical trial is accepted as one of the most effective methods. The ethically relevant reasons to apply this method are: (i) if an experimental drug differentiates from placebo we know for sure that it is efficacious; (ii) placebo control has more statistical power and therefore needs fewer patients; (iii) this advantage reduces not only the duration and costs of the trial but also reduces the number of patients who are exposed to the risk of side-effects of the new drug; (iv) since placebo-controlled trials are more effective, new drugs will be available faster and for less cost; (v) the usual test for equality or non-inferiority of the investigational drug and the control comes to more unequivocal results if the control is a standard therapy the efficacy of which has been established by placebo-controlled trials; nevertheless there remains a gap between the results of clinical trials in highly selected samples and the results of their application in everyday practice (Kienle 2008).

The 2008 amendment of the Helsinki Declaration is said, first, to take into account the complexity of clinical research by calling for the specific determinations of each single trial design; and second, to yield a basis for operationalising criteria for an ethically acceptable use of placebo in clinical trials. As criteria and indications as to their operationalisation the following have been suggested (Carpenter et al. 2003): (i) the probability of a significant clinical advantage of the investigational drug; (ii) the existence of compelling reasons for the use of placebo such as (a) that the standard medication must be assumed to be ineffective for the population from which the trial sample is taken, (b) that the examined condition is prone to considerable fluctuations or to spontaneous remissions, and (c) that a substantial reduction of the exposition risk of the experimental drug is possible; (iii) the selection of probands that reduces the possibility of serious negative consequences and who have full competence to consent; and (iv) a comprehensive benefit-risk-analysis that elucidates the advantages of the use of placebo against its risks for the probands. Accordingly, a placebo control is also seen as acceptable if (a) the placebo rates in the indication range are high, variable, or near the response rates of effective therapies, (b) existing therapies have a high risk of side-effects, or (c) if existing therapies are effective only against single symptoms of the disease. Finally, it should be considered that placebo may be also an effective treatment in many instances at least for a short time.

17.3.2.2 Cons

In general the withdrawal or withholding of an effective medication due to the use of a placebo is not compatible with the ethical primacy of the individual patient's welfare. A pure placebo control is indispensible only for very few questions, and would be less needed if the possibilities of testing for superiority of the experimental drug against the standard therapy or if add-on placebo controls would be used more often.

Even patients with mild diseases and only a small risk of suicidality or relapses or a malignant course of the disorder visit the physician in order to gain help for their complaints. Therefore, the clinical researcher runs two risks: first, informing patients of the inconveniences or disadvantages due to the increasing rigor of scientific methodology may undermine the trust of patients who wish to receive the individually best treatment and not a therapy by chance (randomization) or by blindness of the doctor (double blind) or even without specific efficacy (placebo); second, incomplete information in order to avoid a high primary drop-out rate with the consequence of recruiting a sample with selection bias is unethical. Thus, the ethically adequate information of the patient is a demanding task. Therefore, (i) the recruitment of patients will be difficult and take a long time if they (a) fulfill very strict selection criteria, and (b) will be informed comprehensively that there exists an effective therapy the withdrawal or withholding of which even if for only a short time may prolong their subjective suffering in case they are assigned to the placebo control-group; patients have been shown to be less agreeable to participate in placebo-controlled clinical trials than in trials that use an active comparator (Roberts 1998, Hummer et al. 2003; Fleischhacker et al. 2003); (ii) this may result in a lesser representativeness of placebo-controlled studies; therefore the results are not as unequivocally usable as they are declared; representativeness is also hampered by high drop-out rates. Placebo controlled clinical trials have been demonstrated to have higher drop-out rates – even in the active treatment arms – than trials that use an active comparator (Kemmler et al. 2005); (iii) high drop-out rates – primarily due to refusal to participate, secondarily due to ineffectiveness – may offset some of the potential advantages of placebo controlled clinical trials inasmuch as they may prolong trial duration and cost; (iv) high drop-out rates may also diminish the reliability with which potential therapeutic effects of the investigational drug can be judged; (v) it may be easier to break the blind of a placebo-controlled trial especially if the experimental drug has a reasonably well-known side-effect profile; and (vi) Next to these methodological considerations one also needs to consider whether a clinical trial with limited explanatory power can be justified on ethical grounds.

17.3.2.3 Specific Applications

In general placebo-controlled clinical trials are seen as ethically acceptable in subjectively strongly influenced and *minor* forms of depression, dysthymia, anxiety disorders without major impairments, eating disorders, sexual dysfunctions and sleeping disorders – but only under the condition of a comprehensive information of the patient and his valid consent. Also disorders of cognitive performance or of daily life behaviour in dementias can be added as long as only treatments with questionable or mild effectiveness for these symptoms and the progression of the disease are available. It may be acceptable also in disorders in which an effective therapy exists but is not evidence-based in less severe forms of the disorder.

In the case of *depression* major reasons for the use of placebo controls are taken from a meta-analysis of 75 placebo-controlled trials of antidepressant drugs (Baldwin et al. 2003): a large variation of the placebo response rate between 10 and 50%, and a substantial improvement of 30% in the placebo groups. These results have been explained mainly by the inclusion of mild and reactive depressions into the trials. However, this would justify placebo-controlled trials only in very mild depressions, but not in marked depressions, since for these antidepressant drugs with evidence-based effectiveness can be applied, i.e. a sufficient 'historical evidence' for efficacy is given (Food and Drug Administration 2001). Therefore, in marked depressions the examination of new antidepressant drugs is ethically justifiable only if they will be tested against active controls, at best for superiority (Helmchen 2005), or in cases of therapy resistance against antidepressant standard therapies, or by add-on placebo controls – even if this will require larger samples of patients (Ferriman 2001). Further arguments against the use of placebos in trials of new antidepressant drugs are: (i) the large variation of the placebo response speaks against the use of placebo as a reference for the expected effects of an investigational drug, (ii) in large reviews on studies of antidepressant drugs (Khan et al. 2000, Storosum et al. 2001) both the investigational drug and the standard therapy have been more effective (40%) compared with the placebo rate of 30%, i.e. the test on equality between the investigational and the standard drug would have given evidence for efficacy, and (iii) the standard therapy is effective and thereby the patients of the placebo control group have been withheld an effective therapy, i.e. the ethical principle of the well-being of the patient has been violated (Michels 2000).

Ten years ago in a heated debate reasons were given for the use of placebo controls in trials of new antipsychotic drugs, mainly in *schizophrenic* patients; such reasons were (i) available antipsychotic drugs are only partially effective, and their effectiveness is questionable for symptoms of an unfavorable course such as cognitive disturbances or primary negative symptoms (Carpenter et al. 1997); (ii) available antipsychotics have serious side-effects such as tardive dyskinesias or excessive weight gain, and non-responders are fairly frequent; and (iii) moreover, in placebo-controlled trials every other treatment, e.g. psychosocial interventions, by no means must be withheld. And a careful monitoring as well as a low threshold for an effective medication may reduce the risks for placebo-treated patients at least in the short term. Furthermore, placebo-treated schizophrenic patients seem to have no increased risk for relapses or suicides compared with patients treated with effective antipsychotic drugs (Curson et al. 1986, Khan et al. 2001, Storosum et al. 2003) but more recent findings seem to show an increased risk (Fleischhacker 2009, personal message).

However, these arguments must be examined in each specific case since therapy-resistant patients are probably exactly those who are incompetent to give informed consent; this could lead to a high drop-out rate (Wahlbeck et al. 2001) and prolong the duration of the study and thereby abolish the advantage of the placebo control (Zipursky and Darby 1999). This has to be assumed all the more as the withdrawal of a seemingly inefficient neuroleptic medication before starting a pure placebo-controlled study leads to a change of symptoms for the worse (Pickar and Bartko 2003). Additionally, more than the half of schizophrenic patients in a corresponding investigation refused to participate in a placebo-controlled study (Hummer et al. 2003), and only 30% of European clinicians experienced in trials would perform such a study (Fleischhacker and Burns 2002).

17.3.3 Conclusion

The lack of clarity of the WMA 'Note of Clarification' from 2002 provoked a manifold and intense discussion that made clear and specified the methodological and ethical problems of placebo-controlled clinical trials, particularly with new antidepressant drugs. Positions were differentiated: authors who considered as necessary placebo-controlled trials for evidencing the effectiveness of new antidepressant drugs tried to avoid risks for the patients by excluding those with severe depressions or suicidality (Möller 2005); on the other hand placebo-critical authors no longer assumed as easily that every placebo-controlled study was ethically unjustifiable. Most important for the further development of efficient and ethically acceptable controlled clinical trials will be (i) a reliable methodological and ethical evaluation of every single study protocol as well as (ii) a fair practice of gaining a fully informed consent, and (iii) the final publication of the results including all significant framework conditions. However, there exists some doubt that the FDA will accept the more restrictive formulation of the 2008 amendment of the Helsinki Declaration, and it cannot be excluded that a forthcoming revision of the GCP will fall behind it (Meyer 2008).[2] Reason for such doubt is given by a discrepancy between the cautiously written recommendations of the FDA (Food and Drug Administration 2001) and the EMEA (European Agency for the Evaluation of Medicinal Products 2003) as well and the practice of these authorities to request positive results from at least two placebo-controlled studies before licensing a new drug, i.e. considerably more than the written recommendations demand (Baldwin et al. 2003).

This shadowy discrepancy between theory and practice indicates that the licensing authorities, who are obliged to protect the population, value the scientific clearness of the evidence of effectiveness more highly than the ethical principles that oblige the physician primarily to further the welfare of the individual patient (Michels 2000, Taupitz 2002). This weighting contradicts paragraph 6 of the Helsinki Declaration that 'the well-being of the individual research subject must take precedence over all other interests.' (World Medical Association 2008). Accordingly the Additional Protocol to the Convention on Human Rights and Biomedicine of the Council of Europe, which has been accepted by the Council of Ministers of the European Union and is therefore legally binding, states similarly in article 23.3 that 'the use of placebo is permissible where there are no methods of proven effectiveness, or where withdrawal or withholding of such methods does not present unacceptable risk or burden' (Council of Europe 2003) (see Chapter 5).

Therefore, it remains a task of ethics committees to develop measures for evaluation of the ethical questions that are raised by the complex problems of elaborating an unequivocal evidence of the effectiveness and safety of new drugs (Garattini et al. 2003). Main points of evaluation will be the validity of the consent of patients who are fully informed about the risks and burdens of a delay of an effective therapy, and, second, how acceptable these risks and burdens are. This is valid particularly for placebo controls in disease conditions that are neither life threatening nor are without an established therapy, i.e. that belong to the wide range between a clear acceptability of placebo controls in the case of lack of effective treatments on the one hand and an unequivocal inadmissibility in serious diseases for which an effective treatment exists on the other hand.

The criteria and reasons that have to be considered by ethics committees in study designs with placebo controls are as follows: (i) which objectives will be aimed at, e.g. efficacy or side-effects or quality of life, and (ii) by which measures (active or placebo control, test of superiority, equality or noninferiority) these objectives can be achieved most efficiently. The test measures must be compared by systematic and evidence-based methods both with regard (a) to the previously defined difference (delta) of endpoints between the index and the control group, which is necessary to accept it as of clinical relevance for both potential benefits and risks (Kim 2003), (b) to the needed sample-size as to the number of patients who are burdened by the possibility that they will receive either a non-active or an ineffective treatment with

---

[2] see also the critical comment (Kimmelman et al. 2009) on the October 2008 decision of the FDA (FDA 2008) to discontinue its reliance on the declaration and to substitute the International Conference on Harmonisation's Guideline for Good Clinical Practice (GCP).

potential side-effects (Garattini et al. 2003). If every single study design would be evaluated by such, of course, large-scale procedures, it could be possible in individual cases to achieve a maximum of safety of evidence by a placebo-controlled trial without seriously impairing the well-being of the patient. Only if in the special case the objective of the investigation is undisputed and can by no means be achieved by active or add-on placebo controls (or other methods) but only by a pure placebo control, and, furthermore, if consent can be expected of the fully informed patient because of only light burdens and risks, this procedure may be seen as ethically acceptable. In such cases the assessment of the capacity to consent is specifically important in patients with mental disorders. Because 'ultimately, the best test of the appropriateness of including a placebo arm in a trial design will be whether the patient's consent to participate can be confidently given, and whether the clinician's request for that consent arises from a state of clinical equipoise and of moral equilibrium.' (Editorial 2002) Therefore, a placebo-controlled trial of a new drug will be possible at most in so-called minor mental disorders, whereas at least a pure placebo control will not be ethically acceptable in most cases of depressive and schizophrenic disorders.

References

Addington D, Williams R, Lapierre Y et al (1997) Placebos in clinical trials of psychotropic medication. Can J Psychiatry 42(3):6

Appelbaum P, Redlich A (2006) Use of leverage over patients' money to promote adherence to psychiatric treatment. J Nerv Ment Dis 194:294–302

Baldwin D, Broich K, Fritze J et al (2003) Placebo-controlled studies in depression: necessary, ethical and feasible. Eur Arch Psychiatry Clin Neurosci 253:22–28

Brockmann H (2002) Why is less money spent on health care for the elderly than for the rest of the population? Health care rationing in German hospitals. Soc Sci Med 55(4):593–608

Carpenter WT Jr, Appelbaum PS, Levine RJ (2003) The declaration of Helsinki and clinical trials: a focus on placebo-controlled trials in schizophrenia. Am J Psychiatry 160:356–362

Carpenter WT Jr, Schooler NR, Kane JM (1997) The rationale and ethics of medication-free research in schizophrenia. Arch Gen Psychiatr 54(5):401–407

Council of Europe (2003) Draft additional protocol to the convention on human rights and biomedicine, on biomedical research. Steering Committee on Bioethics, Strasbourg, 23 Aug 2003

Curson DA, Hirsch SR, Platt SD et al (1986) Does short term placebo treatment of chronic schizophrenia produce long term harm? Br Med J (Clin Res Ed) 293(6549):726–728

Editorial (2002) The better-than-nothing idea: debating the use of placebo controls. CMAJ 166(5):573

Emanuel EJ, Miller FG (2001) The ethics of placebo-controlled trials – a middle ground. N Engl J Med 345:915–919

European Agency for the Evaluation of Medicinal Products (EMEA/CPMP)(2003) Position statement on the use of placebo in clinical trials with regard to the revised declaration of Helsinki. http://www.emea.eu.int/pdfs/human/press/pos/174240len.pdf

Ferriman A (2001) World Medical Association clarifies rules on placebo controlled trials. BMJ 323:825

Fleischhacker WW, Burns T (2002) Feasibility of placebo-controlled clinical trials of antipsychotic compounds in Europe. Psychopharmacology (Berl) 162(1):82–84

Fleischhacker WW, Czobor P, Hummer M et al (2003) Placebo or active control trials of antipsychotic drugs? Arch Gen Psychiatr 60:458–464

Food and Drug Adminstration (2001) Guidance for industry: E 10 Choice of control group and related issues in clinical trials. http://www.fda.gov/cder/guidance/index.htm

Food and Drug Administration (2008) Foreign clinical studies not conducted under an investigational new drug application. Fed Regist 73(82):22800–22816. http://www.fda.gov/cber/rules/forclinstud.pdf. 28 Apr 2008

Garattini S, Bertele V, Bassi LL (2003) How can research ethics committees protect patients better? BMJ 326:1199–1201

Grohmann R, Hippius H, Helmchen H et al (2004) The AMUP study for drug surveillance in psychiatry – a summary of inpatient data. Pharmacopsychiatry 37:16–26

Grohmann R, Ströbel C, Rüther E et al (1993) Adverse psychic reactions to psychotropic drugs – a report from the AMÜP study. Pharmacopsychiatry 26:84–93

Hamann J, Langer B, Winkler V et al (2006) Shared decision making for in-patients with schizophrenia. Acta Psychiatr Scand 114:265–273

Helmchen H (1992) Ethische Aspekte der Pharmakopsychiatrie. In: Riederer P, Laux G, Pöldinger W (eds) Neuro-Psychopharmaka. Springer, Wien

Helmchen H (2001) Ethical implications of assessment and application of therapeutic effectiveness. In: Smelser NJ, Baltes PB (eds) International encyclopedia of the social and behavioral sciences. Elsevier, Amsterdam

Helmchen H (2004) Ethical implications of relationships between psychiatrists and pharmaceutical industry. http://www.wfsbp.org/publications.html

Helmchen H (2005) Ethische Implikationen placebokontrollierter Prüfungen von Psychopharmaka. Nervenarzt 76:1319–1329

Hummer M, Holzmeister R, Kemmler G et al (2003) Attitudes of patients with schizophrenia toward placebo-controlled clinical trials. J Clin Psychiatry 64(3):277–281

Johnston SC, Rootenberg JD, Katrak S et al (2006) Effect of a National Institutes of Health programme of clinical trials on public health and costs. Lancet 367:1319–1327

Kelly M, Dunbar S, Gray JE et al (2002) Treatment delays for involuntary psychiatric patients associated with reviews of treatment capacity. Can J Psychiatry 47:181–185

Kemmler G, Hummer M, Widschwendter C, Fleischhacker WW (2005) Dropout rates in placebo-controlled and active-control clinical trials of antipsychotics – a meta-analysis. Arch Gen Psychiatry 62:1305–1312

Khan A, Khan SR, Leventhal RM et al (2001) symptom reduction and suicide risk among patients treated with placebo in antipsychotic clinical trials: an analysis of the Food and Drug Administration database. Am J Psychiatry 158:1449–1454

Khan A, Warner HA, Brown WA (2000) Symptom reduction and suicide risk in patients treated with placebo in antidepressant clinical trials: an analysis of the Food and Drug Administration database. Arch Gen Psychiatry 57(4):311–317

Kienle GS (2008) Evidenzbasierte Medizin und ärztliche Therapiefreiheit. Vom Durchschnitt zum Individuum. Dtsch Ärztebl 105:1161–1163

Kim SY (2003) Benefits and burdens of placebos in psychiatric research. Psychopharmacology (Berl) 171(1):13–18

Kimmelman J, Weijer C, Meslin EM (2009) Helsinki discords: FDA, ethics, and international drug trials. Lancet 373:13–14

Linden M (1993) Postmarketing surveillance of psychotherapeutic medications: a challenge for the 1990's. Psychopharmacol Bull 29:51–56

Linden M, Baier D, Beitinger H, Kohnen R, Osterheider M, Philipp M, Reimitz DE, Schaaf B, Weber HJ (1997) Guidelines for the implementation of drug utilization observation (DUO) studies in psychopharmacological therapy. The "Phase IV Research" Task-Force of the Association for Neuropsychopharmacology and Pharmacopsychiatry (AGNP). Pharmacopsychiatry 30(1 Suppl):65–70

Linden M, Thiels C (2001) Epidemiology of prescriptions for neuroleptic drugs: tranquilizers rather than antipsychotics. Pharmacopsychiatry 34:150–154

Meltzer HY, McGurk SR (1999) The effects of clozapine, risperidone, and olanzapine on cognitive function in schizophrenia. Schizophr Bull 25:233–255

Meyer R (2008) Deklaration von Helsinki. Besserer Schutz von Patienten in klinischen Studien. Dtsch Ärztebl 105:1964

Michels KB (2000) The placebo problem remains. Arch Gen Psychiatr 57:321–322

Möller HJ (2005) Are placebo-controlled studies required in order to prove efficacy of antidepressants. World J Biol Psychiatry 6:130–131

Pickar D, Bartko JJ (2003) Effect size of symptom status in withdrawal of typical antipsychotics and subsequent clozapine treatment in patients with treatment-resistant schizophrenia. Am J Psychiatry 160:1133–1138

Roberts LW (1998) The ethical basis of psychiatric research: conceptual issues and empirical findings. Compr Psychiatry 39(3):99–110

Roberts LW, Geppert CM (2004) Ethical use of long-acting medications in the treatment of severe and persistent mental illnesses. Compr Psychiatry 45:161–167

Sonntag A, Matschinger H, Angermeyer MC et al (2006) Does the context matter? Utilization of sedative drugs in nursing homes – a multilevel analysis. Pharmacopsychiatry 39:142–149

Storosum JG, van Zwieten BJ, van den Brink W et al (2001) Suicide risk in placebo-controlled studies of major depression. Arch Gen Psychiatr 158(8):1271–1275

Storosum JG, van Zwieten BJ, Wohlfarth T et al (2003) Suicide risk in placebo vs active treatment in placebo-controlled trials for schizophrenia. Arch Gen Psychiatry 60(4):365–368

Taupitz J (2002) Note of Clarification: Kaum zu verantworten. Dtsch Ärztebl 99(7):C-311

Wahlbeck K, Tuunainen A, Ahokas A et al (2001) Dropout rates in randomised antipsychotic drug trials. Psychopharmacology 155:230–233

World Medical Association (2000) Declaration of Helsinki 2000/2002. www.wma.net/e/policy/17-c_e.html.

World Medical Association (2001) WMA clarifies its ethical guidance on the use of placebo-controlled trials. http://www.wma.net/e/policy/. Accessed 4 Jan 2003.

World Medical Association (2008) Declaration of Helsinki. http://www.wma.net/e/policy/b3.htm

Zipursky RB, Darby P (1999) Placebo-controlled studies in schizophrenia – ethical and scientific perspectives: an overview of conference proceedings. Schizophr Res 35:189–200

Chapter 18
Ethical Problems in Psychotherapy

Ulrich Rüger and Christian Reimer

Contents

18.1 Introduction – Treatment Errors and Ethical Problems

S. Freud pointed out a series of ethical problems in the psychotherapeutic treatment of patients (1917, 1938), but a more in-depth discussion of ethical problems of psychotherapy began only in the 1980s (cf. Heigl-Evers and Heigl 1989, Toksoz and Karasu 1980). Tress and Erny (2008) present an overview of the current status of the discussion. Morton and Rapp (1984) feel that there are no basic differences in ethical problems in the varying psychotherapy procedures. In this respect our comments here can be understood as a discussion covering various schools of thought, even when certain ethical problems are given a specific emphasis in individual psychotherapy procedures.

U. Rüger (✉)
University of Göttingen, D-37085 Göttingen, Germany
e-mail: urueger@gwdg.de

H. Helmchen, N. Sartorius (eds.), *Ethics in Psychiatry*, International Library
of Ethics, Law, and the New Medicine 45, DOI 10.1007/978-90-481-8721-8_18,
© Springer Science+Business Media B.V. 2010

In order to assure relevance for the actual practice, the discussion will follow the chronological process of a typical psychotherapy treatment, beginning with the first consultation and working toward the end of the treatment. The guidelines for the further discussion of our topic are the principles of biomedical ethics (see Chapter 8; Beauchamp and Childress 1983, last edition 2008).

First treatment errors in psychotherapy and ethical implications shall be mentioned. Treatment errors in psychotherapy always have the potential to damage. They can cause lasting irritations in the therapeutic relationship up to withdrawal of treatment, e.g. too rapid exposition in behaviour therapeutic treatment with resulting panic attack and impending withdrawal of the patient. Even though in single cases with good therapeutic relationship treatment errors do not necessarily lead to a withdrawal of treatment they cause the risk of needless prolongations of treatment courses. One example might be the speaking about resistance phenomena at the wrong time during analytical psychotherapy. Both examples refer to single interventions with the risk to damage of which a long-term risk-entailing treatment planning has to be distinguished. This is the case e.g. with maleficent treatment errors, e.g. regression promoting methods in the course of psychoanalysis of a structural fragile patient with corresponding danger of psychosis. In case of a psychotic decompensation of a patient during the course of such a treatment the therapist's liability to recourse surely has to be discussed. Severe and always legally punishable errors are e.g. sexual misuse of a patient by a therapist or refusal of neuroleptic medication in the case of threatening psychotic derailment in the course of psychotherapeutic treatment.

Treatment error and violation of ethical principles are different categories. A treatment error is the result of a wrong action/decision of the therapist (violation of state of the art) with legal implications when indicated. Ethical principles rather refer to attitudes (e.g. openness, transparency) and obligations (consideration of the actual state of knowledge). These ethical principles are generally mentioned in the guidelines of the respective associations. Treatment errors are possible when violating the ethical guidelines.

Also in psychotherapy the principles mentioned by Beauchamp and Childress (1983, last edition 2008) have to be applied without restrictions, they are considered in the following chapters. Especially the principle of a professional–patient-relationship plays a role, as this relationship is an essential agent for a successful treatment and disturbances have much more negative results than in other fields of medicine (e.g. surgery, radiology and so on).

18.2 Specific Ethical Implications of Treatment Difficulties and Errors

18.2.1 Diagnosis and Determination of Indication

There is general agreement that psychotherapy uses 'methodically defined interventions that influence in a systematically changing manner mental disturbances that have been diagnosed as illness, interventions that strengthen the individual's ability to overcome these' (cf. Rüger et al. 2009, p. 106). This definition indeed presupposes a diagnosis of a mental disturbance but does not say anything about the kind and extent of the diagnostic procedure. For some therapists the assertion of an ICD diagnosis seems sufficient. However, a careful recording of the pre-history of the illness and a plausible pathogenetic explanation of the situation with a following nosological classification must always be obligatory. A superficial diagnostic procedure and treatment planning based on this do not suffice the obligation of medical or psychotherapeutic care and thus oppose the *principle of avoiding harm.*

A false or incomplete diagnostic procedure before beginning a psychotherapeutic treatment leads to a false indication and also endangers the patient in that an alternative but definitely indicated treatment will not or only with clear delay be started. As opposed to the rest of the medical field, there is hardly any tradition in the field of psychotherapy – especially in psychoanalysis – of differentially diagnostic thinking. The first diagnostic classification of psychopathological phenomena often is barely questioned in the further process of treatment.

However, the diagnostic procedure is not only a nosological classification of a current presentation of illness, but should also include an adequately differentiated recording of the pre-history of the illness and thus of recognizable vulnerabilities of the patient that are stable over time – apparent in relapses of illness. Only in this way can, for example, the indication for a treatment procedure be eliminated that would overburden the patient and create the danger of a psychotic decompensation. In addition this touches on the ethical *principle of avoiding harm,* if, in the framework of the diagnostic procedure, the current family and professional situations of the patient are not adequately recorded and if their implications for the treatment planning are not considered.

With patients with recognized physical illnesses particular care must be taken with the diagnosis. Physically caused psychic disturbances can remain undiscovered if they are classified uncritically in connection with the presentation of disturbances leading to psychotherapy. Thus a careful diagnostic procedure, including other medical findings, is among the ethical obligations of a therapist before beginning a psychotherapeutic treatment, in order to avoid possible damage.

18.2.2 Treatment Planning

A correct diagnostic procedure and a well-founded indication for a psychotherapeutic treatment in principle alone do not assure a helpful treatment planning for the benefit of the patient. Indeed, after a preliminary diagnostic clarification a treatment planning must be developed that takes into consideration the particularities of the individual case in the sense of an 'adaptive treatment indication' (cf. Helmchen et al. 1981). As mentioned above this must adequately take into consideration to what extent the patient can be burdened, but also his current family and social resources. In addition, the goal of the treatment must be realistic, and a completely open treatment planning in the sense of 'psychoanalysis without tendency' in the framework of treatment is ethically not acceptable. Ethic dilemmas can arise in differentiating between seemingly urgently necessary short-term goals (e.g. the symptoms remission of a phobic patient) and desired long-term inner psychic changes (e.g. changes in chronically repetitive pathological relationship conflicts). In connection with the duration of treatment Otto Fenichel's old rule should be given more attention: 'Whenever a therapeutic success seems possible by lesser expenditure, the greater one had better be avoided' (1945, p. 577). Indeed, as early as 1930, as a result of 10 years of therapeutic experience in the Berlin Psychoanalytical Institute, the same Otto Fenichel recommended that we should if possible 'be satisfied with partial success, if this assures the patient freedom from trouble and ability to work' (p. 14).

Especially with patients with serious personality disturbances it is usually more sensible to achieve realistic partial goals in a treatment framework that is limited from the very beginning than to aim for basic structural changes in a year-long process. Most patients with personality disturbances come for treatment for current interpersonal difficulties that make undisturbed participation in the social life of society impossible. Long-term consequences are very unpleasant developments in professional and social areas, which should be avoided in good time.

To postpone short-term successful assistance in connection with an important partial goal (e.g. improvement of socially phobic behavior) with the argument that the patient might then lose his motivation for a long-term treatment considered necessary can be ethically very problematic, especially if the patient's treatment wish is based on described symptoms. At the very least the patient must be included in the decision between two alternative treatment paths (informed consent).

18.2.3 Informed Consent

As in every kind of medical treatment (cf. Faden and Beauchamp 1986), and also before starting a psychotherapeutic treatment, patients are to be informed about the planned treatment method, especially about possible risks and side-effects of the planned treatment. Here, however, there is a serious ethical dilemma that was pointed out years ago by Heigl-Evers and Heigl (1989) and by Helmchen (1998). The customary ambivalence of patients in regard to a psychotherapeutic treatment is usually increased by a listing of general risks and side-effects of psychotherapeutic treatments. A still uncertain motivation for a psychotherapeutic treatment would be endangered again. On the other hand, Heigl-Evers and Heigl felt that it seemed 'morally and ethically not appropriate to begin psychotherapy before the patient has been able to develop motivation for treatment resulting from understanding of the reason, motivation for therapeutic cooperation' (p. 70). In cases like these the authors recommend, before beginning a psychotherapeutic treatment itself, creating a phase in which there is concentration only on the acceptance of the illness. This of course is a challenge to the therapist who must have tolerance for his own impatience! Before beginning the actual treatment, the patient should be informed about undesirable and harmful effects of psychotherapy. According to Lambert and Ogles (2004), this happens in about 10% of the cases, whereby a survey by S. O. Hoffmann et al. (2008) shows that this happens relatively less often with behavior therapy but clearly more often with psychoanalytical treatment.

In every case, before starting the actual psychotherapeutic treatment, the patient should be informed of all diagnostic and therapeutic steps. These include especially (cf. Reimer and Rüger 2006, p. 407):

- An explanation of the planned procedure during treatment and a reason for this;
- Adequate explanation of risks and side-effects;
- Reasons for the choice of the treatment method and adequate comments on possible alternatives;
- Answering all questions in connection with the planned procedure;
- Information that it is possible to terminate the planned treatment at any time; but also information that a premature termination of psychotherapy can be risky.

The adequate answering of these questions is the prerequisite of a good cooperation during the following treatment. The patient's questions and doubts are to be taken seriously and not brushed off as an expression of 'resistance'. *Respect for the autonomy of the patient* requires this.

18.2.4 Concluding a Psychotherapeutic Treatment

In the concluding phase of a psychotherapeutic treatment the *respect for the autonomy of the patient* and *the precept of avoiding harm* play a particular role.

The *respect for the autonomy of the patient* requires timely planning of the concluding phase and definite agreement about this with the patient. This concluding treatment phase requires a realistic discussion about the results achieved during the psychotherapeutic treatment, but also acceptance that the originally planned goals of the treatment might not all have been achieved. In this context narcissistically structured psychotherapists can have a difficult time. At the same time there is the dissolution of a frequently long relationship between the patient and the therapist. If the therapist is not satisfied with his own needs, he can, through his own ambivalence, make it difficult to let go of the patient at the right time.

In addition possible future crises and how to handle these should be anticipated in order to avoid harm to the patient. Strategies of a relapse prophylaxis should be clarified. Maintenance psychotherapy, which can make sense in individual cases, should be discussed (Frank et al. 1991, Miller et al. 2003, Schauenburg and Clarkin 2003). In the conclusion phase too the precept of an adaptive procedure is particularly important that sufficiently takes the particularities of the individual patient into consideration. In this way the therapist shows *respect for the autonomy of the patient* and helps *to avoid harm in the future*. Rigid rules are obsolete here and not ethical.

18.2.5 Surrounding Economic Conditions

Before beginning any psychotherapy, the psychotherapist is obligated to clarify the financing of the treatment with the patient. With privately financed treatment the therapist must take into consideration not only the current financial situation of the patient, but must also anticipate its development during the course of treatment (Knapp and VandeCreek 2008). It would be ethically problematic if a patient would be strongly economically disadvantaged by financing his psychotherapy. Following up on the behavioral progress made with a successful psychotherapeutic treatment would then be considerably handicapped.

If the psychotherapeutic treatment is to be financed by health insurance, further ethical problems arise: the economic means of the health insurance company are limited and are contributed by the community of insured persons themselves. Therefore the resources of the insured community must be used sensibly and purposefully. For this reason psychotherapy methods must have proven their effectiveness in regard to the present illness. If there are several equally effective treatment methods for a certain illness, that method should be chosen that is least expensive. Any other procedure would not be acceptable for social and ethical reasons, because, in that case, there would be the risk of a wrong allocation of money belonging to the community of the insured. This would be against the *principle of justice*.

These social and ethical aspects should always be taken into consideration in the discussion of the expense of psychotherapeutic treatment. Not doing so surely contributed to the crisis of classical psychoanalysis in the Anglo-American region in the 1980s (cf. Wilkinson 1986). In Germany, however, with the acceptance of psychotherapy by health insurance companies the principles of a sensible and purposeful usage of insurance money in the area of psychotherapy were acknowledged. This happened because, beside behavioral therapy, not 'psychoanalysis' but rather limited psychoanalytically reasoned treatment methods with appropriate proof of effectiveness were introduced into general health care. A prerequisite for the expense acceptance of psychotherapy by the legal health insurance companies is a treatment plan that gives hope for treatment success within the quite generous payment limits. Therapists who do not take into consideration the financial possibilities even during the planning of the treatment act against the ethical *principle of avoiding harm*, if the given limits cannot be respected from the very beginning and the patient, dependent on his therapist, is forced to continue the treatment by privately financing it or else by ending the treatment before it is completed. This kind of treatment planning would also mean acting against the *principle of justice*, because the means of the insured community would not be used appropriately, and, at the same time, therapeutic resources could not be used for the treatment of other patients.

An individually ethical dilemma should be considered in a long, drawn-out treatment that can cover a number of years: a psychotherapeutic treatment that ties up the psychic energy of a person during a very intensive and long-term process can, after a certain point in time, prevent this person from using this energy for an independent mastery of his life. Not only the economic interest of the insured community gives reason to limit the expense of psychotherapeutic treatment, but also, in the treatment of the individual patient, his own 'personal economics' must be considered. This will also be influenced by the finiteness of time and certain regularities in the phases of our lives.

The *principle of justice,* just mentioned, is also touched socially and ethically in regard to another just distribution of resources: members of higher educational classes still have an advantage in the realization of psychotherapy. This is true for the treatment indication itself as well as for the choice of a (usually more expensive) treatment method and also for the duration of treatment (cf. Rüger and Leibing 1999). Surely different influence variables are responsible for this statistical distribution, and these are also discussed by the authors. Here too actions against basic principles of ethical behavior are not easily proven in individual cases. Nonetheless the question is valid whether or not this situation contradicts ethical requirements of medical action, as were established by the World Medical Association in 1994.

18.3 Ethical Requirements of the Therapist

18.3.1 Different Standards in the Profession of Psychotherapists

Despite all differences medical education has a comparable international standard. This is not the case for the profession of psychotherapists. The required qualification to work independently and responsibly as psychotherapist differ greatly between countries, not so much in regard of the core competence in the applied method but rather concerning the further obligatory clinical experience of psychotherapists (e.g. in psychiatry). The ethical problems are obvious. Therapists without sufficient experience in the whole range of psychiatric disorders are in certain cases a risk for the patient.

A more detailed discussion of this topic with consideration of all different regulations for the profession of psychotherapists cannot be accomplished here.

18.3.2 Professional Discretion and Protection of Intimacy

As early as the Hippocratic Oath the obligation of the physician to keep professional confidentiality was stated. According to general understanding of the law, the sense and purpose of the professional discretion of the physician is twofold: confidentiality serves to protect the intimate area of the patient and, at the same time, serves the general interest in an effective health service, 'because only if one can depend upon the discretion of the physician can one expect that patients will consult with physicians' (Bockelmann 1968). The interests of third persons must be particularly protected, even with appropriate permission from the patient. Whereas in general medicine information about third persons is usually given by the patient casually and generally not as the main subject of medical examinations and treatments, in psychiatry and psychotherapy this is different: without information from the patient about third persons, e.g. partner, family, and social surroundings, a psychiatric or psychotherapeutic treatment is totally impossible. Thus this area must be especially protected. In this connection see the guidelines by the 'International Committee of Medical Journal Editors' (Vollmann and Helmchen 1996, Vollmann 1996).

It can however be seen again and again that sometimes patient data, even in psychotherapy, are handled insensitively. Some patients recognize themselves in study reports published by their therapist. Some of these study reports are hardly coded, and written permission from the patient was not obtained before the reports were published. The injuries to the patient resulting from this are expressions of unethical behavior and represent an offence against professional confidentiality.

The responsibility of protecting intimacy is valid toward third persons as well. The naiveté with which reports about patients are discussed in study seminars shows again and again how insensitive even psychotherapists can be about this area.

Even when a partner might be called in temporarily during individual psychotherapy, the psychotherapist must be careful not to mention in conversation any information that the patient has given him in the individual situation. Otherwise the patient feels betrayed, rightly so, and the individual therapy can be burdened because of this.

18.3.3 Difficulties in the Psychotherapeutic Practice

What problems can arise in the psychotherapist's professional field, a jungle of complicated subjectivity? The following list gives only a few examples:

- The effect of sympathy, closeness, continuity on persons who have missed experiences in this area and have thus developed a great need for this. This need will create particular forms of closeness and wishes toward the psychotherapist.
- The effect of psychotherapy with a patient on the rest of the patient's life, e.g. on his life partner.
- What effect do the psychotherapist's explanations, commentaries, positions have on the patient? How far does he feel obligated to go, to act, etc. How far does a psychotherapist move into the real life of the patient and prepare fateful decisions, e.g. separation?

The therapist must ask himself these and other questions constantly if he wants to use empathy effectively, the main instrument of his work. This requires of course that his empathy for the patient is really present without interruption.

Everyone who works in the area of psychotherapy knows from his own experience how stressful the therapist–patient relationship can be, e.g. in

- The permanent threat to the limits and integrity of the psychotherapist by disturbed patients;
- The obligation to preserve a friendly, strong working relationship even in the face of inward and outward resistance;
- The confrontation with one's own memories and unpleasant biographical details, which are set free by considering the biography of the patient;
- At the same time the repeated recognition of at least partial aspects in common with the patient and the feelings resulting from that;
- Relatively little success in therapy, at least in regard to healing, as well as the resulting disappointments and insults all the way to resigned and pessimistic feelings and behavior toward patients.

The question must thus be asked what such a permanent emotional overburdening does to the therapist himself. Often there are also guilt feelings that not enough was done for the patient. Thus therapists often find themselves in the role of mothers and fathers who worry permanently about their deficient children and often feel an obligation to be available for consultation basically around the clock.

The permanent confrontation with psychically disturbed patients can lead to somatic and/or psychic symptoms (e.g. tension, exhaustion, fatigue, and sleep disturbances) in psychotherapists themselves, if they do not have secure inner stability at their disposal. This in turn can impair the quality of the work with patients.

18.3.4 Misbehaviour of the Psychotherapist

It is not difficult to distance oneself from a serious legally punishable error or to criticize a serious irresponsible error. However, the most frequent treatment errors with ethical implications are probably the result of 'lighter' acts against ethical principles. This is often an unreflected functionalisation of the patient by his therapist, not an error in spite of knowing better! In these cases the patient takes on an important function in preserving the psychic balance of the therapist. He can become the therapist's own object in the regulation of his narcissistic balance or can contribute to the 'solution' of inner psychic conflicts or deficits of the therapist (cf. Rüger 2009). Because of their great meaning various forms of treatment errors and their ethical implications will be discussed at length in the following pages.

18.3.4.1 Exploiting Behavior

Annemarie Dührssen (1969) pointed out frequent subconscious tendencies of therapists who, for economic reasons, extended treatments of well-paying patients, especially if these treatments ran their course in a peaceful and positively colored relationship. This is ethically not acceptable if patients must finance this themselves and thus reach the limits of their financial possibilities. However it is just as ethically unacceptable if the treatment is extended unnecessarily at the expense of the health insurance.

After completing the psychotherapy financed by the health insurance, a therapist demanded that a patient herself finance further analysis. The patient, a single mother of a child, had to refuse for economic reasons. She reacted with strong guilt feelings and, for many years, avoided reestablishing a necessary psychotherapeutic treatment. A retrospective reconstruction of initial situation and treatment course clearly showed that a satisfactory result could have been achieved with the benefits provided by the Health Insurance (detailed discussion in Rüger 2009).

Indeed, many psychotherapists, especially psychoanalysts, seem to have a kind of blind spot in regard to the economic situation of their patients; this was shown in a study by Rüger et al. (1996). This surely has a long tradition: psychoanalysis and the psychoanalytical illness model were originally developed for a clientele with no serious economic problems.

18.3.4.2 Acting Out One's Own Needs

Therapists are always in danger of satisfying their own, non-resolved relationship needs by 'long-term treatments' of their patients. Whenever we therapists, in spite of longer treatment, consider a patient 'not yet ready' or when we say 'not everything has been worked out sufficiently', we must examine this verdict to determine whether we can perhaps not yet let go of this or that patient. We have to consider the function of this patient in the context of our own needs.

It goes without saying that, during very short-term treatments, we must always consider whether the patient's pseudoautonomous behavior leads us to leave him alone too early. When, however, after 300 treatment hours of analytic psychotherapy one can read in the report, 'A premature (sic!) end of treatment at this time would lead to a retraumatisation of the patient', then one must ask what kind of treatment planning can lead to this kind of situation.

As Reimer and Rüger reported in 2006, a patient had to try for years to be released finally from her therapist after long years of psychoanalysis. Even after the patient left, the single, childless analyst bombarded the patient with letters for another year, in which she demanded that the former patient return to analysis, be*cause important things had not yet been worked out* (p. 400).

18.3.4.3 Violating the Rule of Abstinence – Sexual Misuse

The sexual misuse of a patient by a therapist is an extreme form of unreflected therapeutic behavior. This violation of the rule of abstinence is the expression of a serious disturbance of the therapist's empathy (Reimer and Rüger 2006). Sexual misuse is a serious example of unethical behavior, and legal punishment here is only right. Although, in questionnaires, therapists agree completely that this kind of behavior is unacceptable (cf. Conte et al. 1989), up to 10% of all psychotherapists admit to their own violation of the rule of abstinence (Bouhoutsos et al. 1983, Gartrell et al. 1986, Herman et al. 1987).

The corresponding investigations show that most of these therapists seem to be able to legitimize their actions to themselves. They do not call it transference but rather 'true love'. The behavior is seen as a 'one-time occurrence', but at the same time there is the denial of the fact that there is always a necessary preliminary development. The concept of 'time out' is named with the strange implication that the rule of abstinence is no longer valid after the therapy session or after conclusion of the therapy!

The consequences of this kind of border trespassing by the therapist are extremely serious, because the confidence wishes of a human being are exploited by a paternal figure of protection. The constellation is similar to the sexual misuse of a child, because it is not a relationship between two more or less autonomous adult persons. S. Freud warned decades ago of the misusing exploitation of a transference relationship by unscrupulous physicians (1917, p. 482).

Patients are usually psychically seriously damaged after sexual misuse by a therapist. The consequences are often long-term depressive developments with suicidal tendencies as well as alcohol and drug misuse along with many psychic and psychosomatic symptoms (Schoener et al. 1984).

18.3.4.4 Acting Out One's Own Need for Dominance

In these cases the psychotherapist knows what is good for the patient; he determines the directions of important life choices, without reflecting on the reasons for his own preference for a particular one of several decision possibilities for the patient.

> A therapist encouraged his patient to expand his hotel in grand style, in spite of total debt and the following bankruptcy. The therapist completely misestimated the patient's economic condition and his sense of reality. After completing the treatment the condition of the patient was worse than beforehand!

In 1938 Sigmund Freud warned in his *Outline of Psychoanalysis* against misusing one's influence on the patient. 'Regardless of how tempting it might be for the analyst to become the teacher, example, and ideal for others and to create human beings according to his own example, he should never forget that this is not his role in the analytical relationship. Indeed, he would be untrue to his task if he let himself be carried away by his own wishes' (GW XVII, p. 101). To the contrary, in all of his efforts the analyst should 'respect the patient's own personality'.

In this connection, a concept formulated by Harald Schultz-Hencke (1951) can be quite helpful, but only if it is not misunderstood but rather used critically. Schultz-Hencke recommends that, during treatment, the therapist should keep a mental image of his future patient, as the patient would be if he were freed from his neurotic handicaps. He recommends that the therapist use this as an inner guideline for the treatment. He writes, 'During the first part of the analysis the analyst should keep, not binding, of course, a mental image, an idea of his patient' (1951, p. 243 ff.). This concept of the *image of the patient, the idea of the patient* must leave room for an autonomous development and in no way legitimizes an influence on specific life choices. Clarification and recognition of possibly irrational choice determinants are the responsibility of the therapist – but not the recommendation of specific choices!

This is also true for the area of partnership and family. Here there should be a clarification of conflict constellations and negative interpersonal behaviour patterns – but not suggestions of separation, divorce, etc. and certainly not pressure to make certain decisions.

Here, different models of the Physician–Patient-Relationship (Emanuel and Emanuel 1992) can compete: in the paternalistic relationship model the therapist knows what is good for the patient, but does not show sufficient respect for the right of autonomy. In contrast, the therapist as lawyer for the benefit of the patient is also obliged to protect him from severe self-damaging behaviour and decisions. Usually, corresponding ethical predicaments can only be solved for the individual case.

18.3.4.5 The Patient as the Therapist's Own Object in the Regulation of His Narcissistic Balance

In this case the patient becomes the therapist's own object that is supposed to elevate the therapist in a certain way, i.e. the patient is to help regulate the therapist's narcissistic balance. However, several questions arise here immediately: is it narcissistic misuse if a therapist is proud of his patient's success? Is a therapist not allowed to be happy about the development of his patient? Every time a patient becomes a substitute for something missing in the therapist's life, whenever the therapist delegates his own impulses onto the patient, and whenever the therapist does not take these factors into consideration, there is a danger of harming the patient.

This is indeed an important aspect of therapeutic action: the profession of the psychotherapist involves more reflection and less activity. But just those patients with strongly active and decision-making professions run the risk of becoming a substitute for the therapist for something missing in his own life. This kind of situation becomes harmful for the patient if the therapist does not reflect on the situation adequately and if the patient thus possibly becomes a long-lasting function in the regulation of the therapist's own narcissistic balance. Prominent patients seem to be in special danger of contributing to the regulation of the therapist's own narcissistic balance – often not benefiting the patient! (cf. Mecacci 2002).

18.3.4.6 The Narcissistic Overrating of the Chosen Treatment Method

Therapists must be able to value and to view positively the treatment methods that they use. Every surgeon must do the same. However, the idealization and the overrating of a treatment method have negative consequences. In these cases the treatment method used is highly narcissistic and is often the prerequisite for membership in a professional society that is considered important. In the individual case other possible treatment methods are often looked down upon or not considered at all.

Overrating one's own treatment methods often leads to ignorance – or in the best case to obsolete knowledge – of alternative treatment possibilities and especially of possibly necessary treatment measures to accompany psychotherapy, e.g. drug therapy.

Hoffmann et al. (2008) place the responsibility for therapy damage occurring more frequently in classical psychoanalyses in comparison with other forms of psychotherapy onto frequently non adaptive procedures in classic psychoanalysis. In such cases 'being stuck in one's own conclusive logic of psychoanalysis' seems, when a critical development is reached, to be particularly dangerous for further development. In such crises usually 'more of the same' follows (more transference interpretation, more regression, more frequent sessions, etc.); less frequently a change of setting, of the therapy method, or of the analyst. In the individual case it must be decided how seriously the borderline to an error has been crossed. An internist would surely have to justify putting a patient onto a medication that would, taking all of the findings into consideration, surely lead to a toxic reaction. He would certainly be called to justify himself if he did not eliminate the medication after this kind of reaction but rather increased the dosage!

Sometimes there is a missionary element in treatment methods without critical indication. This is usually well-meaning, but the conviction of the therapist involved is often combined with a helper attitude resistant to criticism. Thus, according to investigations by Mayou et al. (2000), so-called debriefing interventions after accidents and catastrophes more frequently lead to negative results than to helpful ones. Lieberei and Linden (2008) describe the very desolate development of a patient after such an invasive treatment, who was completely healthy before the accident and, after the accident, showed no signs at all of a negative resolution of the accident (at which she was not even present).

The overrated choice of a treatment method apparently prevents too easily a sufficiently critical indication determination related to the individual case. In regard to the avoidance of a necessary psychopharmacological treatment within psychotherapy we can refer to the very detailed discussion of the Osheroff vs. Chestnut-Lodge case (cf. Stone 1991, Thiel et al. 1998). The present status of knowledge considers it no longer acceptable to reject in principle treatment modifications and combinations. This will be decided by the peculiarities of the individual case.

> A patient with a borderline personality disturbance looks back at his serious social difficulties prior to medication with a low dosage of a neurolepticum. "Previously I could only be angry, destructive, could not think about conflicts, always had to act immediately! Nonetheless, I always knew somehow that something was wrong, but I could not stop it; it just happened." Correspondingly, previous psychotherapeutic treatments had been concluded too soon and had resulted in no longer controlled, destructive actions. In the framework of a combined psychotherapy and psychopharmacological therapy the patient became more and more able to control his highly destructive impulses and to recognize and reflect on them as inner fantasies.

18.3.4.7 Value Concepts and Ideologies

We must differentiate between two areas of values and norms for therapists:

- Values and norms related to health
- Values and norms related to ideology

The idea of man and the ideas of the therapist about what is necessary for the healthy and well-balanced soul of a human being determine to a great extent the goals of a treatment. This is disadvantageous for the individual patient if these ideals flow too rigidly into the therapeutic procedure and thus the peculiarities of the individual case are not sufficiently taken into consideration and there is possibly too much stress on the patient. The patient then feels pressure to fit in and accepts solutions that might demand too much of him.

An example of this is the case when therapists have ideal notions about possibilities of autonomy development of their patients – without considering the opposite pole of human existence, the need to belong and to be dependent. Autonomy and dependence are in conflict with each other in every human being during his entire life; even healthy persons cannot solve this completely. This is part of human existence. The very different ways of resolving this conflict tension are a part of our personality. In spite of this many psychotherapists seem to value 'autonomy' highly, whereas 'dependence' is seen as a negative factor. However, in most cases the goal should be to enable the patient to find a balance in this human conflict tension that is better and perhaps more suitable to his age. The patient must be allowed to come to his own solutions that are possibly also different from those that the therapist has found for his own life.

The *ideological values and norms* represented by the psychotherapist himself can collide with those of the patient and can possibly negatively influence the treatment. The Geneva Statement of the World Medical Association is absolutely clear in its statement about a physician's obligation not to be influenced in one's medical responsibilities by a patient's ideology and 'political membership'. But previous life convictions are questioned in psychotherapeutic treatment much more often than in the rest of the medical field, at least as far as they play a role in the presentation of the current illness. However, the treatment wish of the patient is always (at least at first) related to a therapy of the current illness as well as its pathogenetic conditions – not necessarily a review of his entire life history.

There is general agreement that a therapist's ideological influence of the patient is obsolete (Reimer and Rüger 2006, p. 409). This probably is seldom consciously intended. However, the therapeutic cooperation with a patient becomes difficult, if he has ideological convictions that are widely diverse from those of the therapist. According to Dührssen (1972), at least a partial identification with the norms and values of a patient must be possible in order to understand his inner world. If there is strong incompatibility here, it would probably be better to recommend a patient to another therapist. But this does not solve the basic ethic dilemma: the therapist has a conflict between his *obligation to help* and a possibly serious inner rejection of the patient on the other hand. In solving this dilemma a recommendation by Annelise Heigl-Evers can be helpful; she recommends, in the case of 'hardly bearable patients', an attitude of pity and respect for the person of the patient in 'his life's history existence along with its fateful involvements, especially the elements of guilt and innocence' (Heigl-Evers et al. 1997, p. 172).

18.4 Ethical Consequences for the Therapeutic Practice

During their training therapists are usually taught to reflect carefully on every exchange and the feelings and fantasies thus caused by the patient. At the same time inappropriate involvement in a therapeutic relationship cannot always be avoided. It is important to recognize such a situation in good time. Supervision (with leader) and intervision (inter-cooperative supervision without leader) are helpful here; these assist in preserving one's self-reflective capabilities as well as recognizing the background and possible rationalizations of borderline behavior in good time. The occasional repetition of an in-depth self-examination even after completing psychotherapeutic training is also recommended.

Establishing the subject of ethics in the framework of psychotherapeutic training is generally commendable. Sensible topics of such ethics seminars could be, for example:

- Professional stress on the part of the psychotherapist and prevention of this stress;
- Ethical problems in dealing with seemingly difficult patients;
- Framework conditions of psychotherapeutic work in connection with ethical aspects;
- Prevention of misuse tendencies, especially in consideration of the frequent taboo of sexual transference feelings on the part of therapists.

After completing the self-examination, which is usually very demanding time-wise, financially, and emotionally, some psychotherapists seem to find it difficult to carry out regularly or at least occasionally their own consultation, supervision, or intervision. The tendency toward 'splendid isolation' is surely the greatest danger for the psychotherapist himself, for his work, and thus, finally, for the patient.

A psychotherapist, who professionally works continually in a relationship conflict of complicated subjectivity, has a certain ethical obligation toward himself and his patient to create his own satisfactory private life. A therapist can, in the long run, only be a good therapist, if he has a good private balance and does not have to involve his patients to satisfy his own personal needs.

References

Beauchamp TL, Childress JF (1983) Principles of biomedical ethics, last edition 2008. Oxford University Press, New York, NY, Oxford
Bockelmann P (1968) Strafrecht des Arztes. Thieme, Stuttgart
Bouhoutsos J, Holroyd J, Herman H, Forer BR, Greenberg M (1983) Sexual intimacy between psychotherapists and patients. Prof Psychol Res Pr 14:185–196
Conte HR, Plutchik R, Picard S, Karasu TB (1989) Ethics in the practice of psychotherapry: a survey. Am J Psychother 43:32–42
Dührssen A (1969) Möglichkeiten und Probleme der Kurztherapie. Z Psychosom Med 15:229–238

Dührssen A (1972) Analytische Psychotherapie in Theorie, Praxis und Ergebnissen. Vandenhoeck und Ruprecht, Göttingen

Emanuel EJ, Emanuel LL (1992) Four models of the physician–patient-relationship. J Am Med Assoc 267:2221–2226

Faden R, Beauchamp TL (1986) A history and theory of informed consent. Oxford University Press, Oxford

Fenichel O (1930) Statistischer Bericht über die therapeutische Tätigkeit 1920 bis 1930. In: Deutsche Psychoanalytische Gesellschaft (Hrsg) Zehn Jahre Berliner Psychoanalytisches Institut. Internationaler Psychoanalytischer, Wien, S 13–19

Fenichel O (1945) The psychoanalytic theory of neurosis. Norton, New York, NY

Frank E, Kupfer DJ, Wagner EF, McEachran AB, Cornes C (1991) Efficacy of interpersonal psychotherapy as a maintenance treatment of recurrent depression: contributing factors. Arch Gen Psychiatry 48:1053–1059

Freud S (1917) Vorlesungen zur Einführung in die Psychoanalyse, GW XI 1-482. Fischer, Frankfurt am Main

Freud S (1938) Abriss der Psychoanalyse, GW XVII 63-138. Fischer, Frankfurt am Main

Gartrell N, Herman J, Olarte S, Feldstein M, Localio R (1986) Psychiatrist–patient sexual contact: results of a national survey. I: prevalence. Am J Psychiatry 143:1126–1131

Heigl-Evers A, Heigl FS (1989) Ethik in der Psychotherapie. Psychother Med Psychol 39:68–74

Heigl-Evers A, Heigl FS, Ott J, Rüger U (1997) Lehrbuch der Psychotherapie, 3. Aufl. Gustav Fischer, Stuttgart

Helmchen H (1998) Ethische Implikationen von Psychotherapie. Nervenarzt 69:78–80

Helmchen H, Linden M, Rüger U (Hrsg) (1981) Psychotherapie in der Psychiatrie. Springer, Heidelberg, New York, NY

Herman J, Gartrell N, Olarte S, Feldstein M, Localio R (1987) Psychiatrist–patient sexual contact: results of a national survey. II: psychiastrists' attitudes. Am J Psychiatry 144:164–169

Hoffmann SO, Rudolf G, Strauss B (2008) Unerwünschte und schädliche Wirkungen von Psychotherapie. Psychotherapeut 53:4–16

Knapp S, VandeCreek L (2008) The ethics of advertising, billing and finances in psychotherapy. J Clin Psychol 64:613–625

Lambert MJ, Ogles BM (2004) The efficacy and effectiveness of psychotherapy. In: Lambert MJ (ed) Bergin u. Garfield's Handbook of Psychotherapy and Behaviour Change, 5. Aufl. Wiley, New York, NY, S 139–193

Lieberei B, Linden M (2008) Unerwünschte Effekte, Nebenwirkungen und Behandlungsfehler in der Psychotherapie. Z Evid Fortbild Qual Gesundhwes 102:558–562

Mayou RA, Ehlers A, Hobbs M (2000) Psychological debriefing for road traffic accident victims. 3-Year follow-up of a randomized controlled trial. Br J Psychiatry 176:589–593

Mecacci L (2002) Il caso Marilyn M. e altri disastri della psicoanalisi. Gius. Laterza & Figli Spa, Roma-Bari. Deutsch: Mecacci L, Der Fall Marilyn Monroe und andere Desaster der Psychoanalyse. Btb, München 2004

Miller M, Frank E, Cornes C, Houck P, Reynolds C (2003) The value of maintenance interpersonal psychotherapy (IPT) in older adults with different IPT foci. Am J Geriatr Psychiatry 11:97–102

Morton S, Rapp MD (1984) Ethics in behaviour therapy: historical aspects and current status. Can J Psychiatry 29:547–550

Reimer C, Rüger U (2006) Ethische Aspekte der Psychotherapie. In: Reimer C, Rüger U (Hrsg) Psychodynamische Psychotherapien, 3. Aufl. Springer, Heidelberg, S 391–412

Rüger U (2009) Über unreflektiertes Funktionalisieren von Patienten in der Psychotherapie und seine schädlichen Auswirkungen. PTT-Persönlichkeitsstörungen 13:31–41

Rüger U, Dahm A, Kallinke D (eds) (2009) Faber-Haarstrick Kommentar Psychotherapie-Richtlinien, 8. Aufl. Urban und Fischer, München

Rüger U, Haase J, Fassel K (1996) Was Psychotherapeuten vom Leben ihrer Patienten (nicht) wissen. Z Psychosom Med 42:329–342

Rüger U, Leibing E (1999) Bildungsstand und Psychotherapieindikation – der Einfluss auf die Wahl des Behandlungsverfahrens und die Behandlungsdauer. Psychotherapeut 44:214–219

Schauenburg H, Clarkin J (2003) Relaps in depressive disorders – is there a need of maintenance psychotherapy. Z Psychosom Med Psychother 49:377–390

Schoener G, Milgrom JH, Gonsiorek J (1984) Sexual exploitation of clients by psychotherapists. Women Ther 3:63–69

Schultz-Hencke H (1951) Lehrbuch der analytischen Psychotherapie. Thieme, Stuttgart

Stone A (1991) The Osheroff debate: final (letter). Am J Psychiatry 148:388–390

Thiel A, Freyberger HJ, Schneider W, Schüssler G (1998) Psychotherapie versus Pharmako-therapie. Der Fall Osheroff vs. Chestnut Lodge. Psychotherapeut 43:39–45

Toksoz BY, Karasu B (1980) The ethics of psychotherapy. Am J Psychiatry 137:1502–1512

Tress W, Erny N (2008) Ethik in der Psychotherapie. Psychotherapeut 53:328–337

Vollmann J (1996) „Informed Consent" des Patienten zur Publikation von Kausistiken. Nervenarzt 67:422–426

Vollmann J, Helmchen H (1996) Publishing information about patients. BMJ 312:578

Wilkinson G (1986) Psychoanalysis and analytic psychotherapy in the NHS: a problem for medical ethics. J Med Ethics 12:87–90

Chapter 19
Neuromodulation – ECT, rTMS, DBS

Matthis Synofzik and Thomas E. Schlaepfer

Contents

Abbreviations

CT   Computed Tomography
DBS   Deep-Brain Stimulation
ECT   Electroconvulsive Therapy
MD   Major Depression
OCD   Obsessive-Compulsive Disorder
rTMS   Repetitive Transcranial Magnetic Stimulation
STN   Subthalamic Nucleus
TMS   Transcranial Magnetic Stimulation

T.E. Schlaepfer (✉)
Department of Psychiatry and Psychotherapy, University Hospital, Sigmund-Freud-Strasse 25,
53105 Bonn, Germany
e-mail: schlaepf@jhmi.edu

H. Helmchen, N. Sartorius (eds.), *Ethics in Psychiatry*, International Library
of Ethics, Law, and the New Medicine 45, DOI 10.1007/978-90-481-8721-8_19,
© Springer Science+Business Media B.V. 2010

19.1 Introduction

Electrical techniques that modulate brain processes have already been applied in psychiatry for many decades but the ethical debate surrounding its clinical application has still not calmed down: Electroconvulsive therapy (ECT), available for more than 60 years now and still the most acutely effective antidepressant treatment available (Eranti et al. 2007), raises genuine public and professional concerns about its nature as a medical treatment, receives negative and stigmatizing perceptions and is subject to large variability in standards and practice (Eranti and McLoughlin 2003). Transcranial magnetic stimulation (rTMS) has just been approved for treatment-resistant depression in countries like Canada or the European Union, but effectiveness for other psychiatric diseases and potential subtle brain damage still remain controversial. The most recent neuromodulation technology in clinical application, deep brain stimulation (DBS), evokes reminiscences to early psychosurgery and – due to its invasive nature – is skeptically scrutinized for its putative side effects, especially unwanted modifications of feelings, experiences, or even personal identity.

Ethical discussion of these neuromodulation techniques has often been hampered by misleading arguments and unclear ethical criteria. We demonstrate here (i) some of those misleading arguments which should be avoided in the ethical discussion of neuromodulation for neuropsychiatric indications, and suggest, as a productive alternative, (ii) ethical criteria than can be applied to perform a technique-specific, evidence-based ethical analysis of ECT, rTMS and DBS.

19.2 Misleading Criteria

19.2.1 Invasiveness

Neuromodulation techniques might be ethically evaluated by their degree of 'invasiveness': The more invasive a neuromodulation technique, the more ethically problematic its application. Based on such an argumentation one would come to the conclusion that, for example, psychotherapy is less ethically problematic than psychoactive drugs, which would in turn be less problematic than rTMS and ECT and that these applications would again imply less ethical concerns than DBS. Although this criterion seems to capture our intuitions, it does not bear up against stringent conceptual analysis: applications of a seemingly non-invasive method, e.g. wrongly conducted psychotherapy, or of a transient, minor invasive method, e.g. rTMS of limbic areas in an emotionally instable person, might lead to more harmful effects for a certain person than certain major invasive applications, e.g. correctly inserted, optimally adjusted DBS (Synofzik 2005). Thus, it is not normatively decisive whether a neuromodulation technique is per se invasive or not, but which benefits and risks it involves.

In fact, there do not seem to be such things as 'non-invasive' psychiatric treatments – this would be an oxymoron. Even psychotherapy presents an invasive method: by intervening in psychological states and processing, it alters brain metabolism (Linden 2006). The assumption of psychotherapy as a non-invasive method is completely misleading in that it seems to build on an outdated dualism between psyche and brain. Psychotherapy and TMS present methods of external neuromodulation which – like brain-internal neuromodulation – might also adversely affect circuits other than those targeted by the procedure, and as some of these techniques involve both excitation of some neurons and inhibition of others, the effects of external stimulation might likewise be difficult to control (Glannon 2006).

19.2.2 Identity

In a similar vein, it is not helpful to refer to the concept of 'identity' or 'personality' to obtain a (negative) criterion for ethically evaluating neuromodulation techniques. Recent analyses have warned that for psychiatric patients who are effectively treated with neuromodulation procedures 'the cure may come at the cost of their identities, their selves' (Glannon 2006, p. 46). But what is meant if one sees patients' identity endangered by neuromodulation techniques? Conceptually, two notions of identity should be distinguished which can both be applied to persons (Birnbacher 2006): The *numerical identity* of a person consists in a specific logical relation that is fulfilled if a person can be considered as one and the same at two different points in time – even if it might have undergone some changes in-between. Numerical identity is strictly dichotomous: If a person can be considered to be one and the same person at t_2 as at t_1, it is identical; if it cannot be considered the same, a new person exists. This logical relation is not specific to persons, but can be applied to each and every entity insofar as it can be counted, identified and re-identified. What is special about persons is that there is a second established way of talking about their identity which can be captured by the notion of *psychological identity*. This term – which is roughly equivalent to the term *personality* – depicts the dynamic and organized set of individual characteristics that uniquely influences a person's cognitions, motivations, and behaviors in various situations. The psychological identity (and thus the personality of a person) is not dichotomous, but constantly changes on a broad gradual continuum with many different nuances. It is hardly conceivable that neuromodulation techniques will in fact delete a person's numerical identity, that is, delete the person him- or herself by turning a human being from one person into another one. As an exception, Galert and co-authors illustrate scenarios where a person's memories are completely wiped out by retrograde amnesia (Merkel et al. 2007, p. 260). However, the neuromodulation techniques discussed here lead – at the most – to rather limited amnestic effects (e.g. ECT), and major amnesia might be expected only in very rare situations, e.g. in case of severe hemorrhage due to DBS electrode insertion. Thus, although the concept of numerical identity might indeed serve as a negative criterion against neuromodulation techniques, its use is limited to very specific, exceptional scenarios.

Neuromodulation techniques will, however, certainly affect a person's psychological identity – and in fact, on the first glance, this seems to be one of the most fundamental ethical questions at stake (Abbott 2005, Fuchs 2006). In particular, one might tend to ask whether psychiatric neuromodulation alters personality or not to draw ethical conclusions from this fact. This can be done explicitly by taking personality change as a negative criterion for psychiatric modulation, claiming for example that 'it should not be used to modify a person's individual character traits' (Hildt 2006); or it can be done implicitly, for example by assessing neuromodulation-induced changes of personality variables mainly under the category of 'risks' (Ford and Kubu 2006).

This approach, however, seems misleading in that it neglects the fact that affecting personality traits is not only an unwanted, coincidental side effect of psychiatric neuromodulation, but also *the main intended outcome*: If mood and cognitive behavior – and thus fundamental aspects of the psychological identity of a person – would not change e.g. in DBS in a patient with obsessive-compulsive disorder or with major depression, it could not be considered an effective treatment (for a more extended discussion of the personality argument see Synofzik and Schlaepfer 2008). Indeed, both psychopharmacological and psychotherapeutic interventions have exactly the same goal in aiming to positively influence and thereby alter aspects of personality such as mood and cognition in psychiatric patients. Thus, the decisive question is not whether neuromodulation alters personality or not, but whether it does so in *a good or bad way* from both the patient's and societal perspectives. Criteria that might guide this evaluation will be proposed below.

19.2.3 Early Psychosurgery

Historic allusions to the infamous term of psychosurgery should be largely avoided in discussions about the legitimacy of psychiatric neuromodulation. For several reasons procedures such as 'frontal lobotomy', popularized in the 1930s and 1940s by psychosurgery pioneers such as Egas Moniz, Walter Freeman and James W. Watts (Fins 2003, Heller et al. 2006), clearly differ from modern stereotactic DBS with respect to its indication, the precision of the surgical intervention, the hypothesis guided approach, reversibility, the potential to optimize treatment, the consenting process and careful follow-up (Synofzik and Schlaepfer 2008). Thus, the term 'psychosurgery' is associated with misleading, negative historical and cultural connotations and rather blurs than clarifies ethical and factual issues at stake. At best, it is of some minor heuristic use: The allusion to early psychosurgery offers an opportunity to point out the striking differences between its crude practices and the modern neuromodulation techniques – but this contrast will be of only little help when being confronted with the current challenge to identify productive and comprehensive ethical criteria for current neuromodulation techniques.

19.3 Ethical Criteria for Psychiatric Neuromodulation

To obtain ethically coherent and clinically applicable criteria for neuromodulation in psychiatric diseases, there is no need for specific criteria: the same criteria as for neuromodulation in neurological disorders or as for any other biomedical intervention apply (Beauchamp and Childress 2008), i.e., the neuromodulation application has to (i) benefit the patient (principle of beneficence), (ii) do no harm to the patient (principle of non-maleficence), and (iii) reflect his preferences (principle of respect of autonomy).

19.3.1 ECT

Since its beginning, use of ECT has been surrounded by various myths (Fink 1977) which are continued to be published in ethical assessments until today, holding that it would cause e.g. 'brain damage, death, and brainwashing' (Frank 2002). Consequently, critiques deliberately use the outdated term 'shock therapy' (Frank 2002, Friedberg 1977) to evoke pejorative connotations and view ECT as another 'great and desperate cure' akin to early psychosurgery (Sterling 2000, p. 242). The three ethical criteria proposed here allow to scrutinize such misleading perspectives and to adopt a much more stringent, evidence-based ethical approach of the ethical implications of ECT.

19.3.1.1 Benefits of ECT

A large body of evidence demonstrates that ECT in its current application technique still serves as a highly effective treatment option for specific psychiatric indications, including major depression, delusional depression, bipolar disorder, manic delirium, schizophrenia, malignant catatonia, and neuroleptic malignant syndrome (Fink 2001, Lisanby 2007, McCall 2001). For example in severe depression, it can partially overcome the disadvantages of psychopharmacological treatment, namely the delay until clinical improvement can be achieved and the considerable rate of non-response and non-remission, thus rendering it an effective short-term treatment for depression, that might under certain circumstances be more effective than drug therapy (The UK ECT Review Group 2003). In schizophrenia, a combined treatment of ECT and antipsychotic drugs seems to be effective in particular when rapid global improvement and reduction of symptoms is desired (Tharyan and Adams 2005).

To provide an actual *benefit* to the individual patients, however, ECT must not only be effective, i.e., improve scores in symptom rating scales, but also demonstrate that these operationalized, quantitative changes indeed are associated with an actual improvement of the individual patient's abilities to achieve *personally valuable goals*, i.e., goals that are valuable in light of his or her individual psychosocial situation and on the basis of his or her particular evaluative concept of a good life. Even if clinical trials would demonstrate that ECT is an effective treatment for a certain indication, common measures of effectiveness do not take into account all the factors that may lead patients to perceive it as beneficial or otherwise. This might explain why a systematic review on patients' perspectives on ECT found high variation in levels of perceived benefit depending on the source of research, where patient-led studies report lower rates of perceived benefit than clinical studies (Rose et al. 2003). In other words: The high rates of ECT effectiveness reported in clinical studies need not necessarily correspond with patient satisfaction. The symptom scales completed by mental health professionals might not conform to perceptions of symptom relief by patients and the simple response categories presented within brief questionnaires might not allow to capture the patients' complex trade-off between risks and benefits of treatment (Rose et al. 2003). Future studies of ECT treatment need to include a range of outcomes valued by patients (as emphasized e.g. by Vaughan McCall and colleagues (McCall et al. 2004, Rosenquist et al. 2006)) and need to complement quantitative methods with qualitative methodology in order to capture the true benefit as perceived by patients themselves.

19.3.1.2 Potential Harms of ECT

Notwithstanding its effectiveness, use of ECT has remained somewhat controversial mainly in the lay audience because of potential risks such as seizure induction, cognitive side effects, memory dysfunction and putative effects on cerebral physiology. However, the claim by many ECT opponents that it would cause brain damage (Breggin 1991, Friedberg 1977, Sterling 2000), has been convincingly refuted by an overwhelming amount of evidence (Devanand et al. 1994, Scalia et al. 2007, Zachrisson et al. 2000). Likewise, the death rate is comparable to that of general anesthesia in other indications (Shiwach et al. 2001) and is less than that associated with uncomplicated pregnancies (Fink 2001). The main risks of ECT remain cognitive side effects, in particular both anterograde and retrograde memory impairments (Ingram et al. 2008). Interestingly, objective measures found memory loss to be relatively short term (<6 months post treatment), whereas subjective accounts report amnesia to be more persistent (>6 months post-ECT) (Fraser et al. 2008). In fact, at least one third of patients report persistent memory loss, despite the fact that routine neuropsychological tests using objective measures do not find evidence of persistent memory loss (Rose et al. 2003). Several explanatory factors might contribute to this weak relationship between objective findings and subjective memory assessment. For example, patients' subjective reports of cognitive function might be strongly influenced by mood state (Prudic et al. 2000), thus reflecting rather their depressive mood and hypersensitivity (Reisner 2003, Squire and Slater 1983) to minor, maybe even pre-ECT memory disturbances than true ECT-induced memory impairments. However, it might also be that current batteries of objective neuropsychological tests of memory do not include components that are most affected in reports about subjective memory or that the group effects reported in the literature even out individual experiences of severe memory impairment following ECT. More research needs to be done specifically in those individuals who believe that ECT has had a markedly negative effect on their memory functioning, thus identifying the exact components of presumed memory loss, the predisposing factors and the influence of certain mood states in perceiving memory disturbances.

As long as the discrepancy between objective and subjective findings is not better understood, it would be wrong to focus only on the objective findings and to consider memory loss to be a clinically unimportant side effect of ECT, as claimed e.g. in an earlier version of the ECT fact sheet issued by the British Royal College of Psychiatrists (Rose et al. 2003).[1] The patients' negative subjective perception of memory loss has to be taken seriously when assessing the harms of ECT since the goal of any medical treatment is not to optimize objective measures, but to improve patients' subjective well-being[2] and to reduce their perceived impairments. The risk of subjectively persistent memory disturbances also has to be part of the individual patient information process. Moreover, each ECT treatment course has to be accompanied by an early monitoring of cognitive side effects that would allow modifying the treatment to reduce their severity. A recent small pilot study suggests that significant changes in memory function can be detected as early as after three treatments of ECT and that such monitoring can indeed be done in routine clinical practice (Porter et al. 2008). Finally, future research has to concentrate more also on the non-memory cognitive functions after ECT in order to avoid a bias in the assessment of ECT-induced cognitive side effects. Again, qualitative methodology might be of some help to get a first impression of the various cognitive (and non-cognitive) side effects patients experience during and after ECT.

19.3.1.3 Respecting the Patient's Preferences in ECT

Application of any treatment is only legitimated if it is in agreement with the patient's preferences. Optimally, these preferences are assessed within an informed consent process. Consent to treatment is valid only under two conditions: the patient has been adequately informed of risks and benefits and she freely chooses to undergo treatment. Based on a recent systematic review of studies investigating patients' retrospective views of informed consent to ECT (Rose et al. 2005) it needs to be questioned whether these two criteria are indeed fulfilled in current ECT practice. First, half the respondents reported that they had not received sufficient information about ECT and side effects.[3] Second, approximately a third did not feel they had freely consented to ECT even when they had signed a consent form.

One might argue that these numbers might just reflect the patients' trust in their treating physicians so that they agree to a treatment even though they haven't

---

[1] The current information sheet which was updated in 2008 by the British Royal College of Psychiatrists offers a much more balanced account of potential side-effects of ECT (http://www.rcpsych.ac.uk/mentalhealthinfoforall/treatments/ect.aspx).

[2] Based on the concept that treatment decision-making should be embedded in a deliberative, participative physician–patient relationship (Emanuel and Emanuel 1992), where the psychiatrist is explicitly asked to scrutinize the patient's will, to give recommendations and to mentor his formation of will, the notation of 'patient's well-being' which is used here does of course also imply beneficence-based considerations which assess, inter alia, also the long-term aspects of a patient's subjective well-being. Therefore it is for example also a legitimate treatment goal to overcome a manic state with transient subjective well-being and no insight but socially disastrous behaviour and consequences about which the mentally ill is unhappy after remission.

[3] Of course, one might criticize that this study did not actually investigate informed consent per se, but the *memory* thereof. Nevertheless, even in this most critical reading, this study would clearly demonstrate that provision of information was at least so unremarkable – and thus inadequate – that half the respondents forgot it. To further support the established findings, additional studies are needed which examine the practice of informed consent more immediately and directly.

received sufficient information or aren't completely convinced of it. A study by Freeman et al. (1980) has shown, however, that patients don't put their faith in a doctor and ECT treatment from a sense of trust but rather from a sense of despair and powerlessness: Those who had signed a consent form either stated that they were so desperate they would have tried anything or expressed a sense of powerlessness when faced by a medical professional so confident in the proposed treatment (Freeman et al. 1980). One might also point out that the lack of sufficient knowledge when giving informed consent is not specific to ECT. This claim is right. For example, in an investigation of informed consent to psychopharmacological treatments among long stay psychiatric in-patients, Brown and colleagues (Brown et al. 2001) discovered a lack of knowledge about medications and the reasons for giving them in two-thirds of the patients. However, the mere reference to other bad practices of inadequate informed consent in psychiatry can certainly not justify the inappropriate practice of informed consent with respect to ECT. Rather, it illustrates the large extent of the ethical problem pinpointed here and the need to develop better procedural safeguards to ensure valid informed consent in psychiatry.

19.3.1.4 The Ethical Problem of ECT Undertreatment

Acceptance of ECT as a therapeutic modality, both within the medical community and in public opinion, is still hampered by concerns about putative brain damage and by myths conveyed by historical antecedents, the media, and movies, respectively (for more detailed discussion see Ottosson and Fink 2004). This might explain the high variability in ECT administration (Eranti and McLoughlin 2003, Glen and Scott 2000, Hermann et al. 1995, Philpot et al. 2002) and the common finding of its underuse (Kalinowsky 1982, Sienaert et al. 2006, van der Wurff et al. 2004). Given that ECT is an efficacious treatment for several psychiatric diseases if applied according to the state of the art, one might hold off important benefits from patients by not doing it in the most effective way and, in particular, by not offering it to one's patients at all. The often chronic and partly even progressive course of treatment-refractory psychiatric diseases like depression and schizophrenia implies a constant increase in psychological suffering, work disintegration, social withdrawal, and partnership and family relation problems. Thus, also by *not performing* ECT one might cause harm to a psychiatric patient. For the same reason it would be unjustified to use ECT only as a last resort intervention since earlier consideration of ECT might reduce the rate of those patients who convert into a chronic and difficult-to-treat phase of their disease.

19.3.2 rTMS

Repetitive transcranial magnetic stimulation (rTMS) is a 20-year-old technique originally introduced to noninvasively investigate nervous propagation along the corticospinal tract, spinal roots, and peripheral nerves in humans. Nowadays, rTMS is also used to evaluate excitatory/inhibitory intracortical circuits, to provide information on brain physiology and pathophysiology of various neuropsychiatric diseases and even to treat certain neurological and psychiatric conditions (Rossini and Rossi 2007). It has already been approved for treatment-resistant depression in some countries and is subject to clinical trials for treating a variety of heterogeneous diseases, from obsessive compulsive disorder and schizophrenia to ataxia, optic atrophy and Pelizaeus-Merzbacher disease (Illes et al. 2006). This spread of clinical application and experimental testing warrants an ethical analysis to guide research and clinical decision-making in the next years.

19.3.2.1 Benefit of rTMS

rTMS has demonstrated efficacy in particular in the treatment of major depression where it can be employed in various ways, e.g. as the only treatment modality (Avery et al. 2006, O'Reardon et al. 2007), as an add-on therapy (Bretlau et al. 2008) or to accelerate the effect of antidepressants (Fitzgerald 2004, Rumi et al. 2005, Schlaepfer and Kosel 2004, Schlaepfer et al. 2003). Empirical support for clinical utility is much more rare for other psychiatric diseases: there is some empirical evidence for efficacy in posttraumatic stress disorder (Cohen et al. 2004) which needs further confirmation. In schizophrenia efficacy seems to be restricted to the treatment of persistent auditory hallucinations so far (Lopez-Ibor et al. 2008). Contradictory results in obsessive compulsive disorder reveal some of the main methodological problems of psychiatric rTMS – and maybe of neuromodulation techniques in general: to identify the adequate brain target, to find the best modulation strategy and to account for the heterogeneity of samples even from one and the same psychiatric disease (Lopez-Ibor et al. 2008).

But even if evidence were growing in the next years that rTMS is effective in treating e.g. posttraumatic stress disorder or certain schizophrenia symptoms, this fact per se would not suffice to legitimate its clinical application. As in other treatments, there is often a high interindividual (and partly even intraindividual) variability in the magnitude of rTMS treatment response (as shown e.g. for rTMS in depression (Lisanby et al. 2009)). Thus, individual treatment decisions might often be characterized by uncertainty about the actual benefit that rTMS may bring about, and the mere proof of statistical effectiveness on some group level will not free individual physicians from the burden to constantly perform a strict assessment whether his very individual patient does indeed receive a recognizable benefit from rTMS rather than just a change in a surrogate parameter on a certain depression score or behavioral scale (Synofzik 2006). In particular in the next years, when we will be confronted with rTMS studies in an increasing number of very heterogeneous diseases where effectiveness will often be only marginal, we will need to ask ourselves not only what is effective, but what is truly beneficial for our patients, that is, helping him to achieve goals that are really important in his everyday life. It is highly unlikely that rTMS will be good for all patients of one disease and for all the diseases for which effectiveness is currently tested.

But if there is proven benefit of rTMS for certain diseases and individual profiles, then it might be offered as treatment not only to patients treatment-refractory to medications, but maybe even more so to those who have not been on medication for longer time and whose brain systems and social and physical life are not yet devastated by disease. First analyses have revealed that younger age, less resistance to treatments (Fregni et al. 2006) and shorter duration of current illness (Lisanby et al. 2009) confer an increased likelihood of beneficial response to rTMS. In other words: The criterion of beneficence implies that effective neuromodulation techniques should not only be considered as applications of last resort.

19.3.2.2 Potential Harms of rTMS

While there was a fear initially that rTMS might produce short- or long-term damages to the brain by kindling effects (Rossini and Rossi 2007), several controlled and open studies have shown high levels of safety for single pulse and repeated rTMS (Janicak et al. 2008, O'Reardon et al. 2007). In a study of over 10,000 cumulative treatment sessions there were no deaths or epileptic seizures, no changes in auditory thresholds and cognitive functioning and only mild to moderate adverse events with transient headaches and scalp discomfort being the most common (Janicak et al. 2008). Other side effects reported in the literature are hand weakness, neck and arm pain, and arm tingling (Illes et al. 2006). Treatment discontinuation because of side effects was necessary only in 4.5% (Janicak et al. 2008). Results from neuropediatrics demonstrate a similar safety profile as more than 1,000 children have been treated with repeated rTMS in 84 studies and none of them had any significant side effects (Frye et al. 2008). In fact, safety of rTMS is so high that it might be of special advantage for treating depressive comorbidity in neurological illnesses that may deteriorate with use of antidepressants and even more with ECT (Lopez-Ibor et al. 2008).

Thus, it seems that rTMS is a rather well tolerated, safe treatment. However, ethicists still wonder about some potential theoretical risks. For example, rTMS 'may adversely affect circuits other than those targeted by the procedure' (Glannon 2006, p. 49). Although this risk indeed exists as the size of the stimulated area can only be fairly controlled in rTMS and circuits connected to the stimulated areas are only seldom completely known, this remark is not specific to rTMS or neuromodulation as e.g. both drug therapy and psychotherapy methods affect many unknown and unintended circuits also. Even more importantly, this remark looses its argumentative power by the fact that no major adverse events due to a coincidental rTMS co-stimulation have been reported to our knowledge. Another theoretical risk is that of undetected side-effects: 'It is possible that current tasks are not sensitive enough to uncover deficits that might remain, or that functions that might remain impaired (e.g. attention or speed of information processing) are not tested.' (Illes et al. 2006, p. 151). Such claims are often read in ethical reviews, but their scientific worth is limited in that they can never be completely falsified. Extensive neuropsychological testing does currently not reveal any neuropsychological deficits after rTMS treatment (but rather some improvement of cognitive performance and alleviation of memory complaints (Schulze-Rauschenbach et al. 2005)), yet one might always say that there might still be some subtle deficits not detected by the present methods. However, it is very questionable whether such undetected hypothesized subtle deficits would indeed present a relevant risk of harm to the patient and should thus be part of the ethical assessment.

19.3.2.3 Respecting the Patient's Preferences in rTMS

It has been claimed that the rTMS informed consent process 'must provide full disclosure of all known significant risks and acknowledge the possibility of yet-unknown longitudinal effects' (Illes et al. 2006, p. 151). However, it should also take into account the relative probability of occurrence of these risks and of the longitudinal effects that is very small and most unlikely, respectively. Thus, it might be confusing and distracting for psychiatric patients if they are told about 'yet-unknown longitudinal effects', which have never been observed even in long-term follow-ups and which merely represent a theoretical risk. Therefore, it is necessary to balance principles of autonomy and beneficence, and taking into account this balance it might often be legitimate, it is legitimate to refrain from mentioning this purely theoretical risk in those patients who are already cognitively or emotionally overstrained by the informed consent process in a given situation.

The autonomy principles obligates to draw on patient preferences to decide which benefit-harm ratio is acceptable and which not. It is those preferences – but not the physician's assessment – which should ultimately decide whether rTMS or ECT or nothing at all will be done. For example, if a patient with major depression whishes the most rapidly and maximally effective therapy based on his preference of urgent relief from depressive symptoms, ECT rather than rTMS would currently be the treatment of choice (Eranti et al. 2007, Knapp et al. 2008). However, in a patient with major depression who holds a risk-adverse preference profile, accepting lower effectiveness for less risk potential, rTMS should be recommended due to is lack of adverse memory effects as compared to ECT (Schulze-Rauschenbach et al. 2005).

19.3.2.4 rTMS in the Future

Due to its excellent safety profile and easy use even in out-patient settings, rTMS will be used more widely in psychiatric practice in the future. If, however, more and more professionals not familiar with the optimal technique and brain neurophysiology are using rTMS for therapeutic purposes (Rossini and Rossi 2007), then benefit might be reduced and side-effects more common. Therefore, it is an ethical requirement to establish guidelines of best practice and to implement mechanisms of monitoring standards of care in outpatient settings.

Likewise, naïve misconceptions of rTMS and of the neurocognitive substrates of psychiatric disease must be prevented, as neuromodulation applications will spread through outpatient practices and lay media. In fact, rTMS has already gained the attention of mass media which oversimplify complex neurocognitive states and the benefits and harms of rTMS, claiming e.g. that 'Happiness Is a Magnet' (relating to rTMS application as antidepressant therapy) or that one might become 'Savant for a Day' (cited after Rossini and Rossi 2007). As shown by a recent systematic review print media mostly report in an overly optimistic way, over-emphasizing the potential clinical benefits of neuromodulation techniques but ignoring the main ethical issues (Racine et al. 2007). A more balanced perspective is needed as the despair of treatement-resistant psychiatric patients makes them highly vulnerable to simple therapeutic promises and new techniques.

A balanced perspective is in particular needed, as the future will bring more and more discussions about its potential use as an enhancement tool, i.e. as a tool not only to treat diseases but also to improve normal functioning in healthy people (see Chapter 30). rTMS might be used to decrease or increase sexual desire, disrupt unwanted memory formation or elevate mood in non-depressed consumers (Illes et al. 2006). Some laboratories are already devoted to the development of rTMS methods for enhancing normal cognition (Snyder et al. 2003, 2006). It has been shown in more detail elsewhere that although using neuromodulation techniques for enhancing emotional and cognitive functioning is not intrinsically unethical, time is not yet ripe (Synofzik and Schlaepfer 2008):

- There are no sufficient systematic studies on enhancement effects of rTMS on mood or cognition in healthy subjects to clearly answer whether it is indeed effective and, even more importantly, beneficial in non-lab everyday life circumstances.
- The mild risks of harm in rTMS receive much more weight in an enhancement application than in a disease application since it is less likely that they will be outweighed by the likely benefit: A person with only minor melancholy will be much less ready to accept headache, scalp discomfort, neck and arm pain or arm tingling for some mood elevation than a patient with major depression.

19.3.3 DBS

High-frequency electrical deep brain stimulation (DBS) of specific brain circuits has gained widespread acceptance in treatment of several neurological disorders because of its high effectiveness and – compared to classical ablative neurosurgical interventions – its less invasive, largely reversible, and adjustable features (Perlmutter and Mink 2006). Very recently, efficacy has also been shown in studies of treatment-refractory psychiatric diseases, such as obsessive-compulsive disorder (Abelson et al. 2005, Greenberg et al., 2006b, Nuttin et al. 2003) and major depression (Lozano et al. 2008, Malone et al. 2008, Schlaepfer et al. 2008). In fact, psychiatric DBS is already rapidly moving from experimental to therapeutic – as evidenced by the fact that two large pivotal trials of DBS for major depression have been launched in the last 2 years (Medtronic Inc. 2006, Advanced Neuromodulation Systems (ANS) Inc. 2007). However, the use of DBS for psychiatric indications remains – compared to neurological indications – controversial and is associated with several specific ethical concerns (Fins 2008, Ford 2007, Kubu and Ford 2007).

19.3.3.1 Benefit of DBS

Although recent studies on obsessive-compulsive disorder (OCD) or major depression (MD) clearly demonstrate effectiveness of DBS in some patients resistant to pharmacotherapy and behavior therapy (Greenberg et al., 2006b, Mayberg et al. 2005, Schlaepfer et al. 2008), they also indicate that 50–75% of OCD (Abelson et al. 2005, Greenberg et al., 2006a) and 25–50% of MD patients (Greenberg et al., 2006b, Malone et al. 2008, Mayberg et al. 2005, Schlaepfer et al. 2008) fail to show a long-term response to DBS-treatment, the individual prognostic predictors for enduring therapy response still remaining unclear. In face of this considerable percentage of DBS non-responders, one has to be keep in mind that some patients, even with severe and chronic forms of the disorder, might finally respond after months or years of creative pharmacological and behavioral treatment (Schlaepfer and Lieb 2005). Moreover, effectiveness of DBS is not yet shown to be clearly superior to other treatments, e.g. ablative surgery in case of OCD (here: gamma-knife capsulotomy) (Abelson et al. 2005). First indirect evidence for superior effectiveness with respect to psychopharmacology and psychotherapy might already be reflected by the fact that DBS was the only treatment in the psychiatric patients studied so far that was able to reduce symptom levels of MD and OCD, respectively, after many years of chronic disease and after many different unsuccessful treatment attempts using psychotherapy and psychopharmacotherapy.

Contrary to the common ethical view that psychosurgical methods are 'justified only to treat severe conditions' (Glannon 2006, p. 46), however, we believe that – in analogy to the previous arguments for ECT and rTMS – there might be good *ethical* reasons to abolish treatment refractoriness as a mandatory criterion in the future: DBS might prove to be so superior in OCD and MD that especially patients who have *not* been on medication for longer time and whose social and physical life is *not yet* devastated by disease might benefit more from DBS and thus present the best candidates for DBS. This scenario seems provocative on first glance, but presents a lesson recently learned with respect to DBS in both Parkinson's Disease (Mesnage et al. 2002, Schupbach et al. 2007, Welter et al. 2002) and primary dystonia (Isaias et al. 2008). Since, however, these patients might be also the best candidates for other available treatments, clinical trials need to be well designed in order to establish the superiority of one treatment over the other.

Presently, however, DBS still has to proof that it is both *effective*, i.e. improves scores in OCD or MD rating scales in a larger fraction of patients, and *beneficial*, i.e. demonstrate that these statistical changes are indeed associated with an actual improvement of the very individual patient abilities to achieve personally valuable goals. This ethically highly important difference, which we have already pointed out for ECT and rTMS, has been overlooked for a long time in DBS for movement disorders: both research and clinical practice have focused initially on motor outcome only, but have neglected quality of life independent of motor function and, in particular, normative and psychosocial factors that are easily missed with quantitative outcome parameters (e.g. with movement scores or quality of life scores) (Agid et al. 2006, Schupbach et al. 2006). Only recently the question whether improvement in motor behavior does indeed lead to a relevant overall improvement in quality of life (Drapier et al. 2005) or whether it might be that only certain subgroups receive an overall benefit from the motor improvement (e.g. younger patients) has been asked (Derost et al. 2007).

Even if it was convincingly demonstrated, however, that motor, behavioral and disease-related quality of life variables improve after DBS surgery (e.g. by RCTs (Deuschl et al. 2006)), these measures might still present invalid surrogate parameter for the true benefit of DBS. As shown by a recent open interview study, many Parkinson patients are not happier with their lives, go through tormented periods in their marriages or fail to resume professional activity after surgery – in spite of (or probably even due to) clear improvement in various of these outcome variables after DBS implantation (Schupbach et al. 2006). Since the contributory factors to these psychosocial maladjustments do not seem to be specific to Parkinson's disease, but can be expected after rapid symptom modification in any chronic life-determining disease, the exact same problems can be expected after DBS in OCD or MD. Therefore, clinical studies should not only ask whether DBS is effective, i.e. demonstrate improvement on OCD or depression scores, but whether it indeed allows the very individual patient with OCD or MD to live a more satisfying life including the psychosocial dimension (which might actually be more relevant for a patient's overall quality of life and satisfaction than movement parameters (Burn 2002, Wilson and Cleary 1995)).

19.3.3.2 Potential Harms of DBS

Deep brain stimulation to different targets is associated with severe short-term and long-term risks on both levels, biological and psychosocial. These include: pre- and postoperative complications (in particular intracerebral hemorrhages), dysarthria, worsening of apathy, depression, cognitive impairments, e.g. in verbal fluency, color naming, selective attention, and verbal memory (Smeding et al. 2006), walking disturbances (Kenney et al. 2007), sudden symptom reoccurrence and aggravation in case of battery depletion (occurring as a function of programmed stimulation parameters, usually after 5–13 months in case of higher stimulation current amplitudes such as required for OCD) or of stimulation interruption, risking e.g. exacerbation of depressive symptomatology (Greenberg et al., 2006a). Adverse short- and long-term effects on a psychosocial level might comprise psychosocial maladjustment, suicidality (Appleby et al. 2007), severe disappointment and renewed desperation in case of non-responsiveness to DBS (Abelson et al. 2005).

While some of the neurological adverse effects have only been reported with respect to STN-DBS (e.g. dysarthria), other psychiatric impacts have primarily been associated with DBS of the anterior limb of the internal capsule and nucleus accumbens region (e.g. rapid mood elevation when DBS begins or affective worsening when stimulation was interrupted (Greenberg et al., 2006a)). Thus, a final risk-(and benefit)-ssessment cannot be performed for DBS in general, but needs to be specifically completed for each DBS location separately. Such an assessment of the location-specific efficacy-harm ratio might be of particular importance in psychiatric DBS, as often several very different anatomic targets (e.g., nucleus accumbens, habenula, inferior thalamic peduncle, internal capsule, Brodmann area 25) are proposed for the same condition (depression) and as it is probable that different targets will have different efficacy-harm ratios. However, even if it was shown in the future that adverse psychiatric events are particularly frequent with certain DBS target sites, future studies still have to prove that this effect is not only due to the fact that psychiatric symptoms are better monitored in psychiatric DBS studies that involve – inter alia – psychologically trained staff and psychological testing instruments which are generally not used in movement disorder DBS studies.

Under the assumption that DBS would be an efficacious treatment, one might do harm to patients not only by performing DBS, but also by *not* performing it. The chronic and partly even progressive course of treatment-refractory OCD or MD implies an constant increase in psychological suffering, work disintegration, social withdrawal, and partnership and family problems. Thus, also *not performing* DBS in psychiatric patients might one day demand specific, well-reasoned ethical justifications. Moreover, all pharmacological treatments are associated with significant adverse effects, e.g. agitation, sexual side effects, sedation, sleep disturbances, and night sweats in case of depression treatment, often leading to non-compliance (Keller et al. 2002). Although less recognized, the same holds true for psychotherapy. These adverse effects have to be counterbalanced against those of DBS, in particular as none of these adverse events has been reported in DBS depression treatment so far (Lozano et al. 2008, Malone et al. 2008, Mayberg et al. 2005, Schlaepfer et al. 2008).

19.3.3.3 Respecting the Patient's Preferences in DBS

Personal value preferences based on one's very individual concept of a good life are of special importance for both, taking the decision to undergo a DBS procedure in the first place and determining adequacy of stimulation parameter adjustment in the further course of treatment. For example, evaluations of probability (does a 0.4–3.6% risk of severe perioperative complications seem to be a lot to me or rather negligible?), of benefit (Is tremor improvement beneficial to me, and if yes, does it outweigh a more dysarthric speech?) and of relevance of neurological and psychosocial risks (do I need a clear speech in my daily living? Can my partnership bear a sudden change in disease behavior? Am I at risk of losing my aim of life when fighting the disease is no longer the main purpose in everyday life?) are clearly *value choices* which vary largely with the individual's life style, the psychosocial support system, and conscious value preferences. For instance in neurological disorders, a college teacher with Parkinson's Disease who is mainly involved in teaching and lecturing might consider an even slightly dysarthric speech or minor cognitive disturbance induced by DBS (Dubiel 2006, Kenney et al. 2007) to be so disabling that he prefers an akinetic movement pattern, while a carpenter whose concept of a fulfilling life strongly depends on laborious handcraft might accept some DBS-induced dysarthria and slight cognitive disturbances for being able to move better. Although comparable trade-offs between different capabilities are not yet known for psychiatric DBS, it is highly probable that they are necessitated here as well.

The capacity for autonomous decision-making and especially value choices might be more often and more severely compromised in psychiatric patients than in patients suffering from neurological disorders. Desperation is high in chronic, treatment-refractory and potentially deadly mental disorders, thus giving ground to overhasty decisions in favor of DBS which potentially undervalue the fact that (i) individual treatment-response to DBS is highly uncertain, (ii) some adverse effects might be deadly or lead to severe disability (e.g. in case of hemorrhage) and (iii) long-term cognitive, emotional and behavioral effects of psychiatric DBS are still largely unknown. The high desperation of treatment-resistant patients predisposes them to severe disappointment in case of non-responsiveness and to suicidal reactions (Abelson et al. 2005). It thus complicates not only assessment of efficacy, but also patient management in demanding protocols and in the subsequent physician–patient-relationship. Moreover, since psychiatric side effects such as elevated mood or anxiety might be more likely in psychiatric DBS, patients' preferences for or against certain parameter settings might be directly induced by the stimulation per se, but not reflect their general value perspective, unaffected by DBS treatment.

19.3.3.4 Portraying 'DBS Miracles' to the Public

Ethical requirements need to be considered not only in DBS treatment and patient care per se, but also with respect to *portrayal* of DBS in the public or to psychiatric patients contemplating DBS in the future. Public events informing Parkinson or tremor patients, their relatives and the public about DBS often convey one-sided, biased information about treatment benefit by presenting only patients with large motor and quality of life improvements who report happily about treatment success and have not experienced any adverse event ('DBS miracles'). In contrast, short-, medium- or long-term adverse events, e.g. hemorrhages, dysarthria, psychosocial maladjustments or insufficient treatment responses, are hardly ever reported in an equally demonstrative and intriguing way, e.g. by displaying hemorrhages on CT scans or by inviting patients who have experienced complications. Since the heterogeneous outcomes after DBS are not adequately represented, the public and patients are misled and information meetings might turn – deliberately or inadvertently – into promotional events. If the media propagate such miracle stories, they will increase existing pressures for expedited neuromodulation development (Racine et al. 2007) and associated professional conflicts (Ausman 2004, Illes et al. 2006). In view of the higher vulnerability and restricted autonomy capacity of psychiatric patients the call for balanced information is warranted.

19.4 Conclusion

Neuromodulation techniques offer focused and efficacious treatment options for psychiatric patients resistant to pharmacological and psychotherapeutic methods. There are no general ethical objections against neuromodulation techniques. In particular, the conclusiveness of critical statements about their invasiveness, their potential to alter personal identity or their resemblance to early psychosurgery needs to be thoroughly scrutinized. It seems much more productive to apply those ethical criteria to empirical evidence about neuromodulation techniques that are already widely used to assess other biomedical interventions. These criteria, however, pose several specific challenges when applied to psychiatric neuromodulation: ECT brings about its effectiveness with some moderate memory disturbances; rTMS still needs to demonstrate superior beneficial effects in many psychiatric diseases, and DBS is still largely experimental in nature and needs several procedural safeguards. But ethical problems arise not only in the application of these techniques per se, but also in their reception and portrayal in the public: While ECT is still surrounded by (unjustified) pejorative myths, leading to some underuse, rTMS and DBS are sometimes presented in an overly optimistic, simplifying way. This leads to questionable overuse, e.g. the application of DBS in severely progressive atrophic brains of Alzheimer disease patients (Hurley 2008). But if we do our ethical home work conscientiously, investigate the risk-benefit-ratio of neuromodulation techniques per se and against each other, learn from earlier mistakes in psychosurgery and in DBS for neurological indications and implement certain specific procedural safeguards, then psychiatry will experience stimulating decades.

Disclosure Dr. Schlaepfer received limited support for an Investigator Initiated Study on DBS in resistant major depression from Medtronic Inc., a manufacturer of DBS equipment between 2004 and 2008 and a grant of the Volkswagen Stiftung for the development of internationally based guidelines on the responsible use of DBS in neuropsychiatry. Dr. Synofzik was supported by the 'European Platform for Life Sciences, Mind Sciences, and the Humanities' grant by the Volkswagen Stiftung.

References

Abbott A (2005) Neuroscience: deep in thought. Nature 436:18–19

Abelson JL, Curtis GC, Sagher O, Albucher RC, Harrigan M, Taylor SF, Martis B, Giordani B (2005) Deep brain stimulation for refractory obsessive-compulsive disorder. Biol Psychiatry 57:510–516

Advanced Neuromodulation Systems (ANS) Inc (2007) BROADEN clinical study. A study of a non-pharmacological device for depression, Plano, TX

Agid Y, Schupbach M, Gargiulo M, Mallet L, Houeto JL, Behar C, Maltete D, Mesnage V, Welter ML (2006) Neurosurgery in Parkinson's disease: the doctor is happy, the patient less so? J Neural Transm Suppl:409–414

Appleby BS, Duggan PS, Regenberg A, Rabins PV (2007) Psychiatric and neuropsychiatric adverse events associated with deep brain stimulation: a meta-analysis of ten years' experience. Mov Disord 22:1722–1728

Ausman JI (2004) I told you it was going to happen. Surg Neurol 61:313–314

Avery DH, Holtzheimer PE 3rd, Fawaz W, Russo J, Neumaier J, Dunner DL, Haynor DR, Claypoole KH, Wajdik C, Roy-Byrne P (2006) A controlled study of repetitive transcranial magnetic stimulation in medication-resistant major depression. Biol Psychiatry 59:187–194

Beauchamp T, Childress J (2008) Principles of biomedical ethics, 6th edn. Oxford University Press, New York, NY, Oxford

Birnbacher D (2006) Hirngewebstransplantationen und neurobionische Eingriffe – anthropologische und ethische Fragen. In: Birnbacher D (ed) Bioethik zwischen Natur und Interesse. Suhrkamp, Frankfurt, pp 273–293

Breggin P (1991) Toxic psychiatry. St. Martins Press, New York, NY

Bretlau LG, Lunde M, Lindberg L, Unden M, Dissing S, Bech P (2008) Repetitive transcranial magnetic stimulation (rTMS) in combination with escitalopram in patients with treatment-resistant major depression: a double-blind, randomised, sham-controlled trial. Pharmacopsychiatry 41:41–47

Brown KW, Billcliff N, McCabe E (2001) Informed consent to medication in long-term psychiatric in-patients. Psychiatr Bull 25:132–134

Burn DJ (2002) Beyond the iron mask: towards better recognition and treatment of depression associated with Parkinson's disease. Mov Disord 17:445–454

Cohen H, Kaplan Z, Kotler M, Kouperman I, Moisa R, Grisaru N (2004) Repetitive transcranial magnetic stimulation of the right dorsolateral prefrontal cortex in posttraumatic stress disorder: a double-blind, placebo-controlled study. Am J Psychiatry 161:515–524

Derost PP, Ouchchane L, Morand D, Ulla M, Llorca PM, Barget M, Debilly B, Lemaire JJ, Durif F (2007) Is DBS-STN appropriate to treat severe Parkinson disease in an elderly population? Neurology 68:1345–1355

Deuschl G, Schade-Brittinger C, Krack P, Volkmann J, Schafer H, Botzel K, Daniels C, Deutschlander A, Dillmann U, Eisner W, Gruber D, Hamel W, Herzog J, Hilker R, Klebe S, Kloss M, Koy J, Krause M, Kupsch A, Lorenz D, Lorenzl S, Mehdorn HM, Moringlane JR, Oertel W, Pinsker MO, Reichmann H, Reuss A, Schneider GH, Schnitzler A, Steude U, Sturm V, Timmermann L, Tronnier V, Trottenberg T, Wojtecki L, Wolf E, Poewe W, Voges J (2006) A randomized trial of deep-brain stimulation for Parkinson's disease. N Engl J Med 355:896–908

Devanand DP, Dwork AJ, Hutchinson ER, Bolwig TG, Sackeim HA (1994) Does ECT alter brain structure? Am J Psychiatry 151:957–970

Drapier S, Raoul S, Drapier D, Leray E, Lallement F, Rivier I, Sauleau P, Lajat Y, Edan G, Verin M (2005) Only physical aspects of quality of life are significantly improved by bilateral subthalamic stimulation in Parkinson's disease. J Neurol 252:583–588

Dubiel H (2006) Tief im Hirn. Kunstmann, München

Emanuel EJ, Emanuel LL (1992) Four models of the physician–patient relationship. J Am Med Assoc 267:2221–2226

Eranti SV, McLoughlin DM (2003) Electroconvulsive therapy – state of the art. Br J Psychiatry 182:8–9

Eranti S, Mogg A, Pluck G, Landau S, Purvis R, Brown RG, Howard R, Knapp M, Philpot M, Rabe-Hesketh S, Romeo R, Rothwell J, Edwards D, McLoughlin DM (2007) A randomized, controlled trial with 6-month follow-up of repetitive transcranial magnetic stimulation and electroconvulsive therapy for severe depression. Am J Psychiatry 164:73–81

Fink M (1977) Myths of "shock therapy". Am J Psychiatry 134:991–996

Fink M (2001) ECT has much to offer our patients: it should not be ignored. World J Biol Psychiatry 2:1–8

Fins JJ (2003) From psychosurgery to neuromodulation and palliation: history's lessons for the ethical conduct and regulation of neuropsychiatric research. Neurosurg Clin N Am 14:303–319

Fins JJ (2008) A leg to stand on: Sir William Osler and Wilder Penfield's "neuroethics". Am J Bioeth 8:37–46

Fitzgerald P (2004) Repetitive transcranial magnetic stimulation and electroconvulsive therapy: complementary or competitive therapeutic options in depression? Australas Psychiatry 12:234–238

Ford PJ (2007) Neurosurgical implants: clinical protocol considerations. Camb Q Healthc Ethics 16:308–311

Ford PJ, Kubu CS (2006) Stimulating debate: ethics in a multidisciplinary functional neurosurgery committee. J Med Ethics 32:106–109

Frank LR (2002) Electroshock: a crime against the spirit. Ethical Hum Sci Serv 4:63–71

Fraser LM, O'Carroll RE, Ebmeier KP (2008) The effect of electroconvulsive therapy on autobiographical memory: a systematic review. J ECT 24:10–17

Freeman CP, Weeks D, Kendell RE (1980) ECT: II: patients who complain. Br J Psychiatry 137:17–25

Fregni F, Marcolin MA, Myczkowski M, Amiaz R, Hasey G, Rumi DO, Rosa M, Rigonatti SP, Camprodon J, Walpoth M, Heaslip J, Grunhaus L, Hausmann A, Pascual-Leone A (2006) Predictors of antidepressant response in clinical trials of transcranial magnetic stimulation. Int J Neuropsychopharmacol 9:641–654

Friedberg J (1977) Shock treatment, brain damage, and memory loss: a neurological perspective. Am J Psychiatry 134:1010–1014

Frye RE, Rotenberg A, Ousley M, Pascual-Leone A (2008) Transcranial magnetic stimulation in child neurology: current and future directions. J Child Neurol 23:79–96

Fuchs T (2006) Ethical issues in neuroscience. Curr Opin Psychiatry 19:600–607

Glannon W (2006) Neuroethics. Bioethics 20:37–52

Glen T, Scott AI (2000) Variation in rates of electroconvulsive therapy use among consultant teams in Edinburgh (1993–1996). J Affect Disord 58:75–78

Greenberg BD, Malone DA, Friehs GM, Rezai AR, Kubu CS, Malloy PF, Salloway SP, Okun MS, Goodman WK, Rasmussen SA (2006a) Three-year outcomes in deep brain stimulation for highly resistant obsessive-compulsive disorder. Neuropsychopharmacol 31:2384–2393

Greenberg BD, Malone DA, Friehs GM, Rezai AR, Kubu CS, Malloy PF, Salloway SP, Okun MS, Goodman WK, Rasmussen SA (2006b) Three-year outcomes in deep brain stimulation for highly resistant obsessive-compulsive disorder. Neuropsychopharmacol 31:2394

Heller AC, Amar AP, Liu CY, Apuzzo ML (2006) Surgery of the mind and mood: a mosaic of issues in time and evolution. Neurosurgery 59:720–733; discussion 733–729

Hermann RC, Dorwart RA, Hoover CW, Brody J (1995) Variation in ECT use in the United States. Am J Psychiatry 152:869–875

Hildt E (2006) Electrodes in the brain: some anthropological and ethical aspects of deep brain stimulation. Int Rev Inform Ethics 5:33–39

Hurley D (2008) Deep brain stimulation is used to suppress appetite in obese man – and his memory improves. Neurol Today 8:11–12; Clinical Trials. gov identifier: NCT00658125

Illes J, Gallo M, Kirschen MP (2006) An ethics perspective on transcranial magnetic stimulation (TMS) and human neuromodulation. Behav Neurol 17:149–157

Ingram A, Saling MM, Schweitzer I (2008) Cognitive side effects of brief pulse electroconvulsive therapy: a review. J ECT 24:3–9

Isaias IU, Alterman RL, Tagliati M (2008) Outcome predictors of pallidal stimulation in patients with primary dystonia: the role of disease duration. Brain 131:1895–1902

Janicak PG, O'Reardon JP, Sampson SM, Husain MM, Lisanby SH, Rado JT, Heart KL, Demitrack MA (2008) Transcranial magnetic stimulation in the treatment of major depressive disorder: a comprehensive summary of safety experience from acute exposure, extended exposure, and during reintroduction treatment. J Clin Psychiatry 69:222–232

Kalinowsky LB (1982) ECT: underused and misunderstood. Hosp Community Psychiatry 33:425

Keller MB, Hirschfeld RM, Demyttenaere K, Baldwin DS (2002) Optimizing outcomes in depression: focus on antidepressant compliance. Int Clin Psychopharmacol 17: 265–271

Kenney C, Simpson R, Hunter C, Ondo W, Almaguer M, Davidson A, Jankovic J (2007) Short-term and long-term safety of deep brain stimulation in the treatment of movement disorders. J Neurosurg 106:621–625

Knapp M, Romeo R, Mogg A, Eranti S, Pluck G, Purvis R, Brown RG, Howard R, Philpot M, Rothwell J, Edwards D, McLoughlin DM (2008) Cost-effectiveness of transcranial magnetic stimulation vs. electroconvulsive therapy for severe depression: a multi-centre randomised controlled trial. J Affect Disord 109:273–285

Kubu CS, Ford PJ (2007) Ethics in the clinical application of neural implants. Camb Q Healthc Ethics 16:317–321

Linden DE (2006) How psychotherapy changes the brain – the contribution of functional neuroimaging. Mol Psychiatry 11:528–538

Lisanby SH (2007) Electroconvulsive therapy for depression. N Engl J Med 357:1939–1945

Lisanby SH, Husain MM, Rosenquist PB, Maixner D, Gutierrez R, Krystal A, Gilmer W, Marangell LB, Aaronson S, Daskalakis ZJ, Canterbury R, Richelson E, Sackeim HA, George MS (2009) Daily left prefrontal repetitive transcranial magnetic stimulation in the acute treatment of major depression: clinical predictors of outcome in a multisite, randomized controlled clinical trial. Neuropsychopharmacol 34:522–534

Lopez-Ibor JJ, Lopez-Ibor MI, Pastrana JI (2008) Transcranial magnetic stimulation. Curr Opin Psychiatry 21:640–644

Lozano AM, Mayberg HS, Giacobbe P, Hamani C, Craddock RC, Kennedy SH (2008) Subcallosal cingulate gyrus deep brain stimulation for treatment-resistant depression. Biol Psychiatry 64:461–467

Mallet L, Polosan M, Jaafari N, Baup N, Welter ML, Fontaine D, du Montcel ST, Yelnik J, Chereau I, Arbus C, Raoul S, Aouizerate B, Damier P, Chabardes S, Czernecki V, Ardouin C, Krebs MO, Bardinet E, Chaynes P, Burbaud P, Cornu P, Derost P, Bougerol T, Bataille B, Mattei V, Dormont D, Devaux B, Verin M, Houeto JL, Pollak P, Benabid AL, Agid Y, Krack P, Millet B, Pelissolo A (2008) Subthalamic nucleus stimulation in severe obsessive-compulsive disorder. N Engl J Med 359:2121–2134

Mallet L, Schupbach M, N'Diaye K, Remy P, Bardinet E, Czernecki V, Welter ML, Pelissolo A, Ruberg M, Agid Y, Yelnik J (2007) Stimulation of subterritories of the subthalamic nucleus reveals its role in the integration of the emotional and motor aspects of behavior. Proc Natl Acad Sci U S A 104:10661–10666

Malone DA Jr, Dougherty DD, Rezai AR, Carpenter LL, Friehs GM, Eskandar EN, Rauch SL, Rasmussen SA, Machado AG, Kubu CS, Tyrka AR, Price LH, Stypulkowski PH, Giftakis JE, Rise MT, Malloy PF, Salloway SP, Greenberg BD (2008) Deep brain stimulation of the ventral capsule/ventral striatum for treatment-resistant depression. Biol Psychiatry 65(4): 267–275

Mayberg HS, Lozano AM, Voon V, McNeely HE, Seminowicz D, Hamani C, Schwalb JM, Kennedy SH (2005) Deep brain stimulation for treatment-resistant depression. Neuron 45:651–660

McCall WV (2001) Electroconvulsive therapy in the era of modern psychopharmacology. Int J Neuropsychopharmacol 4:315–324

McCall WV, Dunn A, Rosenquist PB (2004) Quality of life and function after electroconvulsive therapy. Br J Psychiatry 185:405–409

Medtronic Inc (2006) Medtronic to pursue major clinical trial of deep brain stimulation as depression treatment. In: Merkel R, Boer G, Fegert J, Galert T, Hartmann D, Nuttin B, Rosahl S (eds) (2007) Intervening in the brain. Changing psyche and society. Springer, Berlin

Mesnage V, Houeto JL, Welter ML, Agid Y, Pidoux B, Dormont D, Cornu P (2002) Parkinson's disease: neurosurgery at an earlier stage? J Neurol Neurosurg Psychiatry 73:778–779

Nuttin BJ, Gabriels LA, Cosyns PR, Meyerson BA, Andreewitch S, Sunaert SG, Maes AF, Dupont PJ, Gybels JM, Gielen F, Demeulemeester HG (2003) Long-term electrical capsular stimulation in patients with obsessive-compulsive disorder. Neurosurgery 52:1263–1272; discussion 1272–1264

O'Reardon JP, Solvason HB, Janicak PG, Sampson S, Isenberg KE, Nahas Z, McDonald WM, Avery D, Fitzgerald PB, Loo C, Demitrack MA, George MS, Sackeim HA (2007) Efficacy and safety of transcranial magnetic stimulation in the acute treatment of major depression: a multisite randomized controlled trial. Biol Psychiatry 62:1208–1216

Ottosson JO, Fink M (2004) Ethics in electroconvulsive therapy. Brunner-Routledge, New York, NY, Hove

Perlmutter JS, Mink JW (2006) Deep brain stimulation. Annu Rev Neurosci 29:229–257

Philpot M, Treloar A, Gormley N, Gustafson L (2002) Barriers to the use of electroconvulsive therapy in the elderly: a European survey. Eur Psychiatry 17:41–45

Porter R, Heenan H, Reeves J (2008) Early effects of electroconvulsive therapy on cognitive function. J ECT 24:35–39

Prudic J, Peyser S, Sackeim HA (2000) Subjective memory complaints: a review of patient self-assessment of memory after electroconvulsive therapy. J ECT 16:121–132

Racine E, Waldman S, Palmour N, Risse D, Illes J (2007) "Currents of hope": neurostimulation techniques in U.S. and U.K. print media. Camb Q Healthc Ethics 16:312–316

Reisner AD (2003) The electroconvulsive therapy controversy: evidence and ethics. Neuropsychol Rev 13:199–219

Rose D, Fleischmann P, Wykes T, Leese M, Bindman J (2003) Patients' perspectives on electroconvulsive therapy: systematic review. BMJ 326:1363

Rose DS, Wykes TH, Bindman JP, Fleischmann PS (2005) Information, consent and perceived coercion: patients' perspectives on electroconvulsive therapy. Br J Psychiatry 186:54–59

Rosenquist PB, Brenes GB, Arnold EM, Kimball J, McCall V (2006) Health-related quality of life and the practice of electroconvulsive therapy. J ECT 22:18–24

Rossini PM, Rossi S (2007) Transcranial magnetic stimulation: diagnostic, therapeutic, and research potential. Neurology 68:484–488

Rumi DO, Gattaz WF, Rigonatti SP, Rosa MA, Fregni F, Rosa MO, Mansur C, Myczkowski ML, Moreno RA, Marcolin MA (2005) Transcranial magnetic stimulation accelerates the antidepressant effect of amitriptyline in severe depression: a double-blind placebo-controlled study. Biol Psychiatry 57:162–166

Scalia J, Lisanby SH, Dwork AJ, Johnson JE, Bernhardt ER, Arango V, McCall WV (2007) Neuropathologic examination after 91 ECT treatments in a 92-year-old woman with late-onset depression. J ECT 23:96–98

Schlaepfer TE, Cohen MX, Frick C, Kosel M, Brodesser D, Axmacher N, Joe AJ, Kreft M, Lenartz D, Sturm V (2008) Deep brain stimulation to reward circuitry alleviates anhedonia in refractory major depression. Neuropschopharmacology 33:368–377

Schlaepfer TE, Kosel M (2004) Novel physical treatments for major depression: vagus nerve stimulation, transcranial magnetic stimulation and magnetic seizure therapy. Curr Opin Psychiatr 17:15–20

Schlaepfer TE, Kosel M, Nemeroff CB (2003) Efficacy of repetitive transcranial magnetic stimulation (rTMS) in the treatment of affective disorders. Neuropsychopharmacol 28:201–205

Schlaepfer TE, Lieb K (2005) Deep brain stimulation for treatment of refractory depression. Lancet 366:1420–1422

Schulze-Rauschenbach SC, Harms U, Schlaepfer TE, Maier W, Falkai P, Wagner M (2005) Distinctive neurocognitive effects of repetitive transcranial magnetic stimulation and electroconvulsive therapy in major depression. Br J Psychiatry 186:410–416

Schupbach M, Gargiulo M, Welter ML, Mallet L, Behar C, Houeto JL, Maltete D, Mesnage V, Agid Y (2006) Neurosurgery in Parkinson disease: a distressed mind in a repaired body? Neurology 66:1811–1816

Schupbach WM, Maltete D, Houeto JL, du Montcel ST, Mallet L, Welter ML, Gargiulo M, Behar C, Bonnet AM, Czernecki V, Pidoux B, Navarro S, Dormont D, Cornu P, Agid Y (2007) Neurosurgery at an earlier stage of Parkinson disease: a randomized, controlled trial. Neurology 68:267–271

Shapira NA, Okun MS, Wint D, Foote KD, Byars JA, Bowers D, Springer US, Lang PJ, Greenberg BD, Haber SN, Goodman WK (2006) Panic and fear induced by deep brain stimulation. J Neurol Neurosurg Psychiatry 77:410–412

Shiwach RS, Reid WH, Carmody TJ (2001) An analysis of reported deaths following electroconvulsive therapy in Texas, 1993–1998. Psychiatr Serv 52:1095–1097

Sienaert P, Dierick M, Degraeve G, Peuskens J (2006) Electroconvulsive therapy in Belgium: a nationwide survey on the practice of electroconvulsive therapy. J Affect Disord 90:67–71

Smeding HM, Speelman JD, Koning-Haanstra M, Schuurman PR, Nijssen P, van Laar T, Schmand B (2006) Neuropsychological effects of bilateral STN stimulation in Parkinson disease: a controlled study. Neurology 66:1830–1836

Snyder A, Bahramali H, Hawker T, Mitchell DJ (2006) Savant-like numerosity skills revealed in normal people by magnetic pulses. Perception 35:837–845

Snyder AW, Mulcahy E, Taylor JL, Mitchell DJ, Sachdev P, Gandevia SC (2003) Savant-like skills exposed in normal people by suppressing the left fronto-temporal lobe. J Integr Neurosci 2:149–158

Squire LR, Slater PC (1983) Electroconvulsive therapy and complaints of memory dysfunction: a prospective three-year follow-up study. Br J Psychiatry 142:1–8

Sterling P (2000) ECT damage is easy to find if you look for it. Nature 403:242

Synofzik M (2005) Die neuen Möglichkeiten der Neurowissenschaften und ihre ethischen Implikationen: Eine Kriteriologie der Neuroethik. Zeitschrift für Ethik in der Medizin 17:206–219

Synofzik M (2006) [Effective, indicated – and yet without benefit? The goals of dementia drug treatment and the well-being of the patient]. Z Gerontol Geriatr 39:301–307

Synofzik M, Schlaepfer TE (2008) Stimulating personality: ethical criteria for deep brain stimulation in psychiatric patients and for enhancement purposes. Biotechnol J 3:1511–1520

Tharyan P, Adams CE (2005) Electroconvulsive therapy for schizophrenia. Cochrane Database Syst Rev (online) (2): CD000076

The UK ECT Review Group (2003) Efficacy and safety of electroconvulsive therapy in depressive disorders: a systematic review and meta-analysis. Lancet 361:799–808

van der Wurff FB, Stek ML, Hoogendijk WJ, Beekman AT (2004) Discrepancy between opinion and attitude on the practice of ECT by psychiatrists specializing in old age in the Netherlands. J ECT 20:37–41

Welter ML, Houeto JL, Tezenas du Montcel S, Mesnage V, Bonnet AM, Pillon B, Arnulf I, Pidoux B, Dormont D, Cornu P, Agid Y (2002) Clinical predictive factors of subthalamic stimulation in Parkinson's disease. Brain 125:575–583

Wilson IB, Cleary PD (1995) Linking clinical variables with health-related quality of life. A conceptual model of patient outcomes. J Am Med Assoc 273:59–65

Zachrisson OC, Balldin J, Ekman R, Naesh O, Rosengren L, Agren H, Blennow K (2000) No evident neuronal damage after electroconvulsive therapy. Psychiatry Res 96:157–165

Chapter 20
'Coercive' Measures

George Szmukler

Contents

G. Szmukler (✉)
Health Service and Population Research Department, King's College London,
Institute of Psychiatry, London SE5 8AF, UK
e-mail: George.Szmukler@iop.kcl.ac.uk

H. Helmchen, N. Sartorius (eds.), *Ethics in Psychiatry*, International Library
of Ethics, Law, and the New Medicine 45, DOI 10.1007/978-90-481-8721-8_20,
© Springer Science+Business Media B.V. 2010

Abbreviations

AS   Advance Statements
CCs   Crisis Cards
F-PAD   Facilitated PAD
IOT   Involuntary Outpatient Treatment
JCPs   Joint Crisis Plans
MCA   Mental Capacity Act 2005
PADs   Psychiatric Advance Directives

20.1 Preliminary Remarks

The aim of this chapter is to examine what might broadly be termed 'coercive' measures in mental health care. I will examine the distinctions between 'compulsion', 'coercion' (used in a more precise sense)[1] and 'inducements' as types of treatment pressure directed at patients who are reluctant to accept treatment. I will then examine justifications for their use, and finally, means that might lead to a reduction in the necessity for their use.

Compulsion, *coercion* and inducements are conceptually distinct and can be considered to lie on a more extended hierarchy of treatment pressures. These range from 'persuasion', through 'interpersonal leverage', 'inducements', '*coercion*', to 'compulsion' (see Szmukler and Appelbaum 2008) for a fuller discussion). I will focus on the last three in this chapter, as persuasion (appeals to reason) and interpersonal leverage (using the patient's emotional dependence on the clinician) would not generally be considered as 'coercive'. As well as falling within a hierarchy, 'coercive' measures are also variably regulated. Compulsion is usually regulated by mental health law; the other forms are governed, less formally, by professional ethics or codes of practice.

20.2 Compulsion

Compulsion, or involuntary treatment, involves the use of force; the patient has no choice but to accept treatment against his or her will. Treatment may be required in hospital as an inpatient or, as a recent innovation in health care, in the community. The latter is variously termed in different jurisdictions: 'community treatment order', 'outpatient commitment', 'involuntary outpatient treatment' or 'mandated community treatment'.

Compulsion is regulated by law. The key criteria for involuntary treatment in most statutes are based on 'risk' – for example, to the health or safety of the patient,

---

[1] I will use the term *coercion* in two senses: first, in a general sense of a morally significant treatment pressure; second, in the sense of a specific, narrowly defined form of treatment pressure. When used in the narrow sense, the word *coercion* will be italicized.

or to the safety of other persons. The two kinds of risk are usually presented together. It is unusual for there to be a criterion based on evidence of impaired decision-making capacity by the patient, that is, that the patient is unable to make a sound judgement about the need for treatment. The Mental Health (Care & Treatment) (Scotland) Act 2003 is an exception in this regard.

Ignoring treatment decision-making capacity in mental health law, as opposed to non-consensual treatment in general medicine where it is a necessary condition, raises ethical problems. It is argued, therefore, that the autonomy of the patient (who has decision-making capacity) is not respected equally. The patient who suffers from a 'physical disorder' is able to refuse treatment, even if life-saving, if he or she has capacity; not so for the person who has a 'mental disorder'. In other words there is discrimination against patients with 'mental disorders' compared to those with 'physical disorders' (Campbell and Heginbotham 1991, Szmukler and Holloway 1998). This point will be considered further when we come to examine justifications for coercive interventions (Section 20.4.1 below).

20.2.1 Involuntary Hospitalization and Inpatient Treatment

Hoyer (2008) has recently reviewed problematic aspects of involuntary treatment. Whether involuntary inpatient treatment is effective or not is not known; nor is it likely to become known since a randomised controlled trial would not be considered ethical. But what is troubling from an ethical point of view is the huge variation in involuntary hospitalization rates from country to country, and even within countries. In Europe in 1998–2000 the rates ranged from 6 per 100,000 in Portugal to 218 per 100,000 in Finland (Salize and Dressing 2004), while a later study of six European countries found rates ranging from 32.1 per 100,000 in Sweden to 190.5 in Germany (Priebe et al. 2005). Different forms of legislation and their practical consequences are difficult to compare between countries, and there is often uncertainty about what is included in official figures as well as their accuracy. There is a suggestion that in countries where patients have stronger legal representation, involuntary hospitalization rates are lower (Salize and Dressing 2004). Variation within a country can be demonstrated by considering, for example, Norway, where involuntary admission rates have varied fivefold (cited in (Hoyer 2008) or Sweden (Kjellin et al. 2008)). The last finding suggests that mental health legislation is not the major factor associated with differences. Variation in services probably contributes, but from an ethical perspective, differences in 'custom and practice', attitudes or 'conventional wisdom' are perhaps more significant. While Hoyer rightly describes the difficulties in comparing figures across nations, he concludes that even allowing for the problems, substantial variation remains.

Just as significant are changes in rates of involuntary hospitalisation over time. In England and Wales, compulsory admissions to hospital increased by 63% between 1984 and 1996 (Hotopf et al. 2000), and by another 29% between 1996 and 2006 (Keown et al. 2008). Contributing to this increase has probably been a decrease in the number of inpatient beds. However, also important have been pressures on clinicians arising from a risk-averse society that fears that community care has failed, with an associated increase in the number of dangerous mentally ill people on the streets. Because risk is so vaguely characterized in mental health legislation, there is enormous scope for interpretation. In a highly politicised domain, such as mental health care, this becomes troubling. In contrast to England, however, the compulsory admission rate in Sweden declined steeply between 1979 and 2002 (Kjellin et al. 2008).

Large variations in rates of application of coercive measures following admission are also evident, especially in the use of seclusion and physical restraint, both nationally and internationally (see for example, Fisher 1994, Janssen et al. 2008). Such measures are not as closely regulated as admission procedures, and may be governed by a specific 'code of practice', as in the UK, or by more generally formulated professional codes or standards.

Variations such as these have major ethical significance; they indicate that local 'custom and practice' is hugely influential. Yet, the ethical underpinnings for local practice seem, if the literature is any guide, to be rarely scrutinized and challenged.

20.2.2 Compulsion in the Community – Involuntary Outpatient Treatment (IOT)

It is widely accepted that mental health law should keep pace with developments in psychiatric practice. In keeping with an increasing focus on mental health care in the community, provisions for involuntary treatment outside hospital have been introduced in a number of jurisdictions.

Some forms of IOT may be considered as a 'less restrictive alternative' by allowing commitment to outpatient treatment instead of inpatient treatment, or by permitting earlier conditional release from inpatient commitment. Other forms of IOT may allow 'preventative commitment'. This is the most controversial since it may allow the compulsory treatment of a patient who is not currently at risk. Based on a proven record of relapse when treatment is discontinued, and of dangers previously demonstrated when relapse has occurred, compulsion is used to avert future risk.

As Dawson and colleagues have shown (Dawson 2005, Mullen et al. 2006), the scope of IOT powers varies widely by jurisdiction. They nearly always include the imposition of a duty on the patient to accept psychiatric treatment (even if that duty is not matched by a power to restrain and medicate the patient in a community setting). They may include a direction that the patient accept visits from clinicians and attend appointments; a direction as to where the patient should reside; an authority for a clinician to enter his residence at reasonable times and for purposes directly related to the enforcement of the treatment regime; the power to recall the patient to hospital; and, to provide involuntary treatment in a hospital or clinic where such treatment may safely and appropriately be given.

However, IOT is controversial. Ethical criticisms include potential violations of human rights, such as privacy; the potential to distract attention from the quality of community services and to undermine the development of non-coercive methods for engaging reluctant patients; the potential to divert resources to a small group of patients perceived as 'risky'; the possibility that fears of involuntary treatment may deter patients from seeking treatment; and the limited range of treatments that can be enforced, some being perhaps sub-optimal (for example relying on drugs that can be given by intramuscular injection). There are also concerns (supported by some evidence (Burgess et al. 2006)) that outpatient orders may lead to an increase in the use of compulsion, since unlike inpatient commitment, there is no 'ceiling' imposed by a finite number of available beds on the numbers on compulsory orders. The elasticity of the concept of 'risk', and its potential to encompass a broad range of 'troubling' behaviours in the community is an added concern. Just as with inpatient commitment, rates of IOT vary widely across jurisdictions, at least fivefold (Dawson 2005), and with similar implications.

Less directly an ethical question, but of significance is the lack of agreement about the effectiveness of IOT in reducing admissions, reducing patient violence, or improving psychosocial outcomes. The same studies, including randomised controlled trials, are interpreted quite differently by different experts (Churchill et al. 2007, Kisely et al. 2005, Swartz and Swanson 2004). The first of these is the most comprehensive, and in my view, the most persuasive. The better controlled the investigation, the less effective IOT turns out to be.

Whether, in principle, ethical objections to IOT could rule the practice out altogether will be considered later (Section 20.4).

20.2.3 Patients' Retrospective Attitudes to Compulsion

If patients who had been treated involuntarily were to retrospectively endorse this action, this would have ethical significance. Studies involving involuntary inpatient treatment indicate that while many patients report health benefits, fewer than 50% of patients later regard their detention as justified (Gardner et al. 1999, Kane et al. 1983, Kjellin et al. 1997). In a recent Swedish study, only 32% of involuntary patients interviewed at discharge stated that they would want to have the same treatment in the future (Kjellin et al. 2004). Similarly, a recent English study found that only 40% of involuntary patients regarded this measure as justified when interviewed 1 year later (Priebe et al. 2009).

Involuntary community treatment seems to be better accepted. IOT may be generally preferred by patients to involuntary hospitalization (Swartz et al. 2003a), but the majority of patients who experienced IOT in an important North Carolina study failed to endorse its benefits (Swartz et al. 2003b). In contrast, a small New Zealand study of patients who had experienced a community treatment order and who agreed to interview, found that the majority regarded it as helpful, even though disadvantages were recognised (Gibbs et al. 2005). A significant proportion of patients believe the possibility of mandated community deters patients from seeking help (Van Dorn et al. 2006).

Overall, then, it appears that the majority of patients retrospectively fail to endorse their involuntary treatment, even though they may acknowledge health benefits.

20.3 'Coercion': Threats and Offers

20.3.1 Offers Versus Threats

An important distinction, necessary to an understanding of 'coercion', is to be made between 'threats' and 'offers'. Here we are dealing not with force, but with conditional (or more accurately, biconditional) propositions – that is: if the patient accepts treatment A, then the clinician will do X; or if the patient does not accept A, then the clinician will not do X (or do Y). An account by Wertheimer (Wertheimer 1987) has been especially influential. He proposes that 'threats' coerce, but 'offers' do not. He states: 'The crux of the distinction between threats and offers is that A makes a 'threat' when B will be worse off than in some relevant base-line position if B does not accept A's proposal, but that A makes an 'offer' when B will be no worse off than in some relevant base-line position if B does not accept A's proposal'.

What is meant by the 'baseline'? Wertheimer argues for a moral baseline. An example will make this clearer. A highly skilled surgeon (S) proposes to a patient (P) with a physical disfigurement that he will only perform a cosmetic operation if P pays £1,000. Has he made a threat or an offer? Under a moral test, the key issue is whether S ought to operate for free, for example under a free healthcare service (or whether P has a right to a free operation). If so, S's proposal is a threat. P is worse-off under this moral baseline if he does not agree to pay the fee. On the other hand, if P is not entitled to a free cosmetic operation, even in a free healthcare system (that is, where there is no moral imperative for S to operate for free) then S's proposal is an offer. If P does not pay, he is no worse-off than if S's proposal had not been made. P has no entitlement to the operation.

A threat, if not acceded to, makes the recipient 'worse off' according to the relevant moral baseline, while an offer, if declined, does not. The person threatened is threatened with the deprivation of a right or obligation, and thus would be made worse off. The proposal: 'Either you accept medication or you will be involuntarily admitted to hospital' is thus a threat, according to the baseline that patients should be entitled to choose whether they will take medication or not. An offer of something in the nature of an extra benefit (a ticket to a football match) made on condition that the patient complies with the treatment would, if rejected, not make the patient worse off compared with the relevant moral baseline. The proposal made by a mental health court, that if the patient accepts supervised treatment, the expected custodial sentence for the offence committed will be waived, is also an offer. The baseline here is the established sentence incurred by the offence. Rejection of the proposition of treatment makes the person no worse-off; a custodial sentence would be imposed even if the offer had never been made.

Defining the moral baseline presents difficulties that space does not allow us to consider in detail here. There is a tendency to slip into equating a 'legal baseline' – what a person is entitled to under the law – with a moral baseline. Bonnie and Monahan (2005) have argued for recognition of a legal baseline standard in mental health care. This has the virtue of precision, but it may be difficult, morally, to accept the relativism that accompanies different laws in different jurisdictions. It may be worth considering the possibility of using a statement of principles, such as the UN's 1991 'Protection of persons with mental illness and the improvement of mental health care', as the basis for a moral baseline.

Threats are thus coercive, while offers are not. This is the narrow technical sense of coercion. Note that while a coercive act entails a moral violation of some kind, it may, as we shall see, be justified under certain conditions.

Some authors (see for example, Rhodes (2000)) have argued for a subjective account of coercion, involving the perceiving of a threat or a sense of the will being unduly pressured. These may have value clinically in understanding how a patient experiences a proposition, but do not comport so well with the kinds of justifications for coercion used in mental health care (Section 20.4).

20.3.2 Deception

Under some circumstances deception can be *coercive* in the narrow sense. A patient on an IOT order may believe, for example, that failure to accept treatment in the community will lead to forced treatment even though this may not be permitted under the terms of the order. If this false belief is not corrected by the clinician, this amounts to a *coercive* threat.

20.3.3 Exploitation

Exploitation, while morally unseemly, differs from coercion (Mayer 2007, Wertheimer 2001). The threat here does not issue from the person making the proposal, but is a 'background' threat. Furthermore, there is no entitlement. Take for example, a variation of the surgeon's offer above. Suppose there is only one surgeon available in the area who can do the operation, which elsewhere costs around £1,000. This surgeon demands £5,000. The patient does not have an entitlement to free healthcare for this procedure, nor is there a fixed fee. Exploitation involves not a threat, but an offer, albeit a problematic one. The offer, if refused, leaves the patient no worse-off under the moral baseline. The moral problem here is of the exploiter taking *unfair advantage* of the person exploited. But note that exploitation can be mutually advantageous – the patient gets the operation, and the surgeon earns a lot of money.[2]

20.3.4 Unwelcome Predictions

If a clinician were to say to a patient, that a refusal of medication will eventually result in admission to hospital involuntarily, is this a threat or something else – an unwelcome 'prediction'? Much depends here on the evidential basis of the prediction. How well is it supported by the facts, concerning, for example, previous outcomes in these circumstances? Whether the clinician will be an agent in the

---

[2] With regard to unreflective or unconscious exploitation of a patient by a narcissistic therapist see Chapter 18.

in the involuntary treatment order is also relevant. The distinction can be unclear (Schramme 2004), but unwelcome predictions that meet sound evidential criteria are not *coercive*.

20.4 Inducements

20.4.1 Inducements in Mental Health Services

If we accept the account of *coercion* in the narrow sense given above, then offers (or inducements) represent a lesser degree of treatment pressure than threats. These offers may involve assistance of various kinds – help with shopping, organising free outings, help with obtaining benefits, and so on. However, a little reflection will indicate that such inducements may still present ethical problems. Under what circumstances should some patients be offered an inducement, and not others? Why should a patient who is prudent in deciding to take medication not receive a valued 'good' that is offered to a patient who declines medication? The issue here is one of 'fairness'. The situation in community mental health services is especially complex as a wide range of helpful activities and assistance might be offered to patients, often in the context of problem solving in the individual case. It could be argued, however, that 'fairness' requires a clear statement about what a mental health service offers, for example, as laid out in an operational policy. Without this, the paradox may arise where the greater the range of provision in a mental health service, the greater the scope for 'threats' where a form of assistance is denied a patient who does not comply with treatment.

20.4.2 Problematic Inducements

Some inducements are met with a strong moral intuition of unacceptability by clinicians. An example is the proposal that patients with a psychosis who are at high risk of relapse and who reject medication should be paid to take it (Claassen 2007, Claassen et al. 2007).

I have addressed the ethical concerns raised by this proposal of 'money for medication' elsewhere (Szmukler, 2009). These include, first, the question of *fairness* (as just noted), and second, the potential for *exploitation*. Although 'small' sums of money (say £10–15 per depot injection) are to be offered, for a patient with a psychosis who may be severely financially disadvantaged, such an amount may be substantial. While taking medication may be advantageous for the patient in improving his or her clinical condition, that does not make the transaction non-exploitative. As noted above, exploitation can be mutually advantageous. But what is the advantage for the clinician? Seeing one's patient improve is undoubtedly rewarding. But there may be other motives. In many mental health services there are now major pressures on inpatient beds. Admissions can be difficult to arrange. Preventing admissions also saves the health services money. The clinician may, in effect, become an agent of the health service in reducing costs (and may be rewarded for doing so by management).

A third ethical problem in money for medication, perhaps the most important, is one of '*incommensurable values*'. This complex idea can be only briefly outlined here. The idea of 'incommensurable values' refers to a transaction characterized by an exchange between 'goods' lying in two domains of valuation, one high and one low, or valuations that cannot be aligned along a single metric. Such an exchange is said to denigrate or corrupt the character of the 'goods' that lies in the higher domain. Stark examples would be the selling of a child or of an electoral vote. In the proposal being made for psychosis, money is offered to induce the patient to change a decision about medication. Let us assume, as proponents seem to, that the patient has treatment decision-making capacity. Thus the patient's refusal is a considered decision about medication that represents what the patient judges to be in his or her best health interests. It is a considered decision because the patient has decision-making capacity and previous attempts at persuasion about the benefits of medication have presumably been unsuccessful. The patient has weighed up these benefits and drawbacks of treatment and decided against. The harm then, is the devaluing or denigration of a decision made by the patient about what is in his or her best health interests. This represents in a significant sense, a failure of *respect for the person*, the devaluing of a fundamental moral value (in western societies at least). Such a devaluation is particularly troubling in the case of those with a mental illness, as respect for this group of patients in most societies has not been at the same level as for other patients. Thus money for medication further promotes an inferior conception of personhood – of diminished agency and autonomy – of those with a mental illness.

Different considerations may apply when the patient lacks capacity, where perhaps a case can be made that respect for the person is not violated in the same way, and that 'money for medication' if judged as being in the 'best interests' of the patient (see Section 20.4.2 below) may be acceptable. Space does not permit further discussion here; the interested reader is referred to Szmukler (2009), where a number of practical difficulties associated with financial incentives for treatment adherence are also outlined.

20.5 Justifications for 'Coercive' Interventions in Mental Health Care

20.5.1 A Framework Applicable Across the Whole Range of Interventions

In most countries, mental health law defines the criteria for the use of compulsion, while there is little or no explicit guidance for decisions concerning the use of coercion, as defined above, or of inducements. The risk-based criteria in mental health law that govern inpatient involuntary treatment do not translocate readily into community settings; such criteria, for instance, as 'the health or safety of the patient' or 'the protection of others' have such broad meaning that their limits cannot be easily discerned. I have argued above (Section 20.1) that the criteria for involuntary treatment in most statutes discriminate against people with mental disorder (Dawson and Szmukler 2006, Szmukler and Holloway 1998). In large part this reflects the absence of a set of coherent underlying principles about when non-consensual treatment is justified.

Szmukler and Appelbaum (2008) have drawn attention to two frameworks for making decisions about treatment pressures, one based on a '*capacity-best interests*' model, the other on '*paternalism*' as elaborated by Culver and Gert (Gert et al. 2006). In this chapter I will focus on the former as it is more developed and already finds expression in various forms of guardianship legislation, and more recently the Mental Capacity Act 2005 (England and Wales). Furthermore its principles can be applied across the entire range of coercive interventions, including compulsory inpatient and community treatment (Dawson and Szmukler 2006, Szmukler et al. 2010). This framework also brings the non-consensual treatment of those suffering from a mental disorder into alignment with the treatment of patients in the rest of medicine and thus with general medical ethics.

20.5.2 A 'Capacity-Best Interests' Framework

Definitions of decision-making *capacity* vary, but common elements are the ability to understand and retain information relevant to the decision (including the consequences of deciding one way or the other) and the ability to use that information to make a decision. The latter includes the ability to 'appreciate' that the information applies to the patient's predicament, the ability to reason with that information, and the ability to express a choice (Grisso and Appelbaum 1998) (see Chapter 3.1 and 4.2.1.2).

Only if the patient lacks an adequate level decision-making capacity, it is argued, and who is thus unable to make sound judgments about their health interests, would treatment against a patient's wishes be considered. A patient who has capacity is more capable of making decisions about his or her health interests than anyone else, and should not be subject to non-consensual treatment. For the incapable patient, a further test must be passed – the proposed treatment must be in the patient's 'best interests.' Definitions of 'best interests' may be difficult (McCubbin and Weisstub 1998) but useful, practical guidance for deciding on the matter is available.

An example is the Mental Capacity Act 2005 (MCA) for England and Wales which states:

• The person making the determination of best interests must consider all the relevant circumstances and, in particular, take the following steps:
• He must consider (a) whether it is likely that the person will at some time have capacity in relation to the matter in question; and (b) if it appears likely that he will, when that is likely to be.

- He must, so far as reasonably practicable, permit and encourage the person to participate, or to improve his ability to participate, as fully as possible in any act or decision affecting him.
- He must consider, so far as is reasonably ascertainable: the person's past and present wishes and feelings (and, in particular, any relevant written statement made by him when he had capacity); the beliefs and values that would be likely to influence his decision if he had capacity; and the other factors that he would be likely to consider if he were able to do so.
- He must take into account, if it is practicable and appropriate to consult them, the views of: anyone named by the person as someone to be consulted on the matter in question or on matters of that kind; anyone engaged in caring for the person or interested in his welfare; any donee of a power of attorney granted by the person; any deputy appointed by the Court, as to what would be in the person's best interests.
- regard must be had to whether the purpose for which the action is needed can be as effectively achieved in a way that is less restrictive of the person's rights and freedom of action.

This framework is highly suitable to making decisions about 'coercive' interventions in a community mental health care setting. One would aim for the least coercive option, and restrict interventions that are rejected by the patient (for example, medication, a home visit) to those patients who lack capacity and where it is judged to be in the patient's best interests. This framework offers a structure for clinical discussions concerning the justification for 'coercive' interventions and fosters greater clarity and transparency in decision-making.

Extension of this framework to involuntary treatment, including detention as an inpatient, requires some additional elements. Capacity-based legislation, as exemplified in guardianship statutes or in the MCA, is strong on respect for patient autonomy, but weak on the regulation of the use of force or detention in hospital. These areas of weakness are precisely those where civil commitment statutes are strong. There the authorization of force and detention – what is permitted by whom, to whom, for how long, and under what conditions – are clearly specified. By 'fusing' the strengths of both types of legislation it is suggested that a single comprehensive statute can be formulated that can deal with all patients who lack decision-making capacity, whatever the cause, whether it be a 'mental disorder' or a 'physical disorder' (Dawson and Szmukler 2006, Szmukler et al. 2010). Apart from a few exceptional instances, the fundamental ethical principles are the restriction of involuntary interventions to patients who lack capacity and for whom they are in the patient's best interests.

A capacity-best interests framework does not, in principle, exclude IOT. If treatment can be safely and effectively provided in the community for patients meeting the criteria for involuntary treatment, there is no reason why this should not occur.

20.5.3 The Protection of Other Persons

It will probably have been noticed that a capacity-best interests framework does not address, directly at least, interventions aimed at the protection of others. If it is in the best interests of the incapable patient to prevent harm to others, then involuntary treatment would be perfectly appropriate. But what should happen when a patient who poses an apparent risk to others retains capacity?

The health interests of the patient and the protection of others comprise distinct and separate ends (Culver and Gert 1982, Large et al. 2008b, Szmukler and Appelbaum 2008). Pursuit of the latter does not necessarily entail the former. The latter turns on the question of the magnitude of the risk to others and the potential harms that may occur. If the patient with a mental disorder has capacity, by definition, he or she appreciates the consequences of the illness and of having or not having treatment. Such a patient makes a capable decision about risk, as does a non-mentally ill person who, for example, decides to drink alcohol excessively before driving. To force treatment or to detain such a person on the basis of risk alone amounts to a form of preventive detention, a measure which, through the agency of conventional mental health law ('the protection of others'), is reserved only for those with a 'mental disorder'. Detention as applied to the rest of us follows only on the commission of an offence. If a capable patient suffering from a mental disorder commits an offence, then it may be appropriate, or even incumbent upon us, to offer voluntary treatment, either instead of, or concurrently with the usual sentence for the offence. In general, all capable persons presenting what is deemed an unacceptable risk to others should be equally liable to detention or coercive interventions, whether they suffer from a mental disorder or not. If detention were to be regarded as acceptable on the basis of putative 'risk' alone (and most societies oppose this), liability to detention should be determined by the level of risk, not by whether the person has a 'mental disorder'. Selectively subjecting only those with a 'mental disorder' to preventive detention on the basis of 'risk' is unfair, discriminatory (and stigmatizing, since it promotes a stereotype of the mental ill as dangerous). I put the term 'mental disorder' in quotes as it is poorly defined and regularly comes under considerable pressure to expand its scope as societies try to find solutions for troublesome people whose processing through the criminal justice system may be cumbersome or unlikely to lead to what may become indeterminate detention. The category of persons having a 'dangerous severe personality disorder', a term originated by a Minister in the British Government, but now in common usage for a range of programmes, is a good example (Maden and Tyrer 2003).

Incapable patients for whom treatment is not in their best interests – in all probability meaning that they have been unresponsive to all treatments thus far – are probably rare. If presenting a serious risk to others, detention in a healthcare institution under humane conditions may be appropriate (Szmukler et al. 2010).

Being clear in our thinking about risk is especially important as psychiatric care moves into the community. Allowing coercive interventions on the basis of risk alone carries the danger of their being used increasingly frequently, especially in a risk-averse society. The notion of what constitutes an unacceptable 'risk' is elastic.

It may come to be applied to a wide range of behaviours that might previously have been considered as merely annoying or undesirable.[3]

There is also a significant statistical problem of ethical importance. Despite the public's fears, incidents of serious violence perpetrated by people with mental disorders are rare and not more common since the advent of community care (Large et al. 2008a). The prediction of rare events inherently lacks accuracy. The number of false positives – those predicted to be violent who are not – far outweighs the true positives. It can be shown that current risk assessment instruments (Buchanan and Leese 2001) will, on average, predict incorrectly approximately 7 times out of 10 for a base rate in the population of 20% per annum for violent incidents, and more than 9 times out of 10 if the rate is 5% (Szmukler 2003). The base rate of severe violence is lower still. Even the best instruments (Monahan et al. 2001), tested under research conditions, do not fare much better, especially when replication on a new population is attempted (Monahan et al. 2005). Mossman (2006) has examined the statistical and ethical aspects of risk assessment. He argues, using a neo-Kantian analysis, that the detention of persons assessed as constituting an 'actuarial' risk to others, but where the majority will not carry out a dangerous act, is wrong. It amounts to treating such people as a 'means' (of avoiding harm to others) and not as an 'end' (acting to further the interests of a patient who is respected as a 'person').

These considerations present a major challenge to mental health services. The potential for an expansion of coercive interventions in response to community fears is substantial. Mental health professionals rightly accept an obligation to act to avert a potential serious harm if there is a clearly expressed threat against another person and a credible means of acting on that threat. However, if abuses are to be avoided, they need to also engage in a dialogue with a community that understands 'risk' poorly, entertains unrealistic ideas about what is possible and demands healthcare responses that are unethical.

20.6 Reducing the Need for Coercive Interventions

20.6.1 Involving Patients in Treatment Decisions

A reasonable hypothesis is that patients are likely to feel less 'coerced' if they are able to play a more active role in making treatment choices. The trend for their increased involvement in determining the shape and nature of services may be helpful in this regard (Rose et al. 2003) but evidence has so far not been adduced that this has an effect on the need for coercive interventions.

On the other hand, initiatives at the individual patient level involving the use of 'advance statements' (AS), have been shown to be effective. An AS allows a patient, when well, to state treatment preferences in anticipation of a time in the future when,

---

[3] See the Chapter 29 Abuse of psychiatry for political purposes by van Voren about the involuntary detention of political dissidents in the former Soviet Union.

as a result of the effects of a mental illness, he or she may not be capable of making treatment decisions. The anticipated loss of decision-making capacity usually occurs during a relapse of a psychosis. An important aim of an AS is to give the patient more influence over his or her treatment at a time of crisis and thus to reduce the need for coerced treatment.

20.6.1.1 Advance Statements

A typology of ASs has been described by (Henderson et al. 2008). The key dimensions along which ASs vary include the following: whether the AS is patient or service provider led; whether it is legally binding or not; whether it is facilitated by a person independent of the clinical team. At least three major types of AS have been described: 'crisis cards' (CCs), 'joint crisis plans' (JCPs) and 'psychiatric advance directives' (PADs).

20.6.1.2 Crisis Cards

In a crisis card (CC) patients state their treatment wishes or nominate a person who should be familiar with their preferences, without reference to their treatment team. It is entirely consumer led. In some jurisdictions, for example, under the MCA in England and Wales stated treatment refusals now have legal force, but can be overriden by mental health legislation. Their uptake has been very limited (Sutherby and Szmukler 1998).

20.6.1.3 Joint Crisis Plans

In contrast to CCs, the joint crisis plan (JCP) is the product of a structured discussion between patient (usually accompanied by a relative, friend or advocate) and the clinical team (usually the psychiatrist and case-manager) that aims to reach an agreement on what measures should be taken if a relapse should occur in the future. A critical element is the involvement of a facilitator independent of the clinical team who ensures that the patient's voice is heard (Henderson et al. 2004, Sutherby et al. 1999). A JCP is not legally-binding (although as noted above specific treatment refusals may have legal force in some jurisdictions but may be overridden by an order under a mental health act). The clinical team makes it explicit that while it will attempt to comply with the terms of the JCP, it cannot guarantee to do so.

While the JCP is service provider initiated, it aims at achieving an advance agreement in which the patient has played at least an equal role. Indeed, what finally appears in the JCP, including the exact wording, is determined by the patient (though agreed by the clinician).

A JCP will usually comprise much more than a statement of treatment preferences or refusals in the event of a relapse. For instance, it may include details of the current treatment plan (including medication and dosages), early signs of relapse; what has proven helpful in the past in arresting a relapse; what treatments have proven successful (or unsuccessful) when relapse has become established;

who should be contacted to help; what symptoms or behaviours would indicate that admission should occur; adverse drug effects or allergies; and, practical issues that need attention, (for example, making sure the patient's home is secure in the event of hospitalization, looking after pets or plants, cancelling milk deliveries, and so on). The JCP's specificity of content, based on an analysis of past illness episodes, is an important advantage since relapse in a patient tends to occur in a stereotyped manner.

A randomized controlled trial of JCPs involving 160 patients has now been conducted involving 8 community mental health teams in South-east England (Henderson et al. 2004). Almost 40% of patients who were eligible took up the opportunity to complete a JCP. Eligible patients were those with a diagnosis of a psychotic illness or bipolar disorder, and who had had at least one admission in the previous 2 years. Compulsory admissions over a 15 month period were halved compared to a control group (13% vs. 27%). There was a non-significant trend for reduced hospitalizations. Although numbers were small, there were also significantly fewer violent incidents in the JCP group. Patients formulating JCPs said their agreements were given freely and without pressure, that they felt more in control of their mental health problems as a result of making a JCP, and that they would recommend JCPs to others (Henderson et al. 2009).

20.6.1.4 Psychiatric Advance Directives

Advance directives are in principle legally binding. The directives in a PAD may be of three kinds: first, specified treatments that are refused or requested; second, statements about personal values, attitudes or general preferences that may be used as a guide to those making decisions about treatment for the patient; and third, nomination of a person to act a 'substitute' or 'proxy' decision-maker. A PAD assumes that the patient had decision-making capacity when it was made, and that the circumstances in which the PAD is triggered are those that were anticipated.

Medical advance directives have legal force in the US. Although specific PAD legislation enacted in some states in the US is ostensibly legally binding, there is currently considerable uncertainty about circumstances in which PADs may be overridden (Swanson et al. 2006a). Civil commitment legislation may do so if it specifically authorizes involuntary treatment; however, if a separate decision concerning consent is required before involuntary treatment can be imposed on a patient following detention in hospital (the detention being based, for example, on a separate dangerousness standard) then a PAD may prevail. The significant case here is *Hargrave v Vermont*, heard by the US Court of Appeals, Second Circuit, which upheld an advance directive refusing all medication for a psychosis on the grounds that failure to do so would amount to discrimination against persons with a mental disorder (Allen 2004, Appelbaum 2004). The court ruled that if advance directives for those with other disorders cannot be overridden, it should also be so for those with a mental disorder. Some PAD statutes, for example, in Pennsylvania, stipulate that the patient's wishes may be overridden if they violate 'accepted standards of care'. How this should be interpreted is not clear. It is noteworthy that the evidence on PADs to date shows that refusal of all treatment alternatives is rare (Swanson et al. 2006b).

A further controversial issue is whether a PAD bearing a treatment refusal very likely to lead to a life-threatening risk should be respected or not. This is too big a subject to be covered here.

A variant of the PAD, termed a facilitated PAD (F-PAD) has been introduced following research revealing an apparently widespread appeal of PADs to many patients, yet few actually making one. Difficulties in understanding the implications of a PAD and the practical complexities in their execution act as deterrents to their adoption. In an F-PAD, an attempt is made to overcome these obstacles by employing a trained facilitator who explains what a PAD involves and, if the patient chooses to opt for one, assists with its completion (Swanson et al. 2006b). The service provider may also be asked to become involved in generating the PAD.

A randomised controlled study of F-PADs has indeed shown that facilitation results in a highly significant increase in the number of patients who decide to make a PAD (61% vs. 3% of controls). Of special interest to the subject of this chapter is the finding that at 1 month follow-up, those with an F-PAD reported a much better working alliance with their clinicians and were more likely to say that they received the services they needed (Swanson et al. 2006b). A later report from this study found that the number of 'coercive interventions' over the succeeding 2 years for those who made a PAD was considerably fewer compared to patients who chose not to make a PAD (Swanson et al. 2008). The interventions assessed were: (1) being picked up by the police and transported to an emergency facility; (2) being placed in handcuffs; (3) being involuntarily committed to a hospital for psychiatric treatment; (4) being placed in seclusion in a locked hospital room; (5) being placed in physical restraints; and (6) receiving forced medications. It was also found that an F-PAD was more effective in reducing such interventions in those patients who had suffered a loss of capacity during the crisis.

Thus there is emerging evidence that some forms of AS can reduce coercion and improve outcomes. Patients may be empowered by the process; from a practical point of view, the information contained in an AS may ensure that the most appropriate treatment is given when information from other sources is unavailable. Independent facilitation, either in a JCP or an F-PAD, may be a critical factor. JCPs show that patients can effectively voice their treatment wishes outside a legal framework. The dialogue between patient and mental health professionals is probably a key ingredient.

20.6.2 Interventions Aimed at Reducing Coercive Interventions on Hospital Wards

A number of non-controlled studies relying on before versus after comparisons have suggested that the use of seclusion on psychiatric wards can be significantly reduced (Gaskin et al. 2007). The interventions have been varied, multiple, and have usually involved system changes – for example, state policy and regulation changes aimed at reducing seclusion, monitoring and analysing episodes, strengthening leadership, staff education, changing the ward environment, increasing staff to patient ratios, creating special emergency response teams, and treating patients as active participants in interventions to reduce seclusion. There is also evidence for similar measures achieving a reduction in the use of restraint (Hellerstein et al. 2007) is a recent example).

A study in which a focussed intervention was an attempt to increase the involvement of inpatients in planning their treatment did not affect the 'perceived coercion' experienced by the patients (Sorgaard 2004).

The dearth of rigorously controlled studies in this field is notable.

20.7 Conclusions

This chapter should have made clear the need for ethical analysis of the range of treatment pressures commonly used in psychiatry. Distinctions having moral relevance can be made between them. Justifications for 'coercive' interventions are not often discussed in the course of everyday practice, yet the new types of practice in community psychiatry and the huge variations due to local 'custom and practice' suggest that this is very necessary. Clinical discussion of justifications can be conducted within a coherent framework. Competence in this domain of practice is surely as important as in the more 'technical' aspects of treatment such as choosing the most appropriate medication or psychological intervention. 'Advance statements' may prove a valuable tool for reducing the need for 'coercion'.

References

Allen M (2004) Hargrave v Vermont and the quality of care. Psychiatr Serv 55:1067
Appelbaum PS (2004) Law & psychiatry: psychiatric advance directives and the treatment of committed patients. Psychiatr Serv 55:751–752
Bonnie RJ, Monahan J (2005) From coercion to contract: reframing the debate on mandated community treatment for people with mental disorders. Law Hum Behav 29:485–503
Buchanan A, Leese M (2001) Detention of people with dangerous severe personality disorders: a systematic review. Lancet 358:1955–1959
Burgess P, Bindman J, Leese M, Henderson C, Szmukler G (2006) Do community treatment orders for mental illness reduce readmission to hospital? An epidemiological study. Soc Psychiatry Psychiatr Epidemiol 41:574–579
Campbell T, Heginbotham C (1991) Mental illness: prejudice, discrimination and the law. Dartmouth, Vermont
Churchill R, Owen G, Singh S, Hotopf M (2007) International experiences of using community treatment orders Department of Health: www.dh.gov.uk/en/Publicationsandstatistics/Publications/PublicationsPolicyAndGuidance/DH_072730
Claassen D (2007) Financial incentives for antipsychotic depot medication: ethical issues. J Med Ethics 33:189–193
Claassen D, Fakhoury W, Ford R, Priebe S (2007) Money for medication – financial incentives to improve medication adherence in Assertive Outreach patients. Psychiatr Bull 31:4–7
Culver C, Gert B (1982) Philosophy in Medicine: conceptual and ethical issues. Oxford University Press, Oxford

Dawson J (2005) Community treatment orders: international comparisons. Otago University Press, Dunedin, New Zealand

Dawson J, Szmukler G (2006) Fusion of mental health and incapacity legislation. Br J Psychiatry 188:504–509

Fisher WA (1994) Restraint and seclusion: a review of the literature. Am J Psychiatry 151:1584–1591

Gardner W, Lidz CW, Hoge SK, Monahan J, Eisenberg MM, Bennett NS, Mulvey EP, Roth LH (1999) Patients' revisions of their beliefs about the need for hospitalization. Am J Psychiatry 156:1385–1391

Gaskin CJ, Elsom SJ, Happell B (2007) Interventions for reducing the use of seclusion in psychiatric facilities: Review of the literature. Br J Psychiatry 191:298–303

Gert B, Culver CM, Clouser KD (2006) Bioethics: a systematic approach, 2nd edn. Oxford University Press, New York, NY

Gibbs A, Dawson J, Ansley C, Mullen R (2005) How patients in New Zealand view community treatment orders. J Ment Health 14:357–368

Grisso T, Appelbaum PS (1998) Assessing competence to consent to treatment: a guide for physicians and other health professionals. Oxford University Press, New York, NY

Hellerstein DJ, Bennett Staub A, Lesquesne E (2007) Decreasing the use of restraint and seclusion among psychiatric inpatients. J Psychiatr Pract 13:308–317

Henderson C, Flood C, Leese M, Thornicroft G, Sutherby K, Szmukler G (2004) Effect of joint crisis plans on use of compulsory treatment in psychiatry: single blind randomised controlled trial. Br Med J 329:136

Henderson C, Flood C, Leese M, Thornicroft G, Sutherby K, Szmukler G (2009) Views of service users and providers on joint crisis plans: single blind randomized controlled trial. Soc Psychiatry Psychiatr Epidemiol 44:369–376

Henderson C, Swanson JW, Szmukler G, Thornicroft G, Zinkler M (2008a) A typology of advance statements in mental health care. Psychiatr Serv 59:63–71

Hotopf M, Wall S, Buchanan A, Wessely S, Churchill R, Hotopf M, Wall S, Buchanan A et al (2000) Changing patterns in the use of the Mental Health Act 1983 in England, 1984–1996. Br J Psychiatry 176:479–484

Hoyer G (2008) Involuntary hospitalization in contemporary mental health care. Some (still) unanswered questions. J Ment Health 17:281–292

Janssen WA, Noorthoorn EO, de Vries WJ, Hutschemeakers GJ, Lendemeijer HH, Widdershoven GA (2008) The use of seclusion in the Netherlands compared to countries in and outside Europe. Int J Law Psychiatry 31:463–470

Kane JM, Quitkin F, Rifkin A, Wegner J, Rosenberg G, Borenstein M (1983) Attitudinal changes of involuntarily committed patients following treatment. Arch Gen Psychiatry 40:374–377

Keown P, Mercer G, Scott J (2008) Retrospective analysis of hospital episode statistics, involuntary admissions under the Mental Health Act 1983, and number of psychiatric beds in England 1996–2006. Br Med J 337:a1837

Kisely S, Campbell LA, Preston N (2005) Compulsory community and involuntary outpatient treatment for people with severe mental disorders. Cochrane Database Syst Rev 3:CD004408

Kjellin L, Andersson K, Bartholdson E, Candefjord IL, Holmstrom H, Jacobsson L, Sandlund M, Wallsten T, Ostman M (2004) Coercion in psychiatric care – patients' and relatives' experiences from four Swedish psychiatric services. Nord J Psychiatry 58:153–159

Kjellin L, Andersson K, Candefjord IL, Palmstierna T, Wallsten T (1997) Ethical benefits and costs of coercion in short-term inpatient psychiatric care. Psychiatr Serv 48:1567–1570

Kjellin L, Ostman O, Ostman M (2008) Compulsory psychiatric care in Sweden: development 1979–2002 and area variation. Int J Law Psychiatry 31:51–59

Large MM, Ryan CJ, Nielssen OB, Hayes RA (2008b) The danger of dangerousness: why we must remove the dangerousness criterion from our mental health acts. J Med Ethics 34:877–881

Large M, Smith G, Swinson N, Shaw J, Nielssen O (2008a) Homicide due to mental disorder in England and Wales over 50 years. Br J Psychiatry 193:130–133

Maden T, Tyrer P (2003) Dangerous and severe personality disorders: a new personality concept from the United Kingdom. J Pers Disord 17:489–496

Mayer R (2007) What's wrong with exploitation? J Appl Philos 24:137–150

McCubbin M, Weisstub DN (1998) Toward a pure best interests model of proxy decision making for incompetent psychiatric patients. Int J Law Psychiatry 21:1–30

Monahan J, Steadman HJ, Robbins PC, Appelbaum P, Banks S, Grisso T, Heilbrun K, Mulvey EP, Roth L, Silver E (2005) An actuarial model of violence risk assessment for persons with mental disorders. Psychiatr Serv 56:810–815

Monahan J, Steadman HJ, Silver E, Appelbaum PS, Robbins PC, Mulvey EP, Roth LH, Grisso T, Banks S (2001) Rethinking risk assessment: the MacArthur study of mental disorder and violence. Oxford University Press, New York, NY

Mossman D (2006) Critique of pure assessment or Kant meets Tarasoff. Univ Cincinnati Law Rev 75:523–609

Mullen R, Dawson J, Gibbs A (2006) Dilemmas for clinicians in use of community treatment orders. Int J Law Psychiatry 29:535–550

Priebe S, Badesconyi A, Fioritti A, Hansson L, Kilian R, Torres-Gonzales F, Turner T, Wiersma D (2005) Reinstitutionalisation in mental health care: comparison of data on service provision from six European countries. Br Med J 330:123–126

Priebe S, Katsakou T, Amos M, Leese M, Morriss R, Rose D, Wykes T, Yeeles K (2009) Patients' views and readmissions 1 year after involuntary hospitalisation. Br J Psychiatry 194:49–54

Rhodes M (2000) The nature of coercion. J Value Inq 34:369–381

Rose D, Fleischman P, Tonkiss F, Campbell P, Wykes T (2003) User and carer involvement in change management in a mental health context: review of the literature. NHS Service Delivery and Organisation R&D Programme, London

Salize HJ, Dressing H (2004) Epidemiology of involuntary placement of mentally ill people across the European Union. Br J Psychiatry 184:163–168

Schramme T (2004) Coercive threats and offers in psychiatry. In: Schramme, T and Thome J (eds) Philosophy and psychiatry. Walter de Gruyter, Berlin, pp 357–369

Sorgaard KW (2004) Patients' perception of coercion in acute psychiatric wards. An intervention study. Nord J Psychiatry 58:299–304

Sutherby K, Szmukler G (1998) Crisis cards and self-help initiatives. Psychiatr Bull 22:3–7

Sutherby K, Szmukler GI, Halpern A, Alexander M, Thornicroft G, Johnson C, Wright S (1999) A study of 'crisis cards' in a community psychiatric service. Acta Psychiatr Scand 100:56–61

Swanson JW, McCrary SV, Swartz MS, Elbogen EB, Van Dorn RA (2006a) Superseding psychiatric advance directives: ethical and legal considerations. J Am Acad Psychiatry Law 34:385–394

Swanson JW, Swartz MS, Elbogen EB, Van Dorn RA, Ferron J, Wagner HR, McCauley BJ, Kim M (2006b) Facilitated psychiatric advance directives: a randomized trial of an intervention to foster advance treatment planning among persons with severe mental illness. Am J Psychiatry 163:1943–1951

Swanson JW, Swartz MS, Elbogen EB, Van Dorn RA, Wagner HR, Moser LA, Wilder C, Gilbert AR (2008) Psychiatric advance directives and reduction of coercive crisis interventions. J Ment Health 17:255–267

Swartz MS, Swanson JW (2004) Involuntary outpatient commitment, community treatment orders, and assisted outpatient treatment: what's in the data? Can J Psychiatry 49:585–591

Swartz MS, Swanson JW, Monahan J (2003b) Endorsement of personal benefit of outpatient commitment among persons with severe mental illness. Psychol Public Policy Law 9:70–93

Swartz MS, Swanson JW, Wagner HR, Hannon MJ, Burns BJ, Shumway M (2003a) Assessment of four stakeholder groups' preferences concerning outpatient commitment for persons with schizophrenia. Am J Psychiatry 160:1139–1146

Szmukler G (2003) Risk assessment: 'numbers' and 'values'. Psychiatr Bull 27:205–207

Szmukler G (2009) Financial incentives for patients in the treatment of psychosis. J Med Ethics 35:224–228

Szmukler G, Appelbaum PS (2008) Treatment pressures, leverage, coercion, and compulsion in mental health care. J Ment Health 17:233–244

Szmukler G, Daw R, Dawson J (2010) A model law fusing incapacity and mental health legislation. J Ment Health Law (in press)

Szmukler G, Holloway F (1998) Mental health legislation is now a harmful anachronism. Psychiatr Bull 22:662–665

Van Dorn RA, Elbogen EB, Redlich AD, Swanson JW, Swartz MS, Mustillo S (2006) The relationship between mandated community treatment and perceived barriers to care in persons with severe mental illness. Int J Law Psychiatry 29:495–506

Wertheimer A (1987) Coercion. Princeton University Press, Princeton, NJ

Wertheimer A (2001) Exploitation. Stanford Encyclopedia of Philosophy. http://plato.stanford.edu/entries/exploitation

Chapter 21
Ethics of Deinstitutionalization

Dirk Claassen and Stefan Priebe

Contents

Abbreviations

AOT   Assertive Outreach Team
CMHT   community mental health team
CR   Crisis Resolution
CTO   Community treatment order
EI   Early Intervention
HT   Home Treatment
PACE   Program for Assertive Community Effort
PACT   Program for Assertive Community Treatment
TAPS   Team for the Assessment of Psychiatric Services

D. Claassen (✉)
Psychiatrist and Psychotherapist, Former Consultant Psychiatrist and Honorary Senior Lecturer at
the East London Mental Health Trust, Psychiatric Practice, D-30161 Hannover, Germany
e-mail: praxis@drclaassen.com

H. Helmchen, N. Sartorius (eds.), *Ethics in Psychiatry*, International Library   341
of Ethics, Law, and the New Medicine 45, DOI 10.1007/978-90-481-8721-8_21,
© Springer Science+Business Media B.V. 2010

21.1 Deinstitutionalization

In psychiatric ethics, discussions are usually focused on the behaviour of individual mental health professionals, the care delivered by specific services, defined therapeutic interventions and various aspects of research (Helmchen 2008), sometimes on the design and implementation of new legislation (Lawton-Smith et al. 2008, Lepping 2008), but rarely on the development of and changes in the general approach of mental health care (Priebe 2004).

The term deinstitutionalization has been widely used in both the public and professional arena when reforms of mental health care were and are discussed. Its exact meaning however can vary. It is sometimes referred to as a general intention to reduce the role of institutions throughout mental health care and prefer the least institutionalized and most autonomous form of care whenever possible. According to this notion of deinstitutionalization, self-help without any professional input is preferable to any form of care through professional services; out-patient care is preferable to partial hospitalization, and partial hospitalization is preferable to full hospitalization. This wide understanding of deinstitutionalization contrasts with a much narrower notion of the term which often equates only the abolishment of old type psychiatric hospitals and possibly their replacement through services in the community (sometimes also called de-hospitalization). We will use the term deinstitutionalization as a general reference to reforms of mental health care that led to the closure or down-sizing of former asylums and established services providing care for patients with severe mental illnesses in the community. As a consequence, some of the discussed issues are more closely linked with specific aspects of mental health services providing care in the community than necessarily with the closure of hospitals.

In this chapter, we analyze ethical aspects of deinstitutionalization, a process that dominated major reforms of mental health care across the Western world since the 1950s. Deinstitutionalization with the closure and/or down sizing of former large asylums led to the establishment of different forms of community mental health care, the ethical dilemmas of which are considered and discussed. We do not aim to provide a full review of the very complex issue of deinstitutionalization and all its potential ethical implications. We rather focus on the application of ethics to the main features of deinstitutionalization and its consequences in psychiatry. This may be seen as applied ethics and thus refers to the more theoretical chapters in this book.

Psychiatry in the nineteenth century was characterized by institutionalization, on the levels of both academia with many chairs in psychiatry at universities and practical care in the large asylums for an increasing number of mentally ill patients across Europe. Ideas for a process of deinstitutionalization and care approaches outside traditional asylums were already outlined by Griesinger in Berlin in the 1860s and became more popular at the beginning of the twentieth century. One may argue that the approaches of open care by Kolb in Germany and of home care by Querido in the Netherlands in the 1930s constituted first models of deinstitutionalization and community psychiatry. However, on a larger scale deinstitutionalization in Europe began after World War II with growing societal support for the closure or downsizing of asylums and the development of alternative forms of care in the community. The political drivers, times of onset, pace, forms and exact outcomes of reforms leading to and implementing deinstitutionalization varied substantially across European countries. In some form however deinstitutionalization occurred in practically all countries across Europe. Asylums were abandoned or downsized and some kind of community care for patients with severe mental illnesses was established in all Western European countries. Eastern European countries followed later or are still involved in developing and implementing a form of deinstitutionalization. In the following we consider three western European countries, namely the United Kingdom (UK), Germany and Italy, more closely to illustrate important aspects of deinstitutionalization. The three countries represent different traditions in psychiatry and very different practical approaches to deinstitutionalization.

The UK had a relatively high level of patients in psychiatric asylums (the number of beds peaked in 1954 at 148,000 (Leff 1997)), which was increasingly regarded as a costly and ineffective form to provide care. One of the largest asylums, Friern Hospital, was a symbol of institutionalized psychiatry with allegedly one of the longest corridors in the world. At its peak in 1952 it contained 2,400 patients (Leff 1997). The futility of efforts to reform Friern Hospital and similar gigantic institutions inspired Enoch Powell's water tower speech of 1961, which set the trend away from the maintenance and improvement of large isolated hospitals towards the reprovision of inpatient care as a part of general hospitals and the transition of services into the community (Powell 1961). In practice, however, the process turned out to be slow and difficult, leading the government in 1975 to publish a White Paper (Better Services for the Mentally Ill), stipulating better reprovision of services for the old long stay population and causing more hospital closures during the 1980s (Carrier and Kendall 1997).

Germany started with a lower number (115,000 in 1970 (Machleidt et al. 2004)) of long stay psychiatric inpatients. Ideas for deinstitutionalization became part of a wider political debate following the student rebellion of 1968 and were driven by a societal interest in providing dignified psychiatric care for all patients. The Psychiatrie Enquete of the German Bundestag (1975) was unanimously endorsed by all political parties represented in the parliament and was the major political turning point in the development towards deinstitutionalization. The following process was supported by the Deutsche Gesellschaft für Soziale Psychiatrie, an organization of pro reform professionals, and led by the Aktion Psychisch Kranke, an association aiming to link psychiatric expertise and political decision making. Whilst asylums, at least initially, were rather downsized than closed, reforms led to the establishment of smaller psychiatric departments at general hospitals and a rich – but often poorly coordinated – range of services in the community, mostly provided by voluntary non-profit organizations.

In Italy, in the late 1960s there were about 100,000 long stay patients in mental hospitals. From the 1960s onwards, the attitude of professional groups and society at large on how care for the mentally ill should be provided changed markedly. The leading figure in this debate and the subsequent, relatively sudden political reform was Franco Basaglia, who headed psychiatric hospitals first in Gorizia and then in Trieste. The debate culminated in the passing of law 180 in 1978 by the Italian Parliament, requesting the closure of all traditional psychiatric hospitals. This initiated the closure of all large state hospitals and halved inpatient bed numbers within a few years (Benaim 1983). Thus, the political decision for deinstitutionalization in Italy was more radical than in other countries and law 180 was widely regarded in Europe as the clearest model of consequent deinstitutionalization. Like in other countries, services providing mental health care in the community were established over time, with considerable regional variation.

What were the ethical drivers for these developments, as opposed to political, economic and judicial reasons for the process of deinstitutionalization?

To address this question, we will refer to an established ethical tool, Beauchamp's ethical grid. We extend it by some more specific ethical values which were not part of the more generally phrased grid categories. We will apply the model also to the discussion of typical ethical dilemmas in community psychiatry, especially recovery, assertive outreach and community treatment orders, before we will consider the more recent debate on possible tendencies to reinstitutionalization in psychiatry.

21.2 Beauchamp's Ethical Grid Extended

Moral decision-making and the wish to make the right decision and take the right action in clinical situations generated various ethical theories over time, amongst them virtue ethics, casuistry, deontological theory, utilitarianism and ethics of care (Green and Bloch 2006), some of them explained elsewhere in this book (see Chapter 8).

The need for a clinically relevant and applicable analytic tool to facilitate ethical decisions prompted Beauchamp und Childress (1994) to develop the Four Principles Approach (principlism), providing the clinician with an ethics toolbox. This has four main categories, namely

Beneficence – the obligation to provide benefits and balance them against risks. This requires positive action rather than merely the omission of harmful activities.

Nonmaleficence – the obligation to avoid the causation of harm. This refers back to the Hippocratic tradition: 'Do no harm'.

Autonomy – the obligation to respect the decision-making capacities of autonomous persons. This concerns freedom, privacy, voluntariness and choice.

Justice – The obligation of fairness in the distribution of benefits and risks (e.g. do poorer people have the same access to treatment as rich people?)

The grid is used here as this is the most popular approach to condense different ethical theories into a practically applicable set of principles. Yet, other options are also available (Claassen 2007).

In addition to the four Beauchamp categories, certain values have developed in psychiatry and the social sciences, which are of relevance to the process of deinstitutionalization and the profound reforms of mental health care over the last five decades. They are partly covered by the more general ethical categories described above. Some, however, have evolved into broader concepts and ethical values in their own right (in the sense of what is good and what is bad) and affected further professional and legal developments.

Antistigmatisation (see Chapter 2) – The concept of stigma means the acknowledgement of the additional social handicaps and distress that people with mental illnesses experience as a result of prejudice but also the self-perception of people suffering from a mental illness, e.g. as being worthless. Antistigmatisation thus tries to counteract the negative image of mental illness in the public. This has led to a campaign by the Royal College of Psychiatrists in the United Kingdom (Changing Minds: Every Family In The Land, Crisp et al. 2000) and, as an ethical value, is at the heart of some important developments in service planning.

Non-discrimination (see Chapter 20) – This signifies specifically that mentally ill people should be treated like any other (physically ill) patients and/or healthy subject. As an ethical value guiding service development, it was mentioned by the UK Expert commission informing the new Mental Health Act (Szmukler and Holloway 2000). It is more narrowly defined than normalisation and social inclusion, and related to Antistigmatisation.

Normalization – This concept that all is good which brings people with a mental illness back into 'normal life' (however problematic the notion of normality is) was originally developed by Wolfensberger (1970) for intellectually disadvantaged people, but has had a significant impact on the development of psychiatric services since. However, because of the problematic and unidirectional concept of 'normality', this is considered outdated now and has been replaced by newer concepts like non-discrimination and recovery.

Empowerment – This has been understood as the personal control over all domains of life, not just mental health care, including decisions related to important areas such as vocation, residence and relationships. The concept stems from community psychology and was developed by Rappaport (1987) and others. As an ethical value, it has helped create user forums, self help groups, the recovery concept, advanced directives and advocacy schemes. It transgresses autonomy as a more active value, and members of this movement have taken a stand against some newer developments in community psychiatry (PACE versus PACT, Ahern and Fisher 2001), see below.

Social inclusion – As a value opposed to social exclusion (Morgan et al. 2007) it stipulates access to resources within a society, e.g. income, housing, schooling, jobs, relationships and is often linked with policies promoting equal opportunities and combating social inequalities (Ware et al. 2007). In Britain, this resulted in the creation of a Social Exclusion Task Force, which over several years set up a number of projects, including some addressing mental health issues (Cabinet Office 2009).

Human rights (see Chapter 5) – The European Convention on Human Rights, issued in 1950, and the European Court of Human Rights, established in 1998, established and promoted a set of values i.e. concerning the rights to life, liberty, a fair trial, and privacy, whilst protecting against torture and discrimination. This development has substantially influenced mental health legislation (Bindman et al.

2003). The newest development in this section, the 2006 UN Convention of the rights of people with disabilities, will probably have a profound effect on further legislation of all member countries. The Enable secretariat of the UN proclaims:

> The Convention marks a "paradigm shift" in attitudes and approaches to persons with disabilities. It takes to a new height the movement from viewing persons with disabilities as "objects" of charity, medical treatment and social protection towards viewing persons with disabilities as "subjects" with rights, who are capable of claiming those rights and making decisions for their lives based on their free and informed consent as well as being active members of society (Enable 2009).

Choice – The concept of choice in mental health care, meaning the power of the individual to make informed decisions about treatment, has recently developed into an important core value of the UK government's efforts to reform health care services in general and mental health services in particular (National Institute of Mental Health 2006). In other countries like Germany, where the right of every patient to choose a general practitioner (GP) or psychiatrist is an integral part of the health care system and the insurance arrangements, this has been less pronounced.

21.3 Ethical Drivers for Deinstitutionalization in Europe

A changed attitude of regarding psychiatric care less as aiming to detain 'mad' people than to treat mentally ill patients was arguably one major ethical driver for deinstitutionalization (Carrier and Kendall 1997). The more optimistic view of psychiatric care may have been aided by the advent of new physical treatments (ECT, insulin coma therapy) and later neuroleptic medication for schizophrenia (Chlorpromazine 1952). The experience that treatment of mental illness was indeed, at least in some cases and to some extent, associated with positive outcomes, encouraged the wish to promote patients' mental health (beneficence) and integrate psychiatric care into the realm of medicine (antidiscrimination).

Goffman (1962) described how institutions of mental health care can strip people of privacy, dignity and identity. His work had a strong and lasting influence on the intention of social psychiatrists in both the United States and Europe to avoid further harm for mentally ill people caused by long periods of hospitalization. Thus, non-maleficence was another powerful ethical driver to close down large remote psychiatric institutions in favour of smaller hospital services within general district hospitals or at least to change the nature of asylums and turn them into more therapeutic services with respect for the dignity and needs of patients.

The human rights movement, promoting autonomy and requesting fairness regardless of gender, race or religious beliefs, resulted in new laws preventing people with mental illness from detention and involuntary treatment only on the basis that they present with severe symptoms of a mental illness. Whilst this movement probably originated in the United States, it also impacted on debates and attitudes in Europe (Dreßing and Salize 2004).

In Germany, the Enquete (1975) outlined grave deficits in the provision of psychiatric care. A radical shift was proposed to prevent and treat mental illness early in order to avoid severe disability at a later stage (non maleficence), to decrease the necessity for hospital treatment, to prevent social exclusion (i.e. to foster social inclusion), and to enable psychiatric hospitals to provide the best possible treatment (beneficence) through improving staffing, buildings and organisational standards.

It is also specified in the Enquete that psychiatric and psychotherapeutic care should be available to all mentally ill patients and that psychiatric patients should have the same rights as patients suffering from a physical disorder (justice) and that all legal, financial, and social disadvantages should be abolished (antidiscrimination). Therefore, the psychiatric system as a whole should be an integral part of the health care system (normalization).

In the UK, Enoch Powell's Water Tower Speech (1961) called for the abolishment not only of mass care in asylums, but also of the architectural structures supporting them. However, the actual process of transition to community care took another 25 years to develop fully (Leff 1997). A series of inquiries into malpractice in British hospitals for the mentally ill provided further critical evaluation of psychiatric institutions. The effect of such inquiries was to reinforce the developing view that the large institutions were self-evidently harmful (Thornicroft and Bebbington 1989). Goffman's analysis was reinforced by Wing and Brown (1970) in the UK, who described institutionalism as in part responsible for many of the presenting symptoms, especially the so-called negative symptoms of schizophrenia, i.e. lack of energy and motivation. The assumption that institutionalized care may be harmful was widely accompanied by the idea that less institutionalized forms of care were preferable as a matter of principle, because they respected the patient's autonomy and independence. This led to a demand for the 'least restrictive' form of care, more or less under any circumstances.

In Italy, pressure by the media, inquiries into services and the charismatic efforts of Franco Basaglia helped pass the law 180 in 1978, more radical than any other mental health legislation in Europe, effectively closing most of the major mental institutions within a very short time frame and discharging some 10,000 formerly detained patients. Law 180 stipulated that no new hospitals were to be built and no new patients admitted to old hospitals. Instead, they were to be treated in the community and only very acute or involuntary cases admitted to psychiatric departments at general hospitals, the capacity of which was limited to a maximum of 15 patients at any one time. Involuntary hospital admissions under the law were made cumbersome and short termed, while at the same time, community mental health centres were established to provide care for the majority of patients. Whilst the situation in Italy has been complicated through the co-existence of public and private in-patient units and a significant regional variation in service provision, care in asylums was regarded by Basaglia as both violating human rights and ineffective. Also, some protagonists of the Italian reforms had close links with anti-psychiatric positions, and the influence of anti-psychiatry on the deinstitutionalization debate was stronger than in Germany and the UK.

Another driver for deinstitutionalization came from the fact that psychiatry increasingly developed into a multiprofessional discipline. Whilst initially nurses and medical doctors were the only therapeutic 'agents', over time other professional disciplines evolved and were involved in the provision of psychiatric care. Depending on the exact understanding of the term psychiatry, in some this led to a change of terminology from psychiatric care into mental health care, a semantic aspect not further addressed in this chapter. Traditional work therapy was complemented by outpatient occupational therapy and 'place and train' initiatives with a strong rehabilitative and empowering approach, involving training of social skills and activities of daily living, stepped support schemes and job preparation. The former social worker became housing support and benefit optimizer, facilitating independent living through regular home visits and support for dealing with difficulties in the family and social relationships. Psychologists entered the picture mainly via providing different forms of psychological treatments.

In Germany, families of patients and their wish to provide a humane and high quality treatment for their ill relatives became a strong focus point initiating a 'trialog' between professionals, service users and their relatives. The trialog, e.g. in form of so-called psychosis seminars (Bock and Priebe 2005) suggests that the three groups can learn from each other outside traditional roles and institutionalized frameworks. It intends to complement institutionalized care provision through a more direct human encounter of people with different backgrounds and experiences. In the UK, calls for user involvement became an integral part of directives and policies, and mostly impacted on the level of service organization with user representatives on most committees and panels.

A discussion of whether and, if so, in what precise form these mixed and often more implicit than explicit ethical motives were achieved, goes beyond the scope of this chapter. With respect to the practical effects of the deinstitutionalization reforms, Leff (1997) for the Team for the Assessment of Psychiatric Services (TAPS) and Thornicroft and Bebbington (1989) provide a summary for the UK, while Hoffmann et al. (2000) gives an overview for Germany. Neukirch (2008) describes the Italian reform process and its pitfalls. Yet, it can be concluded that the ethical problems caused by the asylums in the past were now replaced by a different set of ethical issues in community care. These issues are relevant across Europe although a number of potential problems following deinstitutionalization in the US were avoided in most European countries.

One may argue that the principal ethical dilemma of deinstitutionalization was the danger to replace the institutionalized life and lack of autonomy in the asylum with a more independent life in the community that would be associated with a risk to

(a) become homeless with a need to 'sleep rough',
(b) become the victim of violence (a risk that also existed in the asylum, but to a limited degree),
(c) be mistreated by parts of the community,

(d) get socially isolated without any meaningful activity and an increased danger to
     commit suicide,
(e) possibly being forced to beg or earn money through other degrading activities
     such as trading sexual favours,
(f) being more exposed to street drugs which would lead to an exacerbation of the
     illness, and
(g) get in conflict with the justice system and end up in prison.

Research shows that on a very large scale all these risks have been mostly avoided
in Europe. At the same time each of them certainly became reality for some patients.
For each of the affected individuals, deinstitutionalization might be regarded as a
failure and ethically indefensible. However, research also showed that the individual
fate of patients following discharge after long term hospitalization was practically
impossible to predict. Thus, some potentially poor outcomes had to be expected,
whilst other patients faired well and there is evidence that on average the quality of life of formerly long term hospitalized patients improved significantly after
discharge (Priebe et al. 2002).

Finally, a risk that has not been avoided in Europe is the tendency to create new
clusters of mentally ill patients in the community, who often live together in housing projects, spend much time together in specific services such as drop-in centers,
and may even work together in sheltered workshops. The hope that people with
severe mental illness would be enabled to lead a totally 'normal' life mingling with
non-mentally ill people in all aspects of private, social and professional life did
not materialize. Such hope might have been unrealistic, but the emergence of new
'ghettos' of mentally ill people in the community has been often criticized.

When deinstitutionalization commenced, the political, managerial and professional approach to this gigantic venture was rather paternalistic and based on own
values that living in the community was preferable to a life within the walls of an
asylum. At least initially, it did not consider patient empowerment as central and the
concept of the patients' capacity to make decisions was not discussed as important.
Patients were rarely asked, if they wanted to be discharged from long term hospitalization. On the contrary, it became widely accepted that the abolishment of asylums
would usually meet some resistance among both patients and staff. This resistance
needed to be overcome and managed on a practical level, but did not lead to an ethical debate on the pros and cons of the reforms in general and their implementation
in particular.

However, the dominating view in the scientific community is that – on balance –
deinstitutionalization brought more benefit than harm to mentally ill patients in
Europe, but there have also been questions as to whether deinstitutionalization has
gone too far. At the same time, even the most critical voices have not called for a
return to the former system of asylums. They rather demand more modern forms
of institutionalized care or more investment for better community support or both
(Munk-Jørgensen 1999).

21.4 Ethical Issues in Deinstitutionalized Mental Health Care

One of the main problems following deinstitutionalization was the 'revolving door patient': Someone who had been admitted to the hospital, treated for some time, then discharged into the community, only to be admitted again a short time afterwards because of a relapse of symptoms or inability to live in the community (Geller 1986). A good part of the efforts of community psychiatry have since been directed to prevent relapses in the community and reduce the rate of rehospitalisations, increase the quality of life of patients discharged from hospital, intensify support systems and establish a seamless and graded system, by which a patient could be 'weaned' from hospital care in a stepped and careful way.

Parts of the public and some of the media saw the revolving door patient as a sign of failure of deinstitutionalization. An ethical problem in this context is the question, whether the revolving door phenomenon is – sometimes or usually – a good or bad thing for the patient concerned, their relatives, the involved services and the public in general.

Looking at repeated hospitals admissions from the patient's point of view, the possibility to be admitted whenever this is required certainly seems appealing (in terms of greater autonomy and choice), especially if there was no danger involved to be locked up on a hospital ward indefinitely. It may also help organize hospital services and precious resources around the needs of patients (beneficence) and not vice versa. Taking an analogy from physical medicine, in multiple sclerosis or heart failure, few would find it remarkable if patients are admitted in a acute crisis, only to be discharged again as soon as possible, even if this is followed by a readmission at a later stage.

From the services' and public's perspective it is important to acknowledge that – given current treatment options – there is a proportion of patients requiring repeated readmissions to hospital wards without ever reaching a consistently 'well' condition and an overall acceptable quality of life, whatever the quality of a service and the provided treatment are. If services and mental health professionals are not aware of this or cannot accept it, they may tend to admit patients too frequently and over-stress risk prevention (locking people up indefinitely) or admit too rarely (e.g. denying admission to a suicidal patient), both of which may be harmful to the mental health of the patients, their autonomy and long term outcome. In the first case, the benefit of avoiding potential harm for some patients would be bought by restricting the rights of many others. In the second case, the mere recurrence or even persistence of a symptom (in the example suicidal ideation) would prevent the hospital from admission on the grounds of untreatability, which would in turn possibly bring severe harm to some patients (who might harm themselves if not admitted) while saving resources for society and superficially improving outcome for the many who are treated. The difficult bit here will be to strike a balance between two opposing extremes, which might be successful if one looks at the beneficence for the individual patient in the given situation first and acknowledges the deciding factor of autonomy, i.e. one would rather want to admit the patients who want to be admitted and discharge the ones who do not.

The following description of typical ethical dilemmas in community care reflects issues that regularly occur across European countries. Yet, in the illustration we mostly refer to the situation in the UK, as this country conceptually has the most developed psychiatric outpatient system with strategically different team philosophies catering for the needs of diverse mentally ill subpopulations. For various reasons (political, health system characteristics, historical, financial) the precise forms of services vary across Europe, yet the situation in the UK can be used as an example and the specific ethical problems of different approaches can be highlighted. It would certainly be interesting to compare services in different European countries in regard to their ways of working, their policies and the ethical principles involved. However, this would certainly be beyond the scope of this chapter.

The description of the different teams and therapeutic approaches will be followed by summarizing the ethical problems into one or more typical 'dilemmas'.

21.4.1 Rehabilitation and Recovery

Recovery teams in the UK have now replaced the more traditionally oriented rehabilitation and continuing care teams, which have, together with community mental health teams (CMHTs), supported the process of deinstitutionalization by coordinating housing arrangements, providing leisure and work opportunities and organizing psychiatric aftercare (Priebe et al. in press). While patients in the UK were often transferred in groups into residential care homes accompanied by a regular funding budget ('dowry', Knapp 1991), in Germany, care homes for the elderly were often used to adjust to the needs of new housing provision for long-term patients.

Guiding philosophy was the idea of normalization (Wolfensberger 1970) and the plan to provide a 'home for life' (as this was actually called in the UK).

It soon became clear, however, that a certain amount of control over the discharge process and further housing provision was required to minimize the risk that patients would end up homeless or being severely neglected in private care facilities (the book Spider by (McGrath 1991) and the following movie by Cronenberg (2004) provide a good example of that). Sometimes, hospital facilities were turned into supported housing projects by simply changing the door sign of the former long stay ward.

This led to the formal discharge of the patients who, however, stayed in the same buildings and were disadvantaged as compared to their previous situation. Supported housing schemes are normally less well funded than hospitals and not subjected to similar quality standards and control mechanisms. In such cases, deinstitutionalization was mere rhetoric and helped save costs for the funding body or the provider organization or both. As a result, the ethically founded intention to foster autonomy and welfare of long-stay patients was misused by organizations and authorities for purely economic reasons.

Other difficulties faced by rehabilitation teams were the lack of work opportunities and leisure activities for chronic patients. Whilst in the hospital, these needs were often met by in-house workshops and industrial occupational facilities; after discharge there was at first nowhere to turn to. While sheltered workshops became very popular and a regular feature of outpatient care in Germany, these were considered stigmatizing and excluding patients from mainstream work opportunities in the UK. Since about 2000 the traditional approach of a stepwise and gradual approach of work rehabilitation was challenged first in the USA and then the UK. Empirical studies demonstrated that in the traditional approach of 'train and place' patients tend to get stuck in an institution within the chain of stepwise rehabilitation and never reach regular employment. In the alternative approach of 'place and train' patients are placed within regular employment straight away and receive support there. Such an approach is associated with higher rates of patients achieving regular employment (Burns and Catty 2008).

Another, less frequently considered feature of deinstitutionalization was that many patients, while living independently in the community, became socially isolated, neglected by services and prone to be excluded from mainstream social life; contact with psychiatric services often providing the only stimulation during the week (Abrahamson 2000). On the other hand, the tendency of rehabilitation teams to shelter and protect 'their' patients has been criticized by others as a new form of institutionalization, thereby fostering dependence (anti-autonomy), deskilling and stigmatizing people further.

Recently, the remodeling of rehabilitation services to recovery teams has prompted a more user centred approach, where user representation, users' experience of the illness, advanced directives and advocacy schemes are intended to lead to empowerment and anti-stigmatization. On a strategic level, the aim is to move patients on from highly staffed residential housing to less supported schemes and individual placements (Lyall and Claassen 2007), thereby increasing autonomy and decreasing the stigma of traditional 'care' homes. Usually, a patient will have the choice of two or more alternative placements and make an informed decision.

However, all these rehabilitative efforts can put psychological pressure on patients, with excessive demands on the patient's ability to adjust, thereby leading to treatment termination, symptom exacerbation or even suicidal ideation. In any case, patients may fail to achieve the aspired level of functioning and feel devalued as a result. The recent popularity of 'place and train' schemes can illustrate this. As compared to the traditional 'train and place' approach with a stepwise rehabilitation using different schemes with diminishing degrees of protection, such schemes have been shown to integrate more of those patients who feel able and willing to return to regular employment into such jobs. For the patients concerned this is undoubtedly an important success. However, the method is only for patients who are motivated and sufficiently confident to seek competitive employment, and even among those patients a significant number do not achieve lasting regular employment. All those patients who do not manage to maintain work and drop out of the scheme and the even larger number of those who feel they will not be able to cope with the demands of competitive employment still require opportunities for meaningful work activities. Such opportunities can only be provided in a protected environment.

On the level of service provision a balance is required between pressure and support to achieve the optimal level of functioning for some and protected working schemes for those patients who cannot achieve regular employment. What the best balance is in a given context, depends on a number of factors, one of which is the ethical decision on how much 'rehabilitation pressure' a service may put on a patient. Ideally and with respect for the autonomy of the patient, this would be seen as a positive challenge and not (at worst) as coercion (see Chapter 20). An ethically grounded approach will only work here if patients have some form of choice how to develop their future working 'career' and how quickly they wants this to happen.

In clinical practice, teams are often confronted with the 'empowerment versus negligence' dilemma: What amount of autonomy is healthy and beneficent? What kind of interventions foster dependence? On the practical day to day service level this can be decided on by team discussions using ethical grids like the one mentioned above, going through a list of questions or principles and weighing them in the light of the actual case problem. If there is a severe conflict or basic recurring problem, the counsel of clinical ethics committees could be sought. However, these institutions have only recently moved on from a more research ethics oriented focus to general ethical issues.

21.4.2 Assertive Outreach

Assertive outreach is an established specialized community psychiatric care approach in Anglo-American countries geared especially towards non-engaging and high risk patients. Assertive outreach teams are characterized by focusing on severely mentally ill patients with the highest risk of self harm or violence towards others, who would otherwise have to be admitted to hospital, an active and persistent follow up of a small case load and hands-on support with nearly all sorts of issues ranging from medical to social problems.

Reviewing the literature for an ethical analysis of the service model (Claassen and Priebe 2006), we found that there is only scarce research evidence for beneficence, while the frequency of hospital admissions, violent incidents or social functioning was not influenced. On the other hand, no harmful effects were observed either.

The 'never accept no for an answer' approach has resulted in the accusation that Assertive Outreach impedes patients' rights to privacy and self-determination. Other similar arguments are that this means coercion of non-consenting individuals, that it deskills patients, makes them more helpless and is more focused on medication adherence and social control than improved care (Ahern and Fisher 2001). Newer criticism, which goes beyond the impact on the patient's autonomy, sees a danger of Assertive Outreach becoming a form of reinstitutionalizing patients in the community by making them dependent on the team and recreating an environment that facilitates aspects of institutionalized behaviour as in old fashioned asylums (Priebe 2004).

Assertive Outreach tends to provide care for patients who did not ask for it, at least not explicitly. Yet, the approach is not based on legislation overriding the patient's wish. It is rather justified by the presumed best interest of the patient and the idea that care will never reach real compulsion unless properly legislated through invoking the Mental Health Act. This is a dilemma reflected by patients' experiences

(Priebe et al. 2005b, Watts and Priebe 2002), who are concerned about their autonomy whilst often valuing the high staff commitment, practical support and engaging relationships. Team discussions in Assertive Outreach are often characterized by the question as to whether leaving a patient, who does not really want treatment, alone constitutes respect for the patient's autonomy or irresponsible neglect of a patient in need of care.

In this context, the question of capacity could also be important. Is the person able to understand, retain, use and weigh up the information relevant to the decision in question? Impaired capacity does not necessarily mean coercive action on the side of the service, rather, someone would be needed to support the patient in his decision making. This could be a relative, a friend, an advocate, the service itself or some kind of legal guardianship (which is very popular in Germany).

In addition to the ethical questions pertaining to rehabilitation and recovery teams mentioned above, Assertive Outreach is especially confronted with the 'persistence versus coercion' dilemma and the requirement to provide the least restrictive therapeutic environment for the patient without compromising the team philosophy. There is no easy way out of this as the typical Assertive Outreach patient (by definition) does not want to be treated and thus his autonomy will almost certainly be impacted by the team. However, this is understood rather as a very long (and time limited) leash and not as a straitjacket.

21.4.3 Community Treatment Orders (CTOs)

In several countries (US, Australia, Canada, UK) community treatment orders have been established to authorize involuntary outpatient care, often in the wake of serious incidents involving psychiatric patients killing a relative or member of the public (Kendra's law in New York, The Clunis Inquiry in the UK).

Under a CTO patients are usually required to take medication, accept outpatient appointments and reside at a certain place, with the threat of a return to inpatient care if they do not comply. Community treatment orders are now being introduced into psychiatric practice in the UK via the amended Mental Health Act, while there are softer and less outspoken means in Germany (mainly via a Guardianship order) and no known means in Italy to force patients to comply with prescribed treatments in the community.

Again, there is an inconclusive evidence base for the beneficence of such an intervention (Lawton-Smith et al. 2008). However, there are further ethical concerns. As Lawton-Smith et al. (2008) and Lepping (2008) argue, the new Mental Health act is risk-based, not capacity (= autonomy) based and therefore conflicting with other legislation, e.g. the Human Rights Act 1998 and the Mental Capacity Act 2005. Forced treatment in psychiatry, which is the only medical specialty treating significant numbers of their patients against their will, could also mean discrimination for mentally ill patients and less choice. If psychiatric outpatient care is increasingly coercive in nature rather than respecting the autonomy of patients, it may become more difficult for potential patients to seek or at least accept psychiatric treatment.

This may reduce the overall effectiveness of mental health care and indirectly lead to even more coercion for ensuring that unwilling patients receive care. Altogether, this makes community treatment orders problematic in the light of several, if not all, of the ethical principles cited above.

The main point from the pro-side (Lawton-Smith et al. 2008) is that community treatment orders could possibly prevent compulsory hospital care, thereby acting coercively, but less so than in case of a forced hospital admission with all the known collateral damage. By providing forced treatment, this could also benefit someone suffering from a serious mental illness and lacking insight. Through effective treatment, one could possibly prevent a suicide in the future and also improve the person's general health (Munetz et al. 2006). A clear judicial basis would also prevent abuse by covert and uncontrolled forms of coercion, thereby e.g. giving the individual patient the right to appeal.

In clinical practice, it is often difficult to answer questions like 'How much risk is tolerable?' (public protection versus risk aversion) and balance the possible good coming out of forced treatment against the impediment of free will and loss of autonomy (beneficence versus rights-based approach).

We have not addressed other specialised service approaches here, such as Early Intervention (EI) or Home Treatment and Crisis Resolution (HT/CR) teams. They show in part similar ethical problems as the ones described above, although particular dilemmas (like stigmatisation in Early Intervention) may become apparent when the outlined ethical principles are applied (see Chapter 16).

21.5 Reinstitutionalisation

Whilst deinstitutionalization dominated the debate on mental health care provision and actual reforms since the 1950s, data on mental health service provision in Europe since 1990 seem to indicate a potentially new phenomenon, i.e. reinstitutionalization. Across several European countries, the number of beds in conventional psychiatric hospital departments tended to decrease further between 1990 and 2006. However, during the same period of time there was a substantial increase in the number of forensic psychiatric hospital beds, places in different forms of supporting housing and the prison population with an unknown number of inmates with severe mental illnesses (Priebe 2005, Priebe et al. 2008). These changes are not consistent across all studied countries, but are substantial in most countries and occurred in countries with different traditions, health care systems and economic situations. Whether these changes constitute real reinstitutionalization with an increased overall provision of institutionalized care or rather a trans-institutionalization in which conventional beds are only replaced by other forms of care, remains debated (Priebe et al. 2005a). Also, the reasons for this development are not yet clear. An increased morbidity, a stronger tendency to risk aversion on a societal and clinical level, a diminished support for mentally ill patients in families and a strong lobbying influence of provider organizations have all been discussed as possible factors. At the same time, there is much less public and professional debate on this tendency than there was on deinstitutionalization.

From an ethical perspective, in particular the 'forensification' may reflect a wish for increased security in the society as a whole, reflected by risk aversiveness of psychiatric services. The wish to avoid harm coming out of 'dangerous' patients living unobserved (and maybe under-cared for) in the community could be one motive. In this context it should be noted that there is no evidence for an increase of the frequency of homicides committed by people with mental illnesses. Another driver might be the notion among professionals that some patients can only be effectively treated in a secure environment resembling the old hospital settings. However, this notion is not based on research evidence. There is a possibility, however, that the usually long term treatment in a forensic setting could compromise the patient's autonomy and interfere with wishes and possibly skills to live independently, although there is a scarcity of research in this area. Overall the long-term outcome for former patients from a medium secure unit in one study was poor, with excess mortality, high rates of reconviction and readmission, and few gaining employment (Davies 2007).

A particular ethical problem is linked to the lack of precise data on the proportion of mentally ill people in prisons and the quality of their care. There seems to be very limited interest in the issue, although it may be assumed that the percentage of people with severe mental illnesses among the sharply rising prison population might be significant. Also, the increasing provision of supported housing presents an ethical problem as it might compromise the achievements of deinstitutionalization leaving patients, who would have been long term hospitalized a few decades ago, in long term housing schemes of often questionable quality and with limited options to achieve higher levels of social inclusion (Fakhoury et al. 2005). A recent survey of patients in housing services in the UK (Priebe et al. in press) has shown that a substantial number of patients in supported housing do not receive care from specialized mental health services so that they cannot benefit from specialized programs for rehabilitation or recovery.

There are at least three ethical dilemmas for psychiatry in the phenomenon of re-institutionalization:

(a) an ethically driven research would aim to provide precise quantitative data and conduct qualitative studies to understand the phenomenon, but such research is very scarce and the issue receives little attention from funding bodies and research groups;
(b) it remains unclear whether some of the positive effects of deinstitutionalization are quasi reversed and patients tend to lose autonomy and social inclusion;
(c) if the increasing provision of institutionalized care is driven by provider interests rather than patient needs, the ethical argument would be against reinstitutionalization and require a movement for renewed and patient-oriented forms of deinstitutionalization.

21.6 Case History

John is a 25-year old patient with a difficult childhood who has suffered from schizophrenia since he was 17. He has minimal 'activities of daily living' skills and no support from relatives. He is living in his own flat, after his placement in a residential care home was terminated twice because of behaviour problems. He is not able to clean his flat sufficiently and does not possess much insight into the necessity of it. His hygiene is poor and he does not eat properly, living often from the food offered by the team.

John frequently responds to hallucinations, shows formal thought disorder and has acted in a bizarre manner in the past (e.g. he urinated into bottles and kept these in his flat for weeks). At times, he has been aggressive towards neighbours and staff. He has not been compliant with his medication, although he has accessed the service when he needed something, i.e. lost his keys or ran out of money.

He was admitted to an acute ward last year in a perplexed state and responded partially to neuroleptic medication. The ward then advocated a supported housing project, but John was adamant that he wanted to continue to live independently, as his experience with residential care had been very negative. He was more concerned with his everyday problems, i.e. food, than with the often neglected state of his flat. One of his most used statements, when the team approached him, was: 'I am ok, leave me alone!'

The Assertive Outreach Team supported him over the following year with daily supervised medication (i.e. a member of the team actually watching that he took his medication), help with cleaning his flat (which he often did not allow, shouting abuse at members of staff) and support with his benefits, linking with other support agencies to optimize his living conditions and cash availability.

In the regular team meetings, the following topics were discussed:

How many of John's problems are health related and could possibly benefit form medical interventions (e.g. a depot injection or clozapine)?
Would John benefit from a longer inpatient admission, even if that had to be forced, in terms of his skills of daily living?
What about his decision making capacity, i.e. is he always able to judge his own situation correctly and act on this judgement in the best of his interest? If not, what does this mean for the team's approach?
Is a decent life possible for John, even if not much changes in his condition?
What are the risks for Assertive Outreach staff and members of the public in dealing with John?
Is there a risk of a serious violent incident?

Applying the ethical grid e.g. to the second question, if a forced hospital admission would benefit John in the long term, one could argue that the beneficence of such a measure in terms of a dramatic treatment success and possible recovery is at least questionable (although some improvement of his mental and physical health might be expected), whereas some harm could be done by admitting the patient against his will, with possible immediate danger to John's physical and psychological well-being, but also with some risk to the staff or the supporting police team.

A forced hospital admission would be certainly against the autonomy of the patient and his self declared wish to stay in his environment (human rights). Moreover, such admission might stigmatise him (neighbours watching), disempower him by leaving him no choice and add to his social exclusion.

Leaving him in the flat, on the other hand, whilst monitoring him closely and be available in the case of a break down, will limit harm, respect his autonomy in his flat and give him the choice to comply or not. However, it might also stigmatise him (neighbours watching), disrespect his autonomous will to be left alone and his human right to privacy as well as fostering dependence on the team (food, cleaning, communication).

Certainly, if something serious would happen, e.g. if he stabbed a neighbour, it might be very difficult for the team to prove that 'everything' was done to prevent this, critics might label this attitude negligent and not pro-active enough, even unethical, and accuse the team to leave someone in such dire circumstances (= doing nothing), hallucinating and helpless.

Like in other patients with chronic schizophrenia, decision making capacity is usually fluctuating, especially in people who (by AOT definition) often lack insight, cooperation and motivation for continuous treatment. So the question here would not be, if capacity is impaired, but how much and with what immediate consequence for his life or the life of others. If there is impending danger, then the ethical pendulum would surely swing towards moving in fast and preventing further harm (non-maleficence), even if this is against his will (autonomy).

This example should illustrate that in the clinical practice of community psychiatry, it is often very difficult to walk the tightrope between conflicting ethical priorities. Taking into account several ethical categories certainly does not make the task any easier, however, by running through them, discussing these issues from different perspectives and documenting this can help to achieve a balanced point of view informing clinical action.

Before deinstitutionalization, John probably would have been sectioned to one of the large asylums, possibly never to resurface again in the community. Today, in a deinstitutionalized psychiatric landscape, he will have few, if any, hospital admissions, in between watched relentlessly by an Assertive Outreach team struggling to maintain a foot in the door. The pressure on everyone in this system is immense, with forensic psychiatry providing an outlet, in case anything serious should happen.

21.7 Conclusions and Outlook

While the deinstitutionalisation process was informed mainly by a mixture of mostly implicit traditional ethical motives, the further development of current community psychiatry resources depends on a web of traditional and newer ethical values and human rights based approaches, which are in competition with risk-minimizing efforts. This has led to more freedom of choice and social integration of mentally ill people, while on the other hand promoting coercion in the community and possibly reinstitutionalization.

The ethical problems encountered in the current landscape with largely community based care seem to be more complex and multilayered than in the past.

Then, it was easy to be 'against' the large institutions and the consensus to abolish them was broad. Today however, a detailed ethical analysis is often needed to come to grounded and careful clinical decisions beyond policy directions and practice-based action.

New therapies are visible on the horizon which carry their very own ethical dilemmas (e.g. deep brain stimulation,[1] neuroenhancement,[2] brain reading) adding to this already complex picture. Yet, such new treatments are unlikely to sort the ethical dilemmas outlined in this chapter. As with most ethical problems there is no easy solution and attitudes on what constitutes the ethically most appropriate approach will be disputed and may change over time. An explicit discussion of the underlying values and principles might be first step towards a better understanding of the various ethical implications of providing mental health care in the community. At the heyday of deinstitutionalization reforms in the 1970s such discussions were more common and probably more visible and audible in the professional communities. We would argue that mental health care as a whole and most of the patients concerned benefited greatly from those discussions and the supported reforms.

One may conclude that significant progress in mental health care can be achieved in the future only, if new discussions on the values underpinning mental health care involve not only professional groups, patients and their carers, but also society at large so that they can gain a political momentum as it happened with deinstitutionalization in the past.

References

Abrahamson D (2000) Social networks in community care. Psychiatric Bull 24:354

Ahern L, Fisher D (2001) Recovering at your own pace. J Psychosoc Nurs Ment Health Serv 39(4):22–32

Benaim S (1983) The italian experiment. Psychiatr Bull 7:7–10

Bindman J, Maingay S, Szmukler G (2003) The human rights act and mental health legislation. Br J Psychiatry 182:91–94

Bock T, Priebe S (2005) Psychosis seminars: an unconventional approach. Psychiatr Serv 56(Supplement 11):1441–1443

Burns T, Catty J, EQOLISE GROUP (2008) IPS in Europe: the EQOLISE trial. Psychiatr Rehabil J 31(4):313–317

Cabinet Office – Context for Social Exclusion Work (2009) http://www.cabinetoffice.gov.uk/social_exclusion_task_force/context.aspx

---

[1] See Chapter 19.
[2] See Chapter 30.

Carrier J, Kendall I (1997) Evolution of policy. In: Leff J (ed) Care in the community – Illusion or reality? Part I: the rise and fall of the psychiatric hospital. Wiley, Chichester

Claassen D (2007) Financial incentives for antipsychotic depot medication: ethical issues. J Med Ethics 33(4):189–193

Claassen D, Priebe S (2006) Ethical aspects of assertive outreach. Psychiatry 6(2):45–48

Crisp AH, Gelder MG, Rix S, Meltzer HI, Rowlands OJ (2000) Stigmatisation of people with mental illnesses. Br J Psychiatry 177:4–7

Davies S (2007) Long term outcomes after discharge from medium secure care: a cause for concern. Br J Psychiatry 191:70–74

Dreßing H, Salize HJ (2004) Historische Entwicklung. In: Zwangsunterbringung und Zwangsbehandlung psychisch Kranker. Psychiatrie, Bonn

Enable (2009) Convention on the rights of people with disabilities. http://www.un.org/disabilities

Enquete 1975. Bericht über die Lage der Psychiatrie in der Bundesrepublik Deutschland. http://www.dgppn.de/de_enquete-1975_39.html

Fakhoury W, Priebe S, Quraishi M (2005) Goals of new long-stay patients in supported housing: a UK study. Int J Soc Psychiatry 51:45–54

Geller JL (1986) In again, out again: preliminary evaluation of a state hospital's worst recidivists. Hosp Community Psychiatry 37(4):386–390

Goffman E (1962) Asylums: essays on the social situation of mental patients and other inmates. Aldine Publishing, Chicago, University of Michigan, Ann Arbor, MI

Green SA, Bloch S (2006) Theoretical considerations. In: An anthology of psychiatric ethics. Oxford University Press, Oxford

Helmchen H (2008) Ethical considerations in clinical research with mentally ill people. Nervenarzt 79:1036–1050

Hoffmann K, Isermann M, Kaiser W, Priebe S (2000) Quality of life in the course of deinstitutionalisation – Part IV of the Berlin Deinstitutionalisation study. Psychiatr Praxis 27(4):183–188

Knapp M, Beecham J, Anderson J, Dayson D, Leff J, Margolius O, O'Driscoll J, Wills W (1990) The TAPS project. 3: prediction the community costs of closing psychiatric hospitals. Br J Psychiatry 157:661–670

Lawton-Smith S, Dawson J, Burns T (2008) Community treatment orders are not a good thing. Br J Psychiatry 193:96–100

Leff J (1997) Introduction. In: Leff J (ed) Care in the community – illusion or reality? Wiley, Chichester

Lepping P (2008) Is psychiatry torn in different ethical directions? Psychiatr Bull 32:325–326

Lyall M, Claassen D (2007) Specialist psychiatric rehabilitation teams – a historical anomaly? Psychiatr Bull 31:315–316

Machleidt W, Bauer M, Rose HK, Lamprecht F, Rohde-Dachser C (2004) Psychiatrie, Psychosomatik und Psychotherapie. Thieme, Stuttgart

McGrath P (1991) Spider. Viking, New York, NY

Morgan C, Burns T, Fitzpatrick R, Pinfold V, Priebe S (2007) Social exclusion and mental health: conceptual and methodological review. Br J Psychiatry 191:477–483

Munetz MR, Galon PA, Frese FJ (2006) The ethics of mandatory community treatment. In: An anthology of psychiatric ethics. Oxford University Press, Oxford

Munk-Jørgensen P (1999) Has deinstitutionalization gone too far? Eur Arch Psychiatry Clin Neurosci 249(3):136–143

National Institute of Mental Health (2006). Our choices in mental health. http://www.mhchoice.csip.org.uk

Neukirch S (2008) 30 years of psychiatry reform in Italy – looking back at the reform process and its social and health policy factors of influence. Sozialpsychiatrische Informationen 4:2–11

Powell E (1961) Water tower speech. http://www.mdx.ac.uk/www/study/xpowell.htm

Priebe S (2004) Institutionalisation revisited – with and without walls. Acta Psychiatr Scand 110:81–82

Priebe S (2005) Why compare mental health care in European capitals? Eur Psychiatry 20(Supplement 2):S265

Priebe S, Badesconyi A, Fioritti A, Hansson L, Kilian R, Torres-Gonzales F (2005a) Reinstitution-alisation in mental health care: comparison of data on service provision from six European countries. Br Med J 330:123–126

Priebe S, Frottier P, Gaddini A, Kilian R, Lauber C, Martiney-Leal R, Munk-Jørgensen P, Walsh D, Wiersma D, Wright D (2008) Mental health care institutions in nine European countries, 2002–2006. Psychiatr Serv 59(5):570–573

Priebe S, Hoffmann K, Iserman M, Kaiser W (2002) Do long-term hospitalised patients benefit from discharge into the community? Soc Psychiatry Psychiatr Epidemiol 37:387–392

Priebe S, Saidi M, Want A, Mangalore R, Knapp M (2009) Housing services for people with mental disorders in England: patient characteristics, care provision and costs. Soc Psychiatry Psychiatr Epidemiol 44(10):805–814

Priebe S, Watts J, Chase M, Matanov A (2005b) Processes of disengagement and engagement in assertive outreach patients: qualitative study. Br J Psychiatry 187:438–443

Rappaport J (1987) Terms of empowerment/exemplars of prevention: toward a theory for community psychology. Am J Community Psychol 15(2):121–148

Szmukler G, Holloway F (2000) Reform of the mental health act: health or safety? Br J Psychiatry 177:196–200

Thornicroft G, Bebbington P (1989) Deinstitutionalisation – from hospital closure to service development. Br J Psychiatry 155:739–753

Ware NC, Hopper K, Tugenberg T, Dickey B, Fisher D (2007) Connectedness and citizenship: redefining social integration. Psychiatr Serv 58(4):469–474

Watts J, Priebe S (2002) A phenomenological account of users' experiences of assertive community treatment. Bioethics 16:439–454

Wing J, Brown J (1970) Institutionalism and schizophrenia. Cambridge University Press, Cambridge

Wolfensberger W (1970) The principle of normalization and its implications to psychiatric services. Am J Psychiatry 127(3):291–297

Chapter 22
Ethical Issues in Forensic and Prison Psychiatry

Norbert Konrad and Birgit Völlm

Contents

Abbreviations

EUPRIS   Mentally Disordered Persons in European Prison Systems – Needs, Programmes and Outcome
UK   United Kingdom
US   United States
VRAG   Violence Risk Appraisal Guide
WHO   World Health Organisation

N. Konrad (✉)
Institute of Forensic Psychiatry, Charité – University Medicine, 12203 Berlin, Germany
e-mail: norbert.konrad@charite.de

H. Helmchen, N. Sartorius (eds.), *Ethics in Psychiatry*, International Library of Ethics, Law, and the New Medicine 45, DOI 10.1007/978-90-481-8721-8_22,
© Springer Science+Business Media B.V. 2010

22.1 The Dual-Role Dilemma in Forensic Psychiatry

Forensic psychiatry is a subspecialty of clinical psychiatry which requires special legal and criminological knowledge and experience in the treatment of mentally disordered offenders. Forensic psychiatrists should have solid psychiatric training as well as practical experience in dealing with mentally disordered offenders. The double knowledge in psychiatry and law defines the subspeciality of forensic psychiatry and provides the ethical foundations for its practitioners (Arboleda-Flórez 2006). 'In psychiatric ethics, the dual-role dilemma refers to the tension between psychiatrists' obligations of beneficence towards their patients, and conflicting obligations to the community, third parties, other health-care workers, or the pursuit of knowledge in the field. These conflicting obligations present a conflict of interest in that the expectations of the psychiatrist, other than those related to patients' best interests, are so compelling. This tension illustrates how the discourse in psychiatric ethics is embedded in the social and cultural context of the situations encountered. It appears that as society changes in its approach to the value of liberal autonomy and the "collective good", psychiatrists may also need to change' (Robertson and Walter 2008). This quote reminds us that social and political factors are important drivers of decision making, particularly as in-depth analysis of specific ethical problems in forensic psychiatry are only beginning to emerge.

22.2 European Perspective

Forensic psychiatry operates within a certain legal and societal context which undergoes constant evolution. Laws are rules that guide human behaviour and as such are man-made. This means that concepts such as responsibility or competence are normative rather than clinical issues (Morse 2008) which differ from country to country, sometimes significantly (Salize and Dreßing 2005). Therefore, while the ethical issues facing forensic psychiatrists might be similar across cultures, they do also depend on the specific legal system and service provision within each country. We will thus briefly describe the main differences in the characteristics of the relevant legal frameworks within Europe.

All European legislations recognise the concept of criminal responsibility as a prerequisite for punishment. Individuals who lack responsibility for the act they have committed are therefore exempt from punishment which usually means admission to a treatment facility rather than imprisonment. Most, but not all, countries also recognise the concept of diminished (as opposed to full or lacking) responsibility which may lead to a less severe punishment. Most European countries require some degree of reduced responsibility for entry into the forensic-psychiatric system. However, in some countries, e.g. the UK, access to forensic psychiatric care is independent of criminal responsibility and solely determined on the basis of the mental condition at the time of sentencing. Criteria for involuntary admission to hospital are the same than those applied to civil psychiatric patients, i.e. the presence of a mental disorder that requires hospital treatment due to the risk the individual poses to self or others. Both approaches have ethical implications. Restricting forensic-psychiatric care to those with some level of diminished responsibility emphasises the importance of the likely mental state at the time of the offence but might neglect disorders developing subsequent to this thereby potentially hindering service provision to those who become unwell in prison. Admission to forensic-psychiatric care based on current need has the advantage of potentially delivering treatment to all those who require it. However, given the often lengthy stay in forensic-psychiatric treatment settings, mentally disordered offenders with full criminal responsibility might find themselves incarcerated for significantly longer periods than if they had received a prison sentence.

A number of national laws within Europe provide exclusion criteria with regards to mental health legislation. From a civil liberty perspective the exclusion of certain patient groups (e.g. those with personality disorders, substance use disorders or sexual deviancy) from compulsory psychiatric measures might be seen as desirable; however, it has to be kept in mind that such exclusion might also mean a lack of service provision for those in need. For example, substance abuse disorders are not seen as a mental disorder within the Mental Health Act 1983 of England & Wales (Department of Health 2008). Therefore, even if there is a clear link between substance dependence and an offence committed, the individual will not be admitted to a forensic-psychiatric treatment facility but to prison where the standard of care provided might be inferior.

22.3 The Role of a Forensic Psychiatrist

Forensic psychiatrists deal with some of the most difficult patients in psychiatry. They are concerned with the assessment of complex cases, including risk assessment, and with the treatment of mentally disordered offenders, typically in secure settings such as secure hospitals or prisons. Furthermore, forensic psychiatrists act as expert witness in court, commenting e.g. on issues of criminal responsibility and competency to stand trial. In each of these areas forensic practitioners can face particular ethical challenges.

22.3.1 The Forensic Psychiatrist as Expert Witness

Forensic psychiatrists appear in court as expert witnesses, giving their opinion on specific issues as requested by lawyers or a judge. As such they have to act within the law but also have to accept the authority of the legal profession. Psychiatrists in court only provide an opinion while decisions are made by the judge or jury, a situation that differs from that encountered by the highly skilled forensic psychiatrist in his or her other work context and one that can cause discomfort or even resentfulness. A number of difficulties met by the psychiatric expert witness have been described in the literature. They include harassment by the different parties involved in the trial (Calcedo-Barba 2006), public criticism, difficulties in keeping one's dates, low reward, poor relationships with the legal profession partly due to unfounded attacks, loss of dignity and status as a consequence of the confrontation with sharp-shooting lawyers amongst others. Some psychiatrists avoid expert witness duties altogether because of the number of frustrations they are confronted with in court. They are happy that there are 'masochistic' colleagues who are prepared to be available to act as expert witness (Rasch and Konrad 2004).

22.3.1.1 Ethical Issues

Taking on duties as an expert witness is not only associated with external frustrations but has caused grave soul searching when accepting duties which originally were not core tasks of the psychiatrist (Schneider 1977). The terms 'guilt' or 'criminal responsibility' in legal thinking do not exist as an empirical entity in psychiatry. Even if a medical expert does not comment directly on criminal responsibility – and he should not do as a matter of fact – his expert opinion aims at enabling this finding. Helping to select the criminal irresponsible has a serious side effect as de Smit (1977) has shown: the forensic psychiatrist legitimates the punishment of individuals labelled as responsible. The psychiatrist takes on the, at first sight, humanitarian act of treating those who are not punished due to their mental disorder. However, this action becomes problematic as the psychiatrist does not only undertake treatment but also custodial functions. The forensic psychiatrist is 'changing side' (WHO 1977), he moves from protector of the ill individual to being protector of the society (Leyrie 1977): In Germany, mentally disordered offenders are subject to special legal regulations (Konrad 2002), which are based on the concept of criminal responsibility. If future serious offenses are expected from offenders who are considered to have at least diminished criminal responsibility, they are admitted, regardless of therapeutic prospects, to special forensic psychiatric secure hospitals without limit of time as long as they are assessed to be dangerous. In his role as expert assessing dangerousness, the forensic psychiatrist in therefore involved in this preventive detention measure. In this context the German Constitutional Court decided in 2006 (2 BvR 443/02), diverging from former High Court decisions (which restricted the right of patients to gain access to information about their case pertaining objective findings), that patients housed in a forensic psychiatric hospital have the right to gain access to their complete file, which causes continuing debates.

In cases of psychiatric reports on refugees facing deportation, which bear considerable diagnostic and prognostic difficulties, the psychiatrist has a major impact on an individual's life which can have grave consequences including deterioration of existing mental disorders (Zinkler 2003). The most severe role conflict for psychiatrists exists in countries with capital punishment (like USA or Japan) where forensic experts are used to assess the 'competency to be executed' (Okasha 2002), which could be achieved by treating the mental illness.

An alternative to the difficult and sometimes reluctant engagement in legal work is the total refusal, the retreat from the forensic field. But to go on a 'total strike' is certainly not the ideal way to achieve important changes in legal practice (Schneider 1977). Given the reality of expert witness work some authors have questioned whether it is justified to speak of humanitarian engagement of psychiatrists in this field (Hallek 1974). Besides the mentioned possibilities to cooperate half willingly or to refuse expert duties there is also the option of willing adaptation and over-adaptation. This is what Robert Musil described in his novel »Der Mann ohne Eigenschaften« (»The man without qualities«) some decades ago (1952): the tendency of doctors to conform and to adapt to the expectations of the legal profession.

The forensic psychiatrist is indeed confronted with a double dilemma: Either he legitimates punishment by labelling only a fraction of the accused as disordered and in need of treatment; in this case he might be called a servant of justice. Alternatively, if he offers treatment for a large number of offenders he might be accused of brain washing and treatment tyranny. Psychiatry has been criticised for therapeutic nihilism and revenge; on the other hand, when progressive institutions were developed for the therapy of disordered offenders, therapists were criticised for applying to lengthy treatments based on unclear criteria with doubtful success. In public consciousness outsiders of the society are treated either too softly – as nowadays in many European countries – or too harshly. Forensic psychiatry, acting on behalf of society with the doubly stigmatized, is subjected to double reproach (Rasch and Konrad 2004). There is probably no subspeciality of psychiatry which has been more criticised than forensic psychiatry.

22.3.1.2 Practical Aspects of Forensic-Psychiatric Assessments

When asked to provide an opinion on an offender, the Forensic Psychiatrist should obtain the consent of the examinee to provide the report to the requesting body. This means he has to ensure the examinee has the capacity to consent and understands the purpose of the report and that any relevant information will be included. He should also state whether the assessment is compulsory or not and provide information about possible consequences of both cooperation and non-cooperation. The information on possibly negative consequences of non-cooperation can be (mis)understood by the examinee as enforcing his/her pseudo-voluntary consent. The Forensic Psychiatrist should explain the purpose of the consultation clearly and note that it is not a therapeutic consultation and that no help, suggestions, treatment, and possibly not even feedback, will be offered, with the exception of intervention if the examinee is at immediate and serious risk. The relationship is one of evaluation, and the fact that the evaluator is in no position to reassure the person on matters of confidentiality or privacy could mean that negative findings will endanger the interests and cause harm to the person being evaluated, regardless of this person's health and the evaluator being a physician. Because of this, forensic psychiatrists may even be implicated in the criminalization of mentally ill persons (Arboleda-Flórez 2006).

If the examinee discloses confidential information which could jeopardise his social position (e.g. Case Report No 36 in Carmi et al. 2005), the forensic psychiatric expert witness faces an ethical dilemma: On the one hand he has to report every detail to the court by which he was instructed in order for the court to reach a just decision, on the other hand the accused (despite not having the status of a patient) may expect rules of the patient–doctor relationship to still apply, e.g. medical secrecy. One option for the examiner is to inform the judge privately about the material assembled so to not fail the expert duty in concealing pertinent information. How this material will be used, however, is for the judge to determine.

The compilation of third party information through interviewing family or friends, as is common practise in general psychiatry, can easily be construed as inadmissible investigation activity in forensic psychiatry, particularly if it happens without consultation with the examinee and the court and without explaining the right to refuse the evaluation.

Prison physicians have a responsibility to request from the appropriate authorities (e.g. courts) a forensic-psychiatric assessment in cases where they suspect a disorder such as a psychotic disorder, a severe personality disorder or markedly reduced intelligence that may affect the prisoner's criminal responsibility, competence to stand trial or fitness to be detained. In this context, forensic psychiatrists should:

- not, as a matter of principle, and in order to avoid a conflict of roles, assess their own patients (Reid 2008)
- provide the legal client with their expert knowledge about psychopathology and diagnosis
- present their specialist knowledge in such a manner that it is readily understandable to the legal client, and thus provide a basis for independent decision-making
- contribute to humanizing legal procedures by providing expert information and their special view on the development of delinquent behavior in order so that justice can be done to the accused, while at the same time providing objective information.

In dealing with deviant behavior, criminology and psychiatry have developed two parallel approaches that partly give different definitions of terms in their own specialist language, for example of the term psychopathy. These communication problems, especially translating psychiatric findings and conclusions into the legal coordinate system, are an internationally familiar phenomenon. The forensic psychiatrist should avoid overstepping competences and venturing into normative evaluation which is reserved for the courts. An uncontrolled (over)identification with the offender based on unrecognised counter-transference can lead to adoption of an exaggerated helper role. Assuming the role of a prosecutor or judge, on the other hand, can turn the evaluation situation into a cross-examination and lead to incorrect evaluation findings. What is required from the forensic psychiatrist is the psychopathological analysis of the mental condition of a perpetrator and his personality performed on the basis of empirical knowledge and competent specialist examination. The expectations an offender has of the forensic-psychiatric expert may involve not only the hope of a favorable outcome of the trial but also the need for exploration of the self and the delinquent acts in question. Thus, the expert gets assigned therapeutic functions, if only as a communication partner, which makes the completely neutral attitude of his role appear fictitious. To maintain clarity of roles the expert should not change openly sides from the examiner to the treating physician but inform the legal client about the treatment needs which are in the best interest of the offender.

22.3.2 Risk Assessment

Forensic psychiatrists are crucially involved in the assessment of risk, particularly the risk of future (violent) re-offending, but also the risk of institutional violence. These assessments might determine whether an individual is going to be released or the level of security he will be treated in. In countries where the law allows preventative detention, e.g. detention for the protection of others after serving a time-limited prison sentence, forensic psychiatrists may serve as experts dealing with questions of dangerousness. Therefore, risk assessments have major ethical implications for forensic practitioners.

22.3.2.1 Ethical Issues

The interpretation of risk assessments requires great skill and clinical experience. Training in risk assessment generally as well as in the use of the specific instruments[1] to be administered is therefore essential. Practitioners must be aware of the limitations of the instruments they routinely use. It should be noted, for example, that probability risk estimates using actuarial assessment tools have been developed in a specific population and might not be valid in other groups. Therefore, an instrument validated in male offenders released from a high secure prison in Canada may not be valid in women treated in the community in a European setting, and it should therefore be used with caution in such a population, if at all. Furthermore, all risk assessment instruments produce indications about groups of individuals but are limited in what they state about the individual. They also generally do not differentiate the severity of violent re-occurrence or the timeframe in which to expect such an event. Most risk assessment instruments put more emphasis on recording historical, i.e. unchangeable, information than on change, e.g. as a result of treatment.

No matter how sophisticated our risk assessment instruments are, future behaviour cannot be predicted with any certainty, giving rise to particular challenges in balancing the civil liberties of the patient and the protection of the public from potential future harm by that patient. Buchanan and Leese (2001), reviewing studies in which risk assessments were validated by follow up of actual future violence, concluded that six people would have to be detained for 1 year to prevent one person from acting violent during this time period. However, the question how much risk is acceptable to the society is not a psychiatric one. As Halleck (1984) noted:

---

[1] Actuarial risk assessment tools and structured clinical judgement instruments are the two main approaches. An example of an acturial risk assessment instrument is the Violence Risk Appraisal Guide (VRAG, Webster et al. 1994) which provides probability estimates for violent re-offending. Structured clinical judgement instruments such as the Historical Clinical Risk (HCR) 20 scale (Webster et al. 1997) serve as an aide memoire, providing a structure for the assessment while also relying on clinical expertise.

'Society must decide how many it will restrain unnecessarily to protect us from the one person who might hurt us. This is purely a moral and political issue that must be left to the conscience of the community.' In an increasingly risk aversive society public pressure demands to reduce false negative predictions with the expectation to eliminate risk. However, reducing false negative risk prediction leads to an increase in false positives which means that even more patients might be subjected to severe restrictions of their liberties, many of whom might have never gone on to commit a (further) violent offence.

A range of new measures both in criminal law and mental health legislation have been introduced in recent years to address the issue of 'dangerousness' in offenders. They include, to name but a view, the sexual violent predator laws in the US, the proposed (and later withdrawn) legislation regarding individuals with 'dangerous and severe personality disorders' and new indeterminate sentences for public protection in the UK. Common to all such initiatives is the shift from a retributive to a preventative detention model whereby individuals can only be released if a reduction in risk can be demonstrated. Crucially, the assessment of criteria for such preventative detention schemes does often require the expertise of psychiatrists.

While forensic psychiatrists will probably always be expected to make predictions about future risk, they should not loose sight of their primary duty as clinicians which are towards their patients. In the context of risk assessment, this does not mean that they should withhold information regarding risk from relevant agencies but rather that they should focus on the clinical task of treatment, i.e. managing rather than merely predicting risk. They also have a responsibility in educating policy makers and the population as a whole on what can realistically be expected in terms of risk reduction and the means necessary to achieve this task.

22.3.3 Forensic-Psychiatric Treatment

The debate on whether mentally disordered prisoners should be detained in prison or in forensic-psychiatric institutions is primarily a legal-philosophical and political one. Countries applying the construct of diminished or absent criminal responsibility or incompetence to stand trial (unfitness to plead in the UK) divert certain mentally disordered persons from being imprisoned. Secure confinement in these cases will take place in a forensic psychiatric institution, if necessary. Treatment of mentally disordered offenders can therefore take place in three different types of institutions with completely different legal frameworks, namely general psychiatry, forensic psychiatry and prison psychiatry. Different disposal options give rise to ongoing discussions about confinement errors (Konrad 2002).

Each of the three types of disposal outlined above has distinct advantages and disadvantages. Highly specialized institutions, for example, might be in a better position to provide treatment for mentally disordered offenders with complex needs. On the other hand, such institutions carry with them the risk of isolation and stigmatization of both staff and patients. Nevertheless, it seems that the otherwise widely adopted approach of deinstitutionalization and community care for the mentally ill is slow to be implemented in forensic-psychiatric settings (Salize and Dreßing 2005).

22.3.3.1 Ethical Issues

While prison sentences are time limited, forensic-psychiatric care usually isn't. In order for the patient to be discharged (or moved to conditions of lesser security), it has to be demonstrated that the risk to others has reduced to a significant degree. Procedures and time frames for such assessments differ from country to country. From a human rights point of view fairly regular re-assessments are clearly desirable. Different views have been put forward as to whose responsibility such re-assessments should be. Whilst review by an independent assessor might provide a more objective view of the case, the treating psychiatrist will invariably have the most detailed knowledge of the patient which may or may not lead to a more favorable outcome for the patient. Difficulties arise in those systems (e.g. UK) where the court plays no role in determining final discharge or transfer. The responsibility for this decision then lies with the forensic practioner who might be reluctant to take any risks which could jeopardize his career in an increasingly risk aversive society. Of particular concern is further the reported detention of patients in conditions of higher security than necessary simple because of bed shortages in downstream low to medium secure facilities.

Some countries allow for joint disposals, i.e. a prison sentence as well as a treatment order, raising the question in which order these should be applied. If psychiatric treatment is administered first to good effect in terms of a reduction of risk of re-offending, the purpose of a subsequent prison sentence prolonging further the time spent in custody is questionable. On the other hand, if the opposite sequence is imposed, one might argue that this violates the individual's right to treatment by deferring the same to a later stage. If the psychiatrist has a role in recommending which of these two sequences of disposal to choose, he should be guided by what is in the best interest of the patient. For example, acutely unwell individuals who clearly suffer from their condition should be provided with treatment without delay. The offender's wishes might be another guiding principle. In any case the psychiatrist should endeavour to explain to the patient the implications of the different disposals.

Given that forensic-psychiatric treatment might lead to longer incarceration compared to prison disposal, the question arises whether there is sufficient evidence to support such treatment. Concerns have been expressed particularly in relation to individuals with 'psychopathy' following the publication of evidence suggesting that treatment might not only not help such individuals but actually make them worse (Rice et al. 1992). Even though these findings have been widely disputed by a number of authors (e.g. D'Silva et al. 2004), individuals with 'psychopathy' continue to be excluded from some prison programmes and in some countries (e.g. UK) also, at least partly, from forensic-psychiatric care.

A much under-researched subject is that of restrictions within prison or forensic-psychiatric care. Even if one accepts that incarceration might be necessary for public protection or more generally in order to provide treatment for a mentally disordered offender, it does not follow that further restrictions on patients' liberties should be imposed within this setting. However, such restrictions are commonplace and frequently not regulated by law. Restrictions may be imposed, amongst others, with regard to visiting rights, conjugal visits, contact with children, leave, mail, access to financial means or material goods, relationships with fellow inmates or smoking, often without specific justification relevant to the individual patient. Provision of treatment against the will of the capacitous or non-capacitous patient will be referred to below.

22.4 Prison Psychiatry

Diverting mentally ill offenders to forensic-psychiatric institutions does not prevent people from becoming mentally unwell when imprisoned, nor does the presence or history of mental disorder automatically results in the absence of criminal responsibility. The still high prevalence of mental disorders in prisoners has been impressively demonstrated in more recent surveys. In a systematic review of 62 surveys from 12 different western countries including 22,790 prisoners (mean age 29 years, 81% men), 3.7% of the men had a psychotic illness, 10% major depression, and 65% a personality disorder, while 4% of women had a psychotic illness, 12% major depression, and 42% a personality disorder (Fazel and Danesh 2002).

In comparison to the general population, prisoners have an increased risk of suffering from a mental disorder that transcends countries and diagnoses. The increased consultation of forensic psychiatry in this area reflects the interest of the relevant agencies in reducing the high suicide rate in prisons and jails. Some authors have suggested that the suicide rate among prisoners is a marker of the inadequate or even inhumane treatment in prisons (Konrad 2006).

The attribution of competence in the prevention of recidivism is accompanied by the hitherto unfulfilled expectation that forensic psychiatry can decisively reduce the relapse rate in individuals with personality disorders, especially in offenders with dissocial personality disorders and the subgroup of 'psychopaths' (Hare 1991). There is a special need for research into treatment options for these patient groups that are rejected by many – even forensic – psychiatrists (Konrad 2006).

Follow-up treatment for released inmates should be provided for by community specialised services. It is essential that the prison doctor has ample notice of the forthcoming release of his patient so that he may arrange an outside appointment with all relevant services very shortly after the prisoner's release. It should be ensured that all necessary documentation is dispatched to the providers of such services with the full consent of the patient.

22.4.1 Principle of Equivalence

If one accepts that mentally disordered prisoners should be treated in penal institutions, possibly in a hospital wing/ward within the prison, then the principle of 'equivalence' of care in the community and therapeutic provision for incarcerated mentally disordered persons should prevail (e.g. Konrad et al. 2007). However, it is doubtful whether the majority of prisoners with mental disorders receive appropriate care such as that mandated by the European Convention on Human Rights and other international charters.

This essential principle should also be applied to medical treatment of addicted prisoners and of withdrawal symptoms in prison. However, again this does not appear always to be the case. For example, medication-assisted treatments, endorsed by international health and drug agencies as an integral part of HIV prevention and care strategies for opioid-dependent drug users, are unavailable for most prisoners even if they are available to the general public in a particular country (Bruce and Schleifer 2008). Psychotherapeutic and medico-social programmes developed in prisons should be closely linked to the approach used in the community as a whole with regard to drug-dependant individuals (drugs, alcohol, medication).

Opponents of equivalence of standards of care for mentally disordered prisoners argue that prisoners do not deserve it or should not have (even) better care than outside of prison, where they did not use already existing services or were considered problem patients. Commitment in this area hardly promises politicians votes, but it should be pointed out that imprisonment, imposed by society via the courts, establishes a special social responsibility, especially for the health of prisoners, even if psychiatric intervention does not primarily or indirectly prevent crime.

22.4.2 Ethical Issues

A number of guidance documents by the United Nations International Resolutions (esp. Standard minimum rules for the treatment of prisoners), the Council of Europe (esp. Recommendation No R (98) 7 on the Ethical and organizational aspects of health care in prison), the World Medical Association (esp. Declaration of Tokyo 1975), the World Psychiatric Association (esp. Declaration of Hawaii 1977) as well as the Oath of Athens (International Council of Prison Medical Services 1979) touch upon prison psychiatry but lack more detailed guidelines for dealing with mentally disordered prisoners.

The professional medical role of a psychiatrist and/or psychotherapist working in prison has inherent conflicts within. On the one hand the doctor acts according to the requests and interests of his/her imprisoned patient and, following the Hippocratic oath, assigns the highest priority to the preservation and restoration of the patient's health, yet, on the other hand he/she is an employee of that authority which, in carrying out the punishment required by the state, implements measures which may well damage the prisoner's health (Binswanger 1979). Unlike a surgeon or physician working in prison, who treats illnesses which may be pre-existing or which may have occurred regardless of imprisonment, psychiatrists in prisons deal with large numbers of individuals with 'prison reactions', which have arisen directly as a consequence of imprisonment (Konrad 2001). To a certain extent the function of the psychiatric and psychotherapeutic treatment provided is to keep the prisoner fit for imprisonment, serving a pacifying and mollifying function. Prison psychiatrists find themselves in ethically questionable territory if they carry out psychopharmacological or other medical interventions for which there is no primary medical indication, in order to allow judicial proceedings and the penal system to run smoothly (Binswanger 1979).

A typical conflict arises (e.g. Case Report No 35 in Carmi et al. 2005) when prisoners suffering from anxiety, depression and/or suicidal ideas recover in a hospital setting, but relapse after return to prison. The concern of the treating team is the occurrence of an actual suicide of the patient if he returns to prison. In these circumstances it is of particular importance that the treating doctor provides some follow-up for the patient or at least organizes such ongoing treatment.

Treatment in prison has to address inmate-specific problems and circumstances, including post-release services. This has to be guided by the functional level of the patient and the severity of psychiatric symptoms. The high prevalence of mental disorders support the use of routine application of standardized diagnostic screening instruments as a component of the admission procedure in prison (Konrad et al. 2007). In accordance with the principle of equivalence, every prisoner suffering from a mental disorder should receive appropriate treatment equal to the care that such a patient would receive if he was not in prison. Prisoners suffering from serious mental disorders should be kept and cared for in a hospital facility which is adequately equipped and staffed with appropriately trained personal. Inpatient treatment should not be restricted to the distribution of medication to mentally disordered offenders otherwise locked up 23 h a day in their cell but infers the availability of a multidisciplinary team comprising psychiatrists, psychologists, psychotherapists, occupational therapists and counsellors. That means that the treatment standards within a prison hospital should not be worse than in a community setting.

Many prison conditions bear the risk of violating the human rights of prisoners. An analysis of cases taken to the European Court of Human Rights by mentally disordered offenders, demonstrated the difficulties inherent in ensuring appropriate care to individuals, and safeguarding the public at the same time; the issues raised included problems raised by indeterminate sentences, the use of detention for preventive purposes, and debates about availability of treatment (Prior 2007). As noted by Salize et al. (2007): 'The European Prison Rules (Committee of Ministers 2006) include several recommendations that shall safeguard human rights standards of prisoners and ethical principles during imprisonment, e.g. the right of prisoners to "ample opportunity to make requests or complaints to the director of the prison or to any other competent authority" (Principle 70.1). Prisoners "are entitled to seek legal advice about complaints and appeals procedures and to legal assistance when the interests of justice require" (Principle 70.7).' According to the information collected during the EUPRIS study (Salize et al. 2007), prisoners in Bulgaria, France, Greece, Hungary, Ireland and Italy lack the opportunity to complain to an independent organisation.

Of particular concern are disciplinary measures which are coercive by nature, (see also Chapter 20). Mentally disordered prisoners are more likely to become the subject of disciplinary measures due to misbehaviour that may be caused by the disorder. It is well known that specific coercive measures (e.g. solitary confinement) are likely to aggravate mental disorders. Thus, it is crucial to assess the psychological status of a prisoner prior to implementing such measures in order to avoid any additional harm. In Austria, Belgium, Greece, Italy, Norway, Spain and Sweden, any prisoner known to suffer from a mental disorder will be assessed for fitness to undergo disciplinary measures prior to their implementation. Regulations in Bulgaria, England & Wales, Germany, Greece, Hungary, Italy, Poland and Slovenia go one step further in that psychopathological assessments will be conducted in all cases requiring punitive or disciplinary measures. In other European countries, such an assessment is not stipulated (Salize et al. 2007).

Somewhat surprisingly, in most countries disciplinary or coercive measures during imprisonment must be recorded. Such records or files are an essential tool for assessing the appropriateness of such measures, particularly in the case of mentally disordered prisoners. In those cases where the use of close confinement of mentally disordered patients cannot be avoided, it should be reduced to an absolute minimum and be replaced with one-to-one continuous nursing care as soon as possible.

22.4.3 Consent to Treatment

Consent to treatment should be sought from all patients, including offenders suffering from a mental disorder, provided they have capacity to consent. Furthermore, obtaining the patient's consent, especially in the case of psychiatric pathology, is essential if a 'therapeutic alliance' is to be formed which is likely to make the patient more committed to the treatment offered.

A controversial issue in forensic psychiatry and a classical ethical dilemma is whether an incompetent prisoner has the right to refuse treatment or, framed differently, whether the right to refuse treatment supercedes his right to sanity. Some might argue that it is the duty of the treating psychiatrist to zealously persuade the patient, his guardian and, if need be, the courts, that a proposed treatment is indeed in the best interest of the individual, regardless of an expressed (and sometimes psychotic) wish against it. The question is if the courts recognize a patient's right to receive treatment so as not to remain psychotic (implicitly acknowledging the subjective torment and, at times, sheer terror of the psychotic individual). Abramowitz (2005) suggested that the courts will usually support treatment for an individual lacking capacity as long as it is consistent with professional standards of care, however, without asserting a specific, inalienable right of the individual to receive treatment.

The situation of a non-consenting patient who does not lack capacity also has to be taken into account. Generally, every patient has a right to refuse treatment or to informed 'non-consent'. It is of note, however, that not all mental health laws recognise this right for self-determination. Some laws, e.g. the Mental Health Act 1983 of England and Wales, provide procedures to override the informed consent of capacitous patients through a second opinion doctor. Such measures are only available after transfer of the prisoners to a psychiatric institution.

On occasion the decision to refuse treatment results from a conflict relating to non-medical issues; this is, for example, the case when a prisoner goes on hunger strike to protest against a judicial or administrative decision. In this situation the doctor should assess the state of health of the person concerned and subsequently make a detailed note in the patient's file to document that the individual has capacity to understand the treatment proposed but has refused treatment after being given detailed information.

The need for medical care of prisoners who persistently refuse food in order to make a protest is rare but challenging. Knowledge about the hunger strike quickly spreads and gets into the political arena. Governments want to resist the demands, which often have political overtones, but also do not want prisoners to die because of fear of a backlash of public opinion. Pressure is therefore brought on the prison health care staff, including psychiatrists, to keep the prisoners alive, if necessary, by force feeding. However, a doctor must obtain consent from the patient before applying his skills to assist him. The only exception is in an emergency when the patient is incapable of giving consent. Because the end stage of food refusal in coma, it follows that the patient is then not capable of giving consent and it is possible to argue that the doctor may then intervene by artificial feeding, to save the patient's life. However, according to Wool and Pont (2006) this is not the case if the patient has made it clear beforehand that he refuses intervention to prevent death.

22.4.3.1 Compulsory Treatment

If mentally ill prisoners refuse to accept medication, having made an informed decision not to consent, the problem arises as to whether it can be administered against their wishes. In line with principles of medical ethics a person cannot be forced to undergo treatment unless there is a risk to self or others. (The issue of compulsory treatment is addressed in detail in Chapter 20.)

22.4.4 Confidentiality

Confidentiality is central to the doctor–patient relationship. It enables the patient to develop trust, knowing information he discloses will be held in confidence. The basic principles of confidentiality apply to all doctors, including forensic psychiatrists, and most European countries have laws and/or professional guidance to govern this complex area (e.g. General Medical Council 2004). The doctor must not disclose information about the patient to third parties without the patient's consent except in a limited number of clearly specified circumstances, usually to prevent serious harm to the patient or others. If such a situation arises the patient should be informed about the disclosure and the reasons for disclosure clearly documented. Although this has traditionally received less attention, principles of confidentiality also apply to other professions, e.g. psychologists (Younggren and Harris 2008).

In forensic settings the principles of confidentiality may be threatened in a number of ways. For example, multidisciplinary working and liaison between agencies are the norm and are necessary and benefit the patient. However, the increased sharing of information about the patient makes it more difficult to maintain confidentiality and may be contrary to the patient's expectations. Furthermore, due to the nature of the patient group, the forensic psychiatrist may be under particular pressure to disclose information in pursuance of crime investigation or prevention. In addition to universally recognized exceptions to confidentiality (e.g. Konrad 2007) some exceptions arise uniquely in correctional facilities. For example, psychiatrists may be expected to report to authorities serious inmate rule violations and plans for escapes or disturbances (Appelbaum 2005). They are also required to report to different agencies regarding the progress or otherwise of their patients with potentially far-reaching consequences. It is of utmost importance for the forensic psychiatrist to not loose sight of the fundamental principals of confidentiality and to consider each request for disclosure of information on its merit and to weigh up each time whether such request is justified. If confidentiality is not respected, the patient–physician relationship, difficult as it is in an often therapy-hostile prison environment, may be further jeopardized. In case of unavoidable disclosure the patient should be informed about the disclosure and the reasons for it.

22.5 Research

Each medical discipline depends on some form of research in order to increase knowledge. Different forms of such research exist. Epidemiological research may not require individual patient contact or any interventions which may cause harm to the individual (see Chapters 11 and 26). While such research is generally welcome, including in prison settings, in order to increase the evidence on mental disorders and mental health care in the prison context, clinical research is a different matter. The participation of psychiatric patients, particularly those incarcerated, in clinical research poses significant ethical challenges. Prisoners may be particularly vulnerable to (real or perceived) coercion due to their restrictive living conditions in custody. Those prisoners with mental disorders are 'double vulnerable'; in those individuals both voluntariness and capacity have to be considered. However, administering treatment without sound evidence base, based on thorough research, does not appear to be a desirable alternative. Therefore, guidance with regards to the ethics of human experimentation is crucial.

A number of guidelines have been developed to set standards for biomedical research, most notably the 'Declaration of Helsinki' (World Medical Association 2008). The declaration lies down the fundamental principals of respect for the individual, self-determination and the right to make informed decisions regarding the participation in research. It also stipulates the review of research by independent ethics committees. While these guidelines apply to any individual, specific guidance has been developed for prisoners. The European Prison Rules (Committee of Ministers 2006) clearly state that prisoners shall not be subjected to any experiments without their consent (Principle 48.1) and that experiments involving prisoners that may result in physical injury, mental distress or other damage to health shall be prohibited (Principle 48.2). The passing of these rules may at first sight seem unnecessary as they do not go further than other already established guidance governing research in any patient group. However, it has to be considered that incarcerated individuals are rightly seen as a particularly vulnerable group requiring special protection. For example, some authors have reported earlier practises in the USA where patients were bribed to consent to dangerous research by offering them early release in return (Gunn and Taylor 1993).

More than half of the countries in the EUPRIS study generally prohibit biological and pharmacological research on prisoners. The remaining countries allow such research in principle but emphasize the importance of obtaining informed consent of the prisoners, permission of the responsible authorities, and ethics committee approval (Salize et al. 2007). While the intention to protect vulnerable prisoners from being coerced into participating in research is in principle commendable, it also raises a number of questions. For example, how is progress regarding treatment to be made if relevant research is not permitted and what are therefore the ethical implications of NOT conducting research in prisons? Furthermore, a ban on all biomedical research in prisons appears to stem from a perception of prisoners as incapable of making voluntary informed choices for themselves. This perception discriminates against prisoners in depriving them of the opportunity to engage in research projects, which they may well find an interesting and rewarding experience.

References

Abramowitz MZ (2005) Prisons and the human rights of persons with mental disorders. Curr Opin Psychiatry 18:525–529

Appelbaum KL (2005) Practicing psychiatry in a correctional culture. In: Scott CL, Gerbasi JB (eds) Handbook of correctional mental health. American Psychiatric Publishing, Washington, DC

Arboleda-Flórez J (2006) Forensic psychiatry: contemporary scope, challenges and controversies. World Psychiatry 5:87–91

Binswanger R (1979) Probleme der Gefängnispsychiatrie. Nervenarzt 50:360–365

Bruce RD, Schleifer RA (2008) Ethical and human rights to ensure medication-assisted treatment for opoid dependence in prisons and pre-trial detention. Int J Drug Policy 19:17–23

Buchanan A, Leese M (2001) Detention of people with dangerous severe personality disorders: a systematic review. Lancet 358:1955–1959

Calcedo-Barba A (2006) The ethical implications of forensic psychiatric practice. World Psychiatry 5:93–94

Carmi A, Moussaoui D, Arboleda-Florez J (2005) Teaching ethics in psychiatry: case-vignettes. Unesco Chair in Bioethics, Haifa

Committee of Ministers (2006) Recommendation Rec(2006)2 of the committee of ministers to member states on the European prison rules. Council of Europe

Council of Europe (1999) Ethical and organisational aspects of health care in prison. Recommendation No R (98) 7. Adopted by the committee of ministers on April 1998. Strasbourg

Department of Health (2008) Mental Health Act 1983. The Stationary Office, Norwich

De Smit NW (1977) La double face de la psychiatrie légale. Déviance et Société 1:435–439

D'Silva K, Duggan C, McCarthy L (2004) Does treatment really make psychopaths worse? A review of the evidence. J Pers Disord 18:163–177

Fazel S, Danesh J (2002) Serious mental disorder in 23000 prisoners: a systematic review of 62 surveys. Lancet 349:545–550

General Medical Council (2004) Confidentiality: protecting and providing information. General Medical Council, London

Gunn J, Taylor PJ (1993) Ethics in forensic psychiatry. In: Gunn J, Taylor PJ (eds) Forensic psychiatry. Clinical, legal and ethical issues. Butterworth-Heinemann, Oxford

Hallek SL (1974) A troubled view of current trends in forensic psychiatry. J Psychiatry Law 2: 135–157

Halleck SL (1984) The ethical dilemmas of forensic psychiatry: a utilitarian approach. Bull Am Acad Psychiatry Law 12:279–288

Hare RD (1991) The hare psychopathy checklist-revised. Multi-Health Systems, Toronto, ON

International Council of Prison Medical Services (1979) Oath of Athens. Adopted by the World Medical Assembly, Athens

Konrad N (2001) Psychiatry in custody and prisons. In: Henn F, Sartorius N, Helmchen H, Lauter H (eds) Contemporary psychiatry, vol 2, Psychiatry in special situations. Springer, Heidelberg

Konrad N (2002) Prisons as new asylums. Curr Opin Psychiatry 15:583–587

Konrad N (2006) Forensic psychiatry in dubious ascent. World Psychiatry 5:93

Konrad N, Arboleda-Florez J, Jager AD, Naudts K, Taborda J, Tataru N (2007) Consensus paper: prison psychiatry. Int J Prison Health 2:111–113

Leyrie J (1977) Manuel de psychiatrie légale et de criminologie clinique. Librairie Philosophique, Vrin, Paris

Morse SJ (2008) The ethics of forensic practice: reclaiming the wasteland. J Am Acad Psychiatry Law 36:206–217

Okasha A (2002) Death penalty and madrid declaration, vol 1, Abstracts. XII World Congress of Psychiatry, Yokohama

Prior PM (2007) Mentally disordered offenders and the European court of human rights. Int J Law Psychiatry 30:546–557

Rasch W, Konrad N (2004) Forensische Psychiatrie. 3.Auflage. Kohlhammer, Stuttgart

Reid WH (2008) The treatment-forensic interface. J Psychiatr Pract 14:122–125

Rice ME, Harris GT, Cormier C (1992) An evaluation of a maximum security therapeutic community for psychopaths and other mentally disordered offenders. Law Hum Behav 16:399–412

Robertson MD, Walter G (2008) Many faces of the dual-role dilemma in psychiatric ethics. Aust N Z J Psychiatry 42:228–235

Salize HJ, Dreßing H (2005) Placement and treatment of mentally ill offenders – legislation and practice in EU Member States. Final Report. Central Institute of Mental Health, Mannheim

Salize HJ, Dreßing H, Kief C (2007) Mentally disordered persons in European prison systems – needs, programmes and outcomes (EUPRIS). Final Report. Central Institute of Mental Health, Mannheim

Schneider P-B (1977) Tribune: justice et psychiatrie. La psychiatrie et la justice pénale. Déviance et Société 1:427–434

Webster CD, Douglas KS, Eaves D, Hart SD (1997) HCR-20: assessing risk for violence, version 2. Mental Health Law and Policy Institute, Simon Fraser University, Burnaby

Webster CD, Harris GT, Rice ME, Cormier C, Quinsey VL (1994) The violence prediction schene: assessing dangerousness in high risk men. University of Toronto, Centre of Criminology, Toronto, ON

WHO (1977) Forensic psychiatry. Report on a working group. Copenhagen. Siena 13–17 Oct 1975.

Wool R, Pont J (2006) Prison health. A guide for health care practitioners in prisons. Quay Books, London

World Medical Association (2005) Declaration of Tokyo. Guidelines for medical physicians concerning torture and other cruel, inhuman or degrading treatment or punishment in relation to detention and imprisonment. Adopted by the World Medical Assembly, Tokyo in Oct 1975 and editorially revised Divonee-les-Bains in May 2005. http://www.wma.net/e/policy/c18.htm. Accessed Dec 2008

World Medical Association (2008) Declaration of Helsinki. Ethical principals for medical research involving human subjects. World Medical Association, Ferney-Voltaire Cedex

World Psychiatric Association (1977) Declaration of Hawaii. http://www.wpanet.org/content/ethics-hawaii.shtml. Accessed Dec 2008

United Nations (1987) Standard minimum rules for the treatment of prisoners. http://www.unchr.ch/html/menu3/b/h_comp34.htm. Accessed Dec 2008

Younggren JN, Harris EA (2008) Can you keep a secret? Confidentiality in psychotherapy. J Clin Psychol Sess 64:589–600

Zinkler M (2003) Zur psychiatrischen Begutachtung von Migranten bei drohender Abschiebung. R&P 21:22–24

Chapter 23
Treatment of Substance Dependence

Ambros Uchtenhagen

Contents

Abbreviations

AMA	American Medical Association
APA	American Psychiatric Association
DATOS	Drug Abuse Treatment Outcome Study
DSM-IV	Diagnostic-Statistical Manual of Mental Disorders, 4th revision (APA)
EC	European Commission
EMCDDA	European Monitoring Centre on Drugs and Drug Addiction
EU	European Union
HIV	Human Immunodeficiency Virus
ICD-10	International Classification of Diseases, 10th revision (WHO)
NIDA	National Institute of Drug Abuse

A. Uchtenhagen (✉)
Research Institute for Public Health and Addiction, Zurich University, CH-8005 Zurich, Switzerland
e-mail: uchtenhagen@isgf.uzh.ch

H. Helmchen, N. Sartorius (eds.), *Ethics in Psychiatry*, International Library of Ethics, Law, and the New Medicine 45, DOI 10.1007/978-90-481-8721-8_23, © Springer Science+Business Media B.V. 2010

NTIES   National Treatment Improvement Evaluation Study
NTORS   National Treatment Outcome Research Study
QCT   Quasi-Compulsory Treatment
UN   United Nations
UNODC   United Nations Office on Drugs and Crime
WHO   World Health Organisation

23.1 The Ethical Dimensions

23.1.1 The Ethical Basis

23.1.1.1 Consequential Ethics

The philosophical debate includes a distinction between absolutist and utilitarian positions (Hare 1991). Absolutism includes the acceptance of a conduct code based on absolute, indisputable rights and duties (e.g. abstaining from substance use). Utilitarianism has its focus on the consequences, not on the reasons or motives of conduct; whatever the motives are, moral judgment is based on the consequences of behavior (e.g. of substance use). This utility principle has a longstanding tradition in various forms; a most prominent representative was the English philosopher John Stuart Mill (Mill 1861). A contemporary example of a consequential ethics is found in the works of Hans Jonas (Jonas 1984).

> The absolutist position states: "Right is to be done come what will come. I am not answerable for the consequences of doing right, only of not doing it…". The utilitarian position is summed up by "success is the touchstone; the might of obtaining the reward" (Parker 1940).

The focus on consequences asks for an external criterion how to evaluate those. The answers are not unisono: while Mills uses the criterion of pleasure (behavior must be pleasure-seeking), Jonas formulates the imperative that human activities must be compatible with the permanence of genuine human life. In our context, both these criteria are too broad. We have to deal here with the external criteria of the human rights declaration as a general framework, and with the principles of medical ethics which must fully apply when dealing with substance dependence. Both will allow us an ethical judgment on the results of treatment as well as on the consequences of treatment policy.

Today's drug policy claims to be evidence-based. Evidence means that policy recommendations are based on scientific findings on 'what works' and therefore have a good chance to lead to positive results. Drug policy is based on principles of consequential ethics.

23.1.1.2 Human Rights

The Universal Declaration of Human Rights (UN 1948) contains a number of relevant conditions, such as: no discrimination (art.2), no degrading or inhuman treatment (art.5), right of equal access to medical care and social services (art.25/1), but also that everyone has duties to the community (art.29/1) and that limitations of rights and freedom are admissible on the basis of 'just requirements of morality, public order and the general welfare' (art.29/2).

The European Convention on Human Rights (Council of Europe 1950) further stipulates that a person's liberty may be deprived in case of lawful detention of alcoholics or drug addicts (art.5/1/e), with a right to appeal to a court (art.5/4).

These statements try to establish a balance between protecting the individual rights of the person and respecting the needs of society for public order, general welfare and even morality. There is large room for interpretation, so that every society can decide on the compatibility of addictive behavior with the nature and extent of the above mentioned requirements. Compulsory measures against persons with substance dependence may be admissible on the basis of national laws (see also Section 23.2.2.3).

23.1.1.3 Medical Ethics

Medical ethics apply only if and inasmuch as substance dependence is understood as a medical condition. This has been the official position of the leading medical organizations since decades (see 2). And brain research has identified it as a 'brain disease', thereby giving it a biological basis (Leshner 1997). The substance dependent person is a patient and should enjoy the status and all the rights of patients.

The guiding principles of biomedicine are dealt with extensively in Chapter 8.

In the absence of specific rules for the treatment of substance dependence, the general ethical rules for good medical practice apply. The four major principles are: do no harm, improve the well-being, respect the autonomy and apply justice. It is obvious that even these few principles cannot be followed without creating conflict (Rust 2000).

Involuntary intervention to prevent harm is in conflict with the autonomy of an unwilling patient. Treating all patients as being equal (principle of justice) is impossible where the resources are limited. Also, confidentiality and data protection often are in conflict with administrative and law enforcement interests in case of illicit drug use. All such conflicts must be carefully examined, in the best interest of all concerned. When the patient's interests collide with those of relatives or other third parties, a common solution must be found to the extent possible. It is advisable to recur to an ethical consilium if major consequences are expected from the decision. The principle of respecting the autonomy of the patient must never be overruled in the name of some abstract societal value without the presence of concrete harm implications for others.

It is inevitable therefore to discuss the main ethical aspects of addiction treatment in the second part of this chapter, providing empirical evidence on outcomes in order to satisfy the expectations of a consequential ethics. But first I will look at the types of societal and individual values which are relevant for a discussion on addiction and addiction treatment.

23.1.2 Societal Values at Stake

Human rights as well as medical ethics mention the interests of society which may be in conflict with the interests of the individual. In fact, the ethical justification to interfere with a person's substance dependence resides in negative societal consequences of dependence. Which values have to be considered here?

23.1.2.1 Citizen's Obligations

Every citizen is bound, by law or common understanding, to contribute to the well-being of the community, by leading a self-responsible and law-abiding life, caring for his or her subsistence, paying taxes and limiting his or her behavior so as not to infringe the liberty and rightful interests of others.

These obligations are valid for every member of the community, and only illness and handicap are accepted as limiting factors. Substance dependence can only limit the expectations, if and inasmuch the dependent person is considered to have a chronic medical condition or is otherwise unfit for meeting the obligations. And even then, the condition opened a road to enforced treatment wherever such a law has been in place. Also, a resentment against addiction as a burden to society is still present in many official and unofficial attitudes, resulting in an unwillingness to care for addicts just as for other patients.

23.1.2.2 Public Safety, Public Order

All societies organize themselves in a way which protect to the extent possible the safety of citizens and a public order that allows for an efficient daily life, a framework for the pursuit of activities and for protecting properties. Public substance use, public intoxicated behavior, gatherings of inebriated or drugged people, public drug trafficking, any substance-related public nuisance calls for appropriate measures to restore order and safety and to minimize nuisance.

Among the appropriate measures are all treatments which help to bring a majority of addicts into treatment and into protected environments, and competences to law enforcement to keep substance use off public places. Only a combination of such measures were able to make the unacceptable nuisance and misery of the large open drug scenes ('needle park') in the Swiss cities disappear. The acceptable compromise was to let drug use and trafficking happen in private, as long as the neighborhood is not molested.

23.1.2.3 Socio-Cultural Acceptability of Behavior

Substance use is a factor in many social and cultural events, as a facilitator of social contact, a source of emotional well-being, but also of destructive or aggressive behavior. The acceptable limits of intoxication and behavior, of substances used and of opportunities for consumption are cultural-specific and are to be respected when dealing with substance dependence (Edwards and Arif 1980). For instance: a major difference between western and Asian cultures is the place of the individual within the family system; while the individual's interests and autonomy are a core concept of most western psychotherapies, the integrity and the interests of the family are the higher values in many Asian societies.

23.1.3 Personality Values at Stake

In a paternalistic attitude, substance dependent persons must be protected against wasting their personal resources and potential achievements. However, such an attitude collides with the present position that interference is only justified on the basis of negative consequences of a person's behavior for others (except if psychiatric conditions are present which justify involuntary hospitalisation). Also, many people have lived or live a productive life in spite of their substance dependence, and to interfere with their lifestyle would cause more ethical concerns than to respect their autonomy. Even preventive approaches can only reduce the likelihood of starting substance use, but cannot replace a person's decision to abstain or not.

Nevertheless it shall be briefly described, which personality values are at stake and are considered in psychotherapeutic approaches to personality development, as a contribution to protect against and overcome substance dependence.

23.1.3.1 Self-Fulfilment and Needs Satisfaction

In most Western societies, each individual is expected to develop it's own potential, to find out what is good for him or her, to care about needs and how to optimize need satisfaction in a given social context. In a pluralistic society, a variety of life styles and value orientations are co-existing, although under the same laws, and each person has to find it's individual place within this plurality. A religious obligation not to waist your resources (the 'talents') may be associated to the notion of a unique personality which must be developed.

One of the characteristics of substance dependence is the neglect of other interests and duties. A person occupied with keeping up a substance dependence may be unable to care for developing its personality and therefore to miss the essential of his/her life.

23.1.3.2 Self-Responsibility and Freedom of Choice

Developing one's own resources and shaping one's own life is to be facilitated by education and by societal organization, but it is ultimately in the responsibility of the person itself. Conditional are the freedom of choice – and the freedom of the will. Is addiction leaving any room for choices, or is it a negation of free will and therefore the basis for involuntary intervention? This has been debated extensively.

At present and on the basis of research evidence, we have a more differentiated view. A distinction is made between using substances in a controlled manner, and having developed (not 'choosen') dependence with its craving and compulsive consumption style. The decision of many to change their lifestyle and go to treatment (Bergmark 2008, Ekendahl 2007) or to stop the dependence without professional support (the so-called self-healers, Klingemann and Sobell 2007) demonstrate the ability of many addicts to make a choice and stick to it. Not all addicts will make it, but there is no way to predict a 'point of no return'.

In the light of this position, treatment of substance dependence is expected to help the individual to make it's own choices and to regain self-responsibility, rather than to make the decisions for him.

23.1.3.3 Self-Limitation of Pleasure Seeking Behavior

Using psychoactive substances is basically a pleasure-seeking (or at least displeasure-avoiding) behavior. Substance dependence is seen as an unbalanced motivational system where substance induced pleasure has become disproportionally important in relation to other sources of pleasure. This view has been developed on the basis of psychoanalytic theory (Loose 2005) as well as on the basis of clinical and brain research (West 2006). Pleasure seeking behavior is also a basic assumption of the theory of addiction as a rational behavior: past positive experience induces continued efforts to repeat such experience if its costs are not disproportionate in relation to the expected positive effects (Gruber and Koszegi 2001).

One of the most prominent characteristics of dependence is a loss of control over amount, frequency and circumstances of consumption. Treatment is about regaining self-control, or – in the case of agonist maintenance treatment – about compensating insufficient self-control by external controls (supervised intake of supervised dosage).

The expected learning process is to optimize pleasure by avoiding excessive and untimely consumption with its negative consequences, or to prefer other pleasure-seeking behavior with less risks of losing control. Treatment is designed to support this process.

23.2 Treatment of Substance Dependence – Ethical Aspects

23.2.1 Substance Dependence – Disorder or Misbehavior?

23.2.1.1 The Medical Concept

The current medical science defines substance dependence as a condition with diagnostic criteria, described in the generally acknowledged diagnostic systems International Classification of Diseases (ICD-10) of the World Health Organisation, and Diagnostic-Statistical Manual of Mental Disorders (DSM-IV) of the American Psychiatric Association. New editions of both are in preparation. The criteria include biological symptoms (tolerance, withdrawal symptoms in case of discontinued use), psychopathological symptoms (craving, desire to reduce consumption), behavioral symptoms (loss of control over consumption, consuming in order to overcome withdrawal symptoms, neglect of other interests and of duties) and social symptoms (negative consequences of use). The diagnosis is clear, if 3 criteria are met during the last 12 months. No criterium is always present and is considered to be pathognomonic, although loss of control was determined previously as the main characteristic of addiction (Jellinek 1960, Rush 1772). But the concept of loss of control is strongly related to socio-cultural norms which determine acceptable behaviour, and there is no universally accepted set of criteria and measurements of such behaviour (Room 1980, Room et al. 1996). Therefore, dependence syndrome covers a considerable variability, in phenomenology and course.

The etiology has been clarified by brain research, demonstrating structural changes in specific cerebral regions which are the basis for tolerance and craving. Such changes develop after continued use of an addictive substance and are enhanced by genetic factors and under stress, and they are reversible. Both processes – becoming dependent and restitution – are considered to be learning processes (Volkow et al. 2003).

As long as the substance was considered to be the main reason for developing dependence, the logical consequence was the cure by abstinence (Rush 1772). The more complicated and multi-factorial concept of dependence etiology also fosters the role of (at least temporary) abstinence for a remission of the cerebral structural changes, but life-long abstinence is no longer asked for as the only possible therapeutic approach. This notion is reinforced by previous observations of regaining consumption control by former addicts (Kaya et al. 2004) and that not the substance alone, but also the personality and the social environment had a role in the development of dependence, and that many users keep their consumption under control or regain control under more favorable conditions (Robins 1993, Zinberg 1984).

Another major problem was (and is) the belief of doctors that the use of substances such as heroin and cocaine leads inevitably to dependence, and who therefore have no reason to differentiate between treatment of hazardous use, harmful use and dependence. ICD-10 and DSM-IV make such a difference, and specific interventions for these conditions have been successfully introduced into medical practice, especially motivational interviewing and related approaches against hazardous and harmful use without dependence (Miller and Rollnick 2002, Rollnick and Miller 1995).

In an ethical perspective, the medical concept results in defining the dependent person as an ill person who differs from a non-dependent problematic user and whose condition asks for an appropriate specific treatment.

23.2.1.2 Medicalisation of a Social Problem?

Although the medical concept includes the notion of social factors facilitating the development of substance dependence, the claim of medicine to have the responsibility for dependent persons is regarded to basically medicalize a social problem. Non-medical interventions, such as removing the patient from a pathogenic milieu

(Robins 1993) or self-help approaches (Klingemann and Sobell 2007) demonstrate the limitations of a strictly medical model.

In a Public Health perspective however, where the concern for social risk factors of medical conditions is a major issue, these limitations are compensated conceptually, if not effectively. Public Health research demonstrates the role of economic inequality, illiteracy, misguided urbanization, and loss of social traditions and networks plays a role in the worldwide role of substance dependence for the global burden of disease (Uchtenhagen 2004, WHO 2001). In order to achieve health for all, the scope of activities must include health promotion securing the conditions for health protection (WHO 1986). This intention is at the core of the EU Health Programme 2008–2013 (EU 2008), and its effectiveness has been demonstrated (International Union for Health Promotion and Education 2000).

The ethical problem here is evident: substance dependence is a consequence of social problems, and medical treatment alone will not suffice to reduce effectively the prevalence and incidence rates. Medicine however has the potential to evidence the role of social conditions, is required to raise the visibility of the problem and to contribute to efforts at reducing the problem. The AMA Standards of Conduct (2001) recognize this to be part of good medicine.

23.2.1.3 Medical Versus Moral Treatment

The notion of substance misuse and dependence as a moral weakness grew on the basis of a religious ideal of moderation or even abstinence from pleasure seeking behavior. It was a reaction to abundant substance use and its negative consequences, at the dawn of new competitive economies during the seventeenth and eighteenth centuries in Europe (Schivelbusch 1980) and USA (Warner 2007), and at the end of the nineteenth century again leading to temperance movements (Harding 2008) and finally to alcohol prohibition and the prohibition of illicit drugs (Levine 1981). Economic motives played a role as well; widespread inebriety weakened the economic growth, and the costs of social welfare became a burden. But people continued consumption in illegality, the prohibitive strategies had and have limited successes and a high price in unintended side-effects, and in the consumer-friendly societies a moral stance for abstinence still has its difficulties.

The basic ambivalence – substance dependence as a medical condition or a moral weakness – is reflected again in the opposition of medical and moral treatment. If moral treatment is understood to be educational and admonishing, this approach is nowadays believed to have little impact. The new approach is to help the addict in getting motivated by appropriate empathy and information, while confrontation and reproaches are recognized to reinforce the resistance against any change. The basic attitude of enhancing the patient's motivation for change is to avoid confrontation and preaching; and to give the patient the feeling that he is taking the decisions and doing the necessary steps forward himself. Special methods have been developed for enhancing the motivation for change, and they have become an important element in today's treatment of alcohol and other drug problems (Gossop 2006, Moyer et al. 2002).

23.2.2 Principles of Treatment in an Ethical Perspective

23.2.2.1 A Hierarchy of Objectives

The goals to be reached through treatments of substance dependence (and which therefore are the criteria for measuring outcome in treatment evaluation research) have changed over the last decades. While the goal of abstinence was traditionally on top of the list, the present situation can be summarized as follows: the primary goal is the patients survival, moving to health improvements (or at least prevention of deterioration), to improvements in social integration, to reductions in substance use (moving away from addictive behavior), to improvements in quality of life (as defined subjectively by the patient), ultimately resulting in a responsible and satisfactory life style. Abstinence is not always needed for reaching these objectives, nor does abstinence guarantee to reach them.

A national example of listing the objectives is given in the Report on Models of Care by the UK Department of Health (Department of Health 2002):

- reduction of psychological, social and other problems directly related to drug use;
- reduction of psychological, social or other problems not directly attributable to drug use;
- reduction of harmful or risky behaviours associated with the use of drugs (e.g. sharing injecting equipment);
- attainment of controlled, non-dependent, or non-problematic drug use;
- abstinence from main problem drugs;
- abstinence from all drugs.

This document fully endorses the principle of consequential ethics in prioritizing the reduction of the various forms of drug-related harm, including social, medical, legal and financial problems until the drug dependent is ready and able to come off drugs.

23.2.2.2 Cure or Care?

Treatment is often identified with cure (in the sense of healing an illness), while care means serving a chronic patient without chances for healing. As such, care is an equivalent to harm reduction, meaning all interventions designed to improve the health and social status of a chronic addict who continues a dependent behavior (harm reduction is more than HIV prevention). The present debate on treatment and harm reduction is often a debate on opposing principles of action which cannot be reconciled. In this debate, harm reduction has been disqualified as an approach to prolong dependence, to make substance use acceptable for young people and to undermine the readiness of addicts for treatment.

Today, in the light of the updated treatment objectives, harm reduction is considered an ally rather than an opponent of treatment. Accordingly, the treatment system must be an integrated system that enables abstinence and harm reduction services to work together, in order to provide a continuum of care, including:

- Easily accessible low threshold services that meet the immediate needs of continuing drug users
- Clear processes for motivating users to move away from drug dependent lifestyles
- Clear processes for referring users into structured treatment programmes that promote stabilization or abstinence (quoted from Stevens et al. 2006b).

23.2.2.3 Tailoring Treatment to Individual Needs

Bearing in mind the diversity of etiology, symptoms and stages of substance dependence, it becomes obvious that treatment cannot be uniform for all dependent persons. In addition, treatment needs in different age groups and other target groups (gender, ethnicity, comorbidity etc) may differ considerably as well. Treatment must respond to the specific needs of an individual patient, on the basis of a comprehensive needs assessment and a treatment planning process where patient and therapist work together on a shared understanding of what is needed and what should be done.

This concept of a needs based treatment has been intensively researched. I mention two large studies from USA and one European study. The National Treatment Improvement Evaluation Study (NTIES) documented the results of needs-based treatment planning and found a significant correlation between 1-year outcomes (measured as drug-free urines) and the number of needs included in the treatment plan (Gerstein et al. 1997). A comparison of basic, average and enhanced services for heroin dependence evidenced better retention and outcomes in services where psychiatric and social care was available to patients (McLellan et al. 1997). In both studies, covering the needs for psychiatric care and living conditions (housing, jobs) was found to be especially important. This again has consequences for the hierarchy of objectives: a reduction of substance use is facilitated by improved living conditions and not necessarily their precondition.

"The combination of treatment components and services to be employed must be tailored to meet the needs of the individual, including where he or she is in the recovery process." is therefore one of the principles of addiction treatment (NIDA 1999).

The European study focuses on a set of treatment needs determining which level of care is appropriate in order to meet those needs; this concept is in line with earlier attempts to match patients to specific treatments (Merkx et al. 2007).

23.2.3 Public Health Interventions Versus Individual Care

23.2.3.1 Coverage of Treatment Needs

A new kind of conflict was created while the only adequate response to substance dependence was abstinence, voluntary or enforced. It became obvious that only a minority of drug injectors could be reached and retained effectively by abstinence treatment. On the other hand, reaching the majority of drug injectors became a primary objective in order to slow down the HIV epidemic. The public health priority is to offer treatment to all persons in need of treatment. Individual care is optimized by high quality treatment, but public health cannot accept high quality standards for a few as long as the many are not reached adequately.

This principle includes a monitoring of the treatment needs in a given population and, accordingly, a careful planning of the treatment system as a whole. The responsibilities – ethically and professionally – are well distributed: medical practitioners are responsible for good individual care, service directors are responsible for good practice in their services, health authorities are responsible for good coverage of treatment needs.

23.2.3.2 Cost-Effectiveness of Interventions

When caring for the treatment system as a whole, the next step is to take the responsibility for making best use of the available human and financial resources for treatment. This means to look at how much effective treatment is provided at what costs. It does not suffice to give a priority to treatments with good evidence for effectiveness, but to treatments which provide effectiveness at the lowest costs.

Economic studies in addiction are a relatively new era, and methodology is still developing (Single et al. 1995). When it comes to economic analysis of addiction treatment, there are pitfalls to be avoided: e.g. calculating costs only on a patient-day basis without considering the duration of treatment, or considering only the costs occurring at the treatment center without including the costs of other services providing the ancillary care. Also, the effectiveness criteria must encompass more than the reduction in substance use (see Section 23.2.2.1). The overall cost-effectiveness of addiction treatment is well documented (NIDA 1999).

Comparative studies on treatment approaches provide results which favor out-patient over in-patient treatment, professionally supervised care over intensive professional care. In the case of heroin problems, out-patient agonist maintenance treatment is more cost-effective than long-term residential drug-free treatment (Gerstein et al. 1997). In the case of alcohol problems, early brief interventions are more cost-effective than more intensive treatments of dependence, making the former a public health priority (Moyer et al. 2002).

A recent approach are efforts which intend to make best use of resources through models of stepped care, matching patients to specific treatments, on the basis of their characteristics and needs. They could not find better outcomes in matched patients; however, the models met some problems in being implemented correctly and accepted by professionals and services (Longabaugh and Wirtz 2001, Schippers et al. 2002) Nevertheless, models to reserve intensive and expensive treatments to those who cannot profit from self-help or less intensive interventions, will continue to be experimented.

23.2.3.3 Risk Reduction

An efficient protection of public health goes beyond the prevention of starting substance use and providing treatment for substance dependent persons. It includes all measures which are effective in protecting the health and social status of active users, in the interest of users as well as of the population at large (see Section 23.2.2.2).

Risk reduction (also termed harm reduction) measures are part of a comprehensive drug policy, besides prevention, treatment and law enforcement, as included in the European Drug action plans 2005–2008 (EU 2005) and 2009–2012 (EC 2008). The European Monitoring Centre on Drugs and Drug Addiction EMCDDA has accordingly collected and published data on harm reduction measures (Hedrich 2004), while in north America the acceptance is much lower (Broadhead et al. 2002).

Evaluation studies have demonstrated the usefulness of risk reduction approaches, while the concerns about negative consequences, such as increase of prevalence rates, could not be substantiated. Risk reduction is now an internationally approved approach. The dichotomy of prevention and treatment on one side, of harm reduction in the sense of reducing the adverse health and social consequences of drug use on the other side, is now considered to be 'a false dichotomy: they are complementary' (UNODC 2008).

23.2.4 Motivation, Neglect, Coercion

23.2.4.1 Motivation for Change as a Prerequisite for Treatment

As already mentioned in Section 23.2.1.3, motivation for change is considered to be essential for a good treatment outcome. Without identifying oneself with the aims and elements of the program, and without active participation in the program, behavior change is unlikely. Some of the research even indicates that motivation for change is not so much a result of therapeutic efforts, but rather a precondition for going to treatment (Bergmark 2008).

In an attempt to tailor therapeutic interventions to motivation for change, a stage concept was developed. The transtheoretical model (Prochaska and DiClemente 1982) identifies stages of motivation (from pre-contemplation to contemplation to preparation to action to maintenance to relapse) and recommends stage-specific interventions. The research evidence on the usefulness of the model is still scarce (Gossop et al. 2007). The main critique of the model points out the rapid changes in a person's motivational state which makes it a non reliable basis for treatment planning (West 2006). However, motivational enhancement methods may apply, in order to optimise the patient's readiness for behavior change especially in the Criminal Justice System (Czuchry and Dansereau 2005, Hiller et al. 2002).

The findings have implications: it is the patient who decides if and when the behavior and lifestyle will be changed; this is his responsibility. The responsibility of professionals on the other side is to enable and enhance the motivation for change, through adequate information, attractive treatment programs, listening to patient satisfaction and critique. The responsibility of the treatment system is in facilitating and monitoring the respective service quality, in the interest of an optimal coverage.

23.2.4.2 Accessability and Affordability of Treatment

Neglect of patients in need of treatment can take many forms. For instance, an analysis of national guidelines for agonist maintenance treatment of opiate dependence showed a range of restrictive access criteria, such as minimal age, minimal duration of dependence, polydrug use, lack of confidentiality, and likewise restrictive criteria for a continuation of treatment, such as persistent illicit substance use or bureaucratic limitation of treatment duration (Uchtenhagen et al. 2005). Other examples are the denial of social support and/or medical care to active alcohol or drug users or smokers. Examples are the denial of liver transplants to persons with alcohol problems, in spite of research evidence that survival rates in those who continue to drink are not lower than in other patients (Fireman and Rabkin 2001), and the denial of Hepatitis C treatment with Interferon to active drug injectors on the basis of wrongly presumed lack of compliance (Edlin et al. 2001). Excluding addiction treatment from health insurance and reserving treatment to those who can pay for it are also major limitations for good coverage. Throughout the history of addiction treatment, a frequent form of neglect was the attitude to let those live in misery who are not ready to accept the treatment offers until they come humbly on their knees begging for help.

In many countries, prisons have no adequate infrastructure and resources to offer treatment to substance dependent inmates, or the available treatments are restricted to persons who discontinue their habit (Moeller et al. 2007). This is especially problematic, because the risks for mortality and morbidity is elevated for drug users in the prison milieu.

The ethical implication is obvious: without appropriate accessibility and affordability of treatment, there is no way to reach the 'Health for All' goal of good coverage and of treating all patients as equal. The multi-country study on 'Adequacy of treatment and care to the needs and quality standards' described the expectations on behalf of the European Commission and the World Health Organisation's Regional Office for Europe (Uchtenhagen and Guggenbühl 2000).

23.2.4.3 Quasi-compulsory and Compulsory Treatment

Is it advisable to force addicts into treatment, when they are not motivated to engage in it by themselves, in view of an objective of optimal coverage?

It is a common understanding that addicts do not enter treatment unless there is some external or internal force behind such a decision. The forces are quite different, ranging from health concerns to social pressure by family or employer, to legal pressure in order to avoid losing the drivers licence or going to prison. This is called the 'continuum of coercion' (Weisner 1990). The scientific evidence on the effectiveness of such 'coercion' is contradictory (Gerdner 1998, Wild et al. 2006). A recent multi-country study on quasi-compulsory treatment (QCT, treatment on court order as an alterative to imprisonment, with the consent of the patient) documented that perceived pressure does not translate into higher motivation; no significant differences in outcome were found between QCT patients and control groups of voluntary patients (Stevens et al. 2006b).

Coercion through mutually agreed consequences of substance use during treatment (contingency management) is recognised to be helpful, but more so when using positive reinforcers than sanctions (Miller and Flaherty 2000, Schumacher et al. 2007). From an ethical standpoint, it is reluctantly accepted under the term of 'Ulysses coercion' (Odysseus wanted to be bound to the mast of his ship in order to resist the temptations of the sirens, Tännsjö 1999).

The findings indicate that coercion has a role in supporting patient motivation for change, on the basis of informed consent, but it cannot replace motivation through compulsion without consent. Compulsory psychiatric hospitalisation can prevent suicide and can treat acute psychiatric comorbidity or enforce a temporary moratorium, but is no longer accepted as a long-term treatment of addictive behaviour.

23.2.5 The Case of Agonist Maintenance Treatment

23.2.5.1 Rationale and Origins

At first view, prescribing agonists with dependence liability to persons suffering from substance dependence seems to be paradox. One of the main arguments against agonist prescribing refers to a prolongation of dependence, considered to be unethical. How can we understand the rise and successes of this therapeutic approach (known under the terms substitution or replacement treatment)?

The main objective is to replace uncontrolled use of an addictive substance by the controlled provision of a medication acting on the same receptors as the original substance. The secondary objectives to be reached are: reduction of the uncontrolled substance use, reduction of adverse health and social effects of uncontrolled use (including a reduction in drug-related delinquency), normalization of life-style (including drug-free social contacts, improved housing and employment conditions), These objectives are in line with the hierarchy outlined above (see Section 23.2.2.1). The primary objective is reached by prescribing and controlled intake of an agonist, the secondary objectives by eliminating the need to purchase the original substance of dependence and by ancillary care.

The introduction of agonist maintenance treatment is closely linked to the experience of unsuccessful detoxification and abstinence-based treatment. Maintenance without substitution was practiced in Roman times (the emperor Marc Aurel was maintained on opium by Galenus, the eminent physician), daily dosages of opium were provided to dependent persons in South-East Asia in the nineteenth century (Westermeyer 1982); ironically, after the ban of opium these persons switched to heroin (Westermeyer 2006). First attempts at substitution were to replace opium by alcohol in the sixteenth century (Elliott 1920), morphine by heroin (Kramer 1977) and morphine by cocaine (Freud 1885) in the nineteenth century. These attempts were far from being successful. A first well-designed and scientifically based model was developed by Dole and Nywander: the methadone maintenance scheme (Dole and Nyswander 1965). They used oral methadone, a full opiate agonist, with a longer half-life than injected heroin (controlled intake of one dose per day can block the heroin craving effectively). The model included ancillary care for medical and social conditions of enrolled patients.

The ethical justification of agonist maintenance treatment in the modern form was to provide otherwise treatment-resistant heroin addicts with an effective approach to improve their health and social situation without asking for total abstinence from narcotic substances.

Apart from opiate replacement, only nicotine replacement has been introduced into present medical practice, while former alcohol maintenance for alcoholics is discontinued.

23.2.5.2 Results and Availability

Methadone maintenance is one of the most frequently and best researched therapeutic approaches. Extensive reviews of the pharmacological aspects, of service delivery and of therapeutic outcomes have been published (Arif and Westermeyer 1990, Farrell et al. 2004, Health Canada 2008, Uchtenhagen 2003, Ward et al. 1998). In addition to Methadone, Buprenorphine and retarded morphine have been introduced and researched as replacement medicines in the treatment of opioid dependence, with similar outcomes. The positive findings in terms of health and social improvements, including a massive reduction of delinquency, were complemented by the fact that agonist maintenance treatment has the greatest potential to bring heroin dependent persons into contact with therapy and care. It therefore became one of the most welcome instruments for limiting the spread of blood borne diseases – HIV/Aids and Hepatitis – among drug injectors. The reduction of risky injecting behavior and of seroconversion rates in methadone patients became evidenced, for community based programs (Metzger et al. 1993) and for prison based programs (Stallwitz and Stoever 2007).

Economic evaluation documented the cost-effectiveness of methadone and buprenorphine maintenance treatment (Connock et al. 2007) and also the cost-utility ratio of heroin maintenance treatment (Dijkgraf et al. 2005).

Based on the evidence, Methadone and Buprenorphine have been scheduled by World Health Organisation as essential medicines (WHO 2004a, b). Also, international evidence based scientific guidelines for agonist maintenance treatments have been published (WHO 2008). These are the most visible indicators for an almost universal acceptance of agonist maintenance treatment, with the exception of a few countries (e.g. Russian Federation).

By 2006, 24 Member States of the European Union had introduced agonist maintenance treatments, with an estimated number of half a million patients enrolled (EMCDDA 2007). However, considerable differences exist in the availability between countries, the estimated coverage varying between 5 and 83% in the community, between 0 and 55% in prisons (Stoever et al. 2004).

In a Public Health perspective, agonist maintenance has a unique role in the treatment of opioid dependence, due to its higher attractiveness, better retention and higher reduction of heroin use in comparison to non-pharmacological treatments (Cochrane review, Mattick et al. 2003).

23.2.5.3 Concerns and Limitations

A major concern was a weakening of the motivation to change and therefore a prolongation of addictive behavior through agonist maintenance treatment. There is no evidence for this claim. A multi-site major cohort study showed a relapse rate of methadone maintenance patients to daily opiate use of only 27% at 12 year follow-up (Marsh et al. 1990), The DATOS study from USA and the NTORS study of the UK, both multi-site prospective cohort studies, found significant reductions of heroin use and injecting at 5 year follow-up in patients who were enrolled in methadone maintenance treatment (Gossop et al. 2003, Hubbard et al. 2003).

Other concerns focused on a diversion of prescribed substances into the illegal drug market (which is effectively prevented through an adequate take-out policy), on methadone clinics attracting heroin addicts as a form of tourism, and a negative impact on neighborhoods (prevented effectively by adequate regulations and management).

On the other side, limitations of the therapeutic effectiveness come from patients who continue to inject (an effective answer to methadone-resistant addicts is the provision of heroin-assisted treatment (Uchtenhagen 2008)), or from patients with polydrug dependence (which eventually responds to additional behavioral therapy).

23.3 Implications and Recommendations: A Final Statement

The essential ethical message is well summarized in the WHO international guidelines for the psychosocially assisted pharmacological treatment of opioid dependence (WHO 2008):

> When making clinical decisions for the treatment of people with opioid dependence, ethical principles should be considered, together with evidence from clinical trials; the human rights of opioid dependent individuals should always be respected. Treatment decisions should be based on standard principles of medical care ethics – providing equitable access to treatment and psychosocial support that best meets the needs of the individual patient. Treatment should respect and validate the autonomy of the individual, with patients being fully informed about the risks and benefits of treatment choices. Furthermore, programmes should create supportive environments and relationships to facilitate treatment, provide coordinated treatment of comorbid mental and physical disorders, and address relevant psychosocial factors.

There is not much to be added. The statement applies fully as the main recommendation for all treatments of substance dependence. It is a perpetual agenda for future generations.

References

AMA (2001) Principles of medical ethics. In: Code of medical ethics. American Medical Association, Chicago

Arif A, Westermeyer J (1990) Methadone maintenance in the management of opioid dependence. An international review. Praeger, New York, NY

Bergmark A (2008) Specific and contextual treatment mechanisms. Nordic Stud Alcohol Drugs 25:277–285

Broadhead RS, Kerr TH, Grund JPC, Altice FL (2002) Safer injection facilities in North America: their place in public policy and health initiatives. J Drug Issues 32:329–356

Connock M, Juarez-Garcia A, Jowett S, Frew E, Liu Z, Taylor RJ, Fry-Smith A, Day E, Lintzeris N, Roberts T, Burls A, Taylor RS (2007) Methadone and buprenorphine for the management of opioid dependence: a systematic review and economic evaluation. Health Technol Assess 11:1–171

Council of Europe (1950). Convention for the protection of human rights and dignity of the human being with regard to the application of biology and medicine: convention on human rights and biomedicine

Czuchry M, Dansereau DF (2005) Using motivational activities to facilitate treatment involvement and reduce risk. J Psychoactive Drugs 37:7–13

Department of Health (2002) Models of care for substance misuse treatment. Promoting quality, efficiency and effectiveness in drug misuse treatment services. Department of Health, London

Dijkgraf MGW, van der Zanden BP, de Borgie AJM, Blanken P, van Ree JM, van den Brink W (2005) Cost utility analysis of co-prescribed heroin compared with methadone maintenance treatment in heroin addicts in two randomised trials. Br Med J 330:1297–1303

Dole VP, Nyswander ME (1965) A medical treatment for diacetyl-morphine (heroin) addiction. JAMA 193:646–650

EC (2008). Communication from the commission to the European parliament and council on an EU drugs action plan for 2009–2012. Commission of the European Communities COM(2008) 567/4, Brussels

Edlin BR, Keal KH, Lorwock J, Kral AH, Ciccarone DH, Moore LD, Lo B (2001) Is it justifiable to withhold treatment for hepatitis C from illicit drug users? New Engl J Med 345:211–215

Edwards G, Arif A (1980). Dug problems in a socio-cultural context. A basis for policy and programme planning. Public Health Papers no. 73, World Health Organisation, Geneva

Ekendahl M (2007) Will and Skill – an exploratory study of substance abuser's attitudes towards life-style changes. Eur Addict Res 13:148–155

Elliot HM (1920) Memoirs of Jahangir. Islamic Book Service, Lahore

EU (2005). EU drug action plan. Amtsblatt der Europäischen Union 2005/C168/01, Brussels

EU (2008) Health programme 2008–2013. European Commission, DG Health and Consumer Protection, Brussels

EMCDDA (2007) EU drug action plan, progress report 2006. European Monitoring Centre on Drugs and Drug Addiction, Lisbon

Farrell M, Ward J, Mattick R, Hall W, Stimson GV, des Jarlais D, Gossop M, Strang J (2004) Methadone maintenance treatment in opiate dependence: a review. Br Med J 309:997–1001

Fireman M, Rabkin JM (2001) Outcome of liver transplantation in patients with alcohol and other chemical dependence. Psychosomatics 42:172–173

Freud S (1885) Über die Allgemeinwirkung des Cocaïnes. Medizinisch-chirrgischer Centralblatt 32:374–375

Gerdner A (1998). Patient and programme factors with impact on outcome in Swedish compulsory care of addicts: a systematic review. Lund University

Gerstein DR, Datta AR, Ingels JS, Johnson RA, Rasinski KA, Schildhaus S, Talley K (1997). NTIES. The National Treatment Improvement Evaluation Study. Final Report. National Opinion Research Center, Chicago

Gossop M, Marsden J, Stewart D, Kidd T (2003) The national treatment outcome research study (NTORS): 4–5 year follow-up results. Addiction 98:291–303

Gossop M (2006) Treating drug misuse problems: evidence of effectiveness. National Treatment Agency for Substance Misuse, London

Gossop M, Stewart D, Marsden J (2007) Readiness for change and drug use outcomes after treatment. Addiction 102:301–308

Gruber J, Koszegi B (2001) Is addiction rational? Theory and evidence. Q J Econ 116:1261–1303

Harding G (2008) Constructing addiction as a moral failing. Sociol Health Illn 1:75–85

Hare R (1991) The philosophical basis of psychiatric ethics. In: Bloch S, Chodoff P (eds) Psychiatric ethics. Oxford University Press, Oxford

Health Canada (2008) Literature review – methadone maintenance treatment. Public Works and Government Services Canada, Ottawa

Hedrich D (2004) European report on drug consumption rooms. European Monitoring Centre on Drugs and Drug Addiction, Lisbon

Hiller ML, Knight K, Leukefeld C, Simpson DD (2002) Motivation as a predictor of therapeutic engagement in mandated residential substance abuse treatment. Crim Justice Behav 29:56–75

Hubbard RL, Craddock SG, Anderson J (2003) Overview of 5-year follow-up outcomes in the drug abuse treatment outcome studies (DATOS). J Subst Abuse Treat 25:126–134. http://www.ncbi.nlm.nih.gov/entrez/utils/fref.fcgi?PrId=3048&itool=AbstractPlus-def&uid=14670518&db=pubmed&url=http://linkinghub.elsevier.com/retrieve/pii/S0740547203001302

International Union for Health Promotion and Education (2000) The evidence of health promotion effectiveness: shaping public health in a new Europe. Part two, evidence book. European Commission and IUHPE, Brussels

Jellinek EM (1960) The disease concept of alcoholism. Hillhouse, New Haven, CT

Jonas H (1984) The imperative of responsibility. In search of ethics for the technological age. University of Chicago Press, Chicago

Kaya CY, Tugai Y, Filar JA, Agrawal MR, Ali RL, Gowing LR, Cooke R (2004) Heroin users in Australia: population trends. Drug Alcohol Rev 23:107–116

Klingemann H, Sobell LC (2007) Promoting self-change from addictive behaviors. Practical implications for policy, prevention and treatment. Springer, New York, NY

Kramer JC (1977) Heroin in the treatment of morphine addiction. J Psychedelic Drugs 197:193–197

Leshner A (1997) Addiction is a brain disease, and it matters. Science 278:45–47

Levine HG (1981) Mässigkeitsbewegung und Prohibition in den USA. In: Völger G (Hrsg) Rausch und Realität. Drogen im Kulturvergleich. Rautenstrauch-Joest-Museum, Köln, pp 126–131

Longabaugh RH, Wirtz PW (eds) (2001) Project MATCH hypotheses. Results and causal chain analyses. NIAAA project MATCH monograph series, vol 8. National Institute on Alcohol and Alcohol Abuse, Rockville, MD

Loose R (2005) The subject of addiction. Psychoanalysis and the Administration of Enjoyment, Sage, London

Marsh KL, Joe GW, Simpson DD, Lehmann WEK (1990) Treatment history. In: Simpson DD, Sells SB (eds) Opioid addiction and treatment: a 12-year follow-up. Krieger, Malabar, pp 137–156

Mattick RP, Breen C, Kimber J, Davoli M, Breen R (2003) Methadone maintenance therapy versus no opioid replacement therapy for opioid dependence. Cochrane Database Syst Rev (2):CD002209

McLellan AT, Grissom GR, Zanis D, Randall M, Brill P, O'Brien CP (1997) Problem-service 'matching' in addiction treatment. A prospective study in 4 programs. Arch Gen Psychiatry 54:730–735

Merkx MJ, Schippers GM, Koeter MW, Oudejans S, Vujik PJ, Vries de CQ et al (2007) Allocation of substance use disorder patients to appropriate levels of care: feasibility of matching guidelines in routine practice in Dutch treatment centres. Addiction 102:466–474

Metzger DS, Woody GE, McLellan AT (1993) Human immunodeficiency virus seroconversion among intravenous drug users in- and out-of-treatment: an 18-month prospective follow-up. J Acquir Immune Defic Syndr 6:1049–1056

Mill JS (1861) In: Robson JM (ed) Utilitarianism, Chapter 4. Collected works, vol 10, pp 203–259. Toronto University Press, Toronto, p 1963ff

Miller N, Flaherty JA (2000) Effectiveness of coerced addiction treatment (alternative consequences) a review of the clinical research. J Subst Abuse Treat 18:9–16

Miller WR, Rollnick S (2002) Motivational interviewing: preparing people for change. The Guilford Press, New York, NY

Moeller L, Stoever H, Juergens R, Gatherer A, Nikogosian H (eds) (2007) Health in prisons. A WHO guide to the essentials in prison health. World Health Organisation, Regional Office for Europe, Copenhagen

Moyer A, Finney JW, Swearingen CE, Vergun P (2002) Brief interventions for alcohol problems: a meta-analytic review of controlled investigations into treatment-seeking and non-treatment-seeking populations. Addiction 97:279–292

NIDA (1999) Principles of drug addiction treatment. A research-based guide. National Institute of Drug Abuse, Bethesda, MD

Parker T (1940) Transcendentalism. In: Muelder WG, Sears L (eds) The development of American philosophy. Houghton Mifflin, Boston, MA, pp 130–139

Prochaska JO, DiClemente CC (1982) Transtheoretical therapy: toward a more integrative model of change. Psychother: Theory Res Pract 19:276–288

Robins LN (1993) Vietnam veterans rapid recovery from heroin addiction – a fluke or normal expectation. Addiction 88:1041–1054

Rollnick S, Miller WR (1995) What is motivational interviewing? Behav Cogn Psychother 23: 325–334

Room R (1980) Dependence and society. Br J Addiction 80:133–139

Room R, Janca A, Bennett L, Schmidt L, Sartorius N (1996) WHO cross-cultural applicability research on diagnosis and assessment of substance use disorders. Addiction 91:199–230

Rush B (1772) Sermons to gentleman. John Dunlap, Philadelphia, PA

Rust A (2000) Ethische Aspekte. In: Uchtenhagen A, Zieglgänsberger W (Hrsg) Suchtmedizin. Konzepte, Strategien und therapeutisches Management. Urban & Fischer, München, pp 573–584

Schippers GM, Schramade M, Walburg JA (2002) Reforming Dutch substance abuse treatment services. Addict Behav 27:995–1007

Schivelbusch W (1980) Das Paradies, der Geschmack und die Vernunft. Eine Geschichte der Genussmittel. Hanser, München

Schumacher JE, Milby JB, Wallace D, Meehan DC, Kertesz S, Vuchinich R, Dunning J, Usdan S (2007) Meta-analysis of day treatment and contingency-management dismantling research: Birmingham homeless cocaine studies (1990–2006). J Consult Clin Psychol 75:823–828

Single E, Collins D, Easton B, Harwood H, Lapsley H, Maynard A (1995) Proposed international guidelines for estimating the costs of substance abuse. Canadian Centre on Substance Abuse, Toronto, ON

Stallwitz A, Stoever H (2007) The impact of substitution treatment in prisons – a literature review. Int J Drug Policy 18:464–474

Stevens A, Hallam C, Trace M (2006a) Treatment for dependent drug use. A guide for policymakers. Beckley Foundation, London

Stevens A, Berto D, Frick U, Hunt N, Kerschl V, McSweeney T, Oeuvrey K, Puppo I, Santa Maria A, Schaaf S, Trinkl B, Uchtenhagen A, Werdenich W (2006b) The relationship between legal status, perceived pressure and motivation in treatment for drug dependence: results from a European study of quasi-compulsory treatment. Eur Addict Res 12:197–209

Stoever H, Hennebel JC, Casselmann J (2004) Substitution treatment in European prisons. A study in policies and practices of substitution in prisons in 18 European countries. Cranstoun Drug Services Publishing, London

Tännsjö T (1999) Coercive care. The ethics of choice in health and medicine. Routledge, London

Uchtenhagen A, Guggenbühl L (2000). Adequacy in drug abuse treatment and care in Europe (ADAT). Commissioned by World Health Organisation, Regional Office for Europe. Research Institute for Public Health and Addiction, Zurich

Uchtenhagen A (2003). Substitution management in opioid dependence. J Neurol Trans (Supplememt 66): Fleischhacker WW, Brooks DJ (eds) Addictions, mechanisms, phenomenology and treatment. pp 33–60

Uchtenhagen A (2004) Substance use problems in developing countries. Editorial, Bull World Health Org 82(9):641

Uchtenhagen A, Ladjevic T, Rehm J (2005). Guidelines for psychosocially assisted pharmacological treatment of persons dependent on opioids. Working paper for World Health Organisation. Research Institute for Public Health and Addiction, Zurich

Uchtenhagen A (2008) Heroin assisted treatment in Europe: a safe and effective approach. In: Stevens A (ed) Crossing frontiers. International developments in the treatment of drug dependence. Pavilion, Brighton, pp 53–82

UN (1948) The universal declaration of human rights. United Nations, New York, NY

UNODC (2008) Reducing the adverse health and social consequences of drug abuse: a comprehensive approach. United Nations Office on Drugs and Crime, Vienna

Volkow ND, Fowler JS, Wang GJ (2003) The addicted human brain: insights from imaging studies. J Clin Invest 111:1444–1451

Warner J (2007) Why abstinence matters to Americans. Editorial, Addiction 102:502–505

Weisner C (1990) Coercion in alcohol treatment. In: Institute of Medicine (ed) Broadening the base for the treatment of alcohol problems. The National Academic Press, Washington, DC, pp 579–610

Ward J, Mattick R, Hall W (1998) Methadone maintenance treatment an other opioid replacement therapies. Harwood, Amsterdam

West R (2006) Theory of addiction. Blackwell, Oxford

Westermeyer J (1982) Poppies, pipes and people: opium and it's use in Laos. University of California Press, Berkeley

Westermeyer J (2006) The switch from opium to heroin smoking. Addiction 92:686–687

WHO (1986) The Ottawa charter for health promotion. World Health Organisation, Geneva

WHO (2001) The World Health Report 2001 – mental health, new understanding, new hope. World Health Organisation, Geneva

WHO (2004a) Proposal for the inclusion of methadone in the WHO model list of essential medicines. World Health Organisation, Geneva

WHO (2004b) Proposal for the inclusion of Buprenorphine in the WHO model list of essential medicines. World Health Organisation, Geneva

WHO (2008) Guidelines for the psychosocially supported pharmacological treatment of opioid dependence. World Health Organisation, Geneva

Wild DC, Cunningham JA, Ryan RM (2006) Social pressure, coercion, and client engagement at treatment entry: a self-determination theory perspective. Addict Behav 31:1858–1872

Zinberg NE (1984) Drug, set and setting. Yale University Press, New Haven, CT

Chapter 24
Dementia and End-of-Life Decisions: Ethical Issues – A Perspective from The Netherlands

Ron L.P. Berghmans

Contents

Abbreviations

ANH	artificial nutrition and hydration
KNMG	RDMA
PAS	physician-assisted suicide
RDMA	Royal Dutch Medical Association
WGBO	Wet op de Geneeskundige Behandelingsovereenkomst (Law on the Medical Treatment Contract)

R.L.P. Berghmans (✉)
Department of Health, Ethics and Society, Faculty of Health, Medicine and Life Sciences,
Maastricht University, 6200 MD Maastricht, The Netherlands
e-mail: r.berghmans@hes.unimaas.nl

H. Helmchen, N. Sartorius (eds.), *Ethics in Psychiatry*, International Library
of Ethics, Law, and the New Medicine 45, DOI 10.1007/978-90-481-8721-8_24,
© Springer Science+Business Media B.V. 2010

24.1 Introduction

Alzheimer's disease, the most frequent form of dementia, is a progressive and fatal neurodegenerative disorder which is manifested by cognitive and memory deterioration, progressive impairment of activities of daily living, and a variety of neuropsychiatric symptoms and behavioral disturbances (Cummings 2004). Dementia is primarily a disease of the very elderly. Nearly 1% of 65 year olds suffer from dementia. This figure rises to around 40% in people aged 90 and over. The number of very old people in the Netherlands will increase substantially over the coming decades. If the prevalence figures do not change and curative treatments fail to emerge, the number of dementia sufferers, which in 2002 was around 175,000, will have risen to approximately 207,000 by 2010. By 2050 there are expected to be 412,000 dementia patients in this country. In 2000 it was estimated that 1 in every 93 people in the Netherlands had dementia; in 2010 the prevalence will stand at 1 in 81 and in 2050 it will be 1 in 44. In 2000, 35% of the people suffering from dementia in the Netherlands were residing in a nursing home or in a home for the aged (Rurup et al. 2006).

Dementia has far-reaching consequences for patients and their carers. The process of deterioration is slow and not seldomly years pass before the diagnosis is made. As the dementia process progresses, the person's independent functioning gives way to dependence, and participation in social activities becomes more and more difficult, which puts pressure on family relationships and friendships. Slowly but surely, the patient loses all contact with the here and now and also the ability to recognize things, situations and people, even those who have been closest to him, finally gets lost (Health Council of the Netherlands 2002).

No treatments as yet exist which can cure Alzheimer's disease and other dementias, and it is precisely for this reason that symptom control and interventions that are capable of maintaining the spiritual well-being of the patient and carer at an acceptable level are so relevant. Although the achievement of an integrated diagnostic and therapeutic approach to this complex and devastating disorder seems remote (Selkoe 2001), different medical-therapeutic approaches – including neural (embryonic) stem cell therapy (Feng et al. 2009) and deep brain stimulation (Hamani et al. 2008) - by some are considered to be promising as possible strategies with clinical therapeutic potential. Presently, psychosocial approaches aimed at the improvement of the well-being of dementia patients are gaining more interest and research attention (Kitwood 1997, Kitwood and Bredin 1992).

24.2 Dementia, End of Life, and Ethics

The ethics of dementia care differs culturally (Sowmini and de Vries 2008). Not only what is considered to be an ethical issue or dilemma, but also the moral values and norms which are viewed as appropriate may vary between cultures. How people with dementia are cared for, whether dementia is considered to be a problem, and the role of medicine in this context; these are all issues which are related to cultural values, opinions and rituals (Whitehouse 2008). Awareness of (differences with regard to) cultural values implies that we cannot differentiate 'right' practices from 'wrong' practices as long as we do not know or recognize differences in cultural values and moral points of view. It does, however, mean that we can learn from other cultural contexts, without simply trying to replicate other practices.[1]

This chapter is written against the background of societal opinions, moral values and practices regarding dementia care in the Netherlands. The focus will be on moral issues in the context of end-of-life practices and decision-making with regard to patients who suffer from dementia. The range of medical decisions concerning the end-of-life in dementia varies from refraining from treatment (non-treatment decision) to decisions to actively terminate the life of the patient (i.e. physician assisted suicide and euthanasia). It is interesting to note that recent empirical research shows that cross-culturally differences exist between the way in which treatment decisions, particularly regarding demented nursing home patients who develop pneumonia or other end-of-life decisions are made (Helton et al. 2006, Slaets 2007, Van der Steen et al. 2004, 2007).

24.3 Respecting the Individual

When dealing with dementia patients, the starting point always is respect for the individual person, for his individuality and idiosyncrasies (Health Council of the Netherlands 2002). A (probable) diagnosis of dementia does not imply that the patient is suddenly bereft of wishes and desires, or that he should be considered incapable of speaking for himself. While it is still possible, it is important to take the time *with* him, not just *about* him. Furthermore, his wishes with regard to (medical) treatment and care should always be taken into account.

This applies to all who have dealings with dementia patients: professional carers, non-professional carers, family members and others in the patient's social circle. For information about the patient's state of health and well-being, physicians and other professional carers become increasingly dependent on the patient's relatives and friends. Thus it is essential that professionals make active efforts to achieve a relationship, based on mutual trust, with the patient's main informal carer or contact. This is particularly important when medical decisions concerning the end of life of the patient need to be made. In the following, the different decisions are discussed.

---

[1] Dutch euthanasia practice may be taken as an example. This practice has developed during a period of about 40 years within the specific Dutch societal and health care context (Griffiths et al. 1998). Trying to replicate or import this practice in other societies would ignore cultural specifics and historical backgrounds.

24.4 End of Life Decisions

Different end of life decisions can be distinguished (Van der Heide et al. 2007). Each raise different moral and legal issues, and some may even be illegal and/or considered to be immoral in particular countries and jurisdictions (i.e. assisted suicide and euthanasia, and/or ending life without the explicit request of the patient).

The following can be distinguished:

- refusal of treatment by the patient or his/her legal representative
- not starting or forgoing treatment because of medical futility
- intensified alleviation of symptoms
- palliative sedation
- assisted suicide
- euthanasia (based on the explicit request of the patient)[2]
- ending life without explicit request by the patient

24.4.1 A Philosophical Framework for End-of-Life Decisions

To evaluate and assess the moral quality of end of life decisions in dementia, it is important to have a philosophical and moral framework. What is an adequate philosophical and moral framework for end of life decision making in cases of dementia? In the Netherlands, the KNMG or Royal Dutch Medical Association (RDMA) has developed a helpful moral framework for medical decision making in case of the end of life of mentally incompetent patients (Berghmans 1999, KNMG 1997). One of the basic issues in the approach of the RDMA concerns medical legitimation. 'Medical power' is considered as a synonym of technical possibilities. The question of medical legitimation is meant to test medical possibilities in the light of the goals of medicine. The goal of medicine is not merely to produce effects on human bodies, but to help patients (Jecker 1995).

From this perspective on the philosophy of medicine, not only the decision not to treat (anymore) needs justification, but also the decision to (continue to) treat(ment),

---

[2]In the Netherlands, euthanasia is understood to mean: active termination of life at the request of a patient. Euthanasia in this sense is still a criminal offence, but the Criminal Code has been amended to exempt physicians from criminal liability if they report their actions and show that they have satisfied the due care criteria formulated in the Euthanasia Act. The actions of doctors in such cases are assessed by regional review committees (appointed by the Minister of Justice and the Minister of Health, Welfare and Sport), which focus in particular on the medical and decision-making procedures followed by the doctor. Where a doctor has reported a case and a regional review committee has decided on the basis of his report that he has acted with due care, the Public Prosecution Service will not be informed and no further action will be taken. But where a review committee finds that a doctor has failed to satisfy the statutory due care criteria, the case will be notified to the Public Prosecution Service and the Health Inspectorate. These two bodies will then consider whether the doctor should be prosecuted.

as there are limits to the mere technological possibilities of medicine to preserve life, where this may not be beneficial to the patient.

Two questions are of central importance in assessing the moral acceptability of life-shortening treatment for demented patients:

1. under what circumstances is life-prolonging treatment no longer legitimate? And
2. what is the relevance of the remaining capacity of the patient to participate in the decision-making, and how can the wishes of the patient be ascertained? (Berghmans 1999, Griffiths et al. 1998).

Below, the different medical decisions concerning the end of life of patients suffering from dementia will be further explored and discussed.

24.4.2 Non-Treatment Decisions

Optimising well-being and quality of life are central pillars of medical management in dementia. In the later phases of dementia, policy focuses on pain reduction, limitation of comorbidity (cardiovascular disease and diabetes), treatment of intercurrent disorders (such as bladder infection, pneumonia) and the reduction of complications and symptoms that are associated with dementia (Holman and Brendel 2006). While this approach is not primarily aimed at extending the patient's life, this is nevertheless the usual result of such treatment. There may come a time, however, when extending the patient's life is considered to be pointless and/or undesirable, either from the physician's professional point of view, or the patient's perspective (Health Council of the Netherlands 2002, KNMG 1997).

As persons with dementia develop advanced disease, neurological and functional deterioration predispose them to lower respiratory tract infection including pneumonia, which is often their immediate cause of death. For these patients, the benefit of aggressive care, such as invasive procedures, has been increasingly questioned in the face of high mortality, due to side-effects of treatment (Van der Steen 2002, Van der Steen et al. 2004). Also it must be questioned whether further exposure to the irreversible deteriorating course of dementia is desirable from the patient's perspective (Hertogh and Ribbe 1996, KNMG 1997, Van der Steen et al. 2004).

More specifically, two important types of non-treatment decisions in dementia care can be distinguished: (forgoing) artificial nutrition and hydration (Kruit et al. 1999, Onwuteaka-Philipsen et al. 2001, Pasman et al. 2004, The et al. 2002) and the decision not to treat pneumonia (Helton et al. 2006, Van der Steen et al. 2000a, b, Van der Steen et al. 2001, 2002a, b, 2004, 2007, 2002, 2009, Slaets 2007). Below, the focus is on the decision not to (continue to) give artificial feeding to dementia patients.

24.4.2.1 Forgoing Artificial Nutrition and Hydration

In the care for the elderly in general, and in the care for the elderly demented, whether to start artificial nutrition and hydration (ANH) is a complex matter. Food and fluids are one of the primary necessities of life, and although the consensus in palliative medicine is that withholding ANH causes little discomfort in dying patients, forgoing ANH especially appeals to the imagination of the family (Pasman et al. 2004). Recent empirical research shows that foregoing ANH in patients with severe dementia who scarcely or no longer eat or drink seems, in general, not to be associated with high levels of discomfort. Nonetheless, individual differences emphasize the need for constant attention to possible discomfort (Pasman et al. 2005).

The moral justification of treatment involves ultimately an assessment of the possible benefits, burdens and risks of treatment, as compared to non-treatment.

It is obvious that after a decision not to start or to stop treatment which may prolong life if necessary the provision of adequate symptomatic palliative treatment and care deserves priority attention.

24.4.2.2 Refusal of Treatment by Patient or Representative

In conformity with the Law on the Medical Treatment Contract,[3] a *competent* patient has the right to refuse treatment which is unrestricted. In case of *incompetence*, the representative of the patient has decisional authority with regard to treatment decisions. The treating physician has the duty to respect the decision of the representative, unless this is incompatible with what is required of a competent care provider who acts on the basis of the standards of the profession (Gevers 2006). The medical best interest of the incompetent patient, as perceived by the physician, is guiding.

If the incompetent patient objects to a procedure of a radical nature for which the legal representative has given permission, the procedure may be carried out only if it is manifestly necessary to prevent serious harm to the patient's health. Involvement of a judge or court in such cases is not a requirement. Below, in the paragraph on advance directives, the right to refuse treatment will be further discussed.

24.4.2.3 Medically Futile Treatment

Treatment may be futile. *Medical futility* can be defined as a clinical action serving no useful purpose in attaining a specified goal for a given patient (Jecker 1995, Kasman 2004). In order to answer the question *when* a specific treatment can be considered as 'medically futile', different criteria may be applied (Leenen et al. 2007):

– The action does not contribute to the solution of the medical problem or to the preservation or improvement of the medical condition of the patient, or will not reasonably result from this. Preservation of the medical condition generally will

---

[3] Act of 17 November 1994 amending the Civil Code and other legislation in connection with the incorporation of provisions concerning the contract to provide medical treatment (http://www.healthlaw.nl/wgboeng.html).

happen in view of future improvement or in order to allow the patient to continue living in the same medical condition.

– The means to be used are not reasonable in view of the goal(s) to be reached. So-called instrumental proportionality is absent if a comparatively too heavy medical procedure needs to take place in order to reach a comparatively limited benefit. Also the causation of pain should be taken into account in assessing the instrumental proportionality.

– A particular minimum level cannot be reached anymore. Because of the character of the illness or handicap, the patient is very seriously affected or disabled and medical action cannot contribute to the improvement of that situation. Or there may be a tragic situation as a result of a cumulation of medical conditions and disabilities which with an incomparably heavy effort each separately may be treatable but jointly and in totality not or almost not.

What exactly constitutes 'medical treatment' – as distinguished from basic care – in the case of feeding has been subject of debate. In the Netherlands, it is now clear that artificial administration of food and drink is 'medical treatment' that can be terminated (Griffiths et al. 1998, Leenen et al. 2007).

An important question here is whether in assessing medical futile treatment in individual cases the *quality of life* of the patient may play a role (Harris 1987, Leenen et al. 2007). It is agreed that in case an evaluation of the proportionality of a possible treatment takes place, then quality of life considerations may – and often necessarily do - play a role. The same applies to the third criterion mentioned above: a particular minimum level which cannot be reached anymore. What constitutes such a minimum level is not an 'objective' assessment, but requires an evaluation which is at least partly based on quality of life considerations.

24.4.3 Intensified Alleviation of Symptoms

The intensified alleviation of symptoms (such as pain or shortness of breath) is normal medical practice at the end of life. This may be done while taking into account that probably or certainly the life of the patient will be shortened, or (partly) with the intention to shorten the life of the patient. In the first case, the life-shortening effect is a justifiable 'byproduct' of the policy to intensify for instance the use of opioids in increasing dosages. In the case this is done without a medical need but with the intention to shorten the life of the patient, then this cannot be considered normal and accepted medical practice, but as a form of active life termination 'in disguise' which is unjustifiable (Blijham 1996).

24.4.4 Palliative (or Terminal) Sedation

Palliative sedation (sometimes also referred to as 'terminal sedation' or 'deep sedation') is a form of medical management in which a patient in the final stage of his life receives care to combat unbearable suffering (Gevers 2003, Legemaate et al. 2007, Rietjens et al. 2008, Verkerk et al. 2007). Palliative sedation is - in the wording of a Dutch national guideline – 'the intentional lowering of consciousness of a patient in the last phase of his or her life.' The objective of palliative sedation is to relief suffering (caused by refractory, i.e. untreatable symptoms), and lowering consciousness is a means to achieve this (Verkerk et al. 2007).

Because of debate about the (proper) demarcation between palliative sedation and (forms of) active life termination, in the Netherlands the medical profession has formulated a guideline in 2005, which has been updated in 2009 (Hasselaar et al. 2009, Legemaate et al. 2007, Van Delden 2007, Van Wijlick et al. 2009, Verkerk et al. 2007). As in other international guidelines, the indications for palliative sedation is formulated as follows: the presence of one or more refractory symptoms, which leads to unbearable suffering of the patient. A symptom is or becomes refractory if none of the conventional treatments are effective (within a reasonable time frame) and/or these treatments are accompanied by unacceptable side effects. Pain, dyspnea, and delirium are the most common refractory symptoms that in clinical practice lead to the use of palliative sedation (Verkerk et al. 2007). Another condition – which only applies to deep and continuous sedation until death and not to intermittent sedation – is that death must be expected within 1–2 weeks. The assumption is made that if deep and continuous sedation is administered (and the patient is, therefore, unable to take fluids him- or herself), artificial hydration will not contribute to the relief of suffering and may, in fact, have disadvantages, such as the need for a subcutaneous or intravenous cannula and the possible increase of some symptoms and signs. If life expectancy is less than 2 weeks, it is assumed that withdrawing artificial hydration will not hasten death. There is no evidence that proportionally administered palliative sedation shortens life and – at least in the Netherlands – it is stressed more and more often that palliative sedation is normal medical practice (Legemaate et al. 2007). If life expectancy is longer, then the situation is different because in that case the patient would die sooner due to dehydration than would otherwise be the case (Verkerk et al. 2007).

24.4.5 Active Termination of Life on the Request of the Patient

In exceptional circumstances, in the Netherlands it is considered morally and legally justified if a physician actively terminates or helps to terminate the life of a patient on his or her explicit request. This concerns the practice of euthanasia (in Dutch law and practice this is defined as the termination of the life of the patient on his or her explicit, voluntary and enduring request) and of physician-assisted suicide.

A case: dementia and assisted suicide
In the first month of 2004 it became known that the public prosecutor and the minister of justice had decided not to prosecute a physician who assisted a patient with beginning dementia in committing suicide. It concerned a patient who had experienced a long dementia process of his father and his mother and for whom the prospect of going through such a

process of – in his eyes – loss of dignity, caused "unbearable suffering without prospect". The physician accepted his claim of unbearable suffering and assisted in suicide.

The public prosecutor considering the specific circumstances of this case, decided not to prosecute even though the view was taken that in this case the physician should have fulfilled some additional requirements for a careful procedure. After this decision became public, members of parliament asked questions to the minister of justice regarding this case. In the eyes of these members of parliament this decision implied an extension of accepted practice and jurisprudence regarding euthanasia and assisted suicide, since the suffering was caused by a future condition that the patient foresaw and feared. They argued that when that condition would become reality he would no longer be aware of it in the sense that it would then cause unbearable suffering. In his answer, the minister referred to the Supreme Court decision in the so-called Chabot case (1994) that determined that psychic suffering could provide a justification for euthanasia or physician assisted suicide.

Many people fear dementia and some would – as the Flemish author Hugo Claus did in early 2008 - prefer an earlier death over having to progress into the final stages of Alzheimer's disease. One way to avoid the ravages of Alzheimer's disease is to stop eating and drinking[4] or to commit suicide. Another possible route is to ask for assisted suicide or euthanasia (Berghmans et al. 2009). Only in a few countries assisted suicide and euthanasia may be legal: in the Netherlands as well as Belgium physician-assisted suicide (PAS) may be legal if a number of specific conditions are met (De Haan 2002, Nys 2005), and in the US states of Oregon and Washington terminally ill adults are allowed to obtain and use prescriptions from their physicians for self-administered, lethal medications (Chin et al. 1999); in Switzerland physician- as well as non-physician assisted suicide is passively tolerated (Guillod and Schmidt 2005).

Much controversy exists over the moral legitimacy of assistance in dying, if this involves actively and deliberately shortening the life of the patient (Battin 1994, Cohen-Almagor 2004, Gill 2009, Hendin 2002, Jochemsen and Keown 1999, Keown 1995, Keown and Jochemsen 1999, Van Delden 1999).[5] This is particularly the case if patients do not (or not only) suffer from somatic illnesses, but (also) from mental disturbances such as in case of chronic mental illness or Alzheimer's disease.

Actually, in the Netherlands there is debate over the legitimacy of PAS in cases of dementia (Hertogh et al. 2006, 2007). The prospect of the development of

---

[4]The option of stopping with eating and drinking in the Dutch context has been coined as 'versterven' ('starving'); recently the Dutch psychiatrist Chabot talks of 'auto-euthanasia' to qualify all forms of life shortening acts in which no physician is involved, but in which the person involved is at the same time in charge and in dialogue with one or more other intimate others (Chabot 2007).

[5]Opposition to the legalization of PAS generally refers to the presumed incompatibility between a commitment to good palliative end-of-life care and the legalization of physician-assisted suicide. This encompasses a cluster of different claims: 1. that better end-of-life care eliminates requests for PAS; 2. that requests for PAS are often due to depression, which end-of-life care should aim to treat; 3. that physical pain is the only legitimate reason for PAS, and that good end-of-life care can relieve physical pain; 4. that many requests for PAS are illegitimate because they are based in hopelessness; 5. that PAS is incompatible with hospice care, and 6. that PAS is incompatible with the value end-of-life care ought to place on a natural death (Gill 2009). In the Dutch context, PAS is not considered to be in conflict with the goals of good end-of-life care.

Alzheimer's disease can be a morally and legally legitimate reason to give assistance to persons who request active help in dying because they suffer unbearably or consider their life as being irreversibly devoid of dignity, if above that the following conditions are met (the so-called 'due care criteria').[6] The physician:

1. must be convinced that the patient has made a voluntary and well-considered request to die;
2. must be convinced that the patient's is facing hopeless/interminable and unbearable/unendurable suffering;
3. has informed the patient about his/her situation and prospects;
4. together with the patient, must be convinced that there is no other reasonable solution;
5. has consulted at least one other independent doctor;
6. has seen and given his written assessment of the due care requirements as referred to in points 1 to 4;
7. has helped the patient to die with due medical care.

Particularly the requirement of 'hopeless/interminable and unbearable/unendurable suffering' makes the issue of active aid in dying a complex issue. Physicians recognize that the assessment of unbearable, unendurable suffering, because of its (partially) subjective character, can be difficult. Also, physicians may experience problems relating to the assessment of the first requirement: that the request ought to be voluntarily made and be well considered (Buiting et al. 2008). This may be particularly problematic in case of dementia patients.

24.4.5.1 Hopeless/Interminable Suffering

Dementia involves hopelessness in the sense that it is an incurable illness (Health Council of the Netherlands 2002). In time, the patient's mental abilities will unavoidably deteriorate, as will his physical condition. Dementia ultimately leads to death. Bystanders (i.e. next of kin) sometimes perceive the situation to be hopeless. The question, however, is whether the patient perceives things the same way, as in active aid in dying the well-considered, voluntary and uncoerced request of the patient is a precondition.

Firstly, certain symptoms of dementia – such as depression, agitation, behavioural disturbances and the like – can indeed be eased by the use of appropriate interventions. Above that, optimal psychosocial care may make out for a better quality of life of patients. The hopelessness and interminability of the suffering, as stated, is generally connected to the incurability of the illness. As mentioned, Alzheimer's is incurable and inevitably will lead to the death of the patient.

---

[6]These due care criteria are laid down in the Dutch Euthanasia Act (officially: the Termination of Life on Request and Assisted Suicide (Review Procedures) Act). See: http://english.justitie.nl/themes/euthanasia (accessed April 22, 2009)

Hopelesness, as a criterion for careful practice in euthanasia and PAS, does not involve hopelessness as a mental or emotional state, which may be a sign of depression and thus may – with or without adequate treatment - be a temporary state of mind.

Against the hopelessness of the suffering in case of people who are in the early stage of dementia it may be argued that this suffering may come to an end as the condition progresses. In the more severe stages of Alzheimer's disease the awareness of suffering may disappear.

24.4.5.2 Unbearable/Unendurable Suffering

A crucial issue in every case of euthanasia/assisted suicide is the question of the unbearability and unendurability of the suffering of the patient. In general, in cases of active termination of life when an incurable somatic illness is involved, mostly it is the actual suffering of the patient which is assessed. The suffering of a patient may have different causes, such as pain, increasing dependency, and anxiety, but it should be predominantly result from a medically classifiable disease or disorder: other forms of suffering do not justify euthanasia or assisted suicide. The 'unbearability' of the suffering is assessed from the patient's perspective and the patient's ability to cope with the situation. The patient him- or herself should consider the suffering as unbearable (Buiting et al. 2008).

In cases of dementia, it is the prospect of future decline of (cognitive) capabilities which in most cases is the reason why a request for assisted suicide or euthanasia in the early phase of the dementia process is made. Can this be qualified as 'unbearable suffering'?

Some argue that this cannot be considered as such, because the suffering should be in the present and actual. As Den Hartogh (2000, 184) argues: 'For the person's request in his own sincere view is not motivated by his present suffering at all, but by his expectations for the future.' (Besides, arguing against the unbearable suffering criterion in cases of early dementia does not necessarily imply that one opposes active life termination in such cases; what it does imply is that in such cases another criterion may provide an ethical justification for active life termination – as for instance the value of the authenticity, integrity, or the narrative unity of a person's life (Den Hartogh 2000, Dworkin 1993)).

Nevertheless, in the Netherlands there recently have been a number of cases involving Alzheimer's patients in which the regional euthanasia review committee came to the conclusion that the prospect of further decline of cognitive and other capabilities of the person might be qualified as 'unbearable suffering'.[7]

---

[7]More information about the euthanasia review committees can be found at: http://www.euthanasiecommissie.nl/en In the annual reports of these committees which are published at this website, more information about these cases is given. For an analysis of these cases, see Berghmans (forthcoming). See also: http://english.justitie.nl/themes/euthanasia

24.4.6 Ending Life Without Request of the Patient

In the Netherlands, euthanasia is defined as the active termination of life on the explicit, voluntary request of the patient. The requirements of due care, as formulated in the law, are only applicable in cases of euthanasia and assisted suicide.

In 1991 it was reported that in the Netherlands annually about 1,000 cases (0.8% of all the deaths) involve the decision of a physician to actively ending the life of a patient by administering lethal drugs without his or her explicitly expressed and persistent wish (Pijnenborg et al. 1993, Van der Maas et al. 1991). In more than half of these cases the decision has been discussed with the patient or the patient had in a previous phase of his or her illness expressed a wish for euthanasia should suffering become unbearable. In other cases, possibly with a few exceptions, the patients were near to death and clearly suffering grievously, yet verbal contact had become impossible. The decision to hasten death was then nearly always taken after consultation with the family, nurses, or one or more colleagues. In most cases, the amount of time by which, according to the physician, life had been shortened was a few hours or days only (Van der Maas et al. 1991). It appears that using drugs to end life without an explicit request of the patient occurs not only in the Netherlands, but also in other European countries (Rietjens et al. 2007).

A study comparing Dutch general practitioners and nursing home physicians showed that in both the practices of GP's as in those of nursing home physicians termination of life without request by the patient occurs, but is rare (Muller et al. 1994).

With regard to the ethical justification of active life termination without request of the patient, it is clear that one cannot refer to the morality of euthanasia – in the sense of active termination of life on request of the patient (van der Wal 1993). Here not the autonomy of the patient is involved, but (solely) the principle of the best interest of the patient. In health care, there will always be some situations in which terrible suffering, which can end only if the patient dies, arises when the patient cannot give a clear judgment about the desired course of action (Pijnenborg et al. 1993). In such extreme cases a conflict arises between the duty of the physician to preserve life and the duty to alleviate suffering. It is obvious that this is not part of the normal, accepted practice of medicine and health care, but an exception which is in need of open, transparent debate (Strubbe 2000).

24.5 Advance Directives

An advance directive is a document drawn up by a legally competent individual in order to influence decisions about his future medical treatment and care, in case he is no longer able to do so (Buford 2008).[8]

It is important to distinguish broadly between two types of advance directives: negative and positive advance directives (Berghmans 2000, Vezzoni 2005). In a

---

[8] See also Chapter 10

negative advance directive (or negative treatment directive), basically the person states what kinds of treatment he or she *refuses* in advance. Such a negative advance directive, if the wish of the patient is followed, requires from the physician that he refrains from taking actions towards the patient. A positive advance directive, on the other hand, contains a statement about what (medical) action(s) should be taken in case of incompetence. In such an advance directive, a person might ask for active termination of his life (euthanasia directive), or for treatment as long as there is a prospect of life prolongation. Essential is that such actions require action on the part of the physician.

If a patient wishes to draw up an advance directive, ideally a physician – particularly his general practitioner – should discuss this with the patient in order to avoid misunderstandings and problems of interpretation. This, however, is not a legal requirement. In the Netherlands, (negative) treatment directives (as distinguished from euthanasia directives) received legal recognition in 1995, in the framework of the Law on the Medical Treatment Contract (Wet op de Geneeskundige Behandelingsovereenkomst, WGBO) (Vezzoni 2005). One of the grounds on which the law rests is article 11 of the Constitution, which states the right of all persons to the inviolability of the body. To implement this right, the law provides that any medical treatment requires the informed consent of the patient. If a competent patient refuses to give consent for a treatment, the doctor must comply with the refusal, no matter what the reasons may be underlying the refusal and however dire its consequences. The WGBO provides that a refusal of consent can also be given in advance. Under Article 450, if a patient is no longer competent, refusal of consent can take the form of a treatment directive written while he was still competent.

The relevant passage (Art. 450.3) reads in part as follows:

> In case a patient in the age of sixteen years or older cannot be considered capable of a reasonable assessment of his relevant interests, the health care provider and [the personal representative] shall follow the patient's apparent views laid down in writing when he was still capable of such reasonable assessment and containing a refusal of consent [...]. The health care provider may depart herefrom if he considers that there are well-founded reasons for doing so.

This means that the physician must consider the advance directive but is not bound to it if it does not fit the actual situation. It is important to note that connected to a written advance directive a power of attorney may be given to a person who is trusted by the patient and is given the authority to act on the patient's behalf.

24.5.1 Empirical Data on Advance Directives

In a Dutch retrospective study (Rurup et al. 2005) it was found that in the Netherlands, approximately 2,200 demented patients with an advance euthanasia directive die annually after being treated by a physician who knows about this directive. In 76% of such cases, compliance with the directive was discussed, but euthanasia was seldom performed. In two-thirds of the cases of demented nursing home patients with an advance euthanasia directive, the physician was able to identify during the course of the disease a situation for which the patient had intended the directive. One-quarter of the nursing home physicians thought that their most recent patient suffered unbearably to a (very) high degree, and half of them thought that the patient suffered hopelessly to a (very) high degree. In three-quarters of the cases, the relatives did not want the nursing home physician to comply with the directive but they did want to respect the patient's wishes by forgoing life-prolonging treatment, which occurred in approximately 90% of cases. The authors conclude that most nursing home physicians think that the suffering of patients with dementia can be unbearable and hopeless as a consequence of dementia, but most physicians do not consider dementia to be a ground for euthanasia unless perhaps the patient has an additional illness. These results may be taken as evidence that there is not a negative impact of a more 'liberal' legal stance towards practices of euthanasia and assisted suicide, as is feared by some critics who predict abuse of people who belong to 'vulnerable' groups (Battin et al. 2007).

24.5.2 Dementia, Advance Directives and Personal Identity

Next to other issues – which will be addressed below - an important philosophical question with regard to dementia advance directives concerns the problem of personal identity (Berghmans 2000, Buchanan 1988, DeGrazia 2005, Dresser 1992, 1995, Harvey 2006, Kuhse 1999, Vollmann 2001, Wrigley 2007). By virtue of the disorder the patient's former personality is destroyed to a great extent in the advanced stages (Dresser 1992, 1995, Vollmann 2001). On metaphysical grounds it has been maintained that the formerly competent patient who authored an advance directive and the currently demented patient who now exists are different persons. In the case of the severely demented, advance directives issued by their formerly competent selves simply no longer apply (Buchanan 1988, Dresser 1992, 1995).

Obviously, this metaphysical philosophical debate has far reaching implications for the moral force which can legitimably be attributed to advance directives. As a matter of fact, if one convincefully can argue that in case of serious dementia two different persons exist, than one cannot attribute moral authority to the previously expressed wishes in an advance directive. As this issue remains unresolved, a strong conclusion regarding the moral status and force of dementia advance directives cannot be made. However, a defensible conclusion is that advance statements play a useful role in formulating and negotiating what treatment and care is in a patient's best interests, but ultimately in most cases they do not have sufficient moral force to take precedence over paternalistic best interest judgements concerning a demented individual's care or treatment (Hertogh 2009, van Delde 2004, Widdershoven and Berghmans 2001, Wrigley 2007). This conclusion can partly be grounded on some less philosophical but more practical observations which can be made with regard to dementia advance directives and the differences between 'actual' and 'hypothetical' decisions to be made concerning health care and medical treatment in case of dementia.

24.5.3 'Actual' and 'Hypothetical' Decisions

Advance directives are considered as an extension of the competent patient's moral and legal right to refuse treatment. Nevertheless, this view must be nuanced if we look at a number of differences between 'hypothetical' and 'actual' decisions made by a (future) dementia patient. In case of an advance directive, a person foresees and imagines a possible future situation and states what in that specific situation is the course of action which he or she desires to take place (i.e. non-treatment of pneumonia and/or do-no-recuscitate order). In such a hypothetical scenario the future still is uncertain. In an actual scenario, a decision concerning medical treatment and care takes place in reality, in a situation in which the decision actually is made and concerns the present medical situation and treatment needs of the patient. When comparing choices expressed in an advance directive with contemporaneous decisions of competent patients, several differences can be observed (Berghmans 2000, Buchanan and Brock 1989, Cantor 1993). Firstly, even if at the time the advance directive was issued the individual was well-informed about (treatment) options, therapeutic options – and hence prognosis – may change between the time the advance directive is issued, and the time it is to be implemented. This problem may be (partly) tackled by periodically renewing the advance directive. Secondly, advance directives are often formulated with certain implicit assumptions in mind about the expected future condition. People may fear the condition of being demented because they attribute specific experiences to being demented which may or may not be based on knowledge about other demented people (a parent or other relative), and which may or may not be accurate. This, however, does not necessarily match the actual experiences of the person in a future state of dementia. Of course this does not mean that in all cases such assumptions need to be made. In some cases, it may be simply so that the overriding reason for not wanting to (continue to) be in a situation of dementia is that this demented state as such is not desirable in the light of the total of the life of that person (Dworkin 1993). Thirdly, the assumption that a competent person is the best judge of his or her own interests is weaker in the case of a choice about future contingencies under conditions in which those interests may have changed in radical and unforeseen ways than it is in the case of a competent individual's contemporaneous, actual choice. Inasfar as the mental states and experiences of persons – as in case of dementia – deviate more from what we are used to..., the possibility of 'transpersonal introspection' – with regard to the relationship between the former, non-demented, and the later, demented self of that person – becomes more problematic. Our imagination is limited and constrained by our own individual experiences, and these 'color in' the conscious experiences we attribute to other people or creatures (Nagel 1986). Lastly, in the case of an advance directive, important informal safeguards that tend to restrain imprudent or unreasonable contemporaneous choices by a competent patient are lacking. If a competent patient actually refuses life-sustaining treatment, those around that person can and often do urge the patient to reconsider his or her choice. In case of an advance directive one cannot discuss the former wishes with the now incompetent patient. Above that, persons with Alzheimer's disease at a certain time lose their ability to change their minds.

All these considerations also justify the conclusion that dementia advance directives cannot be given absolute moral authority in case of non-treatment decisions
involving incompetent demented patients (Berghmans 2000).

24.6 How to Decide: Making Decisions for Persons with Dementia

Advance discussions (a practice in Dutch nursing homes) may be helpful in finding out the views of patients and their family concerning treatment and care, and
how they feel about the limits of medical treatment interventions. Above that, such
advance discussions create room for the expression of concerns and fears patients
and their loved one may have with regard to the future and decisions which may
need to be made then. Generally, such discussions will provide helpful information
for both patients and/or their families and the physician who is responsible for the
medical care of the patient.

Obviously, when physicians are confronted with requests for physician-assisted
suicide or euthanasia, they should first work to bolster the patient's sense of control and to educate and reassure the patient regarding the management of future
symptoms (Ganzini et al. 2009). And as far as depressive symptoms or feelings of
demoralization are involved, steps should be taken to try to treat such symptoms
(Kissane 2004, Kissane and Kelly 2000). In those cases were a request for active
life termination continues to exist, in some jurisdictions – as in the Netherlands and
Belgium – this may be taken as a situation in which physician-assisted suicide or
euthanasia can be morally and legally acceptable.

24.7 In Conclusion

People suffering from dementia may often not be able to perform abstract tasks,
because they are unable to memorize relevant information. This does not mean,
however, that they are necessarily unable to evaluate their condition and to communicate about is (Jaworska 1999, Widdershoven and Berghmans 2001). More and
more, in research and practice importance and value is given to the *person* with
dementia (Downs 1997, Kitwood 1997, Kitwood and Bredin 1992). Various studies have shown that patients with dementia can share their experience of the care
they receive, if they are carefully listened to (Goldsmith 1996). Many of the problems in communicating with people with dementia come from the fact that for them
responding takes more time. If they are not hurried, but are given the opportunity to
finish the sentences at their own pace, their utterances are much more coherent than
it may appear at first sight (Sabat and Harré 1994, Widdershoven and Berghmans
2006). Anyway, it is important that the patient's perspective is included as much
as possible in the decision-making process, not only at the level of the individual
patient, but also as far as it concerns policy issues (De Boer et al. 2007, Gedge 2004).

In the care for patients who suffer from dementia, ideals regarding the good life play a crucial role (Kalis et al. 2005). These ideals do not disappear at life's terminal stages. On the contrary: in life's final stage decisions should be made in a careful way, with respect for the patient's and the family's perspectives, and a responsible assessment of benefits, burdens, risks and quality of life in each specific treatment decision.

References

Battin MP (1994) A dozen caveats concerning the discussion of euthanasia in the Netherlands. In: Battin MP (ed) The least worst death. Essays in bioethics and the end of life. Oxford University Press, Oxford, pp 130–144

Battin MP, van der Heide A, Ganzini L et al (2007) Legal physician-assisted dying in Oregon and the Netherlands: evidence concerning the impact on patients in "vulnerable" groups. J Med Ethics 33:591–597

Berghmans RLP (1999) Ethics of end-of-life decisions in cases of dementia. Views of the Royal Dutch Medical Association with some critical comments. Alzheimer Dis Assoc Disord 13:91–95

Berghmans RLP (2000) Advance directives and dementia. Ann N Y Acad Sci 913:105–110

Berghmans R, Molewijk B, Widdershoven G (2009) Alzheimer's disease and life termination: the Dutch debate. Bioethica Forum 2:33–34

Berghmans RLP (forthcoming) Decision-making capacity in patients who are in the early stage of Alzheimer's disease and who request physician assisted suicide. In: Youngner S, Kimsma G (eds) Euthanasia in the Netherlands (forthcoming).

Blijham GH (1996) Euthanasia, physician-assisted suicide and palliative care. Ann Oncol 7:879–882

Buchanan AE (1988) Advance directives and the personal identity problem. Philos Public Affairs 17:277–302

Buchanan AE, Brock DE (1989) Deciding for others. The ethics of surrogate decision-making. Cambridge University Press, New York, NY

Buford C (2008) Advancing an advance directive debate. Bioethics 22:423–430

Buiting HM, Gevers JKM, Rietjens JAC et al (2008) Dutch criteria of due care for physician-assisted dying in medical practice: a physician perspective. J Med Ethics 34(9):e12

Cantor N (1993) Advance directives and the pursuit of death with dignity. Indiana University Press, Bloomington and Indianapolis

Chabot B (2007) Auto-euthanasie. Verborgen stervenswegen in gesprek met naasten. [Auto-euthanasia. Hidden roads to death in dialogue with relatives] Uitgeverij Bert Bakker, Amsterdam

Chin AE, Hedberg K, Higginson GK, Fleming DW (1999) Legalized physician-assisted suicide in Oregon – the first year's experience. NEJM 340:577–583

Cohen-Almagor R (2004) Euthanasia in the Netherlands. The policy and practice of merci killing. Kluwer Academic Publishers, Dordrecht/Boston/London

Cummings JL (2004) Alzheimer's disease. N Engl J Med 351:56–67

DeGrazia D (2005) Advance directives, dementia, and the someone else problem. In: DeGrazia D (ed) Human identity and bioethics. Cambridge University Press, Cambridge, pp 159–202

Downs M (1997) The emergence of the person in dementia research. Ageing Soc 17:597–607

Dresser R (1992) Autonomy revisited: the limits of anticipatory choices. In: Binstock RH, Post SG, Whitehouse PJ (eds) Dementia and aging: ethics, values and policy choices. The Johns Hopkins University Press, Baltimore/London, pp 71–85

Dresser R (1995) Dworkin on dementia: elegant theory, questionable practice. Hastings Cent Rep 22(6):32–38

Dworkin R (1993) Life's dominion. An argument about abortion and euthanasia. HarperCollins Publishers, London

Feng Z, Zhao G, Yu L (2009) Neural stem cells and Alzheimer's disease: challenges and hope. Am J Alzheimer's Dis Other Dementias 24:52–57

Ganzini L, Goy ER, Dobscha SK (2009) Oregonians' reasons for requesting physician aid in dying. Arch Intern Med 169:489–492

Gedge EB (2004) Collective moral imagination: making decisions for persons with dementia. J Med Philos 29:435–450

Gevers S (2003) Terminal sedation: a legal approach. Eur J Health Law 10:359–367

Gevers S (2006) Dementia and the law. Eur J Health Law 13:209–217

Gill MB (2009) Is the legalization of physician-assisted suicide compatible with good end-of-life care? J Appl Philos 26:27–45

Goldsmith M (1996) Hearing the voice of people with dementia. Jessica Kingsley Publishers, London

Griffiths J, Bood A, Weijers H (1998) Euthanasia and law in the Netherlands. Amsterdam University Press, Amsterdam

Guillod O, Schmidt A (2005) Assisted suicide under Swiss law. Eur J Health Law 12:25–38

Hamani C, McAndrews MP, Cohn MO et al (2008) Memory enhancement induced by hypothalamic/fornix deep brain stimulation. Ann Neurol 63:119–123

Harris J (1987) Qualifying the value of life. *J Med Ethics* 13:117–123

Harvey M (2006) Advance directives and the severely demented. J Med Philos 31:47–64

Hasselaar JGJ, Verhagen SCAHHVM, Wolff AP et al (2009) Changed patterns in Dutch palliative sedation practices after the introduction of a national guideline. Arch Intern Med 169:430–437

Health Council of the Netherlands (2002) Dementia. Health Council of the Netherlands, The Hague publication no. 2002/04E

Helton MR, Van der Steen JT, Daaleman TP et al (2006) A cross-cultural study of physician treatment decisions for demented nursing home patients who develop pneumonia. Ann Fam Med 4:221–227

Hendin H (2002) The Dutch experience. Issues Law Med 17:223–246

Hertogh CMPM (2009) The role of advance euthanasia directives as an aid to communication and shared decision-making in dementia. J Med Ethics 35:100–103

Hertogh CMPM, Ribbe MW (1996) Ethical aspects of decision-making in demented patients: A report from the Netherlands. Alzheimer Dis Assoc Disord 10:11–19

Hertogh CMPM, de Boer ME, Dröes R-M, Eefsting JA (2006) Would we rather lose our life than lose our self? Lessons from the Dutch debate on euthanasia for patients with dementia. Am J Bioethics 7:48–56

Hertogh CMPM, de Boer ME, Dröes R-M, Eefsting JA (2007) Beyond a Dworkinian view on autonomy and advance directives in dementia. Am J Bioethics 8:4–6

Holman JB, Brendel DH (2006) The ethics of palliative care in psychiatry. J Clin Ethics 17:333–338

Jaworska A (1999) Respecting the margins of agency. Alzheimer's patients and the capacity to value. Phil Public Affairs 28:104–138

Jecker NS (1995) Medical futility and care of dying patients. West J Med 163:287–291

Jochemsen H, Keown J (1999) Voluntary euthanasia under control? Further empirical evidence from the Netherlands. J Med Ethics 25:16–21

KNMG (1997) Medisch handelen rond het levenseinde bij wilsonbekwame patiënten (Commissie Aanvaardbaarheid Levensbeëindigend handelen). [Medical action in regard to the end of life in incompetent patients (Commission on the Acceptability of Medical Behavior that Shortens Life)] Bohn Stafleu Van Loghum, Houten/Diegem (in particular Chapter 7)

Kalis A, Schermer MHN, van Delden JJM (2005) Ideals regarding a good life for nursing home residents with dementia: views of professional caregivers. Nurs Ethics 12:30–42

Kasman DL (2004) When is medical treatment futile? A guide for students, residents, and physicians. J Gen Int Med 19:1053–1056

Keown J (1995) Euthanasia in the Netherlands: sliding down the slippery slope? In: Keown J (ed) Euthanasia examined. Cambridge University Press, Cambridge

Keown J, Jochemsen H (1999) Voluntary euthanasia in the Netherlands. J Med Ethics 25:351–352

Kissane DW (2004) The contribution of demoralization to end of life decisionmaking. Hastings Cent Rep 34(4):21–31

Kissane DW, Kelly BJ (2000) Demoralization, depression and desire for death: problems with the Dutch guidelines for euthanasia of the mentally ill. Aust N Z J Psychiatry 34:325–333

Kitwood T (1997) Dementia reconsidered: the person comes first. Open University Press, Buckingham

Kitwood T, Bredin K (1992) Towards a theory of dementia care: personhood and well-being. Ageing Soc 12:269–287

Kruit A, Ribbe MW, van der Wal G (1999) Afzien van kunstmatige toediening van voeding en vocht bij verpleeghuispatiënten in de laatste levensfase. Ned Tijdschr Geneeskd 143:1401–1404

Kuhse H (1999) Some reflections on the problems of advance directives, personhood and personal identity. Kennedy Inst Ethics J 9:347–364

Leenen HJJ, Gevers JKM, Legemaate J (2007) Handboek gezondheidsrecht. Deel I: Rechten van mensen in de gezondheidszorg. Bohn Stafleu van Loghum, Houten

Legemaate J, Verkerk M, Van Wijlick E, De Graef A (2007) Palliative sedation in the Netherlands: starting points and contents of a national guideline. Eur J Health Law 14:61–73

Muller MT, van der Wal G, Ribbe MW, Eijk JThM (1994) Levensbeëindigend handelen door huisartsen en verpleeghuisartsen zonder verzoek van de patiënt. Ned Tijdschr Geneeskd 138:395–398

Nagel T (1986) The view from nowhere. Oxford University Press, New York, NY

Nys H (2005) Physician assisted suicide in Belgian law. Eur J Health Law 12:39–41

Onwuteaka-Philipsen BD, Pasman HRW, Kruit A et al (2001) Withholding or withdrawing artificial administration of food and fluids in nursing-home patients. Age Ageing 30:459–465

Pasman HRW, Onwuteaka-Philipsen BD, Kriegsman DMW et al (2005) Discomfort in nursing home patients with severe dementia in whom artifical nutrition and hydration is forgone. Arch Intern Med 165:1729–1735

Pasman HRW, Onwuteaka-Philipsen BD, Ooms ME et al (2004) Forgoing artificial nutrition and hydration in nursing home patients with dementia. Patients, decision making, and participants. Alzheimer Dis Assoc Disord 18:154–162

Pijnenborg L, van der Maas PJ, van Delden JJM, Looman CWN (1993) Life-terminating acts without explicit request of patient. Lancet 341:1196–1199

Rietjens JAC, Bilsen J, Fischer S et al (2007) Using drugs to end life without an explicit request of the patient. Death Stud 31:205–221

Rietjens J, van Delden J, Onwuteaka-Philipsen B et al (2008) Continuous deep sedation for patients nearing death in the Netherlands: descriptive study. BMJ 336:810–813

Rurup M et al (2005) Physicians'experiences with demented patients with advance euthanasia directives in the Netherlands. JAGS 53:1138–1144

Rurup ML, Onwuteaka-Philipsen BD, Pasman HRW et al (2006) Attitudes of physicians, nurses and relatives towards end-of-life decisions concerning nursing home patients with dementia. Patient Educ Couns 61:372–380

Sabat SR, Harré R (1994) The Alzheimer's disease sufferer as a semiotic subject. Philos Psychiatr Psychol 1:145–160

Selkoe DJ (2001) Alzheimer's disease: genes, proteins, and therapy. Physiol Rev 81:741–766

Slaets JPJ (2007) 'The old man's friend': verschillen in besluitvorming over behandeling van pneumonie bij demente verpleeghuispatiënten in Nederland en de Verenigde Staten. Ned Tijdschr Geneeskd 151:905–906

Sowmini CV, de Vries R (2008) A cross cultural review of the ethical issues in dementia care in Kerala, India and The Netherlands. Int J Geriatr Psychiatr (published online 2008)

Strubbe E (2000) Toward legal recognition for termination of life without request? Remarks on advice no. 9 of the Belgian Advisory Committee on Bioethics Concerning Termination of Life of Incompetent Patients. Eur J Health Law 7:57–71

The BAM, Pasman HRW, Onwuteaka-Philipsen BD et al (2002) Withholding the artificial administration of fluids and food from elderly patients with dementia: ethnographic study. BMJ 325:1–5

Verkerk M, van Wijlick E, Legemaate J, de Graeff A (2007) A national guideline for palliative sedation in the Netherlands. J Pain Symptom Manage 34:666–670

Vezzoni C (2005) The legal status and social practice of treatment directives in the Netherlands. PhD Thesis, University of Groningen, The Netherlands

Vollmann J (2001) Advance directives in patients with Alzheimer's disease. Med Health Care Phil 4:161–167

Whitehouse PJ (2008) The myth of Alzheimer's. St. Martin's Press, New York, NY

Widdershoven GAM, Berghmans RLP (2001) Advance directives in dementia care: from instructions to instruments. Patient Educ Couns 44:179–186

Widdershoven GAM, Berghmans RLP (2006) Meaning making in dementia: a hermeneutic perspective. In: Hughes JC, Louw SJ, Sabat SR (eds) Dementia. Mind, meaning and the person. Oxford University Press, Oxford, pp 179–191

Wrigley A (2007) Personal identity, autonomy and advance statements. J Appl Philos 24:381–396

de Boer ME, Hertogh CMPM, Dröes R-M et al (2007) Suffering from dementia – the patient's perspective: a review of the literature. Int Psychogeriatr 19:1021–1039

de Haan J (2002) The new Dutch law on euthanasia. Med Law Rev 10:57–75

den Hartogh G (2000) Euthanasia. Reflections on the Dutch discussion. Ann N Y Acad Sci 913:174–187

van Delden JJM (1999) Slippery slopes in flat countries – a response. J Med Ethics 25:22–24

van Delden JJM (2004) The unfeasibility of requests for euthanasia in advance directives. J Med Ethics 30:447–452

van Delden JJM (2007) Terminal sedation: source of a restless ethical debate. J Med Ethics 33:187–188

van Wijlick E, de Graef A, Verkerk M, Legemaate J (2009) Meer houvast voor arts. Herziene KNMG-richtlijn palliatieve sedatie. Med Contact 64(5):194–197

van der Heide A, Onwuteaka-Philipsen BD, Rurup ML et al (2007) End-of-life practices in the Netherlands under the Euthanasia Act. N Engl J Med 356:1957–1965

van der Maas PJ, van Delden JJM, Pijnenborg L, Looman CWN (1991) Euthanasia and other medical decisions concerning the end of life. Lancet 338:669–674

van der Steen JT (2002) Curative or palliative treatment of pneumonia in psychogeriatric nursing home patients: development and evaluation of a guideline, decision-making, and disease course. PhD Dissertation, Free University Amsterdam

van der Steen JT, Helton MR, Ribbe MW (2009) Prognosis is important in decisionmaking in Dutch nursing home patients with dementia and pneumonia. Int J Geriatr Psychiatry (online)

van der Steen JT, Kruse RL, Ooms ME et al (2004) Treatment of nursing home residents with dementia and lower respiratory tract infection in the United States and the Netherlands: an ocean apart. JAGS 52:691–699

van der Steen JT, Kruse RL, van der Wal G et al (2007) Behandeling van pneumonie bij verpleeghuispatiënten met ernstige dementie: terughoudender beleid in Nederland en actiever beleid in de Verenigde Staten naarmate de prognose ongunstiger is. Ned Tijdschr Geneeskd 151:915–919

van der Steen JT, Muller MT, Ooms ME et al (2000a) Decisions to treat or not to treat pneumonia in demented psychogeriatric nursing home patients: development of a guideline. J Med Ethics 26:114–120

van der Steen JT, Ooms MW, Ader HJ et al (2002a) Withholding antibiotic treatment in pneumonia patients with dementia. A quantitative observational study. Arch Intern Med 162:1753–1760

van der Steen JT, Ooms ME, Ribbe MW, van der Wal G (2001) Decisions to treat or not to treat pneumonia in demented psychogeriatric nursing home patients: evaluation of a guideline. Alzheimer Dis Assoc Disord 15:119–128

van der Steen JT, Ooms ME, van der Wal G, Ribbe MW (2002b) Pneumonia: the demented patient's best friend? Discomfort after starting or withholding antibiotic treatment. JAGS 50:1681–1688

van der Steen JT, de Graas T, Ooms ME et al (2000b) When should physicians forgo curative treatment of pneumonia in patients with dementia? West J Med 173:274–277

van der Wal G (1993) Unrequested termination of life: is it permissible? Bioethics 7:330–339

Chapter 25
Ethics of Research with Decisionally Impaired Patients

Giovanni Maio

Contents

25.1 Introduction

Research on decisionally impaired persons is one of the most controversial topics of medical ethics. While for some people, non-therapeutic research on decisionally impaired patients represents a sell-out of human rights, others point out in recourse to the interests of future patients to the necessity of scientific progress. Guaranteeing the protection of human rights whilst at the same time not curbing the growth of knowledge for the treatment of future patients, is the challenge facing all parties involved. These questions refer to philosophy and, for this reason, I would mainly like to explore patterns of reasons and various models of norm justification. The first question is, therefore, what is the basic ethical conflict that arises in conducting research on decisionally impaired persons?

G. Maio (✉)
Institute for Bioethics and History of Medicine, University of Freiburg,
D-79104 Freiburg, Germany
e-mail: maio@ethik.uni-freiburg.de

H. Helmchen, N. Sartorius (eds.), *Ethics in Psychiatry*, International Library
of Ethics, Law, and the New Medicine 45, DOI 10.1007/978-90-481-8721-8_25,
© Springer Science+Business Media B.V. 2010

25.2 Basic Ethical Conflicts of Conducting Research on People

The philosopher Hans Jonas believes that the basic ethical problem of performing research on people lies in the scientific method of experiment, which demands that the subject is objectified, i.e. turned into the object of the experiment. The method of experiment demands further that the subject is instrumentalised, i.e. used for a purpose other than benefiting the subject (Jonas 1969). Thereby, these moments of objectification and instrumentalisation can violate the fundamental unavailability of humans and man's being as an end in himself, and therefore require justification.

If, now, however the heteronomous instrumentalisation of humans on the one hand is pitted against a benefit on the other hand which, despite the instrumentalisation that requires legitimation, cannot simply be classified as futile and irrelevant, otherwise a conflict of norms arises. Ethical conflict is therefore entwined around balancing out individual rights – such as the right to the basic unavailability of the individual and the right to psycho-physical integrity – against the potential benefit future patients could gain from the research. The ethical problem of conducting research on people could therefore be formulated in the polarisation of the rights of present test subjects and the rights of future patients; it could, however, also be captured in the dichotomy of the individual and collective. Both approaches have repeatedly been expressed in the history of the debate on conducting research on people. The following sections aim to explore whether and how this conflict can be resolved.

25.3 Basic Legitimising Factors for Conducting Research on People

If we assume, together with Jonas, that every experiment is attended by an instrumentalisation of man that must be legitimated, the question is raised at this point as to which reasons at all are able to justify the instrumentalisation of using humans in scientific experiments. *Two* central legitimising factors ostensibly present themselves here. The most recognized instrument to legitimise research on humans is the test subject's consent. Philosophically, this can be explained by the fact that by giving their free consent to a trial, test subjects make the purpose of a test their own purpose, thus neutralising the instrumentalising character of the research. It is the principle of autonomy that remains untouched in this way, and hence the test persons are ideally no longer the object. They become subjects that are not only used but participate freely in the undertaking of research. The second legitimising factor of experiments on humans is the benefit gained by the test subject. In the case of a specific benefit to the trial person, the purpose of the experiment is, at least not only, a purpose alien to the subject, but ideally coincides with the test person's purpose, so that in this way the instrumentalisation of the test person, which requires justification, is neutralised.

If, therefore, it is assumed that an experiment on a person is justified by the consent of and the benefit to the test person, it becomes clear why non-therapeutic research on decisionally impaired patients raises special problems, because in this case both legitimising factors for research on humans are absent by definition. Now, however, the question arises as to whether in fact a fundamental illegitimacy of non-therapeutic research on decisionally impaired patients can be derived. In order to clarify this crucial question, I would like to highlight a number of arguments.

25.3.1 Benefit for Present Patients and Groups of Patients

The principle of the benefit is the central consequentially substantiated ethical principle that takes effect in research on humans. If the benefit is described as the legitimising factor for research on people, not much has been gained, however, as long as this benefit is not defined more precisely. In order to clarify this argument, therefore, a precise analysis of the term 'benefit' is required.

Even if there is agreement on the definition of benefit within the meaning of the 'best interest', the *parameters* of these best interests still need to be defined. The parameters of the *probability of occurrence*, for instance, should be considered here. To be specific, this means: how certain must the accomplishment of a benefit be in a research project for the study not to be deemed non-therapeutic? Or in other words: does a potential benefit suffice to justify a study on a decisionally impaired person? What about the fact that all expected benefits in medicine are 'potential' ones? The other parameter would be that of *causality*. This therefore raises the question as to whether an indirect benefit can also act as a legitimising factor, as would be the case, for instance, when the test person himself would not benefit specifically from the study but would certainly gain an advantage from the increased attention or study-related improved diagnostics associated with the study. The British philosopher Richard Ashcroft, for instance, expresses the following: 'But in this case it might be that giving them a measure of control over their situation, which they might not otherwise feel they had, and may be therapeutic of itself.' (Ashcroft 1998, p. 635). Furthermore, the *parameter of objectifiability* should also be considered. It must be clarified here whether classification of a study in the therapeutic category should only be made dependent on the objective interests of the test person, or whether subjective evaluation patterns should also be considered. Some researchers, for instance, interpret the feeling of altruistic help to be a benefit (Ratzan 1980). It remains, however doubtable whether such conceptions of benefit are really legitimising factors for research on decisionally impaired persons.

Moreover, the *parameter of quality* must also be taken into consideration. This refers to the difficulty encountered when endeavouring to further define the term of the 'best interest'. Finally, reference is made to the *parameter of the addressee*, because here, too, we are dealing with a wide range of possibilities of definition. The benefit associated with a study can, for instance, be geared towards at least four different addressees. The most obvious interpretation of benefit as the legitimating factor for carrying out research on humans is the benefit enjoyed by the affected individual test person. While a benefit for future patients can clearly be excluded as a legitimising factor, the evaluation is no longer quite as clear if it involves a benefit relating to the interests of the same group of people, such as the group suffering from the same disease, e.g. in research on dementia patients. The above highlights how important it is to differentiate the term 'benefit'.

25.3.2 Risk Mitigation for Future Patients

A further argument that could be considered for calling on decisionally impaired persons for scientific tests is that, in the case of renouncing such studies, a direct potentiation of risks for future patients would have to be accepted. Such future patients would, for instance, have to take medication that has not been proved to be harmless to this particular group. If, therefore, the obligation of society to mitigate risks is taken seriously, this would also contain the obligation to carry out research on decisionally impaired persons because only then could at the group of decisionally impaired persons be prevented from being subjected to more risky treatments. This line of argument, therefore, avails itself of the instrument of counterbalancing risks and benefits, and it is highly intriguing to discover that almost all official statements on research on decisionally impaired persons argue in this consequentialist manner. Implicitly, at least, this ethical thought pattern is expressed wherever research on decisionally impaired persons is legitimised by the relevance of the expected findings (cf. Frank and Agich 1985) or when the argument of the classic catchphrase of 'therapeutic orphans' is put forward, such as in the 1992 guideline of the Canadian 'National Council on Bioethics in Human Research' (National Council 1992). The problem with this argumentation is that, in this case, the violation of a deontological principle is to be justified by a consequentialist legitimisation. This is a highly controversial type of argumentation because it would not be able to prevent the alienation of basic rights in favour of a positive consequence, however it may be defined (cf. Glass and Speyer-Ofenberg 1992, Veatch 1987).

25.3.3 Minimal Risk as a Protective Measure

The minimal risk is used in many research-ethical regimes as a central condition for carrying out research on decisionally impaired persons. However, even today it is unclear what is meant by the minimal risk. First of all, with the term 'risk', a differentiation must be made between the definition of risk, the determination of risk, the perception of risk and risk assessment. The term 'minimal risk' falls under the category of risk assessment, because the expression 'minimal risk' implicitly simply means 'tolerable risk'. We are not, therefore, dealing with a descriptive term, but primarily with a normative term, because minimal is connotated with negligible or tolerable. Now, the reasonableness of any risk is not identical to the actual social acceptance of a risk, because this acceptance does not correlate with the factual risk, but depends on the perception of a risk. In other words, social acceptance need not coincide with ethical legitimation. This means that not every generally accepted experiment must be, as it were, ethically acceptable and that, vice versa, not every ethically acceptable experiment must also be deemed acceptable in society. The question then arises as to which parameters can be used to exemplify this reasonableness. The 'minimal risk' alone reveals little; its contours only take shape once, in turn, the parameters are determined, based on which this minimal risk is to be defined.

25.3.3.1 The Risk of 'Daily Life' as a Parameter?

The Belmont Report of 1978 defined the minimal risk as

> "the probability and magnitude of physical or psychological harm encountered in the daily lives, or in the routine medical, dental, or psychological examination of healthy persons". Following the Belmont Report, the "Department of Health and Human Services" also gives a similar definition in 1981 oriented towards the term 'daily life':
>> Minimal risk means that the risks anticipated in the proposed research are not greater, considering probability and magnitude, than those ordinarily encountered in daily life or during the performance of routine physical or psychological examinations or tests.

Yet this definition alone does not solve the problem. Once more, the question is raised as to which parameter should now been used to determine *daily life* (Kopelman 1989). Does daily life mean the life of a secretary, a professional football player or a pilot? Despite the fact that the attempt to link the minimal risk to the risks of everyday life may have found its way into a variety of codes, it ultimately appears to be less than helpful, even inadequate, because no logical link can be made between the risks of daily life and the risk of sustaining injury from participating in the study. This can already be seen in the mind game that the term 'daily life' itself and the term 'average person' cannot manage without reference to a particular group, because it is utterly impossible to sum up the whole world and all standards of living as an average. In other words, a certain group or a certain region must always be defined. Depending, however, on how these definitions are made, a minimal risk in Calcutta appears by all means to be higher than in Copenhagen. If one follows the definition of the minimal risk as the risk of daily life, one would have to say that it would be more justifiable to conduct a risky study with the mental ill in Calcutta than with a patient in Copenhagen (cf. Kopelman 1989). And, of course, this does not make any sense.

The second part of the definition compares the minimal risk with the risks that occur in diagnostic or therapeutic methods within the scope of normal curative treatment. For these so-called therapeutic experiments, there would be the possibility to equate the 'minimal risk' to the risk that would be associated with the treatment the patient would be subject to in the event of not participating in the study. But this comparison also comes up against contradictions, because there are so many investigations and therapies in daily hospital life that trigger such anxieties in many people and are also highly risky in parts, so that it appears unjustified to simply call all conceivable stresses of daily hospital life minimal risks (Kopelman 1989). The recourse to 'standard risks' is therefore not particularly helpful.

25.3.3.2 The Term 'Mere Inconvenience' as an Alternative?

Robert Levine proposed taking the term of *'mere inconvenience'* as the legitimation borderline instead of the minimal risk (Levine 1986). The advantage of this terminology is that it would express more strongly the relationship to the special life situation of each individual. Furthermore, with the term 'inconvenience', the most common stresses accompanying scientific tests would be described more accurately than with the term risk or the damage expressed by it. For the term 'minimal risk' semantically always also implies an element of danger, and if one considers that this term can also mean contents such as urine tests, secondary blood tests or even boredom, this connotation of danger does not seem very adequate. However, whether inconvenience should definitively be preferred to the term minimal risk must nevertheless remain debatable. The term minimal risk, on the other hand, reflects two variables: the extent of damage and the probability of its occurrence. Now, it is of considerable importance for the assessment of research risks to stress that we are only dealing here with potentialities of damage and not actual damage. And if a term does not contain this aspect of potentiality, it is a serious semantic displacement that would be inappropriate because it would suggest an inevitability of damage which is not given in any test. For this reason, minimal risk appears to me to be the more complex and hence also the more differentiated term.

25.3.3.3 The Minimal Risk as a Relational Term

The basic problem arising in the determination of the so-called 'minimal', i.e. 'tolerable risk' is the unavoidable ability of subjective elements. This subjectivity takes effect at two levels. The first level affects the orientation of the 'tolerable' to one's own desires or aversions. A risk is then no longer tolerable to the test person if, for instance, it impairs the realisation of certain desires or is linked to specific personal aversions. But there do not exist any objective parameters with which such moments as desires and aversions could be determined. This determination is primarily a subjective process. But subjectivity also extends into a second level of the decision, because not only the definition of desires and aversions is responsible for subjectivity, but quasi the parameters on the basis of which these moments are weighed against one another. The balancing process is not a rationale balancing out; in most cases, moments such as hope, anxieties and personal convictions also flow into this process of balancing in particular, so that not only the preferences as such but also the process of balancing is tainted with a large degree of subjectivity, not only on the side of the research subject but on the side of the researcher, too.

And yet it is not only the subjective elements that make it difficult to outline minimal risk. It is equally difficult to consider that the risk does not only depend on the individual person but quasi on the entire context in which this risk occurs. Taking a sample of blood, for instance, cannot be classified per se as a minimal risk. The subject-related binding nature of the definition results from the fact that a patient with Alzheimer's dementia may well consider having blood taken to be a great strain. The fundamental context-related binding nature of the minimal risk is expressed in this example in that taking blood can indeed be classified as a great risk, depending on the situation. The context alone prescribes to which category the taking of blood samples belongs. For instance, taking a blood sample from a patient on Marcumar is a completely different matter to taking blood from a healthy subject, quite independently of the subjective assessment of taking blood. Equally, taking a blood sample in a hospital is more likely to involve the danger of infection than taking a blood sample in a free practice; and the danger of a haematoma resulting from blood being taken depends, in turn, on the experience or disposition of the person taking the blood as well as on the situation in which the blood is taken. Taking blood under emergency conditions, for instance, is considerably more risky than taking routine blood samples in daily work on the ward.

Minimal risk, in particular, is a relational and context-bound term, and it will be difficult to stipulate minimal risks based on a certain pattern. After all, every threshold, regardless of how it has been defined, from which a risk should be acceptable, will have a limit that will only be defined on the basis of opinions of the individuals participating in a definite situation and not on the basis of figures.

These contemplations are of importance because we can conclude from them that an objective method will never be found in order to define a minimal risk.

One could of course object at this point that, despite all subjectivity, it is nevertheless possible to define certain risks that are not acceptable in any case. One could object that statistics should be able to at least stipulate a certain more or less broad range in which the probability of damage falls, if not a point. Both objections are justified, and both objections highlight the fact that, despite all limitations, risk analysis will remain a valuable tool. It is not the fundamental usefulness of this tool that is under debate, but the degree of validity attributed to statistics.

If, therefore, one would like to determine research risks, one must first clarify which *phenomena* are of importance at all with regard to risk analysis; this preliminary definition is not a natural scientific but a normative definition. But not only the determination of the type of risk or damage, but also the determination of the limit from which damage should be evaluated as such, is the product of a value assessment. For this reason, there cannot be a purely objective and purely unbiased determination of risk. There cannot be such a thing because scientists themselves cannot disengage from subjective decisions. There can also not be such a thing because risk determination and, above all, risk assessment cannot be left to scientists alone. The point of view of a *lay person* is of no less importance than that of the expert in the case of risk assessment, for example.

A more promising possibility of determining minimal risk would be to empirically determine such a scale. One could, for instance, draw up a list of undesirable side-effects and ask certain specific group of the population for their subjective assessment of the side-effects, using standardised questionnaires. Using this psychometric method, the fact that damage can ultimately only be made comprehensible from the subjective view of the individual subject is borne in mind. Such a psychometric investigation would be the first step towards achieving such individualisation, because in this way it could be ascertained how differently a certain method of investigation is perceived by various patient populations. The more specific the population to be asked, the more informative and formative the result would be. After all, tendencies could be made out which, for instance, could highlight the fact that people with a certain disease perceive certain risks or side-effects differently to other groups of people. The limitation of such a process is, of course, that the specific test subjects could have a different conception of damage to that of the determined average. For this reason, a scale created in such a manner can only serve as a preliminary matrix, which would then have to be complemented by the subjective and personal assessment of each individual test subject.

25.3.4 Duty of Solidarity of Decisionally Impaired Patients?

Various regulatory guidelines, such as the Belmont Report of 1978, or the Biomedical Convention of the European Council of 1997 as well as bodies of law such as the French Research Law of 1988, have declared research with no individual benefit on decisionally impaired persons to be permissible in the presence of certain preconditions (Maio 1994, 2002a, b, Taupitz 2002). In reviewing these regulatory guidelines, it becomes noticeable that frequent references are made to the Principle of Justice as an argument for the enlistment of even decisionally impaired patients for scientific experimentation. In its 1977 report on research on psychiatric patients, the Ethics Committee of the American government even ascribes decisionally impaired patients a 'duty of beneficence'. Primarily American bioethicists and theologians have formulated a general duty of solidarity for decisionally impaired persons as justification for non-therapeutic research on these groups of people (cf. Caplan 1984, Engelhardt 1979, McCormick 1974, Veatch 1987). This argument with regard to the principle of justice would demand that all people have a duty of solidarity since otherwise the decisionally impaired subjects would be discriminated unjustified. The argument of Ludger Honnefelder is that of non-discrimination, i.e. not to exclude certain groups of patients from scientific progress – as mentioned in the first chapter of this paper. He advocates non-therapeutic research on decisionally impaired persons with the argument of an 'impending discrimination' of decisionally impaired persons (Honnefelder 1998). The Principle of Justice also resonates in this mode of expression. But how can justice guide decisions in this context? This question shall be explored in recourse to John Rawls.

It goes without saying that John Rawls' 'A Theory of Justice', published in 1971, is one of the most influential philosophical works of the twentieth century; it has greatly influenced a variety of disciplines, from political philosophy and legal and social philosophy to politics and business administration. The work could also be relevant to our question.

The fundamental idea that John Rawls develops in his 'A Theory of Justice' is based on the assumption that whatever free and equal citizens agree upon under fair conditions is just. He uses a thought experiment, which many readers will be familiar with, to justify the grounds for generally applicable Principles of Justice. He proposes a hypothetical decision situation in which people would be put in an original state of equality and would have to name principles of distribution according to which they would like the most important institutions of their society to be ordered. Acting from behind a 'veil of ignorance', these free and rational persons would – as a result of a rational choice – decide in favour of two central normative principles of social justice which would guarantee the greatest possible proportion of basic freedom and primary goods in the event that they should happen to find themselves among the least favoured.

The Two Central Principles of justice are:

(1) Each person is to have an equal right to the most extensive total system of equal basic liberties compatible with a similar system of liberty for all (Rawls 1998, p. 81).
(2) Social and economic inequalities must meet two conditions: first, they must be attached to offices and positions open to all under conditions of fair equal opportunities and, second, they must bring the greatest benefit to the least advantaged members of the society (Rawls 1992, p. 160).

I would like to examine the first basic principle with regard to our question, since Rawls clearly considered the first principle to have priority over the second. To what extent, then, is Rawls' so-called principle of liberty applicable to our problem? Rawls' basic liberties include the following liberties: freedom of political thought, freedom of speech and assembly, freedom of conscience and thought, personal liberty, the right to personal property and protection against arbitrary arrest and imprisonment (Rawls 1998, p. 82). For our problem area, personal liberty is of particular importance, since Rawls would like to see it determined within the meaning of the 'liberty and integrity of the person' (Rawls 1992, p. 161). For this reason, it seems obvious to apply this principle to our problem area.

If it can be assumed that research on a decisionally impaired patient could be interpreted as an infringement of his liberty and integrity, the question arises as to whether there is any constellation that might justify such a restriction of liberty. Rawls clearly states the following on this matter:

A basic liberty that falls under the first principle can only be restricted for the own sake of liberty, i.e. only for the sake of the same liberty of another basic liberty and to optimise the entire system of liberties (Rawls 1998, p. 232).

According to the Rawls' first Principle of Justice, non-therapeutic tests on decisionally impaired persons would only be approved if such approval were to achieve an increase in basic liberty for those affected. According to this, even an individual benefit that only maximises certain primary social goods would not suffice to justify a restriction of the basic liberty. This is precisely where Rawls' main statement lies, which he recorded as follows:

In practice, the priority of liberty implies that a basic liberty may be restricted or denied solely for the sake of one or more other basic liberties, but never for reasons of the greater public good or due to perfectionist values (Rawls 1992, p. 165).

One could object to this formulation, but this methodical individualism is the quintessence of Rawls' Theory of Justice, which transforms justice into a rational choice of prudence rather than a moral decision, and according to this theory, it would only be possible to endorse a non-therapeutic experiment on a decisionally impaired person if an increase in basic liberty, i.e. an increase of capacity to consent can be achieved for the person concerned. Since by definition non-therapeutic tests do not bring any advantages at all to the patient, let alone an increase in basic liberties, no justification of any kind for restricting the freedom of decisionally impaired persons can be derived from Rawls' principle of liberty (Maio 2002c).

Hence although considerations on justice – if one wishes to follow Rawls – are irrelevant to the question of any duty of solidarity decisionally impaired persons may have, it is clear that research on people indeed touches upon the Principle of Justice to a great extent. And it touches upon justice not only in the meaning of a just distribution of research risks but also in the meaning of a just distribution of the opportunities associated with research. This is a crucial point as primarily it is not the research which infringes the liberty and integrity of the subject but the disease itself. Particularly in mentally disturbed patients the basic liberties are destroyed by the disease, e.g. transiently by acute psychosis or irreversibly by dementia (Helmchen 2008). Assuming that research constitutes assistance to decisionally impaired persons in the medium term, an institutionally prescribed global renouncement of all kinds of research on these groups of persons would also be under pressure to legitimise themselves.

In this way, John Rawls makes it clear that, under certain conditions, the institutionally prescribed global renouncement of all kinds of research on these groups of persons could also be ethically not acceptable. The principle of Justice as Fairness therefore points out that a society has to seek ways to also enable decisionally impaired patients to partake in scientific progress, especially in order to give them the feeling of self-respect. Therefore, non-discrimination with regard to not withhold scientific progress from decisionally impaired patients is an important reason for legitimising such research. This was the central conclusion of the Ethics Committee at the German Board of Physicians (see Zentrale Ethikkommission 1997).

But – and this is the decisive question – can this way in fact consist of calling upon decisionally impaired patients for non-therapeutic research? If we now consider the problem area of research on dementia patients, the question already arises as to whether it would not be more obvious to understand the obligation to restore basic liberties as an obligation to provide better social care than as an obligation for more research. For this reason, it would be a reduced assumption of a certain preconception of medicine if one only wanted to define the help to grant basic liberties in the meaning of intensified biological research. As long as, on the other side, the intensification of biological research prospectively now has the possibility for patients suffering from Alzheimer's dementia to be given back part of their basic liberties, research remains an aspect of this obligation. As long as the study remains non-therapeutic, it will not manage to legitimise a restriction of liberty from the perspective of Justice as Fairness. For this reason, it would not correspond to Justice as Fairness if one had the intention of imposing a duty of solidarity for patients with similar diseases on decisionally impaired patients, as long as this obligation involves a loss of liberty. Interestingly, John Rawls also held this in recourse to man's being as an end in himself:

Treating people in the basic plan of society as an end in itself would mean renouncing such advantages that would not improve everyone's prospects. On the other hand, treating people as a means constitutes imposing even worse chances of survival on already disadvantaged people (Rawls 1998, p. 206).

This argumentation in particular highlights the extent to which Rawls' Theory of Justice adheres to Kant and how close the connection is to the principle of **human dignity**. This vicinity to human dignity is expressed in his search for a generalisable norm justification system in the form of the original position which, ultimately, is nothing other than 'general rationality'; it is primarily expressed in his catalogue of basic liberties, which fundamentally represents an interpretation of human rights (Kress 1999).[1]

The principle of human dignity is of great interest to the assessment of research on decisionally impaired patients because human dignity in particular is produced as the most frequent argument against such research. Once, therefore, the attempt has failed to name an ethical principle that could justify the instrumentalisation of humans in the interest of third parties, finally the premise itself should be questioned, according to which non-therapeutic research on decisionally impaired patients is deemed to be a violation of human dignity. The crucial question is therefore not: When can we instrumentalise a person in his core area? Instead, the question is: Is there any non-therapeutic research on decisionally impaired people that does not constitute instrumentalisation in the core area?

25.3.5 Violation of Human Dignity?

The matter of human dignity is a principle of morals used to justify why humans have basic rights. A further possibility of concretisation of human dignity would be to define human dignity negatively. Thus, one could see a common feature of all violations of human dignity in the degradation or debasement of humans (Balzer et al. 1998, p. 27). An alternative would be to draw up a catalogue of obligations related to dignity, simply from a certain natural constellation of human needs. Rawls made a similar attempt with his basic liberties. Hence one could view human dignity as a demand for a minimal area of rights, which could not be violated under any pretence, because these minimal rights are accorded to the bearer of human dignity merely on the basis of his existence and completely independently of his morality or capability (Birnbacher 1987, p. 81). In the so-called purpose formula of the categorical imperative, Kant characterised the term dignity within the meaning of the individual person as an end in himself, according to which every person should be used 'at all times together as a purpose, never simply as a means'. According to

---

[1] The basic liberties include Rawls' following liberties: 'Freedom of political thought and liberty of conscience; freedom of association; and the freedom defined by the liberty and integrity of the person, as well as by the rule of law; and finally the political liberties' (Rawls 1982, p. 162).

this, the aspect of autonomy would be important in the assessment of instrumental-isation, which is respected as long as humans are not treated as a mere disposable quantity. The crucial aspect here is not whether or not the person can perceive this autonomy; the decisive issue is respecting this autonomy when dealing with people, which forbids a person being placed in the service of interests alien to the subject. Human dignity can therefore be interpreted within the meaning of the prohibition of instrumentalisation. But not every instrumentalisation of humans constitutes a violation of the principle of human dignity. If every instrumentalisation were to be assessed as an attack on the core of personal rights of protection, social interaction among humans would be utterly impossible because social roles are unavoidably connected to mutual instrumentalisations that are tacitly tolerated or even accepted.

When considered more closely, several forms of instrumentalisation can be made out, which are qualitatively so varied that they are also subject to different degrees of the need to legitimise themselves. It would be helpful to further differentiate instru-mentalisation, whereby it would lend itself to divide it into essential and partial instrumentalisation. Ultimately, the ethical legitimacy of instrumentalisation stands and falls with the assessment whether the respective test subject would consent to this instrumentalisation if he were responsive. Thus, for our problem area, a non-therapeutic test on a decisionally impaired person would be an *essential instru-mentalisation*, even in the event of the minimum conceivable risk, if the test subject is only selected on the basis of his defencenessless. There is clearly no justification for tests on a decisionally impaired person if they are only carried out because there is the chance to do so or because it is the easiest way, since there would then be no respect of the liberty of others. A *partial instrumentalisation,* on the other hand, would be determined such that it would have to be comparable with the instrumen-talisation of humans in their daily lives, because such forms of instrumentalisation are generally accepted. If, then, it can be assumed that – if the person could be asked – they would consent to such research because they would have no reason not to do so, in the face of the negligible stress, then we can talk of a partial instrumen-talisation. The argument is the probability that the research subject would have given his consent if he hat the ability to do this and not, for instance, the postulate of an obligation to participate in a test, however justified. If, therefore, it were guaranteed that the risk involved in a study would indeed also, and particularly be considered as minimal by the test subject, then in such a case this partial instrumentalisation could appear to be ethically justified, because in the case of a minimal risk, the pre-sumption of the actual consent by the test subject fundamentally appears to be more justified than in the event of a greater risk.

And yet this argumentation also has its weak points. Let us take the case of research associated with minimal risk and which is indeed comparable with the form of instrumentalisation that we would accept as a matter of course in our daily lives. In this case, we can now say that it is a partial instrumentalisation that indeed appears to be legitimisable, and for this reason, it would be possible to also jus-tify such an instrumentalisation with decisionally impaired persons. If, however, such research is declared as generally justified across the board for all decisionally impaired persons, the situation arises that we do not apply the same standard for persons able to give their consent and for decisionally impaired persons. In the case of those able to give their consent, we would naturally nevertheless request their consent, even in the event of the most partial level of instrumentalisation; in other words, we would grant the person able to give his consent the right to reject participating in a study, although it may involve little stress. If we now, in the case of a decisionally impaired person in the same study, allege substituted consent without further proviso, this would constitute an inequality that would not be tolerable, because we would grant liberties to those able to give their consent that we deny decisionally impaired persons. In order to evade this inequality that we cannot ask decisionally impaired patients, a compromise would have to be created, which is where the substitute consent forms and the protective function of an interdisciplinary Ethics Committee have an effect. The central function of such a review board would be the attempt to get a more objective judgment. That means not that the committee has the right to speak for the decisionally impaired persons but that it has to be excluded that the interests of third parties are conditioning the decision of making patient participate in studies.

As a hypothetical construction, a substitutional consent naturally has its limitation with regard to its inherent insecurity. Because ultimately, a decision by proxy – just as with an anticipated decision in the form of a durable power of attorney for health care – can never be a guarantee that the decision will be made in the same way the represented person would de facto have made it. A substitutional decision can always only remain a hypothetical one, and this remaining insecurity robs any decision that accepts considerable damage or a greater risk of its ethical legitimacy. All of the aforementioned official guidelines and conventions have attempted to meet the remaining insecurity on the factual state of wishes with further provisos, including granting the test subject the possibility to reject his participation, as is it never acceptable to restrain someone for research. In practice this would mean that the surrogate had to be with the subject nearly at all times during the research in order to have to possibility to call the study to a halt. Under these provisos, it therefore indeed appears to be justified to also declare non-therapeutic research on decisionally impaired persons permissible.

25.4 Conclusions

If, therefore, a substitutional decision is made obligatory, and the patient would have to retain the possibility to object, and if then also an Ethics Committee were installed as a protective instance, under these provisos it appears that also such non-therapeutic tests on a decisionally impaired person would also only be ethically legitimised in the case of minimal risk. However, the symbolic statement that would emanate from such practice – despite its legitimacy – would have to be taken into consideration. In the Federal Republic of Germany in particular, the absolute inviolability of decisionally impaired patients has become a symbol of their worthiness of protection, and if this inviolability is relativised, regardless of how well founded, one has to accept the consequences that such a practice would have at the symbolic level. For this reason, the actual challenge regarding research on decisionally impaired patients does not lie in the argumentation but in the cautious handling of this understandably sensitive problem area, which has a symbolic connotation.

References

Ashcroft R (1998) Selection of human research sujects. In: Chadwick R (ed) Encyclopedia of applied ethics, vol 2. Academic Press, San Diego, CA, p 627–639

Balzer P, Rippe KP, Schaber P (1998) Menschenwürde versus Würde der Kreatur. Karl Alber, Freiburg/München

Birnbacher D (1987) Gefährdet die moderne Reproduktionsmedizin die menschliche Würde? In: Braun V, Mieth D, Steigleder K (eds) Ethische und rechtliche Fragen der Gentechnologie und der Reproduktionsmedizin. Schweitzer, München, pp 77–88

Caplan AL (1984) Is there a duty to serve as a subject in biomedical research? IRB: Rev Human Sub Res 6:1–5

Engelhardt TH (1979) Basic ethical principles in the conduct of biomedical and behavioral research involving human subjects. Tex Rep Biol Med 38:139–168

Frank S, Agich GJ (1985) Nontherapeutic research on subjects unable to grant consent. Clin Res 33:459–464

Glass KC, Speyer-Ofenberg M (1992) Incompetent persons as research subjects and the ethics of minimal risk. Camb Q Healthc Ethics 5:362–372

Helmchen H (2008) Ethische Erwägungen in der klinischen Forschung mit psychisch Kranken. Nervenarzt 79(7):1036–1050

Honnefelder L (1998) Zur ethischen Beurteilung von Forschung am Menschen unter besonderer Berücksichtigung der Forschung an einwilligungsunfähigen Personen. In: Markus P (ed) Möglichkeiten, Risiken und Grenzen der Technik auf dem Weg in die Zukunft. Friedrich-Ebert-Stiftung, Bonn, pp 131–140

Jonas H (1969) Philosophical reflections on experiments with human subjects. Daedalus 98:219–247

Kopelman LM (1989) When is the risk minimal enough for children to be research subjects? In: Kopelman LM, Moskop JC (eds) Children and health care. Moral and social issues. Kluwer, Dordrecht, pp 89–99

Kress H (1999) Menschenwürde im modernen Pluralismus: Wertedebatte – Ethik der Medizin – Nachhaltigkeit. Lutherisches Verlagshaus, Hannover

Levine RL (1986) Ethics and regulation of clinical research. Urban Schwarzenberg, Baltimore/München

Maio G (1994) Forschung am Menschen. Eine französische Debatte. Ethik in der Medizin 6(3):143–156

Maio G (2002a) Die Forschung am Menschen als ethisches Problem. Philosophische Analyse und historischer Kontext. (Medizin und Philosophie, Bd. 6) Frommann-Holzboog, Stuttgart

Maio G (2002b) The cultural specificity of research ethics – or why ethical debate in France is different. J Med Ethics 28:147–150

Maio G (2002c) The relevance of the principle of justice for research on cognitively impaired patients. Theor Med Bioeth – Philos Med Res Pract 23:45–53

Maio G (2003) Research ethics and the principle of justice as fairness – a restatement. Theor Med Bioeth – Philos Med Res Pract 24:395–406

McCormick RA (1974) Proxy consent in the experimentation situation. Perspect Biol Med 18:2–20

National Commission for the Protection of Human Subjects of Biomedical and Behavioral Research (ed) (1978) The belmont report. Ethical principles and guidelines for the protection of human subjects in research. DHEW Publications No. (OS) 78-0012, Washington, DC

National Council on Bioethics in Human Research (NCBHR) (ed) (1992) Report on research involving children. Prepared by the consent panel task force of the national council of bioethics in human research (NCBHR) with the support of the Canadian paediatric society. NCBHR, Ottawa

Ratzan RM (1980) Being old makes you different: the ethics of research with elderly subjects. Hastings Cent Rep 10:32–46

Rawls J (1971) A theory of justice. Harvard University Press, Cambridge

Rawls J (1982) Social unity and primary goods. In: Sen A, Williams B (eds) Utilitarism and beyond. Cambridge, Cambridge University Press, pp 159–185

Rawls J (1992) Der Vorrang der Grundfreiheiten. In: Hinsch W (ed) Die Idee des politischen Liberalismus. Aufsätze 1978–1989. Suhrkamp, Frankfurt a. M., pp 159–254

Rawls J (1998) Eine Theorie der Gerechtigkeit. Suhrkamp, Frankfurt a. M., 8 Auflage.

Taupitz J (ed) (2002) The convention on human rights an biomedicine of the council of Europe. Springer, Berlin, Heidelberg, New York, NY

Veatch RM (1987) The patient as partner. A theory of human-experimentation ethics. Indiana University Press, Indianapolis, IN

Zentrale Ethikkommission bei der Bundesärztekammer (1997) Stellungnahme "Zum Schutz nicht-einwilligungsfähiger Personen in der medizinischen Forschung". Deutsches Ärzteblatt 94:B811–B812

Chapter 26
Ethical Concerns in Carrying Out Surveys of Psychiatric Morbidity

Howard Meltzer and Traolach S. Brugha

Contents

H. Meltzer (✉)
Department of Health Sciences, University of Leicester, Leicester, LE1 6TP, UK
e-mail: hm74@le.ac.uk

H. Helmchen, N. Sartorius (eds.), *Ethics in Psychiatry*, International Library
of Ethics, Law, and the New Medicine 45, DOI 10.1007/978-90-481-8721-8_26,
© Springer Science+Business Media B.V. 2010

Abbreviations

CIOMS   Council of International Organizations for Medical Science
DSM-IV   Diagnostic and Statistical Manual of Mental Disorders, 4th edition
EPA   Environmental Protection Agency
GHQ12   12-item General Health Questionnaire
GP   General Practitioner
ICD 10   International Classification of Diseases – 10th revision
NSIs   National Statistical Institutes
NSOs   National Statistical Offices
OCD   Obsessive-Compulsive Disorder
PTSD   Post-Traumatic Stress Disorder
SDQ   Strengths and Difficulties Questionnaire

26.1 Introduction

This chapter is primarily concerned with ethical considerations in conducting psychiatric morbidity surveys of the private household population that are carried out by non-medically trained social survey researchers and organisations but may involve medical input. The use of epidemiological surveys among samples identified through health services will also be considered.

In recent years ethical considerations across the psychiatric research community have come to the forefront. This is partly a consequence of legislative change in human rights and data protection, but also a result of increased public concern about the limits of health research in general and mental health research in particular. National Statistical Offices (NSOs), National Statistical Institutes (NSIs) as well as market research organisations which conduct national surveys of psychiatric morbidity on behalf of government continue to monitor the ethical consequences of their research activities. This increased concern for accountability in these spheres has led to the establishment of systems for research governance – ways of discovering and sharing information that are open to public scrutiny and can be seen to be subject to the highest ethical standards. In an era of advanced information and communications technology and concerns about the security of confidential records, access to and the management of data collected from psychiatric morbidity surveys require ethical scrutiny (Social Research Association 2003).

26.2 Ethics Committees

In most countries, a formal review of ethical issues takes place before researchers are allowed to conduct psychiatric morbidity surveys.[1] Even though such surveys involve sampling private household populations and do not involve somatic invasive procedures, there is a growing insistence that an Ethics Committee gives ethical approval before a psychiatric morbidity survey is launched. Some people believe that ethics committees apply only to interventionist research such as medical experiments or pharmaceutical trials. In fact, most generalisable social survey research is interventionist. Interviews and surveys are interventions in the lives of the population studied – quite probing questions are asked – so some form of ethical review for such studies is desirable (Social Research Association 2003).

However, a balance is required so that over-protective and bureaucratic procedures do not restrict valuable, particularly innovative, social research methodology. Medical or health service ethics committees may not fully understand the checks and balances of social survey research. For example, there may be a difference over what precisely constitutes informed consent (Social Research Association 2003). For example, the ethical requirements for recruiting participants are different in clinical trials from those used in sampling members of the public in a national survey. Allowing an interviewer to attempt to persuade people to take part in a study is regarded as coercion in social research and, therefore, not regarded as informed consent. In clinical trials, however, such persuasion is a common feature of subject recruitment for drug trials (see Chapters 9 and 20).

There are some anomalies in the ethical review process which leaves a lot of responsibility in the hands of the researcher – there are no legal penalties or sanctions for not submitting for ethical approval for research on members of the private

---

[1] See Chapter 6

household population nor for failing to fulfil the requirements of the ethics committee even though there may be organisational penalties for doing so. There are some concerns about both the competence and the knowledge of some of these committees which can unnecessarily restrict research activity to the detriment of social scientific progress (Social Research Association 2003).

Once approval has been given for the project to be conducted a follow-up process initiated by the Ethics Committee will confirm whether or not the project has been completed or abandoned or if there were any difficulties with the study which were not anticipated in the original application.

Some commentators suggest that, since ethical decision taking may occur throughout the life of a project, ethics committees should maintain review of the project throughout and not consider their job as merely to cast ethical judgement at the outset. However, to avoid the 'big brother' connotations of such a supervisory model, ethics committees should instead ensure at the outset that researchers have established a system for the maintenance of ethical 'awareness' throughout the project to allow for the occurrence of unanticipated ethical problems, or problems that could not have been foreseen at the outset (Social Research Association 2003).

The ethical concerns in carrying out surveys of psychiatric morbidity are described below in the chronological order of the stages in the survey process. Ethical concerns are relevant to all stages of the survey process from sampling design through fieldwork to archiving the cleaned and anonymised dataset. All of these elements are scrutinised by Ethics Committees.

26.3 Sampling

26.3.1 Coverage

Most counties (without national population registers) which conduct health or psychiatric surveys apply a multi-stage, clustered, random sampling design utilising lists of named individuals or household addresses. The non-private household population, i.e. those living in care or nursing homes, military establishments, nurses homes or education establishments, religious retreats, prisons, hostels for workers or homeless people, are often excluded. However, there is a lot of evidence that a greater proportion of people with the more severe mental disorders live in some of these communal establishments particularly residential care homes for elderly people (Meltzer et al. 1996) prisons (Singleton et al. 1998) and hostels for homeless people (Gill et al. 1996). The ethical issue here is to what extent efforts should be made to include these socially excluded populations in a national survey. Despite the high prevalence of mental disorders in some of these groups, the overall population prevalence will not increase that much (perhaps, an increase of a few decimal places) if the institutional population was included.

26.3.2 *Principles of Sample Selection in Psychiatric Survey Research*

Surveys of psychiatric morbidity are based on the principle of statistical inference: being able to generalise from a sample to a population, for example, to estimate the prevalence of common mental disorders or psychosis in the country by interviewing 5000 or 10000 sampled individuals. Survey researchers use scientific methods to select samples for surveys so that, within the quantifiable limits of sampling error, the sample selected is representative of the population of interest.

Self-selected volunteers therefore would not form a valid scientific sample from which generalisations to the population could be made, nor are substitutes for those originally selected for the survey permissible.

The principle of statistical inference underlying the survey method is in contrast to the basic principle underlying the experimental method on which psychiatric intervention studies or indeed any clinical trials are based. Therefore it may be useful to contrast the two principles. It is intrinsic to the clinical method that prior differences between the groups are controlled by randomisation so that differences in outcome can be attributed solely to the experimental treatment. Although undesirable, it is not detrimental to a clinical trial if a sizeable proportion approached to take part do not participate; what matters is that those who do agree are randomly assigned to experimental and controls groups so that there are no systematic differences between the two groups which could confound the effects of the experimental treatment (internal validity). Hence if people were allowed to choose whether to be in the experimental or control group differences between the groups would invalidate the trial. It is also crucial that once they have agreed to take part and been assigned to a group, those participating continue in the trial. If there is differential loss between the two groups, particularly if the drop-out is related to the nature of the trial treatments, the principle of comparability between the groups is violated.

26.4 Informed Consent

Informed consent involves providing people with sufficient information to make a reasoned decision about whether to take part in the survey: they must know what will be involved, what any consequences to them personally, either positive or negative, might be and thus be in a position to decide whether or not to participate.[2]

The importance of obtaining informed consent from those participating in a clinical trial is clear: it stems from the potential direct personal consequences of taking part, whether in the experimental or the control group. Those participating do not

---

[2]See also Chapter 9

know which group they have been allocated to and certainly cannot choose which to be in. Participants therefore have to be informed on two counts: they need to know what the treatment will involve and what effect it might have on them; and they need to be aware that they run a risk of either not having treatment which subsequently proves to be effective, of having treatment which is not effective, or of suffering some unintended adverse effect of treatment.

Survey interviewers require many of the same skills as those conducting clinical trials. They too are trained to seek consent from people selected to participate in the survey and need to explain what will be involved (e.g. the length of the interview, and any other tasks, such as competing self completion exercises). However, survey research does not normally involve any invasive procedures or any consequences to the individual so there is minimal risk to an individual as a consequence of taking part in a survey.

In survey research, it is essential to maximise response in order to ensure that the generalisability of the results will be valid. Thus survey interviewers are trained to seek co-operation from everyone selected. Since there is little personal benefit in taking part in a survey, interviewers need to motivate the public to take part. However, they are also trained to inform those selected that their participation is voluntary and to accept definite refusals to take part. Agreement to participate has to be obtained based on a clear understanding of what will be involved so interviewers are trained to provide such information and answer any questions or concerns. Even when initial agreement is given, respondents can (and do) refuse to answer certain questions or terminate the interview before the end. Dropping out part way through, although unwelcome, is less of a threat to the validity of the results than not taking part in the first place since at least some information about the respondent will have been collected. The information from a partial interview can be used as consent was given before interviewing began but is often of limited use.

26.5 Ensuring that Consent Is Informed

In their review of the survey process, Ethics Committees are keen to ensure that consent is truly informed. The clinical trial recruiter or interviewer can ensure that those selected have understood what participation will entail and can answer any questions and deal with any concerns. Questions raised on surveys normally relate to what the survey is for, how the respondent has been selected, and what will happen to the information about the respondent. Attempting to seek informed consent by letter (the normal method of obtaining agreement in an opt-in procedure) suffers from a number of serious drawbacks and as such is not generally used in survey research. The choice between opt-in and opt-out procedures is the most frequently discussed issue between researchers conducting mental health surveys and members of ethical committees.

To obtain informed consent by letter a number of conditions need to be fulfilled. First, the letter must reach the person whose consent is sought. This can not always be guaranteed because the first stage of the sampling in some countries is to select addresses. There is no way of knowing the names of those living at the address so that named individuals can be approached. Without a personal visit by an interviewer it is not possible to apply scientific sampling procedures to select at random an individual to take part in the survey so that the consent of the selected person may be sought rather than that of another member of the household. Thus it is not possible to send a letter to the individual who would be selected for the survey and to give them the opportunity of taking part.

Second, the person whose consent is sought must read the letter and fully understand its contents. It is common practice to send advance letters to the selected addresses informing the residents that an interviewer will be visiting and providing brief information about the survey. The aim of such letters is not to seek agreement to participate but to prepare the household for a visit from an interviewer who can explain about the survey, collect sufficient information to sample one individual for the survey and then approach the selected individual for agreement to take part in the survey, providing full information and answering any queries.

Although interviewers find it useful to refer to the advance letter when making an initial call, in order to overcome reluctance to open the door to a stranger, research carried out by the Social Survey Methodology Unit in Office for National Statistics in the UK (White et al. 1998) shows that some households do not even remember receiving such a letter (and certainly not all members of the household have necessarily seen it). Households that normally discard junk mail without opening the envelope may have treated the survey letter in the same way. Even when someone in the household reads the letter they may only glance at it briefly and not study the contents closely. White et al. (1998) showed that most recipients do not recall the contents of advance letters in any detail. They found that a very short letter was therefore more effective. This research (and similar research reported in Groves and Couper 1998) indicates that most recipients would be very unlikely to meet the criterion stated above of reading a detailed letter and fully understanding its contents. They therefore would not fulfil the criterion of being *fully informed* before giving or withholding consent.

Thirdly the person whose consent is sought must be motivated to agree to participate and to convey their agreement. There is no direct benefit to individuals in taking part in a survey yet on most government surveys somewhere between 60 and 80% of those contacted by an interviewer agree to take part. Most research (e.g. Groves and Couper 1998, Groves et al. 1992) indicates that people are motivated by a desire to contribute to the public good, particularly in the case of government surveys. They therefore need to understand the importance of the survey, how the results will be used and why their own cooperation is important (since *they* have been selected for the survey and no substitutes are allowed). Others take part because they like being interviewed or are interested in the subject of the survey. Interviewers emphasise that most people enjoy taking part in surveys. The cost to respondents is mainly their time so interviewers make great efforts to carry out interviews at times convenient to respondents.

Within the research programme of psychiatric morbidity surveys in Great Britain two methods were implemented to obtain informed consent.

26.5.1 Sending an Advance Letter Followed by an Interviewer Visit

This method was used in the series of three adult psychiatric morbidity surveys, 1993, 2000 and 2007 (Meltzer et al. 1995, McManus et al. 2009, Singleton et al. 2001) Sending an advanced letter aimed to encourage participation in the survey without explicitly seeking agreement or not to participate: the letter itself is not intended as a means of obtaining informed consent. The address of the survey organisation and a contact phone number were provided on the letter in case of queries. A small minority (less than 5%) of selected individuals rang to say they did not want to take part in the survey and therefore did not want an interviewer to call. Interviewers then visited the remaining addresses to select the individual to be interviewed, to explain fully about the survey and to seek their agreement to take part. This is the normal understanding of what constitutes informed consent in the survey context. There is no evidence that such standard procedures are viewed by the public as in any way coercive as interviewers always make clear that participation in a survey is voluntary and that even if they start the interview people are informed that they can refuse to answer particular questions.

26.5.2 Opting-Out in Response to a Letter

In the two psychiatric morbidity surveys among children and young people, 1999 and 2004 carried out in Great Britain (Green et al. 2005, Meltzer et al. 2000), children were sampled from an administrative database which contained personal details of nearly all the children under 16 years of age in the population. Because the list was not held by the national survey organisation, the sample selection and advanced letters were sent out by the guardians of the database in order to protect the confidentiality of the children and their contact addresses.

The advanced letter was sent to the parents or primary caregivers asking them to reply only if they did not wish to take part in the research. Failure to reply was taken to mean permission for approach by an interviewer, although, there is no means of ensuring that the appropriate person has received, read and understood the letter so some of the selected respondents may not have been informed adequately about the proposed research. In 1999, 6% opted out and 15% refused when the interviewer made a visit (Meltzer et al. 2000).The corresponding rates in the 2004 survey were 9% and 17% respectively (Green et al. 2005).

Whether individuals are motivated to reply by opting-out depends crucially on the amount of effort replying involves for them. Different procedures have been shown to produce very different rates of opting-out. The main options are: providing a phone number; providing a freephone number (used in the GB child mental health survey); and providing a prepaid return card. The last produces the highest opt-out rates.

26.6 Content of the Advance Letter

Even though few people bother to contact the survey organisation on receipt of the letter, its wording can influence whether or not the selected respondents agree to participate when an interviewer calls. The survey literature (reviewed in Groves and Couper 1998; see also Dillman 2000) has documented good practice with respect to drafting such letters. The main features are that the letter should be short and easy to read; it should stress the importance of the survey, the legitimacy of the survey sponsor and the survey organisation, and the importance of the selected household or person's participation as well as the voluntary nature of their participation. Some key elements of the advanced letter are described below.

26.6.1 The Title of the Survey

Deciding upon the title of a psychiatric morbidity survey can be problematic for several reasons. The expression psychiatric morbidity can not be used as it is jargon – many people will not understand the academic terminology. Any expression with the word 'mental' in it may dissuade people from participating. Therefore, positive expressions such as emotional well-being are used. The ethical question here is whether calling a psychiatric morbidity survey, a survey of health and well-being is misleading or even a deception. Personal relation sensibilities conflict with the requirement of informed consent. However, once the interviewer begins (with, for examples, question on the symptoms of anxiety and depression) it soon becomes apparent what the survey is about. There tends to be two types of reactions. The respondent either says 'you are trying to find out if I am mad, aren't you?' or 'what interesting questions, no-one has asked me about my feelings before'.

26.6.2 Pledge of Non-disclosure

The general public and potential survey respondents may not perceive confidentiality as likely to be so rigorously maintained in surveys as with their medical records. Whether or not survey respondents perceive any danger to themselves of data disclosure, the advance letter should state the organisation's pledge of confidentiality. Loss of data from administrative authorities may have negative repercussions as each occurrence dents public confidence in the security of official statistics and makes the data collectors' job more difficult (Social Research Association 2003).

26.7 On the Doorstep

The interviewer normally goes through the same procedures in introducing a psychiatric morbidity survey as for any other survey – what is the subject of the survey, what are the aims of the survey, who is the sponsor, the name of the survey organisation carrying out the survey, how their household was selected plus the pledge of confidentiality. The issue of a payment for participation is an important ethical concern. There are differing views on incentive payments. In some countries any form of payment is regarded as coercion; in other countries payment is regarded as a justifiable expense for the use of the respondent's time. Sometimes, a small payment or gift is given to the respondent at the end of the interview as a token of gratitude.

26.8 Conducting the Interview

26.8.1 Sampling of Individual Within the Household or Interviewing All Household Members

Once an address has been selected according to the sampling design specifications either everyone in the household is interviewed or one person is sampled at random using, for example, the Kish grid method (Kish 1949). The Kish grid method works very well in a statistical sense but can cause some incredulity if the selected respondent is the partner or carer of a person with mental health problems. 'Why me, you should be interviewing my wife/husband, s/he is the one with mental health problems?'. The interviewer has to explain the random nature of sampling to the respondent. Conversely, if the sampled individual just happens to be someone with mental health problems, s/he becomes suspicious that their medical condition has been disclosed by their doctor or the people providing them with health or social services. Yet again, it is the interviewer's responsibility to explain the random nature of sampling to the potential respondent.

Where all people in the household are interviewed, a different set of ethical issues come to the fore. Issues related to privacy are paramount. The confidentiality pledge should state that apart from the information not being passed to other government departments, or market research organisations or members of the public or press, information will not be passed among family members. The other consideration is respondent burden especially if there are four or five adults in the family.

26.8.2 Competence of the Selected Respondent to Give Consent

A major limitation in gaining informed consent lies with vulnerable populations. Such groups include adults with florid mental states, young adults, those with an intellectual disability and those living in institutions where decision making is taken over by staff. Whereas these restrictions may be evident for these vulnerable groups there are many individuals living in private households, looked after by relatives, who also may not be competent to give consent to participate in a survey, regardless of the topic being researched.

In order to protect the researcher from accusations of failing to secure informed consent a practice has grown of having subjects sign a consent form. While this may serve as some indication that the subject understands some of the implications of their consent to participate it may also compromise principles of confidentiality and anonymity.

Signed consent forms tend to be used in psychiatric morbidity surveys with a longitudinal element or when the researcher wants to link the survey data with administrative databases on for example, mortality or criminal convictions. At a minimum there should be clarity about any opt-in and opt-out arrangements, the length and degree of commitment required of respondents, and the precise goals of the research.

26.8.3 Competence of the Selected Respondent to be Interviewed

Whether the sampled informant is competent to give consent to take part in the survey has to be distinguished from competence to participate in the survey – to understand the meaning of the questions and give considered responses. One way of assessing competence to be able to respond to the whole questionnaire is to start with a module which is relatively straightforward, for example: SF12: a12-Item Short-Form Health Survey (Ware et al. 1996), K10: a ten item screen for serious mental illness in the general population (Kessler et al. 2003), GHQ12: a 12-item General Health Questionnaire (Goldberg and Williams 1988), SDQ: the Strengths and Difficulties Questionnaire (Goodman 1999). If respondents struggle with these, the survey can be terminated. This is also useful for assessing the competence of those where the language in which the survey is conducted is not the mother tongue of the respondent.

26.8.4 Use of Proxy Informants

There are numerous ethical issues surrounding the use of an informant who knows the sampled respondent well, that is a proxy informant. It is standard survey practice to ask a proxy informant for information about a sampled individual who can not participate owing to physical or mental illness, poor language skills, or absence during the field period. The sampled respondent may not be competent to participate in the survey but should be able to indicate whether someone else can answer the questions on his or her behalf. There are repercussions of using proxy informants. One does not know whether the sampled individual would want the information disclosed or whether the questioning may change the relationship between the subject and the proxy. However, the application of strict field procedures would preclude most questions in a psychiatric morbidity survey from being asked. In brief, proxies should only be asked factual questions normally about observable behaviour for example medication taken or services visited. They can not reasonably be asked to evaluate the mood of others. Hence, the survey guidelines about not asking proxy informants attitudinal questions would prohibit the use of most epidemiological instruments based on the ICD10: International Classification of Diseases – tenth revision (World Health Organisation 1992), or the DSM-IV: Diagnostic and Statistical Manual of Mental Disorders – fourth edition (American Psychiatric Association 1994), for research on mental disorders.

26.8.5 Asking Sensitive Questions

Some questions in psychiatric surveys can be regarded as sensitive in the respect that they may cause the respondent to be embarrassed or upset. The three groups of questions which tend to cause most distress are those relating to Post-Traumatic Stress Disorder (PTSD), Obsessive-Compulsive Disorder (OCD) and suicidal ideation or behaviour. Methodological research has shown that in private household surveys it is important to try and establish the traumatic event or the focus of the obsession. The reason for asking for details is that it is difficult to get over to survey respondents what should be included or excluded as a trauma or an obsession in a structured questionnaire. Without the details prevalence estimates tend to exaggerate the real prevalence of PTSD and OCD by including too many false positives. Is it ethical to ask for such details knowing that there is a possibility of upsetting or distressing the respondent? The difficulty in resolving this dilemma is that the consequences of asking the question may not emerge until after the interviewer has left.

26.8.6 Requests for Help

On some occasions, sample respondents ask the lay interviewer for help, particularly if it is the first time that anyone has spoken to them about their thoughts and feelings. Over a 1–2 h period the interviewer can build up an excellent rapport with the respondent. However, the field researcher is a lay interviewer, not a trained counsellor. The interviewer is normally an employee of either a National Statistical Office or a National Statistical Institute or works for a market research organisation subcontracted to the research organisation or even someone on a temporary contract employed solely for the purpose of interviewing people in the national psychiatric morbidity survey. Lay interviewers are told explicitly not to interfere. What they are allowed to do is tell the respondent to seek help from their GP. In addition interviewers carry with them a sheet with the names of ten to twenty societies which provide advice and assistance to people with different types of mental health problems. This information sheet is left with the respondent.

26.8.7 Follow Up Interviews by Clinically Qualified Professionals Such as Psychologists and Psychiatrists

In some psychiatric morbidity surveys mental health professionals may take part in supplementary or follow-up interviews in order to examine complex conditions such as psychosis. Therefore they are participating because of their assessment or diagnostic expertise. Because of their professional role and codes of ethical standards this places them in an unusual position because in a survey of the private household population they are not examining someone who is consulting as their patient. However, they may become aware of mental health issues or related issues such as self harm or risk of harm to others that in their professional role with patients, in a clinical service, they would be expected to deal with in a particular way. In the survey context they only have consent to collect information for the purposes of the research. Any other action will require the consent of the respondent, which cannot be presumed (see Chapter 3, Section 3.3). It is important in preparing health professionals taking part in surveys of the lay public that they understand that their role is essentially the same as that of a lay survey interviewer, which means that the pledge of confidentiality applies equally to the clinically qualified interviewer. However, there are rare exceptions in which the pledge of confidentiality is superseded by other considerations (see later sections on suicide issues, on *revisions to the pledge of confidentiality* and Chapter 11).

26.8.8 Requests to Change or Destroy Part of the Interview

Having spent 2 h in a person's home asking questions about their mental health and usually gathering information on several other related topics: risk factors, use of services, lifestyle behaviours, the interviewer can be asked by the respondents, without giving any reason, for the answers they have given to be deleted or destroyed. Respondents are quite entitled to ask for this to be done. This occurs very rarely and tends to happen within a few days when the respondent becomes concerned that they may have said things that they felt they should not have said. Reiteration of the pledges of confidentiality has little effect on those requesting deletion to change their minds.

26.9 Effect of Survey Participation

26.9.1 Effect on the Survey Respondent

Whether or not respondents have given verbal or written consent to participate in a survey of psychiatric morbidity, the social researcher still has an obligation to protect the subject as far as possible against potentially harmful effects of participating. Interviewers are briefed that if the respondent starts to become distressed or tearful that they should stop the interview and perhaps ask whether the respondent wants to continue, wants a break or wants them to come back another day. Interviewers can say at the beginning of the interview that some questions may be upsetting so at least the person interviewed know what to expect.

Cassell (1982) argues, people can feel wronged without being harmed by research: they may feel they have been treated as objects of measurement without respect for their individual values and sense of privacy. In many of the social enquiries that have caused controversy, the issue has had more to do with intrusion into subjects' private and personal domains, or by overburdening subjects by collecting 'too much' information, rather than with whether or not subjects have been harmed. In some cases a researcher's attitudes, demeanour or even their latent theoretical or methodological perspective – medical versus social model of health and illness – can be interpreted as doing an injustice to respondents. Examples include an offhand manner on the part of a survey interviewer. By exposing subjects to a sense of being wronged, perhaps by such attitudes, by such approaches, by the methods of selection or by causing them to acquire self knowledge that they did not seek or want, social researchers are vulnerable to criticism. Participants' resistance to future social enquiries in general may also increase as a consequence of such perceived lack of consideration.

On the positive side, many respondents find participating in a psychiatric morbidity survey a cathartic exercise. The majority of people who participate in surveys of psychiatric morbidity say that they enjoy the experience (Henderson and Jorm 1990), particularly mothers talking about the emotional or conduct problems of their children. Some respondents find the experience so positive that they want to stay in contact with the interviewer. There is anecdotal evidence of one interviewer being telephoned every week, another was stalked. Again the interviewer has to make it clear that they can not provide any help or service and encourage the respondent to go through the normal helping channels. Nevertheless, respondents can view the results of the survey when the report is published but they can not be told their own results of their psychiatric morbidity assessment.

26.9.2 Safety of Interviewer

Social researchers have a moral obligation to attempt to minimise the risk of physical and/or mental harm to their field staff when conducting an interview. The principal investigator may also have a legal obligation in terms of health and safety regulations to ensure that risk to lay interviewers is minimised. Structured questionnaires of psychiatric morbidity can cause the respondent to become aggressive but well-trained lay interviewers are briefed that if they feel slightly uncomfortable during an interview they must terminate the interview immediately.

In some areas, interviewers visit households in pairs but this is more to do with feeling safe working in a particular area regardless of the topic of the survey.

26.9.3 Effect on the Interviewer

Psychiatrists are accustomed to listening to the sometimes harrowing stories recounted by their patients with concern but also with a sense of detachment. Most lay interviewers do not have that capacity. Within the field period an interviewer may be given a quota of twenty or thirty interviews to carry out and some interviewers will have more than one quota. Distressing accounts also tend to cluster among households in particular areas. To cope with this potential additional stress, interviewers are forewarned about the upsetting situations they may encounter and have access to a support group of other interviewers and their regional field managers. Sometimes group supervision is wanted and helpful. Of course, they can not talk about specific individuals and their problems.

26.10 Confidentiality of Respondent's Answers

What does the interviewer do if the respondent says that s/he has been subjected to severe domestic violence or the interviewer sees that the respondent is dealing in drugs or it is evident that all the answers given by the respondent are untrue. The answer to all of these is clear from the pledge of confidentiality given to the respondent, the interviewer must not take any action or pass any information on to anyone else.

26.10.1 Questions Relating to Suicide

Most psychiatric assessment instruments include a section of questions on suicidal ideation and suicidal attempts. The standard structured survey questionnaire include questions on tedium vitae, death wishes, suicidal ideation and suicidal attempts in the past week, the past year or lifetime (Lewis et al. 1992). It is not the interviewers' job to evaluate the suicidal proclivity of the respondent. However, if the respondent says that they have made plans to commit suicide then their public duty is to report such cases.

26.11 Psychiatric Morbidity Survey Data

26.11.1 Anonymisation and Confidentiality

Epidemiological research is not concerned with identifiable individuals. Data are collected to answer questions such as 'how many people have psychosis?' or 'what proportion of the population have anxiety or depression?' not 'who has anxiety, depression or psychosis?'. Most large scale, social surveys do not record names or addresses on the survey database – hence this database is anonymised. The names or addresses of co-operating (or non-cooperating) households are kept confidential in a separate database.

Survey organisations make sure that even anonymised datasets are stored safely with restricted access. The requirements of data protection and human rights legislation together with modern computer technology make this principle harder to maintain with complete security. Social survey organisations are especially vigilant to take all reasonable steps to prevent the disclosure of identities.

26.12 Preventing Disclosure from Published Data

Social survey organisations take extensive measures to prevent their data from being published or otherwise released in a form that would allow any subject's identity to be disclosed or inferred. The disclosure of identity in itself represents a potential risk to survey respondents. Researchers cannot however be held responsible for any subject that freely chooses to reveal their participation in a study.

Respondents are usually informed if their data is to be deposited in a data archive. Data deposited with data archives are usually subject to specific conditions for deposit and release. Many methods exist for lessening the likelihood of confidentiality breaches, the first of which is anonymity. Anonymous data should be distinct from non-disclosive data. Anonymity helps to prevent unwitting breaches of confidentiality

Social researchers take steps to minimise the opportunities for others to infer identities from archived databases. Data are grouped in such as way as to disguise identities or to employ a variety of available measures that seek to impede the detection of identities without inflicting very serious damage to the aggregate dataset. These include amalgamating the data for groups of minority ethnic populations, or not including four digit codes for occupation and industry. The collapsing of variables undoubtedly diminish analysis possibilities but it is unavoidable in the circumstances.

As a consequence of data base enhancements and the 'matching' or 'fusion' of data sets the probabilities of disclosure of participants' identities has been increased in recent years so that it becomes harder to guarantee anonymity. The release of non-anonymised data, such as in sharing data between governmental agencies when the identities of individuals could be discovered needs to be agreed with participants in advance. This may not be necessary when there are adequate safeguards to ensure that confidentiality is ensured.

26.13 Surveys on the Mental Health of Children

When children are the focus of mental health survey, even greater care has to be taken in applying sound ethical practice. There may be interviews with the parent or primary caregiver, the young person over a particular age and a teacher. Despite the source of information normal confidentiality rules apply. What needs to be stressed is nothing any one of them said will be passed on to any other respondent. Thus, what the parent said would not be divulged to the child nor the teacher. Interviewers normally follow the same instructions as if they were interviewing the parents about their own mental health if asked for help – they should talk to their social worker or to their GP or the child's teachers, as appropriate. A leaflet containing 'helpful contacts' can be prepared to give to parents in this situation.

26.13.1 Revised Pledge of Confidentiality

In the survey on the mental health of children looked after by local authorities in England. (Meltzer et al. 2003) ethical approval for the survey was only given on condition that, in the exceptional circumstances of a child reporting that s/he is being physically or sexually abused and is in a situation where serious harm is being done to him/her, there was an obligation to pass this information on to an appropriate authority. Exceptionally therefore, for this survey only, the confidentiality pledge was revised for the child.

This stated that:

> Nothing you say or write will be passed on to anyone else except if you mention that someone is harming you in some way. In such a case what you said will be passed to child health experts working on your behalf and concerned for your health and happiness.

The child was reassured that answers to all the questions in the survey were confidential, i.e. that their answers would not be passed on to their carers, the local authority or school. It was only if the child reported serious harm being done to him/her that this information would be passed on to child health experts.

26.13.2 Child Consent Form

Both the revised confidentiality pledge and a request to audio-tape record the interview were included in the Child Consent form which needed to be signed by the child before starting the interview. If the child reported serious abuse, the comments would be on the tape and would be forwarded to child experts attached to the survey. They would listen to what was said and assess whether the information needed to be passed on to the Director of Social Services.

This procedure ensured that responsibility for reporting abuse rested with specially recruited experts and not with the lay interviewer. The interviewers' role was solely to send the tape back with a comment that it needs to be assessed. Interviewers were instructed not to contact the local authority nor the child's social worker themselves.

26.13.3 Cases of Abuse

Although it was unlikely that a case where the child reported on-going abuse would occur, there was a greater likelihood that the child might talk about abuse in the past which has led to the child's current difficulties. If any abuse was reported or if the child mentioned problems s/he was experiencing which s/he found difficult and distressing, interviewers asked whether they had been able to talk to anyone else about these problems. If they had, interviewers encouraged them to speak to this person again if the problems were still ongoing. Interviewers were also able to give the children a sheet containing a list of organisations which offer help to children in different circumstances.

Lepine and Smolla (2000) undertook a review of ethical issues concerning participants in community surveys of child and adolescent mental disorders. They stated that the same ethical principles apply for research on children as on adults – the benefits of the research to science and society should be maximised while the harm resulting from participation should be minimised. But what does harm mean for non-therapeutic research on children? Psychological harm is not only the most frequent type of harm but the most difficult to assess (Sieber 1992). Another key question is whether any harm is associated with the child's psychopathology.

It is taken for granted that children can reveal without embarrassment or shame the deviant, negative or abnormal dimensions of their experiences or that they do not experience guilt when they choose to remain quiet about, or falsify, such experiences (Thompson 1992). There is also an assumption that children experience no conflict of loyalty when their deviant behaviour involves other family members (Sieber 1992).

The review by Lepine and Smolla (2000) cites three studies which have assessed the reactions of participants to a child psychiatric interview for research purposes. The researchers concluded that the research was innocuous or had no deleterious effect but that does not mean that some questions were not upsetting. Lewis et al. (1985) assessed the reactions of 612 community children aged 9–18 years to a structured interview experience covering measures of vocabulary, neighbourhood and school environment, peer and family relationships. The most upsetting questions involved the police or illegal acts, regardless of the child's age or psychopathology. However, parents questioned 6–9 months after the interview reported no adverse effects on their child. Researchers therefore concluded that a trained lay interviewer could administer the psychiatric assessment instrument without any major risk of negative effects for the child.

26.14 Psychiatric Morbidity Surveys Among Elderly People

There are particular ethical problems in conducting research among elderly people but it is important to include them in national psychiatric morbidity surveys as they are increasingly larger proportions of the total population and are less frequently involved than the rest of the population in clinical trials. Dementia, depression, alcoholism, and suicide are some of the most important mental health issues for the aging population. Many elderly people, even those living in private households may have memory and concentration difficulties. Because elderly people are more likely to be cognitively impaired, vulnerable and dependent on others, capacity to give consent may require additional consideration. Questionnaires have been developed specifically for epidemiological use and survey techniques have been developed to try and remove the barriers to their participation, for example, the Geriatric Mental State Examination (McWilliam et al. 1988).

26.15 Psychiatric Morbidity Surveys Among Prisoners

A survey on psychiatric morbidity among prisoners was conducted in 1997 in Great Britain (Singleton et al. 1998). Despite the restricted rights of prisoners, it was felt that the ethical guidelines observed when the private household population was conducted should also be enforced in penal institutions.

26.15.1 Type and Place of Interviews

Interviewers had to liaise with prison officers to discuss the organisation of the interviewing: when they should interview prisoners and where the interview should take place. Arrangements varied considerable within prisons. Some prisons offered interviewing in their health centres; in other cases interviewing took place on the wings. Irrespective of where the interview took place, instructions were given to interviewers that they should only interview a prisoner in a room with no other person present to meet the pledge of confidentiality. If this was considered to pose a risk to the interviewer, the interview could be conducted in the presence of another interviewer or if too dangerous, the interview was recorded as non-response.

The timing of the interview was dependent on other prisoner activities, such as work, sport, visits from family and friends or court appearances. It was arranged that prisoners be excused work without loss of pay or bonuses, to avoid a reduction in response which could have biased the prevalence estimates.

26.15.2 Confidentiality

A pledge of confidentiality was given to prisoners – nothing said in the interview would be passed to any prison staff. Prisoners appreciated the fact that the interviewers were not employees of the Prison Service. Interviewers stressed that they worked for an independent body and that no data would be released in any way in which an individual's responses could be identified.

26.15.3 Safety

Interviewers were instructed to implement a number of measures to ensure their safety and security, for example checking the location of the panic button in the interview room, carrying alarms and following the advice of the prison staff.

26.16 Psychiatric Morbidity Surveys Among Homeless People

Survey research on homeless people presents a multitude of methodological challenges – most notably, sampling and interviewing. If one is concerned with the mental health and well-being of the whole population then there is a social obligation to conduct research with particular 'hard-to-survey' groups to avoid cutting them off from providing information which could improve the services that they receive. All the challenges in following ethical principles are exacerbated in the homeless population, for example, respondents may have serious alcohol and drug problems with limited periods of lucidity. This may compromise informed consent and the validity of responses. The field staff need extra protection from verbal and physical abuse. The Ethics Committee who reviewed the application to carry out a survey of psychiatric morbidity among homeless people requested that all field staff should have been inoculated for tuberculosis and Hepatitis B.

26.17 Psychiatric Morbidity Surveys Among Samples Identified Through Health Services

Whereas ethical issues in carrying out national surveys of the private household are the main concern of this chapter, surveys can also be administered in which respondents are randomly sampled from lists held by health services, e.g. lists of patients in contact with psychiatric services. In general the same principles set out in this chapter apply in such cases. The key difference is that respondents are being identified using information about their identity previously given by them to the service for other reasons. The chapter by McClelland deals with the release of information on the identity of service users for various purposes such as administrative and audit reasons as well as for research reasons and in the public interest.

26.18 Conclusions

Psychiatric morbidity surveys carried out by lay interviewers in the private household population play a key role in informing government about the mental health of the nation, identifying the barriers to participation in major life activities, and optimising resources allocated to the treatment of mental disorders. The ethical issues concerning participants in adult and child psychiatric morbidity surveys are not widely covered in the literature although most studies have to be scrutinised by an Ethics Committee. Therefore, ethical concerns are not neglected but reconsidered on a study by study basis. However, guidelines have been produced for proper conduct of epidemiological research in general, for instance, those developed by the International Epidemiological Association have been used in many countries (EPA 2004) and more specifically, International Ethical Guidelines for Epidemiological Studies (CIOMS 2008).

Lepine and Smolla (2000) suggest why there are so little formal guidelines for studies of children but the issues they put forward also apply to adult surveys.

First, ethical issues are difficult to formalise because they emerge from case-by-case discussions. It may be that epidemiologists, who are interested primarily in group data, find little relevance in the publication of the treatment of individual cases.

Second, it may be that child (and adult) psychiatric epidemiology has not yet developed a tradition of evaluating ethical issues concerning participants, such as is found in clinical research. The complexity of evaluating such issues may exceed the capability of research teams mobilised to solve the numerous methodological and practical challenges inherent in a survey. In this regard, medical ethicists could become valuable collaborators.

There is a danger of seeking solutions to ethical issues through a legal approach at the expense of an open debate involving all the societal stakeholders. The law only defines minimal protection criteria, the violation of which results in penalties. Solutions must come from those who design research and those who authorize it. Improving academic training for researchers on ethical matters relevant to their domain of research is mandatory. We also need to recognize that only a reflection initiated at the time of protocol design and based on knowledge of the research domain's methodology could ensure cogent answers to problems.

References

American Psychiatric Association (1994) Diagnostic and statistical manual of mental disorders. DSM-IV-TR 4th

Cassell J (1982) Harms, benefits, wrongs and rights in fieldwork. In: Sieber JE (ed) The ethics of social research: surveys and experiments. Springer, New York, NY

Council for International Organizations of Medical Sciences (CIOMS) in collaboration with the World Health Organization (2008) International ethical guidelines for epidemiological studies. CIOMS, Geneva

Dillman D (2000) Mail and internet surveys: the tailored design method. Wiley, New York, NY

European Epidemiology Federation (2004) Good epidemiological practice (GEP): proper conduct in epidemiological research

Gill B, Meltzer H, Hinds K, Petticrew M (1996) OPCS surveys of psychiatric morbidity in Great Britain, report 7: psychiatric morbidity among homeless people. London, HMSO

Goldberg D, Williams P (1988) A user's guide to the general health questionnaire. NFER-Nelson, Windsor, UK

Goodman R (1999) The extended version of the strengths and difficulties questionnaire as a guide to child psychiatric caseness and consequent burden. J Child Psychol Psychiatry 40:791–801

Green H, Maginnity A, Meltzer H, Goodman R, Ford T (2005) Mental health of young people in Great Britain. TSO, London

Groves RM, Couper MP (1998) Non-response in household interview surveys. Wiley, New York, NY

Groves R, Cialdini RB, Couper MP (1992) Understanding the decision to participate in a survey. Pub Opin Q 65(4):475–495

Henderson AS, Jorm AF (1990) Do mental health surveys disturb? Psychol Med 20:721–724

Kessler RC, Barker PR, Colpe LJ, Epstein JF, Gfroerer JC, Hiripi EC, Howes MJ, Normand ST, Manderscheid RW, Walters EE, Zaslavsky AM (2003) Screening for serious mental illness in the general population. Arch Gen Psychiatry 60:184–189

Kish L (1949) A procedure for objective respondent selection within the household. J Am Stat Assoc 44:380–387

Lépine S, Smolla N (2000) Ethical issues concerning participants in community surveys of child and adolescent mental disorders. Can J Psychiatry 45(1):48–54

Lewis SA, Gorsky A, Cohen P, Hartmark C (1985) The reactions of youth to diagnostic interviews. J Am Acad Child Psychiatry 24:750–755

Lewis G, Pelosi AJ, Araya RC, Dunn G (1992) Measuring psychiatric disorder in the community: a standardised instrument for use by lay interviewers. Psychol Med 22:465–486

McManus S, Meltzer H, Brugha T, Bebbington P, Jenkins R (2009) Adult psychiatric morbidity in England 2007: results of a household survey. National Centre for Social Research. www.ic.nhs.uk/pubs/psychiatricmorbidity07

McWilliam C, Copeland JR, Dewey ME, Wood N (1988) The geriatric mental state examination as a case-finding instrument in the community. Br J Psychiatry 152:205–208

Meltzer H, Corbin T, Gatward R, Goodman R, Ford T (2003) The mental health of young people looked after by local authorities in England: summary report. The Office for National Statistics, HMSO, London

Meltzer H, Gatward R, Goodman R, Ford T (2000) Mental health of children and adolescents in Great Britain. Stationery Office, London

Meltzer H, Gill B, Hinds K, Petticrew M (1996) OPCS surveys of psychiatric morbidity in Great Britain, report 4: the prevalence of psychiatric morbidity among adults living in institutions. London, HMSO

Meltzer H, Gill B, Petticrew M, Hinds K (1995) OPCS surveys of psychiatric morbidity in Great Britain, report 1: the prevalence of psychiatric morbidity among adults living in private households. HMSO, London

Sieber J (1992) Planning ethically responsible research. Sage Publications, Newbury Park, CA

Singleton N, Bumpstead R, O'Brien M, Lee A, Meltzer H (2001) Psychiatric morbidity among adults living in private households, 2000. TSO, London

Singleton N, Meltzer H, Gatward R, Coid J, Deasy D (1998) Psychiatric morbidity among prisoners in England and Wales. Stationery Office, London

Social Research Association (2003) Ethical guidelines. http://www.the-sra.org.uk/documents/pdfs/ethics03.pdf

Thompson RA (1992) Developmental changes in research risk and benefit. In: Stanley B, Sieber JE (eds) Social research on children and adolescents: ethical issues. Sage Publications, Newbury Park, CA, pp 31–64

Ware JE, Kosinski M, Keller SD (1996) A 12-item short-form health survey: construction of scales and preliminary tests of reliability and validity. Med Care 34(3):220–233

White A, Martin J, Bennett N, Freeth S (1998) Improving advance letters for major government surveys: final results. Survey Methodol Bull 43:July 1998

World Health Organisation (1992) The ICD-10 classification of mental and behavioural disorders: clinical descriptions and diagnostic guidelines. WHO, Geneva

Chapter 27
Genetics – Ethical Implications of Research, Diagnostics and Counseling

Peter Propping

Contents

P. Propping (✉)
Institute of Human Genetics, University of Bonn, D-53105 Bonn, Germany
e-mail: propping@uni-bonn.de

H. Helmchen, N. Sartorius (eds.), *Ethics in Psychiatry*, International Library
of Ethics, Law, and the New Medicine 45, DOI 10.1007/978-90-481-8721-8_27,
© Springer Science+Business Media B.V. 2010

Abbreviations

CAG repeat	Cytosin-Arginin-Guanin
CNS	central nervous system
CNV	copy number variant
DNA	desoxyribonucleic acid
ELSI	ethical, legal and social issues
ICSI	intracytoplasmatic sperm injection
IQ	intelligence quotient
IVF	in-vitro-fertilization
LDL	low density lipoproteins
OMIM	Online Mendelian Inheritance in Man
PD	prenatal diagnosis
PGD	pre-implantation genetic diagnosis
PPV	positive predictive value
SNP	single nucleotide polymorphism
STR	short tandem repeats, so-called microsatellites

27.1 Introduction

Genetics has become the leading discipline in improving our understanding of disease mechanisms for two main reasons:

- Information concerning the structure and function of the entire organism is encoded in the nucleotide sequence of DNA. Due to the advent of molecular genetics, the sequence of the 3.2×10^9 nucleotides of the haploid human genome is now known. A somatic cell is diploid, and thus contains around 6.4×10^9 nucleotides. Around one in every thousand nucleotides differs from the 'reference genome' which is the consensus nucleotide sequence established by an international collaborative effort. In addition, an unknown number of structural variants such as deletions, duplications, repeats and rearrangements render the genome of each human being unique.
- Genetic factors are involved to a varying extent in the etiology of most diseases. This conclusion is based on findings from classical family, twin, and adoption studies. Through knowledge of the human genome and the advent of molecular genetics, it has become possible to systematically examine patients and their families for genetic variants. It is a central task of medical research to correlate variation at the DNA level to clinical phenomena. A major justification for this

type of research is the elucidation of disease etiology. Once the genotypes contributing to a disease have been identified, the pathophysiology of a disease can be studied, and specific preventive options may then become available. Knowledge of the functional pathways involved in the pathophysiology of a disease may enable the development of new pharmacological strategies.

27.1.1 Hopes and Expectations

The human brain is extremely inaccessible to direct biochemical or molecular research. The etiology of psychiatric diseases is particularly poorly understood. Current hypotheses concerning the pathophysiology of mental diseases are mainly based on indirect evidence such as pharmacological effects, brain imaging, or – even more remotely – peripheral biochemistry. Genetic approaches assume that DNA is the same in all cells of the body and that variation in genetic material is one aspect of a person's individuality. A major goal of human genetic research is to find genotype–phenotype relationships. Once such a relationship has been established, it may be used for diagnostic purposes and for the identification of functional pathways. The identification of specific pharmacological targets for the treatment of mental disorders achieved through the application of genetic approaches would represent an important breakthrough in this field of research.

27.1.2 Fears and Challenges

From its inception, the study of inheritance in humans has been closely correlated with an interest in brain function. In his research into the influence of inheritance in humans, Francis Galton (1865) used professional achievement among the relatives of 'successful men' as an indicator of a genetic influence on brain function. It was Galton who coined the term 'eugenics' to describe the concept of improving the quality of the human species or any human population by preventing the birth of children with 'undesirable' inherited traits while encouraging reproduction in those with 'desirable' traits. In the decade following the rediscovery of Mendel's laws in 1900, mental disorders became the focus of particular interest upon the publication of Ernst Rüdin's impressive review of the potential value of family studies in achieving a better understanding of such diseases (Rüdin 1911). It is not an overstatement to claim that this article represented the birth of modern psychiatric genetics. Beginning with Galton and Rüdin, interest in the genetic control of brain function and mental disorders was interwoven with eugenics and racism. Eugenic ideology flourished in all industrialised countries in the first decades of the twentieth century. Many people, including scientists, were afraid that the relaxation in natural selection introduced by modern medicine would lead to deterioration in the genetic quality of the population, especially in highly developed countries. Although these ideas were held with particular conviction among political 'conservatives,' they were also propagated by some representatives of the political 'left.' The German physician Alfred Grotjahn, famous for introducing public health in Germany and later a Socialist member of the German parliament, advocated the sterilization of the 'dregs of society' ('Bodensatz') (Grotjahn 1912). As a consequence, many countries introduced legislation for enforced sterilisation. This was the case as early as 1907 in Indiana, and then later in other US states (Paul 2001). In Germany, such legislation had been proposed in the final years of the Weimar Republic, but was only formally introduced in 1933 when Hitler and his Nazi party had come to power. Doctors were obliged to report all patients suffering from any of eight particular diseases (mental retardation, schizophrenia, manic depression, epilepsy, Huntington's chorea, hereditary blindness, hereditary deafness, severe hereditary malformation, and severe alcohol dependence) to the public health authorities. These cases were then brought before a court which determined whether or not the patient should be sterilised. With the outbreak of the Second World War in 1939 patients with mental disorders were secretly and systematically killed in special extermination hospitals with the active involvement of medical staff. As we now know, this was the prelude to the Holocaust.

This inhuman episode of history is firmly established within the collective memory of society. It is thus an eternal burden to the formal study of human genetics, and the study of psychiatric genetics in particular. Many fear that the steadily increasing potential of genetic diagnostics may lead to the 'geneticisation' of society (Lippman 1992) and promote a revival of inhuman practices. It is feared that the adoption of a blueprint of individual genotypes as a metaphor for the biological constitution of an individual would result in restricted concepts of health and illness. There is a fear that the widespread application of genetics may continue a tradition of reductionism and determinism in the social arena as well as within medicine.

The following section outlines the fears that are predominant within modern society. These issues will be discussed in the later parts of this chapter.

Predictive genetic diagnosis. A number of genetic diseases follow a simple i.e. monogenic mode of inheritance and become manifest later in life. Depending on the mode of inheritance, the offspring or siblings of a patient, respectively, may be considered to be at risk for the disease. In cases where the responsible gene is known and the specific mutation has been identified within the family, predictive diagnosis is available for any person at risk. Chorea Huntington – an incurable disease – is the 'classical' example. There is a widespread fear that predictive diagnostics will become possible on a larger scale and pose a threat to society.

Prenatal and pre-implantation genetic diagnosis. With the advent of chromosome analysis and the introduction of amniocentesis and chorionic villous biopsy, prenatal diagnosis has become a widespread practice in older pregnant women. The risk for the birth of a child with a newly arisen chromosome mutation increases with the age of the mother. Trisomy 21, which leads to Down syndrome, is the most frequent chromosome aberration in the children of older women. Prenatal diagnosis (PD) may be used in cases in which there is an increased risk for any testable inherited disease. PD is only possible, of course, if the increased specific risk in the couple has been recognized and the genetic basis of the disease in the respective families is known. In addition to various chromosome aberrations, it is possible to detect thousands of monogenic diseases in the at risk fetus. These include metabolic defects that lead to mental retardation and other symptoms, severe muscle diseases, and complex malformation syndromes. All of these diseases are rare in the population. If the disease follows an autosomal recessive mode of inheritance, the recurrence risk for the siblings of a patient is 25%. Many parents request prenatal diagnosis in this situation in order to preserve the option of an elective abortion of an affected fetus.

The earliest method of establishing the diagnosis of a monogenic disease is pre-implantation genetic diagnosis (PGD). This procedure necessitates extracorporal insemination through intracytoplasmic sperm injection (ICSI). Several egg cells are harvested following hormonal treatment of the woman, and several of these are then fertilized. The embryos are examined at the eight-cell stage for the genetic defect present in the older sibling. If the embryo proves to be free of the defect, it will be transferred to the womb of the mother; it will otherwise be sacrificed.

Two major fears exist concerning prenatal diagnosis. One concern is that society might impose pressure on women to undergo prenatal diagnosis and to abort an affected child in order to avoid the costs of future treatment. The other concern is that parents might begin to use PD and PGD in an increasingly liberal fashion ('bottom-up eugenics'). Only those pregnancies with a disease-free fetus might be allowed to progress to term. It is even feared that the concept of the perfect 'designer baby' might be realised through PGD.

Genetic determinism. The capacity to predict an individual's biological destiny will increase with the continued identification of genes and their functional contexts. There is concern that the free will of the individual may become endangered by the potential of genetic testing to predict the biological future. An individual's genetic make-up would become a highly significant factor in determining the course of their life, a consequence which would have profound implications for a democratic society.

Social stigmatisation. The revelation of genetically determined propensities or predispositions might lead to discrimination, reduced social solidarity and an augmentation of social inequality. This is a particular concern with respect to the diagnosis and genetic-based prediction of mental disorders (see Chapter 2).

Genetic testing and underwriting. The closer the correlation between genetic test results and disease process, and the greater the ability of such tests to predict the onset of a disease, the greater their importance becomes within the context of underwriting. There is a widespread concern that private insurers might ask for predictive genetic testing before a contract can be agreed. This would undermine societal solidarity.

Genetics and data protection. Genetic data are particularly sensitive since they may reflect the biological constitution, the predisposition to disease, or even the mental health status of an individual. Genetic data are contained in many medical case files, and may also be generated within research projects. It is feared that the protection of such data cannot be ensured, and that the potential exists for insurers or employers to gain access to the genetic data of clients or applicants (see Chapter 11).

Complete genome sequencing. Molecular genetic methods are continually being perfected. It may even become possible in the next few years to determine the complete sequence of the personal genome at affordable financial cost. Neither society nor the medical system is sufficiently prepared to cope with this potential development: we lack the capacity to interpret the overwhelming wealth of data that would be generated, and our knowledge of the phenotypic correlates is inadequate. This ominous constellation might result in incompetent and harmful clinical management.

27.2 The Genetic Background

In order to make this chapter understandable to the non-geneticist, it is necessary to briefly review the fundamentals of modern human genetics.

27.2.1 Variation Within the Human Genome

The correlate of the human genome that can be examined under the microscope at a 1,000-fold magnification is the set of 23 pairs of chromosomes that are localized in the nucleus of each cell (cf. Strachan and Read 2004). The carrier of the genetic information is the DNA strand. This is densely packed within each chromosome. The genetic information is encoded twofold in the human genome, and humans are thus diploid. The haploid genome contains 3.2×10^9 nucleotides, the sequence of which is known. Four different nucleotides exist: adenine, guanine, thymine, and cytosine. The nucleotide sequence determines the sequence of the amino acids in the proteins, but only 1.5% of the DNA encodes proteins. The majority of the DNA is non-coding. In humans, 99.9% of the nucleotide sequence is identical in all individuals.

At the population level, the human genome is dynamic: new mutations (variants) arise that may remain in the human gene pool, or may disappear slowly if harmful to the individual. In any given individual, around one in every thousand nucleotides is a variant compared to the reference genome. A variant such as this may be responsible for a genetic disease. The vast majority of variants, however, are single nucleotide polymorphisms (SNPs) that have no functional consequences. The human genome contains different classes of non-coding repeats that may also be polymorphic. One class of repeats that has gained importance in the context of genetic mapping is short tandem repeats (STR, so-called microsatellites). A class of variants that has received particular attention in recent years is copy number variants (CNVs). These may appear as deletions or duplications and may result from new mutations, but they may also be inherited. Certain deletions at CNV loci are correlated with mental disorders such as autism, schizophrenia, mental retardation, and epilepsy. It is not yet understood how different diseases may result from presumably identical deletions. Carriers of these same deletions may even be completely healthy.

27.2.2 Genetic Mechanisms, Simple Modes of Inheritance

If a mutation in the DNA significantly impairs the function of a gene product, this will have implications for the mode of inheritance of the defect. Transmission is termed dominant if a mutation only has to be present in one gene copy i.e. on one of the chromosomes of a pair. Mutation carriers are thus heterozygous. The auto-somal dominant mode of inheritance is typically characterised by transmission of the genetic trait from one parent to 50% of all offspring ('vertical transmission'). Inheritance is termed autosomal recessive if mutations of both copies of a gene are necessary for phenotypic manifestation. In this case, both parents carry the mutation in a heterozygous form, but they themselves are unaffected. 25% of the offspring are homozygous for the mutation and are phenotypically affected ('horizontal trans-mission'). Two thirds of the offspring are heterozygous mutation carriers, and the remaining 25% are genetically normal.

The status of the sex chromosomes is different in males and females. Males have one X- and one Y-chromosome, while females have two X-chromosomes. If a mutation responsible for a monogenic trait is located on the X-chromosome, an X-chromosomal mode of inheritance results. If the mutation has recessive characteristics, a male will be affected if his X-chromosome carries the mutation. A heterozygous female is usually unaffected: she is a 'carrier'.

When the heterozygous (if autosomal dominant) or homozygous (if autosomal recessive) state leads to a phenotypic expression in 100% of cases, the genotype is said to be completely 'penetrant.' In many cases, only a certain percentage of mutation carriers will express the phenotype, and this is termed 'reduced penetrance' of a genotype. The majority of mutations have only a reduced penetrance. There is, in fact, a continuous gradient of very low to complete penetrance between these two extremes.

Thousands of monogenic traits – many of them diseases – have been defined, and usually on the basis of formal genetic studies. A characteristic phenotype occurring in the presence of a simple i.e. autosomal dominant, recessive, or X-chromosomal mode of inheritance can be taken as proof of the existence of a mutation in a specific gene. Current knowledge about all monogenic traits, the spectrum of mutations in the corresponding genes, genotype–phenotype relationships and pathophysiology are documented in a continuously updated data bank, the Online Mendelian Inheritance in Man data bank (OMIM, Gene Tests). At present, diagnostic tests are available for around 2,000 monogenic disorders. The traditional brief description of Mendelian modes of inheritance is, in fact, simplistic. Molecular and formal genetic studies have established the involvement of a range of other phenomena (cf. Strachan and Read 2004): maternal inheritance through mutations in the mitochondrial genome, autosomal recessive inheritance leading to somatic mutations in the mitochondrial genome, germline mosaics, and imprinting. Imprinting is a sex-determined mechanism of transmission which is dependent upon whether a particular mutation has been inherited from the mother or the father. The Angelman syndrome is one example of a disease that may result from 'imprinting' mechanisms.

27.2.3 Multifactorial Inheritance

There are many normal traits and diseases that run in families for which the pattern of familial occurrence is incompatible with a Mendelian mode of inheritance. Examples of normal traits are height, IQ and blood pressure. Examples of diseases are hypertension, idiopathic epilepsy, and unipolar depression. For these traits, the concordance rate in identical (monozygotic) twins, although higher than for fraternal (dizygotic) twins, is far below 100%. This is in contrast to Mendelian traits with complete penetrance, for which identical twins are consistently concordant. The classical genetic model applied to this type of familiality is polygenic inheritance. It has been hypothesized that a large number of genotypes, each making a small contribution alongside environmental or even stochastic factors, contribute to the expression of a certain phenotype. Recent findings from molecular genetic studies have shown that this hypothesis is in principle correct, but that the degree to which any particular genotype contributes may vary.

In fact, most 'multifactorial' traits, particularly diseases, result from a spectrum of genetic mechanisms. These include dominant and recessive inheritance, polygenic mechanisms, new mutations and phenocopies (a phenocopy mimics the phenotype of an otherwise genetic trait). The term 'genetically complex disease' has been introduced to describe the phenomenon by which a disease may result from various genetic or non-genetic mechanisms.

Most of the genotypes predisposing to a multifactorial disorder make only a limited contribution to the development of the disease. The relative risk for carriers of a predisposing genotype may be as low as 1.5, or even lower. The 'positive predictive value' (PPV) depends on the frequency of the predisposing genotype in patients and in healthy controls. If the genotype is common in patients and rare in controls, the PPV is high. In the reverse situation, the PPV is low.

In the public debate about genetic testing (predictive genetic testing in particular), Huntington's chorea is the usual paradigm. It must be remembered, however, that the great majority of mental disorders are multifactorial in origin. A monogenic mode of inheritance is only observed in two types of mental disorder: mental retardation and storage diseases (Propping 2005). The CNS is characterized by a high degree of functional redundancy and neuronal plasticity. The fine structure of the human brain develops under the influence of both genetic factors and environmental stimuli, which results in an unusual degree of flexibility. Only primary developmental defects or the accumulation of un-metabolizable residual products can override this.

27.2.4 Genotype–Phenotype Relationship

Although most monogenic disorders have a characteristic phenotype, phenotypic variation does exist. Familial adenomatous polyposis, for example, is an inherited cancer predisposition syndrome that is caused by a mutation in the *APC* gene. The severity of the disease varies depending upon which codon is mutated. A remarkable range in the age of onset is observed (Friedl et al. 2001). This particular phenotypic variation is presumed to be due to a second somatic hit that occurs stochastically.

Huntington's chorea also has an autosomal dominant mode of inheritance. It is a neurodegenerative disorder, and the age of onset is usually between 40 and 60 years of age. A polymorphism of a CAG repeat is present within the coding region of the disease-relevant (huntingtin) gene in the population. This produces a polyglutamine stretch of variable length. If one of the two alleles contains more than 37 repeats, Huntington's chorea develops. There is a clear negative statistical correlation between the number of repeats and the age of onset. However, the age of onset also varies in carriers of the same number of repeats. This effect can be at least partly explained by the involvement of a modifier gene, the product of which interacts with the polyglutamine stretch (Metzger et al. 2008).

These two examples show that different mechanisms may explain a variation in the phenotype.

27.2.5 Chromosome Disorders

A large number of chromosome disorders are known. Autosomal aberrations generally lead to more severe phenotypic consequences than gonosomal (i. e. X- or Y-chromosomal) aberrations. Autosomal aberrations may be due to new numerical mutations, such as trisomy 21 with its typical phenotype, or structural aberrations. Structural aberrations only have phenotypic consequences if they are unbalanced i.e. when they are characterized by a loss or a gain of genetic material. The presence of a balanced structural aberration in a parent may lead to transmission from one generation to the next.

Gonosomal aberrations may affect the X or the Y chromosome. The XXY, XXX, and XYY karyotypes result from new germline mutations. The karyotype 45, X0, which leads to the Ullrich-Turner syndrome, is caused by the postzygotic loss of a gonosome (i.e. X or Y chromosome). This syndrome is therefore the consequence of a somatic mutation event that occurs in early embryogenesis. The XXY and 45, XO karyotypes lead to infertility. Cognitive functions are only marginally impaired in these four syndromes, and the average IQ deficit ranges between 5 and 10 IQ points. Each gonosomal aberration may also lead to particular psychological problems. For the majority of numerical gonosomal aberrations, the individuals concerned are otherwise healthy; individuals with the 45, X0 karyotype, however, frequently present with a syndromic phenotype.

Over the last decade, molecular cytogenetic methods have succeeded in identifying a large number of chromosomal microdeletions, most of which correlate with a characteristic clinical presentation. The methodological gap existing between chromosome disorders of the 'classical' period and monogenic traits identified through molecular genetic methods is gradually disappearing.

27.2.6 The Future of Genetic Diagnostics

Since many monogenic diseases are highly heterogenous, assays for particular groups of genetic diseases (e. g. diseases of the retina, neurodegenerative diseases, and inherited cancer syndromes) will be developed that can detect all potential mutations in the respective genes. It is probable that sequencing of the personal genome at an affordable cost will become feasible within the next decade. It would then only be necessary to sequence each individual once during their entire lifetime. Genetic diagnostics would become computer-based ('in silico'). The ethical implications of these technical advances must still be clarified.

27.3 Genetic Counseling

The concept of eugenics is older than the genetic paradigm. Scientists of previous generations were not able to quantify genetic effects at the population level, and did not take the psychological and societal consequences of eugenic goals into account. In the early, pre-Second World war phase of human genetic research, genetic counseling was an integral aspect of eugenics in the US and most European countries including Germany. Famous research institutions were founded by proponents of eugenic ideology: The Galton Institute in London (Francis Galton coined the term 'eugenics'), the Cold Spring Harbor Laboratory on Long Island/New York (founded by Charles B. Davenport, an ardent supporter of eugenics), the Dight Institute in Minneapolis (founded by Charles Fremont Dight, a physician who, in his will of 1927, left his estate to the University 'to promote biological race betterment') (Reed 1974). The aphorism 'good judgement comes from experience, and experience comes from bad judgement' (Levy 2003) is certainly pertinent to the paradigm shift involved in applying eugenics to help people cope with the burden of an inherited disease.

At the time when the first genetic counseling facilities were being established in the US following the Second Word War, enthusiasm for eugenics had not completely disappeared. Over time, the interests of the individual gradually became the main focus of genetic counseling, as evidenced, for example, at the First International Congress of Human Genetics in Copenhagen in 1956. In 1975, the modern concept of genetic counseling was formulated by a committee of the American Society of Human Genetics (American Society of Human Genetics 1975): 'Genetic counseling is a communication process which deals with the human problems associated with the occurrence, or the risk of occurrence, of a genetic disorder in a family. This process involves an attempt by one or more appropriately trained persons to help the individual or family to (1) comprehend the medical facts, including the diagnosis, probable course of the disorder, and the available management; (2) appreciate the way heredity contributes to the disorder, and the risk of recurrence in specified relatives; (3) understand the alternatives for dealing with the risk of recurrence; (4) choose the course of action which seems to them appropriate in view of their risk, their family goals, and their ethical and religious standards, and to act in accordance with that decisison; and (5) to make the best possible adjustment to the disorder in an affected family member and/or to the risk of recurrence of that disorder.' Although it is mainly oriented towards procreation, and does not specifically address mental disorders, this definition encompasses the major elements of the counseling process. The Council of Europe (2009) has briefly summarized existing practice in cases where a genetic test is indicated: 'When a genetic test is envisaged, the person concerned shall be provided with prior appropriate information in particular on the purpose and the nature of the test, as well as the implications of its results.' Furthermore, it is a requirement that genetic counseling should be conducted in a non-directive manner, and that the individual concerned must provide willing and informed consent to any genetic test.

27.4 Important Opinions on the Ethics of Psychiatric Genetics

The history of psychiatric genetics reflects the fatal misapprehensions of applied genetics (Schulze et al. 2004). The goals of genetic counseling have developed in a stepwise manner. The original ideology of eugenics has been replaced by an emphasis on fundamental values of medical ethics such as autonomy, non-malevolence, dignity, and confidentiality. Many statements have been formulated over recent decades concerning the management of genetic information. All of them are in agreement as regards the essentials of human rights. The central element of all regulatory proposals is the importance of self-determination.

The ethical aspects of research into mental disorders and genetics have been specifically debated by the Nuffield Council on Bioethics (1998). In addition to clinical and methodological aspects of research, their recommendations address the ethical and legal issues involved in the clinical application of genetic findings and in psychiatric genetic research. In 1995, The German Society of Human Genetics summarized all ethical aspects of the study of human genetics, with revisions in 2007 (Deutsche Gesellschaft für Humangenetik 2007). The ethical aspects of the use of biobanks for research purposes have been addressed by the German National Ethics Council (2004). The Council of Europe (2009) has recently published an Additional Protocol to the Convention on Human Rights and Biomedicine which deals specifically with genetic testing for health purposes. Particular emphasis is given to the conditions under which genetic tests may be conducted in persons who are not able to provide consent.

These statements will be referred to later in this chapter.

27.5 Ethical Issues in the Application of Genetics to Mental Disorders

The ethical aspects of genetic diagnostics for mental disorders will now be briefly discussed.

27.5.1 Genetics as a Diagnostic Tool

Facts and examples. When the responsible gene is known, a monogenic disease can be diagnosed using molecular genetic methods. A hypothesis based on clinical methods is usually required. If, for example, Chorea Huntington is suspected, a genetic test can confirm or exclude the disease. Parkinson disease, in contrast, is genetically heterogeneous. Genetic investigations may allow identification of a particular hereditary subtype of the disease, but cannot exclude another genetic form. The same is true for mental retardation, which has an extremely heterogeneous phenotype. Many monogenic forms of mental retardation exist, as well as forms with a multifactorial inheritance and others that are clearly exogenous. Deletions in various regions of the genome may be responsible for cases of autism and even schizophrenia (see Section 27.2.1).

Ethics. Any patient affected by a disease that is presumed to be genetic should be informed of this in an appropriate manner. Before being tested, the patient should be informed about all of the pros and cons of genetic testing, the right not to know, and all other medical implications. There is general agreement that there should be no objection to genetic testing for diagnostic purposes if informed consent has been obtained. Nevertheless, one important aspect must be taken into account when the patient is asked to provide consent: there are essential differences in conducting genetic tests for a defined disorder such as Chorea Huntington, for which the prognosis is more or less clear, and for identifying one disease type from among a heterogenous group of disorders. The latter case is true for Parkinson disease, for example. The diagnosis of any one particular type of Parkinson disease has important implications for the individual concerned as regards the course of the disease, the prognosis and the recurrence risk for relatives.

27.5.2 Predictive Diagnosis

Facts and examples. When a disorder with a late onset follows a simple mode of inheritance and the relevant mutation has been identified in an affected individual, the offspring, and potentially also the siblings, of a patient are at an increased risk for the disease. Chorea Huntington may again serve as an example. Although the first symptoms of this disease may develop in early adulthood, onset may only occur when the patient is in their late sixties. Most patients develop their first symptoms between the ages of 40 and 60. The a priori risk that the child of a patient with Chorea Huntington carries the mutation and will develop the disease is 50%. The offspring of a patient with this late onset disorder must therefore endure a long period of uncertainty until he or she can be sure that they have not inherited the mutation. Chorea Huntington is due to an extension of the CAG repeat motif that codes for a polyglutamine stretch. There is a statistical correlation between the number of repeats and the age of onset of the disease. The more repeats – beyond the critical threshold of 37 – the earlier the onset. Although the correlation is clear, it should not be used for the purposes of individual counseling since there is appreciable variation, even for carriers of the same repeat length.

Once the mutation has been identified, the children and other relatives of the patient have the option of undergoing predictive testing. If the child does not carry the mutation, he or she will not be affected by Chorea Huntington. If the child does possess the mutation, however, he or she will definitely develop the disease, since the penetrance of the mutation is 100%.

Ethics. Chorea Huntington has become the paradigm for genetic counseling since it is a disorder which represents all of the major issues of modern human genetics: (i) it was the first genetic disease for which the genetic locus was mapped with polymorphic genetic markers (Gusella et al. 1983), (ii) it is a late-onset disorder with a long period of uncertainty, (iii) predictive genetic diagnosis is available for suitable families, (iv) the disease is ultimately fatal, since no prevention or therapy is available, and (v) genetic research into Chorea Huntington was championed by a member of a family affected by the disease (N. Wexler). Soon after the discovery of the genetic locus, geneticists, neurologists and representatives of the self-help support group formulated guidelines to be taken into account when predictive testing is being considered (Went 1990). The major recommendations of these guidelines are: (i) the person at risk should have the right to request testing or to refuse it, (ii) genetic and psychological counseling of a person at risk for Chorea Huntington is necessary both before and after genetic testing, (iii) there should be an interval of at least 3 months between the decision to take the test and blood sampling, (iv) predictive genetic testing in minors should be strongly discouraged, since the result would have no implications for clinical intervention, (v) predictive testing should not be considered if there is evidence of suicidality, (vi) predictive testing should not be conducted in a pregnant woman, and (vii) the counselee has the right to resign from predictive testing at any time during the procedure.

It is a common for predictive genetic testing to be requested by only 10–20% of persons at risk. This is a much lower figure than is the case for treatable diseases such as inherited breast or colon cancer.

The guidelines developed for Chorea Huntington have become a model for other devastating genetic diseases of late onset such as monogenic forms of Alzheimer dementia and the inherited ataxias.

27.5.3 Genetic Testing in Children

Facts and examples. The performance of genetic tests in children is technically as straightforward as in adults. The ethical issues involved are nevertheless very different.

Ethics. Parents may want their child to be tested for a genetic disorder in order to resolve uncertainty. Older children may wish to be tested on their own initiative, but their parents may object. The degree to which a child is capable of judging the implications of a genetic test is questionable. Whatever the particulars of the case may be, the wishes of the child and the parents may not always coincide. Even by the age of 16, it is not certain that a child has reached a sufficient level of understanding and intelligence to be capable of formulating a reasonable decision (Nuffield Council on Bioethics 1998).

A medical or genetic examination should only be conducted in the interests of the individual concerned and with appropriate informed consent. The Nuffield Council on Bioethics (1998) and the guidelines of the German Society of Human Genetics (Deutsche Gesellschaft für Humangenetik 1995, 2007) recommend that predictive genetic testing should be strongly discouraged in children, who are unable to give consent, unless there are implications for clinical intervention during childhood. The guidelines of the German Society of Human Genetics also address the issue of heterozygosity testing for the purpose of making future decisions regarding a decision to have children. It is recommended that such testing should be postponed until the child has reached adulthood, and then performed within the context of genetic counseling.

27.5.4 Examination of a Child in the Interests of a Healthy Relative

Facts and examples. The parents of a child affected by a heritable genetic disorder may wish to know whether any future children are at an increased risk for the disease. Down syndrome may serve as an example. In 95% of cases, the disorder is due to trisomy 21, a newly arisen numerical chromosome aberration. In 2–3% of cases, chromosomal mosaicism for trisomy 21 exists, and in a further 2–3% of cases, the phenotype is due to a centric fusion of chromosome 21 with another acrocentric chromosome, typically chromosome 13, 14 or 15. The translocations may have been transmitted from a parent. In a case such as this, the respective parent is a 'balanced' carrier of a translocation. If the parents have normal karyotypes, a new mutation in a parental germ cell must have occurred, and the respective parent may possess a chromosomal mosaic.

In order to calculate the risk of Down syndrome in future offspring, it is necessary to examine the chromosomes of the affected child. The genetic test is therefore being conducted in the interests of the parents, and not of the child.

A comparable situation is one in which the child of healthy parents suffers from a severe, and presumably genetic, malformation syndrome. A precise diagnosis is necessary for the parents to be able to make an informed decision regarding further children. The diagnosis may potentially be established through genetic testing of the affected child.

Ethics. According to the guidelines of the German Society of Human Genetics (Deutsche Gesellschaft für Humangenetik 1995, 2007), the right of the parents to know the recurrence risk in future children carries more weight than the rights of the affected child. In this situation, genetic examination of the affected child in the interests of the parents would be possible.

27.5.5 Prenatal Diagnosis (Amniocentesis, Chorionic Villous Biopsy)

Facts and examples. Prenatal diagnosis is a method applied to avoid the birth of a severely handicapped child. In the majority of ultrasound surveillance examinations, no abnormality is detected. Performance of an invasive diagnostic approach requires a clear indication. There are two major indications for invasive prenatal diagnosis within the context of genetic disorders: (i) advanced maternal age, which is correlated with an increased risk for numerical chromosome aberrations, particularly Down syndrome, and (ii) the unexpected birth of a child with an autosomal recessive or a X-chromosomal recessive disease. The parents may fear the birth of a second affected child. If it is established that the embryo or fetus has a genotype that leads to a severe disease, the mother may wish to consider an elective abortion.

In the vast majority of cases, prenatal diagnosis is carried out in order to identify diseases with an early onset. Among disorders of the CNS, syndromes associated with mental retardation are a frequent indication. These disorders may be due to an autosomal recessive, or a X-chromosomal mode of inheritance, or to new mutations. Deletions at polymorphic loci for copy-number-variants (CNV) have recently been found to be responsible for some childhood cases of mental retardation and autism (Cook and Scherer 2008, Koolen et al. 2009). In the future, it is possible that prenatal CNV testing will be requested in cases for which a molecular diagnosis is available. New mutations cannot be identified through currently available methods of prenatal diagnosis since their manifestation cannot be foreseen. .

Prenatal diagnosis is not appropriate for genetically complex diseases such as schizophrenia or bipolar affective disorder. These diseases are genetically very heterogenous, and many genotypes will only contribute to a limited extent to the manifestation of disease. The situation may change once empirical data have established that a major factor, such as a microdeletion, is more frequent in patients than in healthy individuals (Stefansson et al. 2008). It is likely, however, that this will only be the case for a small percentage of all schizophrenia and bipolar affective disorder cases. A crucial point will be the degree of relative risk. The issue of performing prenatal diagnosis for a disorder with a late onset also remains contentious, particularly in the case of diseases such as these which are treatable. Even for a monogenic disease with a fatal prognosis such as Huntington' chorea, prenatal diagnosis is rarely requested.

The earliest form of prenatal diagnosis is pre-implantation genetic diagnosis (PGD). In-vitro-fertilisation (IVF, ICSI) and subsequent short-time culture of the fertilized ovum are both necessary for this procedure. In the 8-cell stage, one or two cells are taken from the early embryo and examined for the respective mutation. Only those embryos that do not carry the mutation are subsequently transferred to the uterus. In a prospective clinical study, an established pregnancy was achieved in 22% of all cycles with oocyte retrieval. Following embryo transfer, a 27.8% chance of a live birth was reported. The rate of perinatal death and stillbirth was relatively high, but no excess in major congenital anomalies was observed in the surviving children (Keymolen et al. 2007).

In Europe, PGD is permitted and practiced in Belgium, Spain, Holland, France, the UK, the Czech Republic, and is prohibited in Ireland, Germany, and Switzerland. The major reason for prohibition is the toti-/pluripotential status of the 8-cell stage. It is argued that a toti-/pluripotent cell is a potential human being who should not be rendered an object in the interests of another human being.

PGD has been successfully applied to many monogenic diseases, including those with a late onset such as Chorea Huntington and hereditary breast cancer. It seems that it is easier for a couple at risk for the birth of a child with a late-onset disease to opt for PGD than for prenatal diagnosis, which has the potential for rendering termination of pregnancy necessary.

Ethics. Prenatal diagnosis is practiced in nearly all European countries. There is widespread consensus that a mother has the right to opt for an abortion if she has a high risk of giving birth to a severely handicapped child (e. g. Deutsche Gesellschaft für Humangenetik 2007). The issue of whether disorders with a late onset should be an indication for prenatal diagnosis is a matter of ongoing debate. In accordance with medical practice in European countries, most ethical statements do not exclude the option of prenatal diagnosis for genetic disorders with a late onset, if adequate genetic counseling is provided.

The situation in Germany is different. Following intense debate in the German parliament, recent legislation (April 2009) on genetic diagnostics has been introduced prohibiting prenatal diagnosis for late-onset disorders (however defined).

Whether PGD should be allowed or prohibited depends on the decision made regarding the status of the preimplantation embryo.

27.5.6 Genetic Screening in Newborns and Children

Facts and examples. The screening of newborns for rare genetic diseases that, without specific treatment, would lead to devastating diseases with an early onset is common practice in all European countries. This is undoubtedly a success story. Where no therapy or prevention is available, screening has been shown to have negative psychological effects (Levy 2003). In Germany, newborn screening is strictly limited to diseases for which therapy is available, e. g. phenylketonuria and aplasia of the thyroid gland. Laboratory data obtained through tandem mass spectrometry that indicate the diagnosis of untreatable diseases are systematically blinded.

Ethics. A screening program for health purposes should enable the early identification or exclusion of a genetic disease or a genetic predisposition to a disease, where appropriate preventive or treatment measures are available to the persons concerned (Council of Europe 2009). Thus, for minors who are at increased risk for a genetic disorder that manifests before adulthood, genetic tests should only be performed with the authorisation of his or her representative, and when therapeutic or preventive measures are available. At present, this mainly applies to certain inherited cancer syndromes.

27.5.7 Screening for Late Onset Genetic Disorders

Facts and examples. Progress in medical genetics and the availability of ever improving molecular methods will eventually make it possible to introduce genetic screening programs for late-onset disorders in adults, in a similar fashion to the screening programs offered for newborns. Cascade screening for familial hypercholesterolemia in Holland may serve as an example. Individuals who are heterozygous for a mutation in the LDL receptor gene (frequency in the population 1: 500) develop early atherosclerosis followed by coronary heart disease and other disorders resulting from impaired perfusion. Early and systematic treatment can, however, effectively counteract the development of atherosclerosis. This screening was introduced in 1994 (Umans-Eckenhausen et al. 2001, Wonderling et al. 2004). Cholesterol levels are measured, and individuals with a cholesterol level above a certain limit are offered genetic testing. Participation in the screening procedure is voluntary. Guidelines have been introduced to regulate insurance contracts in individuals who have been shown to have increased cholesterol levels (Homsma et al. 2008).

Ethics. The Nuffield Council (1998) has stated that genetic testing for susceptibility genes of predictive or diagnostic value should be discouraged unless and until the information can be put to effective preventive or therapeutic use. The same should apply, of course, to the offering of screening. The testing for genotypes that predispose to a neuropsychiatric disease with a late onset, such as ApoE, should therefore be discouraged until the Nuffield criteria are met. The same conclusion is warranted for screening for Chorea Huntington or any other triplet repeat disease.

Approval by a competent, State approved body is required before the implementation of a screening program for health purposes. The scientific validity and effectiveness of the program, as well as the sensitivity, specificity, and reliability of any proposed test must be confirmed (Council of Europe 2009).

27.5.8 Heterozygosity Screening

Facts and examples. Premarital heterozygosity testing for autosomal recessive mutations that lead to devastating disease in the homozygous state has been successfully introduced in a number of populations and geographical regions. Well-known examples of such diseases are Tay-Sachs disease in Ashkenazi Jews (Kaback et al. 1993) and beta-thalassemia in the populations of Sardinia (Cao 2002) and Cyprus. Voluntary genetic testing is propagated in Sardinia and Cyprus because the responsible mutations have a high frequency in the respective populations. Once both partners have been shown to be heterozygous, the couple is informed that the risk for the birth of a homozygous child is 25%. They may decide to undergo prenatal diagnosis, and the woman is given the option of early abortion of any affected fetus. The entire procedure is supported by the public health system. The number of newborns with Tay-Sachs disease and beta-thalassemia has dropped considerably in the respective societies. In Israel, heterozygosity screening has been extended to include other autosomal recessive diseases that are common in Ashkenazi Jews.

Ethics. In these examples, heterozygosity screening has proved to be successful for three reasons: (i) the genetic disease in question is devastating and treatment is either unavailable (Tay-Sachs) or is effective for only a few weeks (blood transfusions in beta-thalassemia), (ii) a clear management option exists following a positive test result, and (iii) the whole procedure is voluntary and is supported by the State. These experiences should be drawn upon in the future if screening for late-onset disorders is ever considered.

27.5.9 'Over the Counter' Genetic Testing

Facts and examples. In genetic testing for health purposes, DNA is usually extracted from blood samples, but saliva may also be used since only tiny amounts of DNA are required. This is the basis of market-driven, cross-border offers of anonymous genetic testing over the internet. Companies such as deCodeme, 23andme, and Navigenics offer screening for an ever increasing number of genetic diseases and predispositions to disease. Clients are required to send a sample of saliva. Following receipt of payment, they are granted access to a computer-based interpretation of the identified genotypes. A safe prediction is that such internet-based companies will continue to extend the range of tests offered, perhaps even up to the level of complete sequencing of the personal genome.

Ethics. Two aspects of such testing must be considered: (i) the technical quality of the test, and (ii) the ethical considerations. These companies usually examine various genes implicated in disease predisposition. Results are forwarded over the internet, without the necessary medical advice and genetic counseling that should accompany this process. The potential exists for such devastating scenarios as a woman being informed electronically that she has an 80% risk of developing breast cancer.

Since locus and allele heterogeneity are characteristic of human genetics, the genotypes examined will only represent a selection of all of the possible genotypes. The relevance of the tested genotypes is therefore unclear. It is also not possible to exclude a genetic disease or predisposition without taking the family history into account. An adequate interpretation of any genetic test result should take into account all of the available medical information. 'Over the counter' tests therefore provide incomplete and inadequate results.

In statements on the ethics of genetic testing, there is broad agreement that the personal data obtained from such testing belong to the sphere of informational self-determination. Special precautions are necessary to safeguard this data. Anonymous testing raises the possibility of DNA being collected without the knowledge of the donor (e. g. residual saliva on a cup, or used tissue). Even minors or persons not able to consent could be tested without any medical indication and without adequate authorisation. The statement of the Council of Europe (2009) requires that a genetic test for health purposes is only performed under individualised medical supervision. 'Over the counter' genetic testing does not comply with this fundamental ethical standard.

27.5.10 Unexpected Findings

Facts and examples. As in other fields of medical diagnostics, genetic testing may produce unexpected findings. There has been extensive experience of this in chromosome diagnostics. For example, the chromosomes of a healthy male are examined at his request to see whether his children have a risk for a genetic disease, on the grounds that a relative was affected by an undiagnosed malformation syndrome. It may transpire that this healthy male carries a complex chromosome aberration involving the short arm of the X-chromosome with a deletion in Xp21.2. It can thus be predicted that the currently healthy male will eventually develop Becker Muscular Dystrophy. It is usual practice to inform the person about the unexpected finding and the prognosis of the disease. In view of the potential for an unforeseen diagnosis, a counselee should always be informed about the possibility of unexpected findings.

In medical genetics and genetic counseling, molecular genetic diagnostics usually focuses on a specific locus or gene. Due to the enormous heterogeneity of many genetic diseases, it will eventually become routine to use one assay to screen many genes simultaneously for a mutation. Subtelomer deletion screening is a recent example of a method used to determine etiology in patients with mental retardation. Use of this method and similar approaches will undoubtedly lead to the identification in patients of unexpected genetic diagnoses or predispositions. The more that is known about the significance of mutations, the higher the chance that genetic diagnostics will generate unexpected findings. Once the personal genome can be completely sequenced at affordable costs, thousands of mutations with proven or potential disease implications will be revealed.

Ethics. The issue of unexpected findings – a problem that might occur in other fields of medicine as well – has not yet been systematically examined. Before the performance of any genetic diagnostic procedure, the patient should be informed of the possibility of unexpected findings. While the patient has the right to know everything about his or her genetic endowment and predisposition to disease, he or she also has the right not to know. However, the patient may ask that he or she is only given additional information in the case that the finding is medically relevant. The patient may even request that he or she is only informed of unexpected findings if appropriate therapeutic or preventive measures exist. The responsible doctor must then decide. This situation is difficult because many cases will be borderline. The more genetic information is generated, the more likely it is that unexpected data will be produced. It is questionable whether this scenario would be manageable in practice.

27.6 Research in Psychiatric Genetics

It was agreed at the inception of the human genome project in 1990 that 5% of the invested finances should be used for the investigation of ethical, legal, and social issues (ELSI). Similar rules were later adopted by genetic programs at the European and national level. It has always been acknowledged that genetic information may be sensitive. It is therefore important that the conditions and requirements of genetic research and diagnostics should be systematically analysed.

It has been shown that single genotypes only slightly increase the relative risk for psychiatric disorders. Whether or not a profile of predisposing mutations leads to a stronger correlation with the disease has not yet been clarified. New mutations might have a much higher relative risk, but they are rare. Large patient samples are necessary to determine a reliable association estimate. In many instances, international cross-border collaborations are the only means of providing studies with sufficient power (e. g. O'Donovan et al. 2009).

27.6.1 Biobanks

Facts and examples. Biobanks are an important infrastructure in human genetic research. They are collections of samples of human materials (e. g. cells, tissue, blood, or DNA as the physical medium of genetic information) that are, or can be, associated with personal information about their donors. Biobanks have a twofold nature, as collections of both samples and data (German National Ethics Council 2004). When the genetic contribution to any neuropsychiatric disease is to be examined, large numbers of samples must be collected in order to establish a biobank. Clinical information is collected separately and in parallel. Data are usually pseudonymized (coded) or anonymized. Researchers usually prefer to code the samples. If not prohibited by the donor, researchers like to maintain the option of re-contacting subjects at a later time-point. Biobanks can maintain their scientific value for a long time (see also Chapters 11 and 26).

Ethics. There should be a balance between the opportunities for and requirements of research and the protection of the donor. The use of human material and the gathering of personal data for subsequent use in a biobank for the purposes of medical research must be subject to the donor's consent. Consent must be given voluntarily and the donor must be informed appropriately of the purpose, nature, significance, and implications of the sample collection and its intended use. Donors should also be able to consent to the transfer of samples and data from biobanks to third parties for the purposes of medical research. The transfer should only take place in anonymized or coded form; in the latter case, the recipient should have no access to the code. Organizational measures should be in place to ensure that the code and the encrypted data are stored and administered separately (Nationaler Ethikrat 2004). In addition, the German National Ethics Council has argued that donors should be able to give general consent to the use of their samples and data for the purposes of genetic research, since individual research projects are not always specifiable in advance.

It has become standard practice to provide study volunteers with information as part of the consent procedure ('Informed consent'). This information includes details of all possible side effects and consequences. The situation concerning biobanks established for genetic research is different. Since individual research projects cannot always be specified at the time of sample collection, a proband or patient cannot be informed about all of the goals and details of future studies. An alternative to the option of general consent advocated by the German National Ethics Council has been proposed (Mascalzoni et al. 2008). The authors advance the view that research initiated following the initial storage of samples should also be explained to the proband. Informed consent should thus be seen as an ongoing process between researcher and participant. If this proposal were adopted, the researcher would have to re-contact each individual donor every time a new project was being considered. This is not feasible. It may, however, be a possible within the context of prospective cohort studies.

In genome-wide association studies (GWAS), correlation between mutations with small effects and the phenotype can only be demonstrated in large samples of patients and controls. Up to one million SNPs may be examined in each individual. Since such a study is labor- and cost-intensive, it would be helpful if the complete genotype-set were made available to the scientific community. Data from different studies could then be combined. It has been shown, however, that even when genotypes are made public in an aggregated form the presence of a specific individual can be identified (Homer et al. 2008). If the genotype profile of an individual were made public, it would be possible for an external scientist to establish whether the person was a patient or a healthy control. It could therefore become known that the individual is affected with the disease under investigation in the study. The risk of this occurring is undeniably low, but the threat cannot be ignored. Several funding organisations have removed the genotype data of dozens of genetic studies from their websites (Couzin 2008). Scientific exchange of genetic mass data should only be permitted under the conditions of a contract that respects the privacy of each individual included in the study.

27.6.2 Prospective Cohort Studies at the Population Level

Life course and life expectancy are influenced by a host of factors including genetic factors. Large scale prospective studies are necessary to identify genes that are medically relevant. It is hoped that the establishment of the genotype–phenotype relationship will make it possible to formulate preventive approaches. Cohort studies are being planned in various countries. The UK Biobank is the first project of this kind. For this project, 500,000 blood samples are being collected together with basic medical information from all probands. A certain proportion of the probands will be re-contacted years later. It is likely that this study will uncover genetic information of relevance to mental development. A detailed and transparent protocol has been established for this project that takes ethical issues into account. Participants are made familiar with the protocol, and a call-center is available to answer all questions.

27.6.3 Systematic Disease-Oriented Genetic Research – The Danish Paradigm

In Denmark – a country with a population of seven million – conditions for medical genetic research are unusually good due to balanced legislation (Mors 2009, nielmors@rm.dk, personal communication). The country has population-based registries for psychiatric patients and twins that can be used for research purposes. Access to the registry is strictly regulated by the National Board of Health, the Data Protection Agency, and the Scientific Committees. The Ethics Committee examines the research protocol and decides whether a researcher may, for example, access patient files, or approach a patient through the general practitioner, the psychiatric department, or even directly. The patient may, of course, refuse to participate in a study. The majority of the population agree that patient data should be made available to researchers in order to advance science for the benefit of the common good. The experience in Denmark shows that in a free and liberal society, the protection of personal rights can be compatible with the needs of research. Similar legislation exists in the other Scandinavian countries.

27.6.4 Experience from Iceland

An unusual event was the establishment of deCODE Genetics in Iceland. In 1998, the Icelandic parliament passed a national database law which awarded the company deCODE exclusive rights to create a centralized National Health Sector Database (HSD) of the Icelandic population (Koay 2004). The intention was that the database should contain three centralized sources of information: genealogical, medical/health, and genetic. It was envisaged that the database would enable cross-referencing of genotypic information with phenotypic and genealogical data. It was unusual for legislation to grant a private corporation an exclusive licence to create, operate, and commercialize a centralized national health database. Equally unusual was the fact that the legislation permitted the State to provide the 'presumed consent' of all citizens (past, present, and future) to inclusion in the database. This particular aspect of the legislation only included an 'opt-out' clause to the presumed consent at the initiation of the project. Researchers were permitted to use information from the database concerning any individual without any obligation to notify them (Koay 2004). Unsurprisingly, the project was heavily criticized by bioethicists, physicians, and others. Only 1 year after its implementation, more than 11,000 Icelanders had already opted out of the HSD database (Hauksson 1999).

27.6.5 The Opinion of the General Population and Patients with Mental Disorders

An important issue in this context is the opinion of the general population and patients with major mental disorders and their relatives towards research into psychiatric genetics. An informative study has been conducted in Germany (Illes et al. 2003). The authors interviewed 3,077 members of the general population by telephone, and administered a questionnaire to 316 patients with psychiatric disorders and 163 relatives in a personal face to face interview. The majority of those interviewed were in favor of psychiatric genetic research. The patients and the relatives were even more positive than the general population. Among patients and relatives, the hope prevailed that treatment options might improve through genetic research. Ethical concerns were also apparent, however. There was overall agreement that results from genetic tests should be treated confidentially. Interestingly, a generally positive attitude to predictive genetic testing was observed across the entire sample, which was more marked in patients than in the general population. It is essential for researchers to take the views of patients and the general population into account.

27.7 Outlook and Conclusions

Through the advent of molecular genetics, our understanding of human life, the predisposition to disease and the issue of nature-nurture is ever increasing. The riddle of psychiatric diseases will ultimately be solved (at least in part) through the collaborative efforts of clinicians, neurobiologists and geneticists. This development will increase the scope for the prevention and early treatment of mental disease. However, both the individual and society will have to cope with the challenge of rational decision making within this context.

Research involving medical and (more particularly) genetic data carries a certain risk of the infringement of personal rights. However, this risk can be minimized by appropriate legislation, the involvement of ethics committees, and the professional integrity of individual researchers. A society that wishes to benefit from progress in medical research must also contribute to this research. It is worrying, if not blatantly unethical, for developed societies to prefer to merely import knowledge obtained from basic or translational research conducted abroad, particularly when this involves research conducted in developing nations. The UK Biobank and disease-oriented genetic research programs in Denmark have demonstrated that it is possible to achieve a balance between the protection of personal rights and the needs of research.

Two widespread fears mentioned above (Section 27.1.2) remain to be prevalent in society. One is that our ever increasing knowledge of genetics may ultimately reveal an underlying biological determinism (cf. Propping 2002). This would relegate humans to the status of mere machines that are under the control of their genes. In fact, the fate of individuals who carry a mutation that leads to a monogenic and devastating disease is only biologically determined in the absence of appropriate therapy. However, the more a phenotype is in the 'normal' range, the higher is its polygenic determination. This also applies to multifactorial diseases (see Section 27.2.3). For all multifactorial diseases, the concordance rates for monozygotic twins are around 40–60%, a figure far below 100%. Both genetic and environmental mechanisms contribute to these figures. The predisposition to a multifactorial disease will not progress to manifest disease in a deterministic fashion. Deeper insight into the processes underlying multifactorial diseases will lead to developments in both treatment and prevention. It is thus certain that a general genetic determinism does not exist.

The other fear is that society will witness a 'geneticisation' of all aspects of human life and a revival of eugenics as a consequence of large scale genetic screening,. It is, however, unlikely that the imprecise and unrealistic goals of early eugenics will experience a renaissance. It is perhaps possible, however, that the pursuit of healthy offspring will lead to an increased use of prenatal diagnosis and the elective abortion of affected embryos. This could be achieved by a systematic preconceptual screening program for heterozygosity for autosomal recessive diseases (see Section 27.5.8). This might be termed 'bottom-up' eugenics, in contrast to the eugenics of the twentieth century. It is unlikely that couples will request prenatal diagnosis in order to prevent the birth of offspring with traits carrying minor health implications, and prenatal diagnosis is not suitable for multifactorial diseases. It is also possible that health economy arguments will assume greater importance in society. Insurance companies may be interested in avoiding the birth of handicapped children, and insurance companies or employers may be interested in using the positive predictive values (PPVs) of genotypes to evaluate clients or applicants.

Geneticists are well aware that public acceptance of their work is dependent upon a strict adherence to those ethical standards that are in place to avoid ethically questionable results or strategies. Society and legislators must engage in ongoing and transparent discussion of all developments in this field of research in order to ensure the integrity of human rights and the maintenance of autonomy, non-malevolence, dignity, and confidentiality within the context of human genetic research.

References

American Society of Human Genetics (1975) Ad-Hoc-Committee on genetic counseling. Am J Hum Genet 27:240–242

Cao A (2002) Carrier screening and genetic counselling in beta-thalassemia. Int J Hematol 76(Supplement 2):105–113

Cook EH, Scherer SW (2008) Copy-number variations associated with neuropsychiatric conditions. Nature 455:919–923

Council of Europe (2009) Additional protocol to the convention of human rights and biomedicine, concerning genetic testing for health purposes. Council of Europe Treaty Series No. 203 (1102/2009). Also: http://conventions.coe.int/Treaty/Commun/QueVoulezVous. asp?NT=203&CM=8&DF=4/8/2009&CL=ENG

Couzin J (2008) Whole-genome data not anonymus, challenging assumptions. Science 321:1278

Deutsche Gesellschaft für Humangenetik (1995, 2007) Leitlinie – Genetische Diagnostik bei Kindern und Jugendlichen. http://gfhev.de/de/leitlinien/LL_und_Stellungnahmen/2007

Deutsche Gesellschaft für Humangenetik (2007) Positionspapier der Deutschen Gesellschaft für Humangenetik. http://www.medgenetik.de/sonderdruck/2007_gfh_positionspapier.pdf

Friedl W, Caspari R, Sengteller M, Uhlhaas S, Lamberti C, Jungck M, Kadmon M, Wolf M, Fahnenstich J, Gebert J, Möslein G, Mangold E, Propping P (2001) Can APC mutation analysis contribute to therapeutic decisions in familial adenomatous polyposis? Experience from 680 FAP families. Gut 48:515–521

Galton F (1865) Hereditary talent and character. Cited after Vogel F (1961) Lehrbuch der allgemeinen Humangenetik. Springer, Berlin, Heidelberg, New York, NY

Gene Tests http://www.genetests.org/

German National Ethics Council (2004) Biobanks for research. Nationaler Ethikrat, Berlin http://www.ethikrat.org/_english/publications/Opinion_Biobanks-for-research.pdf

Grotjahn A (1912) Soziale Pathologie. von August Hirschwald, Berlin

Gusella JF, Wexler NS, Conneally PM, Naylor SL, Anderson MA, Tanzi RE, Watkins PC, Ottina K, Wallace MR, Sakaguchi AY et al (1983) A polymorphic DNA marker genetically linked to Huntington's disease. Nature 306:234–238

Hauksson P (1999) Icelanders opt out of genetic database. Nature 400:707

Homer N, Szelinger S, Redman M, Duggan D, Tembe W, Muehling J, Pearson JV, Stephan DA, Nelson SF, Craig DW (2008) Resolving individuals contributing trace amounts of DNA to highly complex mixtures using high-density SNP genotyping microarrays. PLoS Genet 4(8):e1000167

Homsma SJ, Huijgen R, Middeldorp S, Sijbrands EJ, Kastelein JJ (2008) Molecular screening for familial hypercholesterolaemia: consequences for life and disability insurance. Eur J Hum Genet 16:14–17

Illes F, Rietz C, Fuchs M, Ohlraun S, Prell K, Rudinger G, Maier W, Rietschel M (2003) Einstellung zu psychiatrisch-genetischer Forschung und prädiktiver Diagnositk. Ethik Med 15:268–281

Kaback M, Lim-Steele J, Dabholkar D, Brown D, Levy N, Zeiger K (1993) Tay-Sachs disease–carrier screening, prenatal diagnosis, and the molecular era. An international perspective, 1970 to 1993. The International TSD Data Collection Network. JAMA 270:2307–2315

Keymolen K, Goossens V, De Rycke M, Sermon K, Boelart K, Bonduelle M, Van Steirteghem A, Liebaers I (2007) Clinical outcome of preimplantation genetic diagnosis for cystic fibrosis: the Brussels' experience. Eur J Hum Genet 15:752–758

Koay PP (2004) An icelandic (ad-)venture: New research? New subjects? New ethics? In: Roelcke V, Maio G (eds) Twentieth century ethics of human subjects research. Franz Steiner, Stuttgart, pp 335–348

Koolen DA, Pfundt R, de Leeuw N, Hehir-Kwa JY, Nillesen WM, Neefs I, Scheltinga I, Sistermans E, Smeets D, Brunner HG, van Kessel AG, Veltman JA, de Vries BB (2009) Genomic microarrays in mental retardation: a practical workflow for diagnostic applications. Hum Mutation 30:283–292

Levy HL (2003) Lessons from the past – looking to the future. Pediatr Ann 32:505–508

Lippman A (1992) Led (astray) by genetic maps: the cartography of the human genome and healthcare. Soc Sci Med 35:1469–1476

Mascalzoni D, Hicks A, Pramstaller P, Wjst M (2008) Informed consent in the genomics era. PLoS Med 5(9):e192

Metzger S, Rong J, Nguyen HP, Cape A, Tomiuk J, Soehn AS, Propping P, Freudenberg-Hua Y, Freudenberg J, Tong L, Li SH, Li XJ, Riess O (2008) Huntingtin-associated protein-1 is a modifier of the age-at-onset of Huntington's disease. Hum Mol Genet 11:1137–1146

Nuffield Council on Bioethics (1998) Mental disorders and genetics: the ethical context. Nuffield Council on Bioethics, London

O'Donovan MC, Norton N, Williams H, Peirce T, Moskvina V, Nikolov I, Hamshere M, Carroll L, Georgieva L, Dwyer S, Holmans P, Marchini JL, Spencer CC, Howie B, Leung HT,

Giegling I, Hartmann AM, Möller HJ, Morris DW, Shi Y, Feng G, Hoffmann P, Propping P, Vasilescu C, Maier W, Rietschel M, Zammit S, Schumacher J, Quinn EM, Schulze TG, Iwata N, Ikeda M, Darvasi A, Shifman S, He L, Duan J, Sanders AR, Levinson DF, Adolfsson R, Osby U, Terenius L, Jönsson EG, Cichon S, Nöthen MM, Gill M, Corvin AP, Rujescu D, Gejman PV, Kirov G, Craddock N, Williams NM, Owen MJ, Molecular Genetics of Schizophrenia Collaboration (2009) Analysis of 10 independent samples provides evidence for association between schizophrenia and a SNP flanking fibroblast growth factor receptor 2. Mol Psychiatry 14:30–36

OMIM – Online Mendelian Inheritance in Man. http://www.ncbi.nlm.nih.gov/sites/entrez?db=omim

Paul D (2001) History of eugenics. In: Smelser NJ, Baltes PB (eds) International encyclopedia of the social & behavioral sciences. Elsevier, Amsterdam, pp 4896–4901

Propping P (2002) Die Freiheit des Menschen im Zeitalter der Genetik. In: Elsner N and Schreiber H-L (Hrsg) Was ist der Mensch?. Wallstein, Göttingen, pp 127–142

Propping P (2005) The biography of psychiatric genetics: from early achievements to historical burden, from an anxious society to critical geneticists. Am J Med Genet B Neuropsychiatr Genet 136B:2–7

Reed S (1974) History of genetic counseling. Soc Biol 21:332–339

Rüdin E (1911) Einige Wege und Ziele der Familienforschung, mit Rücksicht auf die Psychiatrie. Z Ges Neurol Psychiat 7:487–585

Schulze TG, Fangerau H, Propping P (2004) From degeneration to genetic susceptibility, from eugenics to genethics, from Bezugsziffer to LOD score: the history of psychiatric genetics. Int Rev Psychiatry 16:246–259

Stefansson H, Rujescu D, Cichon S, Pietiläinen OP, Ingason A, Steinberg S, Fossdal R, Sigurdsson E, Sigmundsson T, Buizer-Voskamp JE, Hansen T, Jakobsen KD, Muglia P, Francks C, Matthews PM, Gylfason A, Halldorsson BV, Gudbjartsson D, Thorgeirsson TE, Sigurdsson A, Jonasdottir A, Jonasdottir A, Bjornsson A, Mattiasdottir S, Blondal T, Haraldsson M, Magnusdottir BB, Giegling I, Möller HJ, Hartmann A, Shianna KV, Ge D, Need AC, Crombie C, Fraser G, Walker N, Lonnqvist J, Suvisaari J, Tuulio-Henriksson A, Paunio T, Toulopoulou T, Bramon E, Di Forti M, Murray R, Ruggeri M, Vassos E, Tosato S, Walshe M, Li T, Vasilescu C, Mühleisen TW, Wang AG, Ullum H, Djurovic S, Melle I, Olesen J, Kiemeney LA, Franke B; GROUP, Sabatti C, Freimer NB, Gulcher JR, Thorsteinsdottir U, Kong A, Andreassen OA, Ophoff RA, Georgi A, Rietschel M, Werge T, Petursson H, Goldstein DB, Nöthen MM, Peltonen L, Collier DA, St Clair D, Stefansson K (2008) Large recurrent microdeletions associated with schizophrenia. Nature 455:232–236

Strachan T, Read AP (2004) Human molecular genetics. Garland Science, London

Biobank UK. http://www.ukbiobank.ac.uk/

Umans-Eckenhausen MA, Sijbrands EJ, Kastelein JJ, Defesche JC (2001) Review of first 5 years of screening for familial hypercholesterolaemia in The Netherlands. Circulation 106:3031–3036

Wonderling D, Umans-Eckenhausen MA, Marks D, Defesche JC, Kastelein JJ, Thorogood M (2004) Cost-effectiveness analysis of the genetic screening program for familial hypercholesterolemia in The Netherlands. Semin Vasc Med 4:97–104

Went L (1990) Ethical issues policy statement on Huntington's disease molecular genetics predictive test. International Huntington Association. World Federation of Neurology. J Med Genet 27:34–38

Part IV
Non-medical Uses of Psychiatry

Chapter 28
Political Abuse of Psychiatry

Norman Sartorius and Hanfried Helmchen

Introductory Comment by the Editors

The definition of political abuse of psychiatry is not simple. What is usually meant by those terms is that an intervention (e.g. coercion, forced admission to a psychiatric institution, medically not indicated application of psychopharmacological medications) designed for use in dealing with persons suffering from a mental disorder is used on persons who are not mentally ill – and that this is done because the government has wanted it to be done. Psychiatrists usually participate in the process: but it is also possible to imagine the use of psychiatric means without psychiatrists (e.g. use sedatives for the extraction of information by the some governments' secret services). Bringing a person to a mental hospital by force – even if that person is not found ill and is immediately discharged by psychiatrists – will be harmful to such persons and diminish their credibility in most countries because of the stigma that is accompanying mental illness. In such an instance the governments' officials are using stigmatization related to mental illness which unfortunately does harm to the person concerned regardless of psychiatrists' assurance that the victim is not suffering from a mental disorder.

Other forms of governments' action negatively affecting the mentally ill – e.g. economically or ideologically based but nevertheless political decisions against psychiatry by negligence due to discrimination and stigmatisation of the mentally ill by political powers – are evident in many countries. Letting large numbers of people with a mental illness starve by providing insufficient funds for food to a mental health facility is not uncommon. The funds that mental hospitals receive in many countries often do not allow the treatment of intermittent physical illness – and inmates of mental hospital may die of pneumonia, tuberculosis or a variety of other diseases that could be effectively treated if the government were to provide adequate supplies of medications to mental hospitals and if it undertook what is necessary to make the hospitals a place that are appropriate for human habitation and treatment of an illness.

N. Sartorius (✉)
Association for the Improvement of Mental Health Programmes, CH-1209 Geneva, Switzerland
e-mail: sartorius@normansartorius.com

H. Helmchen, N. Sartorius (eds.), *Ethics in Psychiatry*, International Library
of Ethics, Law, and the New Medicine 45, DOI 10.1007/978-90-481-8721-8_28,
© Springer Science+Business Media B.V. 2010

Similarly, the governments have abused psychiatry by using a psychiatric diagnosis as the justification for sterilization without consent. This 'eugenic' measure has been introduced 100 years ago and has still been used in several European countries (and elsewhere) until relatively recently – abandoning the practice of sterilizing people with a mental disorder, intellectual disability and other problems (sometimes leaving the law in power) in the 1960s and 1970s. Experts from countries in which this measure at the time has been legal did not hesitate to criticize other countries in which the measure had been or still was in use.

Political abuse of psychiatry is also happening when mental health facilities are run by individuals who have obtained their position – say of director of a mental hospital – on the grounds of their allegiance to a political party or on grounds of battlefield merit. The harm that such people can produce for all the persons in the facility is huge and governments in these instances are also engaged in political abuse by having neglected the need to ensure that persons who run governmental medical facilities have the qualifications that are necessary to do so.

Recent reports alleged abuses of psychiatry to silence opponents of government in a number of countries including China, Cuba, Romania, Paraguay and others. It would require a separate volume to describe all forms of abuse and instances of direct or indirect political abuse of psychiatry. It would take even more space to describe the ways in which national and international organizations and well-meaning individuals have fought the political abuse of psychiatry. Such space is not available in this volume and we have therefore opted for a simpler solution and included an example of the abuse of psychiatry in one country, written by Mr. van Voren an activist who has spent many years fighting against it. We hope that example will stimulate the reader to look for other examples of abuse of psychiatry and ways in which it was stopped: in these examples there are many lessons about dangers of abuse of psychiatry (and medicine) and about ways to prevent them and ensure that the rights of citizens – with a mental illness or without it – are respected and protected.

Chapter 29
Abuse of Psychiatry for Political Purposes in the USSR: A Case-Study and Personal Account of the Efforts to Bring Them to an End

Robert Van Voren

Contents

Abbreviations

AUSPN	All Union Society of Psychiatrists and Neuropathologists (of the USSR)
DSM-IV	Diagnostic and Statistic Manual (of Mental Disorders), 4th Revision (American Psychiatric Association)
IALMH	International Academy of Law and Mental Health
IAPUP	International Association on the Political Use of Psychiatry
ICD	International Classification of Diseases (World Health Organisation)
KGB	Komitet Gosudarstvennoi Besopasnosti
RCP	Royal College of Psychiatrists
STASI	Ministerium für Staatssicherheit", i.e. Secret Service of the German Democratic Republic
USSR	Union of Soviet Socialist Republics

R. Van Voren (✉)
Global Initiative on Psychiatry, 1282 BG Hilversum, The Netherlands
e-mail: rvvoren@gip-global.org

H. Helmchen, N. Sartorius (eds.), *Ethics in Psychiatry*, International Library
of Ethics, Law, and the New Medicine 45, DOI 10.1007/978-90-481-8721-8_29,
© Springer Science+Business Media B.V. 2010

29.1 Introduction

There are few issues that have so much dominated the agenda of the world psychiatric community over the past 40 years as the issue of political abuse of psychiatry. Starting in 1971, when Soviet dissident Vladimir Bukovsky sent a collection of documents on the alleged political abuses to the West with the request to consider them, until this very day when the political abuse of psychiatry in China is still on the agenda, the issue has on one hand resulted in fierce debates and outrage, yet on the other hand it has stimulated an ongoing debate on human rights and professional ethics. During those years the World Psychiatric Association (WPA), around which most of the discussions evolved, adopted a series of ethical codes and declarations on human rights that condemn the use of psychiatry for non-medical purposes, and also installed mechanisms to investigate complains of violations of these regulations. Yet at the same time, critics believe that the organization itself has not always implemented the regulations it imposed on its member societies, thereby triggering further debates on the issue.

In this chapter, the case of political abuse of psychiatry in the Soviet Union is used as the main example, specifically because it was a well-documented case, because it strongly influenced the concept of medical ethics and its application internationally, and because with maybe a few exceptions it is generally accepted that psychiatry in the Soviet Union was abused for political purposes in a systematic manner in the course of several decades.

However, the Soviet Union is certainly not the only country where these abuses took place. Over the past decades my colleagues and myself have seen a lot of documentation on other countries.[1] One of the countries where systematic political abuse of psychiatry seemed to have taken place was Romania; in 1997 we organized an investigative committee to research what actually happened.[2] We received information on cases in Czechoslovakia, Hungary and Bulgaria, but all these cases were individual and there was no evidence that any systematic abuse took place. An extensive research on the situation in Eastern Germany came to the same conclusion, although in this socialist country politics and psychiatry appeared to be very closely intermingled (Süss 1998). Later, information appeared on the political abuse of psychiatry in Cuba, which was however short-lived (Brown and Lago 1991). In the 1990s Global Initiative on Psychiatry was involved in a case of political abuse of psychiatry in The Netherlands, in the course of which the Ministry of Defense

---

[1] The author has been involved in the struggle against the political abuse of psychiatry since the late 1970s and was one of the founding members of the International Association on the Political Use of Psychiatry. This organization was later renamed in Geneva Initiative on Psychiatry and since 2005 it is called Global Initiative on Psychiatry. Van Vorens memoirs *On Dissidents and Madness* were published by Rodopi, Amsterdam/New York, in 2009. A book on the issue of Soviet psychiatric abuse and the WPA in 1983–1989, titled *Cold War in Psychiatry*, will appear with the same publisher in 2010

[2] See *Psychiatry under Tyranny, An Assessment of the Political Abuse of Romanian Psychiatry During the Ceaucescu Years*, Amsterdam, IAPUP, 1998.

tried to silence a social worker by falsifying several psychiatric diagnoses. The case took many years to be resolved, and although the victim was compensated and even knighted by the Dutch Queen, it is still not fully resolved (Nijeboer 2006). And, finally, since the beginning of this century the issue of political abuse of psychiatry in the People's Republic of China is again high on the agenda and has caused repeated debates within the international psychiatric community.

During the years of our existence, we were regularly approached with requests to deal with abusive situations in psychiatry in countries such as South Africa, Chile and Argentine. However, on basis of research we concluded that in these cases one could not speak of *political* abuse of psychiatry. In the case of South Africa severe abuses were the result of the policy of Apartheid, which resulted in very different conditions in mental health services for the white ruling class and the black majority of the population. Claims that psychiatry was abused as a means of political or religious repression were never confirmed. In South America the abuse concerned psychiatrists themselves, not psychiatry as such: psychiatrists were used to determine which forms of torture were the most effective, and although these abuses clearly constituted a serious violation of the Hippocratic Oath, they could not be classified as political abuse of psychiatry.

Admittedly, those involved in the struggle against political abuse of psychiatry never reached full consensus on what the exact boundaries were between political abuse of psychiatry and more general misuse of psychiatric practice. In the course of the years, many individual cases were discussed extensively, determining whether it should be considered as one of political abuse of psychiatry or not. The issue continues to be discussed, in particular because recent cases are often more complex and involve a less overt government involvement. A current position paper of the Global Initiative on Psychiatry states that political abuse of psychiatry refers to the misuse of psychiatric diagnosis, treatment and detention for the purposes of obstructing the fundamental human rights of certain individuals and groups in a given society.

The fact that the use of psychiatry for political purposes is reported from so many diverse countries, shows that there is an ongoing tension between politics and psychiatry, and that the opportunity to use psychiatry as a means to stifle opponents or solve conflicts is an appealing one, not only to dictatorial regimes but also to well-established democratic societies. At the same time, it is clear that using psychiatry as a means of repression has been a favorite of socialist oriented regimes. An explanation might be found in the fact that socialist ideology says to be targeted at the establishment of the ideal society, where all are equal and all will be happy, and thus those who are against must be mentally ill. In fact, this second part seemed to have had the strongest influence, because even in the Soviet Union of the 1970s where many were no happy and society was far from ideal, many psychiatrists still believed that those who turned against the regime must be mad.[3] Also in that sense the Soviet case is the best example.

---

[3] Article 62 of the Soviet Constitution (1977) stated: 'Each citizen is obliged to protect the interests of the Soviet State and to contribute to its power and authority' (Helmchen 1986).

29.2 Why Was Soviet Psychiatry Abused Politically?

The political abuse of psychiatry in the Soviet Union originated from the concept that persons who opposed the Soviet regime were mentally ill, as there was no other logical explanation why one would oppose the best socio-political system in the world. Soviet leader Nikita Khrushchev worded this in a speech: 'A crime is a deviation from the generally recognized standards of behavior frequently caused by mental disorder. Can there be diseases, nervous disorders among certain people in Communist society? Evidently yes. If that is so, then there will also be offences that are characteristic for people with abnormal minds [. . .]. To those who might start calling for opposition to Communism on this basis, we can say that.[. . .] clearly the mental state of such people is not normal.'[4]

The diagnosis 'sluggish schizophrenia', developed by the Moscow School of Psychiatry and in particular by Andrei Snezhnevsky, provided a very handy framework to explain this behavior. According to the theories of Snezhnevsky and his colleagues, schizophrenia was much more prevalent than previously thought because the illness could be present with relatively mild symptoms and only progress later. As a result, schizophrenia was diagnosed much more frequently in Moscow than in other countries in the World Health Organization Pilot Study on Schizophrenia reported (WHO 1973). And in particular sluggish schizophrenia broadened the scope, because according to Snezhnevsky patients with this diagnosis were able to function almost normally in the social sense. Their symptoms could resemble those of a neurosis or could take on a paranoid quality. The patient with paranoid symptoms retained some insight in his condition, but overvalued his own importance and might exhibit grandiose ideas of reforming society. Thus symptoms of sluggish schizophrenia could be 'reform delusions', 'struggle for the truth', and 'perseverance' (Bloch 1989). In an interview with the Soviet newspaper *Komsomolskaya Pravda* two Soviet psychiatrists, Professor Marat Vartanyan and Dr. Andrei Mukhin, explained how it was possible that a person could be mentally ill while those around him did not notice it, as could happen in case of 'sluggish schizophrenia'. What did mentally ill then mean? Vartanyan: '. . . When a person is obsessively occupied with something. If you discuss another subject with him, he is a normal person who is healthy, and who may be your superior in intelligence, knowledge and eloquence. But as soon as you mention his favourite subject, his pathological obsessions flare up wildly.' Vartanyan confirmed that hundreds of persons with this diagnosis were hospitalized in the Soviet Union. According to Dr. Mukhin this was because 'they disseminate their pathological reformist ideas among the masses' (*Komsomolskaya Pravda*, July 15, 1987). A few months later the same newspaper listed a number of symptoms 'a la Snezhnevsky', including 'an exceptional interest in philosophical systems, religion and art.' The paper quoted from a 1985 *Manual on Psychiatry* of Snezhnevsky's Moscow School and subsequently concluded: 'In this way any – normally considered sane – person can be

---

[4]Speech published by Pravda on May 24, 1959.

diagnosed as a "sluggish schizophrenic" ' (*Komsomolskaya Pravda*, November 18, 1987).

While most experts agree that the core group of psychiatrists that developed this concept on the orders of the Party and the KGB knew very well what they were doing, for many Soviet psychiatrists this seemed a very logical explanation, because they could not explain to themselves otherwise why somebody would be willing to give up his career, family and happiness for an idea or conviction that was so different from what most people believed or forced themselves to believe. In a way, the concept was also very welcome, as it excluded the need to put difficult questions to oneself and one's own behavior. And difficult questions could lead to difficult conclusions, which in turn could have caused problems with the authorities for the psychiatrist himself.

On basis of the available data, one can confidently conclude that thousands of dissenters were hospitalized for political reasons. The archives of the International Association on the Political Use of Psychiatry (IAPUP) contained over a thousand names of victims of whom we had multiple data (name, date of birth, type of offense, place of hospitalization), all information that had reached the West via the dissident movement. However, this number excluded the vast 'grey zone', people who were hospitalized usually for shorter periods of time because of a complaint to lower officials, conflicts with local authorities or because of unorthodox behavior. It is estimated that this group was much larger. Their names were, however, not known to the dissident movement and thus nor recorded in the West. A biographical dictionary published by IAPUP in 1990 listed 340 victims of political abuse of psychiatry as well as more than 250 psychiatrists involved in these practices (Koppers 1990).

As indicated before, many of these psychiatrists were probably unaware that they engaged in unethical behavior and that they were part of a governmental repressive machinery. For example, Ukrainian psychiatrist Ada Korotenko found out only in the mid-1990s that former colleagues of her had been involved in the political abuse of psychiatry when she participated in a Ukrainian study into the origins of political abuse of psychiatry and in the course of that study examined sixty former victims. Under the original Soviet diagnoses she found the names not only of former colleagues but even of some of her friends. While interviewing the former victims and comparing their state of mind with the original diagnoses, she not only realized they had been hospitalized for non-medical reasons, but also that she could have authored the original diagnoses herself (Korotenko and Alkina 2002).[5] When visiting the Special Psychiatric Hospital in Chernyakhovsk (Kaliningrad region) for the first time in 2006, I was introduced to a nurse who had been on duty when the famous dissident General Pyotr Grigorenko was held there in 1970–1973 (Koppers 1990, pp. 97–98). She remembered Grigorenko well, and was praising his intellect, his concern for his family and his gentleman behavior. Her description of him

---

[5]and private conversations of the author with Dr. Korotenko. Other former Soviet psychiatrists confirmed this dilemma, see Van Voren, R., *On Dissidents and Madness*, Rodopi, 2009, pp. 168–171.

very much fitted my own recollections of the man, yet the question apparently never crossed her mind why he had been hospitalized if he was such a wonderful person: she just obeyed orders, no questions asked. Undoubtedly, this counted for the majority of those who encountered political 'patients' in their psychiatric departments.[6]

The onset of political psychiatry can probably best be seen as the result of a combination of factors that were only possible to mature under a totalitarian regime. The decision in 1950 to give monopoly over psychiatry to the Pavlovian school of Professor Andrei Snezhnevsky was one crucial factor. Here we had a scientist who had a vision, who believed he could make history by proving his view of psychiatry, and the totalitarian climate made it possible for him to implement his plans unobstructed. Well-known psychiatrists who disagreed with him lost their jobs, some were even exiled to Siberia. It was unwise, to say the least, to oppose Snezhnevsky, and he cleverly made use of the sense of terror that reigned the Soviet Union during the last years of Stalin (Bloch and Reddaway 1977, pp. 220–223).

Secondly, Soviet society had become a centrally ruled totalitarian State. Doctors had been subordinated to the Party by having them swear the Oath of the Soviet Doctor instead of the Hippocratic Oath. And the Oath of the Soviet Doctor was very clear: the ultimate responsibility was before the Communist Party, and not before medical ethics.

Thirdly, the Soviet Union had become a closed society, a society that was cut off from the rest of the world. World psychiatric literature was unavailable, except to the politically correct psychiatric elite. The power of the Party seemed endless, whether you believed in their ideals or not. And thus any person who decided to voice dissent openly ran a high risk of being considered mentally ill. And thus the political abuse of psychiatry, that initially mostly effected intellectuals and artistic circles, grew into a important form of the repression with approximately one-third of the dissidents in the 1970s and early 1980s being sent to a psychiatric hospital, rather than to a camp, prison or exile.

Dissident psychiatrist Dr. Anatoly Koryagin, who served 6 years out of a total sentence of 14 years of camp and exile for having been a member of a 'Working Commission to Investigate the use of psychiatry for political purposes', examined over forty victims or potential victims of political psychiatry. His diagnoses were used as a defense against being declared insane, or as a means to show the outside world that a hospitalized dissident had been incarcerated for non-medical reasons. On basis of his sample, Koryagin came to the interesting conclusion that the length of hospitalization seemed to correspond to the length of the sentence a political prisoner otherwise would have got. In other words, a political prisoner charged with 'slandering the Soviet state' usually stayed hospitalized for about 3 years (the maximum term under that article of the USSR Criminal Code) while a person accused of anti-Soviet agitation and propaganda usually stayed in for much longer, 7 years

---

[6]The same bibliographical dictionary of A. Koppers lists over 150 institutions where these abuses took place.

or more (again the maximum sentence under that article). Cynically, one could say that the more crazy a person was, the more serious his damage to the Soviet state! (Koryagin 1987, pp. 43–50; Korotenko and Alkina 2002).

In other cases, dissidents had the feeling that mentally weaker persons were more quickly sent to camps, while the mentally strong and unbreakable faced an uncertain future in a psychiatric hospital, while not having a sentence and being tortured with neuroleptics and other means. All in all, it is safe to conclude that the victims of political repression were carefully selected, and that this form of punishment seemed to be the most fitting to them.

29.3   The Campaign Against Soviet 'Political Psychiatry'

Generally speaking, the systematic use of psychiatry to incarcerate dissidents in psychiatric hospitals started in the late 1950s and early 1960s. However, there are cases of political abuse of psychiatry known from a much earlier date such as the case of the Russian philosopher Pyotr Chaadayev from the times of Tsar Nicholas I (Bloch and Reddaway 1977, pp. 48–50). Also in early Soviet times some attempts to use psychiatry for political purposes took place, yet these cases, as well as the Chaadayev case, can be compared to the Spijkers case in The Netherlands: sticking on a psychiatric diagnosis seemed to be the easiest option.

In the 1930s the political abuse of psychiatry took on a more systematic form. According to a series of letters published by a Soviet psychiatrist in *The American Journal of Psychiatry*, it was one of the leaders of the Soviet secret police, Andrei Vyshinsky, who ordered to use psychiatry as a means of repression.[7] According to the author of the letters, whose name was known to the editor but otherwise remained anonymous, the first Special Psychiatric Hospital in Kazan was used exclusively for political cases. Half of the cases were persons who indeed were mentally ill, but the other half were persons without any mental illness, such as the former Estonian President Päts who was held in Kazan from 1941 till 1956 for political reasons.[8]

Also the Serbski Institute for Forensic and General Psychiatry in Moscow had a political department, headed by Professor Khaletsky. However, according to Soviet poet Naum Korzhavin the Serbsky was at that time a relatively humane institution with a benevolent staff (Bloch and Reddaway 1977, pp. 53–54). However, the atmosphere changed almost overnight when Dr. Daniil Lunts was appointed head of the Fourth Department, which was later usually referred to as the Political Department. Before psychiatric departments had been considered a 'refuge' against being sent to the Gulag, but from that moment onwards this policy changed (Van Voren

---

[7] American Journal of Psychiatry, 1970, vol 126, pp. 1327–1328; vol. 127, pp. 842–843; 1971, vol. 127, pp. 1575–1576, and 1974, vol. 131, p. 474.

[8] *Kaznimye sumasshestviem*, Frankfurt, Possev, 1971, p. 479.

1978, Daniil Lunts, Psychiatrist of the Devil, unpublished manuscript; van Voren 1989, p. 16).

More cases of political abuse of psychiatry are known from the 1940s and 1950s, including that of a Party official Sergei Pisarev who was arrested after criticizing the work of the Soviet secret police in connection with the so-called Doctor's Plot, a anti-Semitic campaign developed at Stalin's orders that should have led to a new wave of terror in the USSR and probably to the annihilation of the remaining Jewish communities that had survived the Second World War. Pisarev was hospitalized in the Special Psychiatric Hospital in Leningrad, which together with a similar hospital in Sychevka had been opened after the Second World War. After his release in 1955, Pisarev initiated a campaign against the political abuse of psychiatry, concentrating himself on the Serbsky Institute that he considered to be the root of all evil. As a result of his activity the Central Committee of the Communist Party established a committee that investigated the situation and concluded that the political abuse of psychiatry was indeed taking place. However, the report disappeared in a desk drawer and never resulted in any action take (Pisarev 1970, pp. 175–180).

Until the mid-1960s the political abuse of psychiatry in the USSR went mostly unnoticed, and also among Soviet dissidents the notion that a dangerous new form of repression threatened them remained absent. In his memoirs Vladimir Bukovsky writes about his stay in the Serbsky Institute: 'We were absolutely not afraid to be called lunatics – to the contrary, we rejoiced: let these idiots think that we are lunatics if they like or, rather, let these lunatics think were are idiots. We remembered all the stories on lunatics by Chekhov, Gogol, Akatugawa and of course also *The Good Soldier Schweik*. We roared with laughter at our doctors and ourselves' (Bukovsky 1978, p. 199). But it was only later that they realized that the old woman who cleaned the ward told everything to the doctors, who used the information to prove their mental illness. In 1974, Bukovsky wrote together with the imprisoned psychiatrist Semyon Gluzman a *Manual on Psychiatry for Dissenters*, in which they gave guidelines to potential future victims of political psychiatry how to behave during investigation in order to avoid being diagnosed as being mentally ill (Bloch and Reddaway 1977, pp. 419–440).

On basis of the available evidence one can conclude that in the course of the 1960s the political abuse of psychiatry in the Soviet Union became one of the main methods of repression. By the end of that decade many well-known dissidents were diagnosed as being mentally ill.

29.4 The International Community Becomes Involved

As a result of the growing numbers of dissidents winding up in psychiatric hospitals the protests in the West grew and eventually culminated into a campaign to end this abuse of the psychiatric profession. In 1971, Vladimir Bukovsky sent a file of 150 pages documenting the political abuse of psychiatry to the West. For the first time Western psychiatrists could study copies of the psychiatric diagnoses

by Soviet psychiatrists involved in the abuse and learn the details of their diagnostic methods. The documents were accompanied by a letter by Bukovsky asking
Western psychiatrists to study the six cases documented in the file and say whether
these people should be hospitalized or not. A group of British psychiatrists examined the file and concluded: 'It seems to us that the diagnoses on the six people
were made purely in consequence of actions in which they were exercising fundamental freedoms. . .' (The Times November 16, 1971) They suggested to discuss
the issue during the upcoming World Congress of the WPA in November 1971 in
Mexico.

However, such discussion was not to take place. Although the President of the
Congress, Dr. Ramon de la Fuente, referred to documents that had been received
about some places in the world where political opposition was treated as mental illness, and he argued that 'to keep silent about such an ignominious situation
would weigh heavily on our conscience' (Mexico City News, November 13, 1971),
his words found no echo in the WPA General Secretary, Dr. Denis Leigh. Leigh
had already informed Snezhnevsky of the complaints and had sent the latter the
'Bukovsky Papers', and laid out his position with regard to the WPA's obligations:
'Nowhere in the statutes is there any mention of the WPA making itself responsible for the ethical aspects of psychiatry, nor is there any relevant statute or by-law
relating to complaints made by one member society against another member society. I think it is legally quite clear that the WPA is under no obligation to accept
complaints from one member society directed against another member society'
(WPA 1971). According to him the only thing the WPA could do was to refer the
cases to the relevant member society, in this case the Soviet All-Union Society of
Psychiatrists and Neuropathologists, which is exactly what he had done.

Leigh's interpretation of the WPA Statutes was tendentious, to say the least,
because one of the purposes of the organization as set out in its statutes was 'to promote activities designed to lead to increased knowledge in the field of illness and
better care for the mentally ill.' However, the Committee did not dispute Leigh's
interpretation of the statutes, and as a result it was clear that there would be no
debate in the WPA's General Assembly. Three days later Leigh suggested establishing a committee to consider the ethical aspects of psychiatric practice, but also in
this case no mention was made of the issue of political abuse of psychiatry in the
Soviet Union. Soviet psychiatrist Marat Vartanyan, by then already one of the main
apologists of Soviet psychiatric abuse, was even elected as associate secretary of
the Executive Committee. A day after the Mexico Congress he stated publicly 'the
nature of our [socio-political] system is such that this could not possibly happen'
(Reuter 1971).

The failure to discuss the issue opened the door for the Soviet authorities to
sentence Vladimir Bukovsky to 12 years in camp and exile, and to increase the use
of psychiatry as a means of repression.[9]

---

[9]Vladimir Bukovsky was eventually exchanged for the Chilean Communist Party leader Luis
Corvalan at Zürich airport on December 18, 1976.

In the period between the World Congresses in Mexico in 1971 and in Honolulu in 1977 a growing number of national psychiatric associations expressed their concern over the issue, but not more than that. The World Psychiatric Association did not study any of the evidence it received, nor did it interview former victims of Soviet psychiatric abuse. At the same time, however, they continued to maintain friendly relations with the Soviet psychiatrists that were closely involved in the political abuse of psychiatry. In November 1972, Secretary General Denis Leigh and Treasurer Professor Rees even accepted an honorary membership of the Soviet All-Union Society. Professor Rees would later change his position and became an active opponent of Soviet psychiatric abuse.

The Soviets continued to win their Western colleagues over. In October 1973 a WPA Congress on schizophrenia was held in Moscow, with among the main speakers Andrei Snezhnevsky and Georgi Morozov, director of the Serbski Institute since 1957 and heavily involved in psychiatric abuse. After the conference a group of psychiatrists were invited over to the Serbski Institute, where they were shown the case histories of six dissidents including the well known victims General Pyotr Grigorenko, Ukrainian dissident Leonid Plyushch and Moscow biologist Zhores Medvedev, accompanied by a short English summary. Subsequently they were shown an examination by a person who was said to be a dissident, and they ascertained that the man was indeed suffering from schizophrenia. The foreign visitors refused to sign any document, but this did not stop Marat Vartanyan from issuing a statement that five of the six cases that had undergone forensic psychiatric examination were, in the opinion of the WPA Committee, suffering 'from a mental illness at the time of their respective commissions of enquiry' (Bloch and Reddaway 1977, p. 317). It was to be repeated several times in the following years.

A month after the meeting at the Serbsky the British Royal College of Psychiatrists adopted a motion in which it deplored the political abuse of psychiatry and condemned the doctors who participated in it. For the first time the College discussed whether it should withdraw from the WPA if the Soviets would remain among its membership. This caused Dr. Leigh to 'associate himself and the WPA [. . .] with the decision to consider more deeply than hitherto the whole matter of psychiatric abuse and to seek ways of bring pressure to bear on countries where abuses occur' (Bloch and Reddaway 1977, p. 335). Although he voiced his opinion that the WPA did not have the resources to examine complaints about misuse of the psychiatric profession, he suggested that the WPA should make a declaration at the next General Assembly of the organization on 'the general principles underlying the ethical practice of psychiatry', and then it would be up to national associations 'whether or not to draw up a detailed code on matters affecting practice in its own country. Thus we avoid problems connected with religion, national policies, forms of political belief and so forth, and can concentrate on the principles' (WPA 1975). Dr. Leigh's evasive moves triggered a response from the Royal College, that it would do everything possible to have the next General Assembly of the WPA condemn the systematic political abuse of psychiatry in the Soviet Union. This was quickly followed by a request from the American Psychiatric Association that a special session be held on concrete abuses of psychiatry at the World Congress (RCP 1976).

29.5 The WPA World Congress in Honolulu

The move by the British and Americans had been supported by other societies as well, and thus there was no way the issue could be kept off the agenda. The first plenary session of the Congress saw the introduction of the Declaration of Hawaii, a statement of the ethical principles of psychiatry that had been drawn up by the Ethical Sub-Committee of the Executive Committee set up in 1973. One of the principles stated in the Declaration was that a psychiatrist must not participate in compulsory psychiatric treatment in the absence of psychiatric illness, and also other clauses could be seen as having a bearing on the political abuse of psychiatry. The Declaration of Hawaii was accepted by the General Assembly without difficulty 3 days later. Also an Ethics Committee was established, chaired by Prof. Costas Stefanis from Greece; one of the members was Marat Vartanyan from the Soviet Union (Bloch and Reddaway 1984, pp. 54–71).

The Soviet issue passed the General Assembly less easily. Prior to the meeting both sides had held press-conferences, and during the Assembly the discussions continued. Two motions were put to the vote, a British one condemning the systematic political abuse of psychiatry in the Soviet Union, and an American one calling on the WPA to establish a Review Committee to examine the allegations of political abuse of psychiatry. The British resolution passed with 90 votes against 88, but only after long debate about not only about the issue itself but also about the allotment of votes and other procedural issues. The American resolution asking for the establishment of a Review Committee received a larger majority of votes, 121–66 votes, but was also contested, among others by the Greek delegate and later WPA President Professor Costas Stefanis, who said he could not see the point of such a committee and that, just as the All-Union Society had been condemned on what he considered inadequate evidence, so the Review Committee might act in the same way (Bloch and Reddaway 1984, p. 69).

From the very first day the Soviets refused to acknowledge the existence of the Review Committee and to cooperate with it. By early 1981, 27 complaints had been made by nine member societies of the WPA. However, the Soviets remained mute and no response was received to any of the inquiries. Then, although not in line with procedures, the Executive Committee itself contacted the Soviets urging them to respond, initially also without success. In October 1981, when the Executive Committee received a report by Review Committee Chairman Dr. Jean-Yves Gosselin in which he expressed his frustration about the inability to move ahead, it became clear that patience with the Soviets was wearing thin (Shaw 1989, pp. 44–46). A month later the British College of Psychiatrists adopted a resolution calling on the General Assembly of the WPA to expel the Soviet society from its membership.

More societies joined the campaign, either by adopting resolutions calling on the suspension or expulsion of the Soviet society, or in support of dissident psychiatrists such as Dr. Semyon Gluzman and Dr. Anatoly Koryagin. By the end of 1982 it became clear that the General Assembly of the WPA would almost certainly vote for suspension or expulsion of the Soviet All-Union Society.

Interestingly, in the course of 1982 the Soviets had actually started to respond to requests for information from the Executive Committee (deliberately bypassing the Review Committee they refused to acknowledge), albeit in a very limited form and in Russian, not one of the official languages. In a report titled *The Issue of Abuse 1970–1983* (WPA 1983), WPA President Prof. Pierre Pichot and General Secretary Prof. Peter Berner tried to accommodate the Soviets maximally, even by providing an overly optimistic message to the WPA that the Soviets were willing to collaborate fully.

29.6 The Soviets Leave the WPA

It was too little and too late: on January 31, 1983, the All-Union Society resigned from the WPA. In their letter of resignation, the Soviets complained of a 'slanderous campaign, blatantly political in nature [...] directed against Soviet psychiatry in the spirit of the 'cold war' against the Soviet Union. [...] The leadership of the WPA, instead of taking the road to uniting psychiatrists, has embarked on the path of splitting them, and has turned into an obedient tool in the hands of the forces which are using psychiatry for their own political goals, aimed at fanning up contradictions and enmity among psychiatrists of different countries.'(Bloch and Reddaway 1984, pp. 249–252).

In the months between the Soviet's resignation and the World Congress of the WPA in Vienna in July 1983, accusations on who and what had caused the crisis went back and forth.

In the official report of Prof. Berner to the General Assembly this worry about the effect of the Soviets' withdrawal on the future of the WPA was highlighted: 'The officers that have administered the WPA during the last election period have been obliged [...] to countenance open confrontation between the national member societies, and have thus been unable to prevent the ultimate outcome of that confrontation, namely the withdrawal of psychiatric societies from membership of the Association. In consequence the Association's influence has decreased. A most dangerous precedent has been set, which does not augur well for the future of the Association' (WPA 1983).[10]

The General Assembly of the WPA in Vienna was probably one of the most disorganized and tense meetings in its existence. After long, sometimes highly emotional and confusing discussions (some delegates didn't even know which resolution they were asked to vote upon) eventually a resolution drafted by the British delegate Prof. Kenneth Rawnsley was adopted with a large majority of 174 votes in favor and 18 against, with 27 abstentions. The resolution was remarkably conciliatory in tone: 'The World Psychiatric Association would welcome the return of the All-Union

---

[10]*Report of the Secretary-General, 1983*. WPA, Vienna July 1983. Following the Soviet resignation, several other Eastern European societies withdrew from the WPA, as well as the Cuban Association.

Society [. . .] to membership of the Association, but would expect sincere cooperation and concrete evidence beforehand of amelioration of the political abuse of psychiatry in the Soviet Union' (Bloch and Reddaway 1984, p. 218). Subsequently, Soviet dissident psychiatrist and political prisoner Dr. Anatoly Koryagin was elected Honorary Member of the WPA, with 119 votes for and 58 against. A resolution calling on the WPA to take up the defense of opponents of political abuse of psychiatry was voted upon by a show of hands, 21 societies in favor and five against. And finally, the General Assembly agreed that the work of the Review Committee was important and should be continued, but also agreed that the mandate should be widened to include also other forms of abuse of psychiatry.

The General Assembly was concluded with the election of the new Executive committee. Interestingly, a Hungarian psychiatrist, Dr. Pal Juhasz, was elected to the Executive Committee, but his membership would not last a year. Immediately following the World Congress he was called in for questioning by his authorities. What exactly happened has never become fully known. Stasi archives, published after the German reunification, showed that he was immediately pressured by his government and called in for questioning on several occasions. His main 'crime' had been that he did not stick to an agreement reached between the socialist member societies and had not condemned the attacks on the Soviet society. Instead, he had agreed to be elected to the WPA Executive Committee. He was disciplined by the Party, then also by his Ministry of Health. Merely 7 months after the World Congress, on February 27, 1984, he died of cancer of the pancreas (Süss 1998, pp. 648–649). His place was taken by an East German psychiatrist, Jochen Neumann. After 1989 it would turn out that the latter was at the same time working for the Stasi as an unofficial agent and kept his bosses in East Berlin (and via them the relevant authorities in Moscow) informed of everything that happened within the WPA leadership. His reports are fascinating reading (Süss 1998, pp. 648–653).[11]

29.7 'Secret' Negotiations

During most of the years between the Vienna Congress and the World Congress in Athens in October 1989 the relations between Soviet psychiatry and the WPA were limited to four persons: on the side of the WPA President Costas Stefanis and General Secretary Fini Schulsinger, and on the Soviet side the President of the All-Union Society Dr. Georgi Morozov, also Director of the infamous Serbski Institute in Moscow, and Professor Marat Vartanyan, Director of the All-Union Center for Mental Health. In particular the latter maintained close relations with Prof. Stefanis,

---

[11] This book is based on extensive investigation of Stasi archives and shows how Prof. Juhasz was put under maximum pressure by the Hungarian authorities until his untimely death on February 27, 1984. It also explains how Prof. Jochen Neumann was proposed as successor by the East German authorities and managed to get himself elected by the WPA Executive Committee. The author has extenssively interviewed Jochen Neumann and former Hungarian colleagues of Pal Juhasz on the matter in preparation of the book *Cold War in Psychiatry*.

as becomes clear from Stasi archives and from reports by Jochen Neumann. For instance, immediately following the death of Prof. Juhasz, the East German secret service suggests to nominate Jochen Neumann as his successor, and to ask Prof. Vartanyan to inform Costas Stefanis of his nomination. The report then adds: 'As far as I know the Committee for state security works with Prof. Vartanyan with regard to the World Psychiatric Association' (Süss 1998, p. 650).

Indeed, from all information available it is clear that Profs. Stefanis and Schulsinger saw a return of the Soviet society as a primary objective. From the very beginning of their tenure contacts were maintained with the Soviets, and the two were able to operate quite solitarily because only few other members on the Executive Committee showed any interest in the matter. In fact, in their endeavors they were supported by a Stasi agent, Jochen Neumann, and only opposed by Dr. Melvin Sabshin, at the same time Medical Director of the American Psychiatric Association. How intensive the contacts were can be derived from that fact that between November 1986 and November 1987 at least six meetings took place between Stefanis/Schulsinger and the Soviets, once even involving Soviet Minister of Health Evgeni Chazov.[12]

The first signs that the Soviets were actively preparing the grounds for a return to the WPA came in late 1984 and early 1985. In the mean time, secret communications and negotiations between the Soviets and part of the WPA leadership continued, as is clear from the reports of Prof. Jochen Neumann to the Stasi and the GDR leadership. Interestingly, the Stasi archives also show that Jochen Neumann himself became increasingly critical of the Soviets, who seemed to be paralyzed because of a deep conflict between Georgi Morozov, who was seeking a return to the WPA but who was facing increasing opposition within the All-Union Society for his lack of flexibility and late understanding that the world had changed, and Marat Vartanyan who tried to block Morozov in any possible way (Süss 1998, pp. 656–670).

The worries about the Soviet intransigence both among Soviet psychiatrists and their Western supporters is not strange, because Soviet society was indeed changing rapidly and more and more Soviet publications reported on cases of political abuse of psychiatry and individual psychiatrists being involved in these practices, and on corruption among Soviet psychiatrists. The press increasingly called on Soviet psychiatrists to come clean and acknowledge what happened, yet they seemed to be stuck on old denials. The film studios in Sverdlovsk decided to make a film on political abuse of psychiatry in the Soviet Union, and the film crew interviewed dozens of victims both in the Soviet Union and in Amsterdam, where they gathered at the offices of the International Association on the Political Use of Psychiatry.[13] It was as if the leadership of Soviet psychiatry was unable to adapt to the quickly changing political climate in the country.

---

[12]data based on examination of both WPA Executive Committee minutes and Stasi archives.

[13]*Blazhenny Izgnannye* (Beatified Exilees) (1989). The film was eventually broadcasted in many countries, including the USSR, and remains a powerful document of the political abuse of psychiatry in the Soviet Union.

Eventually, in 1988, the US State Department became involved as well, probably because the ongoing reports on political abuse of psychiatry increasingly form an obstacle to détente. Most political prisoners had been released, but somehow the prisoners in psychiatric hospitals were the last to go, and only one by one. The State Department decided to institute an official investigation in an attempt to establish whether the political abuse of psychiatry in the Soviet Union indeed took place, had ended or still continued.

Long negotiations preceded the visit of the delegation. They were outlined in minute detail and included plans for sending the names of psychiatric prisoners or former psychiatric prisoners to be examined, conditions as to the examinations, including taking urine samples from each person interviewed. In addition, there was to be a hospital visit team. Eventually the Soviet authorities (the Ministry of Health and the Ministry of Foreign Affairs) agreed on virtually all points and in 1989 the group of about 25 people traveled to the USSR. The delegation visited a number of psychiatric institutions, examined a number of alleged victims of Soviet psychiatric abuse together with Soviet colleagues and subsequently compared joint conclusions with those reached by Soviet psychiatrists in the past. For this purpose, a long list of persons was handed over whom the Americans wanted to invite for such an examination. Apart from that several meetings with dissidents were planned. The Soviets had admitted to most issues, but continued to obstruct the delegation's work both before and during it's stay in the USSR.

After a visit of more than 2 weeks the delegation returned home and wrote its report. The report, although published much later, was quite damaging to the Soviet authorities. Not only had the delegation established that there had been systematic political abuse of psychiatry, but also that this abuse had not ended, that there were still victims of the political abuse in psychiatric hospitals and that the Soviet authorities – in particular the Soviet All-Union Society of Psychiatrists and Neuropathologists – still denied that psychiatry had been used as a method of repression (Schizophrenia Bulletin 1989).[14]

29.8 The Soviets Return to the WPA – While the Country Disintegrates

The WPA World Congress in Athens attracted almost ten thousand participants, among them a sizeable Soviet delegation. The Soviets benefited from the political climate; the democratization in the Soviet Union was, for many delegates, reason to vote in favor of a return of the Soviet All-Union Society. That was also clear to the WPA Executive Committee, who decided to organize an Extraordinary General

---

[14]The US State Department visit was not the only attempt to assess the state of affairs in Soviet psychiatry and whether Soviet psychiatric abuse was indeed coming to an end. Also organizations such as the International Academy of Law and Mental Health (IALMH) sent teams to the USSR to meet with Soviet authorities and psychiatrists, yet the reports were mostly conflicting and showed an intransigence on the part of the Soviet psychiatric leadership to adjust to the new political situation in the country.

Assembly where a debate would be held between the Soviet dissident psychiatrist Dr. Semyon Gluzman and representatives of the Soviet delegation. The WPA possibly hoped that the debate would cause the opinions to change in favor of the Soviets. The opposite happened. It strengthened the opinion of the opponents that too little had changed in Soviet psychiatry to allow a return of the Soviet Society and that their statements were still dominated by lies.

The case soon reached a climax. Virtually all Soviet psychiatrists, including Marat Vartanyan, were banned from the Congress by the Soviet authorities[15] and the leadership role of the Soviet delegation was now openly taken up by a diplomat rather than by a psychiatrist: Yuri Reshetov, deputy Minister of Foreign Affairs of the Soviet Union. It was clear that the game was now being played at the highest possible level, with direct involvement of the political top in Moscow. On the other side, a small group of negotiators was formed, led by the British delegate and President of the Royal College of Psychiatrists Dr. Jim Birley. The situation that developed was unique. While the World Congress continued, the press waited with suspense; the WPA Executive Committee had been mostly pushed to the sideline; and the four delegates negotiated with Yuri Reshetov, who was in continuous contact with Moscow in order to receive his instructions.

Even during the General Assembly negotiations with the Soviets continued. They were offered their last chance: if they wanted to return they would have to read a statement and admit their guilt; otherwise, they would not make it. The communication with Moscow was intense, the negotiations started about the content of the statement. Every word was debated. Eventually the Soviets acknowledged that systematic political abuse of psychiatry had taken place, promised that all political prisoners would be released and that democratic changes would be carried out within the Soviet society. The text had actually been written by the American delegate Dr. Harold Vysotsky, since the Soviets couldn't or didn't want to put anything to paper.

The die was cast, the Soviets were allowed to return to the WPA. Dr. Anatoly Koryagin was deeply shocked. He hadn't thought that the Soviets would be allowed to return and considered the statement by the Soviets as completely insincere and hypocritical. Out of anger he renounced his honorary membership of the WPA on the spot, in spite of attempts by many to change his mind.[16]

The film by the Sverdlovsk filmstudio's showed a jubilant Soviet delegation returning to Moscow. Dr. Marat Vartanyan was interviewed, happily smiling, and asserting that the All-Union Society was welcomed with open arms in Athens and that the All-Union Society was accepted into the World Psychiatric Association. Not a word about the conditions, not a word about the fact that also a new 'Independent Psychiatric Association'[17] from the Soviet Union had been admitted to the WPA.

---

[15]Information from Dr. Yuri Yudin, member of the Soviet delegation.

[16]Dr.Koryagin would remain in Switzerland for a few more years and then returned to Russia. He now lives in a small town east of Moscow.

[17]The Independent Psychiatric Association was founded in Moscow in March 1989 and was the first challenge to the monopoly over Soviet psychiatry of the AUSPN.

29.9 The Soviet Lesson

Looking back, the issue of Soviet political abuse of psychiatry had a lasting impact on world psychiatry, as well as on the World Psychiatric Association. The most positive conclusion is that the issue triggered the discussions on medical ethics and the professional responsibilities of physicians (including psychiatrists), resulting in the Declaration of Hawaii and subsequent updated versions. Also many national psychiatric associations adopted such codes, even though adherence was often merely a formality and sanctions for violating the code remained absent.

The years 1983–1989 made absolutely clear that psychiatry had become completely politicized. The WPA leadership said they tried to keep politics out of psychiatry, yet the result of their actions and their secret negotiations with the Moscow psychiatric leadership was exactly the opposite: it opened the door to carefully orchestrated interventions by the political leadership in Moscow, supported by active involvement of the secret agencies Stasi and KGB. It is not unthinkable that other secret agencies were also having their share in this game. At the same time, the goal of the opponents of political abuse of psychiatry to take politics out of psychiatry was equally unsuccessful. Their work was, whether they wanted or not, an element in the Cold War between East and West, and also in their case 'higher forces' undoubtedly had their influence.[18] In the end, it was the Soviet political leadership that negotiated a return in Athens, and a US State Department that was directly involved in the process of bringing the Soviets back in. The WPA had become a political arena for the two superpowers.[19]

With the fall of communism in Eastern Europe in the late 1980s most of the regular practice of using psychiatry to suppress political opponents ceased to exist. Some cases surfaced in Central Asia, notably in Turkmenistan and, more recently, in Uzbekistan. Also in Russia individual cases of political abuse of psychiatry continue to take place. The ranks of the victims over the last years have included women divorcing powerful husbands, people locked in business disputes and citizens who have become a nuisance by filing numerous legal challenges against local politicians and judges or lodging appeals against government agencies to uphold their rights. However, there appears to be no systematic governmental repression of dissidents through the mental health system. Instead, citizens today fall victim to regional authorities in localized disputes, or to private antagonists who have the means, as so many in Russia do, to bribe their way through the courts. Finally, many of the current leaders of Russian psychiatry, especially those who already belonged to the establishment in Soviet times, have revoked the earlier confession read at the 1989

---

[18]For instance, from the book by Sonja Süss it becomes clear that the main actor against the political abuse of psychiatry in the USSR, the International Association on the Political Use of Psychiatry, was unknowingly infiltrated by the Stasi. The available reports run until 1983; the files of a later date seem to have been destroyed when Eastern German totalitarian rule disintegrated.

[19]The negotiations in Athens were led by Yuri Reshetov of the Soviet Ministry of Foreign Affairs, with direct and constant communication with the Soviet leadership in Moscow.

WPA General Assembly that psychiatry in the Soviet Union had been abused systematically for political purposes. They now preferred to refer to individual cases of 'hyper-diagnosis' or 'academic differences of opinion' (Dimitrieva 2001, pp. 116–130).

29.10 Conclusion

It is beyond doubt that much progress had been made since Vladimir Bukovsky sent his documentation to the West in 1971. Medical ethics, human rights and the professional responsibility of physicians are now a cornerstone of mental health policy, whether adhered to or not. Most societies, at least in developed nations, have an Ethical Code, which clearly specifies what a psychiatrist can and cannot do. However, not all have mechanisms in place to monitor the adherence to the Code and to sanction those who violate it.

As to the World Psychiatric Association, one can say there is more openness to human rights issues and a general understanding that one cannot close one's eyes when the psychiatric profession is abused for non-medical purposes. It has adopted clear regulations that specify the Association's positions on these matters. However, at the same time it is clear that the wish to be a truly global association sometimes hinders the WPA from taking action, thereby risking that the documents itself adopted become a hollow sequence of nice phrases.

References

BlazhennyIzgnannye (Beatified Exilees) (1989).
Bloch S, Reddaway P (1977) Russia's political hospitals, Gollancz, London
Bloch S, Reddaway P (1984) Soviet psychiatric abuse, Gollancz, London
Bloch S (1989) Soviet psychiatry and Snezhnevskyism. In: Van Voren R (ed) Soviet psychiatric abuse in the Gorbachev era. IAPUP, Amsterdam
Brown CA, Lago A (1991) The politics of psychiatry in revolutionary Cuba. Transaction Publishers, New York, NY
Bukovsky V (1978) To build a castle, my life as a dissenter. André Deutsch Publishers, London
Dmitrieva D (2001) Alyans Prava i Miloserdiya. Nauka, Moscow
Helmchen H (1986) Ethische Fragen in der Psychiatrie. In: Kisker KP, Lauter H, Meyer JE, Müller C, Strömgren E (eds) Psychiatrie der Gegenwart. Bd 2 Krisenintervention, Suizid, Konsiliarpsychiatrie. Springer, Berlin, Heidelberg, New York, NY, pp 310–368
Kaznimye sumasshestviem (1971)
Komsomolskaya Pravda (15 July 1987)
Komsomolskaya Pravda (18 Nov 1987)
Koppers A (1990) A biographical dictionary on the political abuse of psychiatry in the USSR. IAPUP, Amsterdam
Korotenko A, Alkina N (2002) Sovietskaya Psikhiatriya – Zabluzhdeniya I Umysl. Sphera, Kiev
Koryagin A (1987) Unwilling patients. In: Van Voren R (ed) Koryagin: a man struggling for human dignity. IAPUP, Amsterdam
Mexico City News (13 Nov 1971)
Nijeboer A (2006) Een man tegen de Staat. Papieren Tijger, Breda

Pisarev S (1970) Soviet mental prisons. Survey, London
Pravda (24 May 1959)
Psychiatric News (1 Oct 1976)
Reuter report, Mexico City (2 Dec 1971)
Royal College of Psychiatrists (Nov 1976) News and notes
Schizophrenia Bulletin (1989) Supplement to 15(4):
Shaw C (1989) The world psychiatric association and Soviet psychiatry. In: Van Voren R (ed) Soviet psychiatric abuse in the Gorbachev era. IAPUP, Amsterdam
Süss S (1998) Politisch Missbraucht? Psychiatrie und Staatssicherheit in der DDR, Ch. Linke, Berlin
The Times (16 Nov 1971)
Van Voren R (ed) (1987) Koryagin: a man struggling for human dignity. IAPUP, Amsterdam
Van Voren R (ed) (1989) Soviet psychiatric abuse in the Gorbachev era, IAPUP, Amsterdam
Van Voren R (2009) On dissidents and madness. Rodopi, Amsterdam, New York, NY
WPA (Nov 28 1971) Minutes of the committee meeting
WPA Newsletter (31 Oct 1975)

Documents

Human Rights Watch/Geneva Initiative on Psychiatry (2002) Dangerous minds, New York
Kaznimye sumasshestviem (1971) Possev, Frankfurt
Psychiatry under Tyranny, An Assessment of the Political Abuse of Romanian Psychiatry During the Ceaucescu Years (1998) IAPUP, Amsterdam
Schizophrenia Bulletin (1989) Supplement to 15(4)
World Health Organization (1973) The international pilot study on schizophrenia
WPA (1983) The issue of abuse 1970–1983, Jan 1983
WPA (1983) Report of the secretary-general, Vienna, July 1983

Chapter 30
(Neuro-)Enhancement

Bettina Schöne-Seifert and Davinia Talbot

Contents

Abbreviations

ADS	Attention-Deficit-Syndrome
AMPA	α-Amino-3-Hydroxyl-5-Methyl-4-Isoxazole-Propionate
CREB	*cycloAMP Response Element Binding Protein*
DAK	Deutsche Angestellten-Krankenkasse
DBS	Deep Brain Stimulation
IQWIG	Institut für Qualität und Wirtschaftlichkeit im Gesundheitswesen
NICE	National Institute for Health and Clinical Excellence
SAS	Sleep-Apnoe-Syndrome
SSRI	Selective Serotonin Reuptake Inhibitor
TMS	Transcranial Magnetic Stimulation

B. Schöne-Seifert (✉)
Institut für Ethik, Geschichte und Theorie der Medizin, Von-Esmarch-Str. 62, 48149 Münster,
Germany
e-mail: schoeneb@ukmuenster.de

H. Helmchen, N. Sartorius (eds.), *Ethics in Psychiatry*, International Library
of Ethics, Law, and the New Medicine 45, DOI 10.1007/978-90-481-8721-8_30,
© Springer Science+Business Media B.V. 2010

30.1 Preliminaries

In recent years, biomedical ethics has dealt with a new subject, namely the use of drugs and other medical devices in order to improve healthy people, e.g. in terms of cognition, sexual performance, or mood. Although many of these attempts of so-called 'enhancement' are not yet proven to be both safe and effective, increasingly more people seem to use them. Ethical concerns, however, do not stop at issues of risks and effectiveness: Even under the hypothetical assumption that in the future these devices turn out to work well and without substantial risk, many people frown at what they consider a blatant 'abuse' of medicine. A whole array of deeper ethical criticism has been forwarded, both on the individual and the societal level. At the same time, there are other voices that consider such improvements ethically permissible or even desirable. These ethical controversies are particularly fierce when the attempted improvement aims at mental functions. In the following, we want to provide a systematic overview of the ethical debates around such 'neuro-enhancements'. To get there, some introductory remarks about concepts, debates, and methods of (neuro-)enhancement are called for.

30.1.1 Conceptual Questions

Eric Juengst, US-American bioethicist and one of the pioneers in the enhancement-debate, introduced the following definition: 'The term *enhancement* is usually used in bioethics to characterize interventions designed to improve human form or functioning beyond what is necessary to sustain or restore good health' (Juengst 1998, p. 29).

Obviously this understanding with its neutrality regarding means encompasses a very wide range of 'interventions' – including bodily exercise, education, mind jogging, or common life style 'kicks' such as caffeine or nicotine. Some authors go even further and include preventive medicine – such as vaccinations – in their concept of enhancement (Brock 1998, see also Merkel et al. 2007, 297ff). Thus considering a number of unanimously accepted interventions as enhancement, proponents of a wide or very wide concept of enhancement deliver it from a possible a priori negative connotation. At the same time, they have been accused of promoting acceptance for enhancement by (conceptual) fraud. Other authors, therefore, doubly limit the definition of enhancement to the narrow realm of interventions that (a) are, as Juengst points out, clearly *not* disease-related which also excludes prevention and in addition (b) use *biomedical* means. This narrow, both goal- and means-specific conceptualization of enhancement is pragmatically motivated: it implies that disease-unrelated interventions with biomedical tools raise a number of unprecedented *new* questions, not the least in ethics. When we, in the following, adopt this narrow concept of enhancement as *disease-unrelated improvements of human form or function by biomedical means,* we do, however, not presuppose that the grounding distinction between health and disease is always unambiguous. Nor do we, by choice of concept, want to suggest that enhancement is eo ipso ethically impermissible or questionable, nor that there is a normative difference between e.g. drinking a pot of tea and taking a smart pill. Rather, these and other questions need to be subjected to ethical scrutiny.

Finally, we like to make a brief remark on the treatment-enhancement distinction – one of the topics many authors have elaborated on (see Daniels 2000, Lenk 2002). Whether one contrasts enhancement with therapy only or with both therapy *and* prevention: in each case, the differentiation dwells on the notoriously contested distinction between health and disease. Again, we think that we can pragmatically circumvent these conceptual quandaries, accept a grey area in between, and rather turn to clear cut cases of enhancement: e.g. memory improvement of a mentally fit manager or mood elevation in a person without any symptoms of clinical depression. It is in these clear cases, that ethical concerns can readily be exemplified and analysed.

30.1.2 A Short History of the Recent Enhancement Debate

Attempts of self-improvement are probably as old as mankind itself. Human beings have always struggled to become better in many respects: running faster, knowing more, looking prettier, living longer, being 'better' humans. Especially the last idea is particularly prominent in Christian religious contexts (Gordijn 2004, 233ff). Improving oneself – and sometimes taking an edge – is not a new phenomenon at all.

'Doping' one's athletic capabilities and cosmetic surgery are two widely practised areas of attempted self-improvement. Due to progress in medical genetics in the 1980s, prenatal gene enhancement of one's offspring has become an imaginable, though not yet a realizable option. All three have been the subject of ongoing ethical discussions. But only recently have they been embedded in a more general systematic debate on ethical and social implications of improving human nature – triggered by the emergence of new biotechnological means for such improvement strategies. This intense and interdisciplinary dispute has been labelled the 'enhancement debate' (Parens 1998, for an early discussion see Glover 1984). The broad range of 'modern' biotechnological enhancement possibilities comprises the aforementioned 'sports doping' and cosmetic surgery as well as modification of sexual potency, body length and aging by hormones and other biomedical means.

And of course not only the physical, but also the mental, particularly the cognitive and emotional functions have become targets of attempted improvement: *neuro-enhancement*. Some authors see a close connection with the ancestors of this practice, i.e. certain mind-widening psychedelic drugs used in antiquity and undergoing a renaissance in the 1960s,[1] but the recent history of mental modification

---

[1] See the work of the contended psychologist Timothy Leary, e.g. *The Politics of Ecstasy* (1968).

is rather young and comprises modern psychopharmacological drugs[2] (see Section 3.3.4) and advanced technical means. Surgical neuro-enhancement is a topic in the discussion but has not yet been applied to humans. Most of the substances and appliances have a therapeutic background, i.e. they usually have been developed and are administered to treat diseases.

When in 1993 the US-American psychiatrist Peter Kramer published his book *Listening to Prozac* in which he described that certain antidepressants (SSRIs) were used by individuals who were not in need of a therapy but who desired to further improve their well-being, he coined this practice 'cosmetic psychopharmacology'. Some of the clients reported that with Prozac® they felt more optimistic, more energetic and less shy, they had the impression of feeling 'better than well' (Elliott 2003).[3]

Parallel to that, small, self-report studies were conducted suggesting that the stimulant Ritalin®[4] was popular among students, who used the substance off-label to improve their ability to concentrate (Babcock 2000). There are, nevertheless, no reliable data as to how widespread the phenomenon of neuro-enhancement already is, because systematic epidemiological data are missing. However, recent larger surveys – conducted mostly online – suggest that in the US (McCabe 2005) and in Germany (DAK-Gesundheitsreport 2009) psychostimulants and SSRIs are used by students and academics as well as by middle-leveled employees.[5]

Enhancing one's mood and cognition seems a tempting but controversial topic and thus there has been an explosion in the number of ethics publications in this field (e.g. Parens 1998, Harris 2007, Sandel 2007, Savulescu and Bostrom 2009, Schöne-Seifert and Talbot 2009, Schöne-Seifert). No doubt, neuro-enhancement has by now become one of the 'hottest' issues in 'neuroethics' – a recently evolved subspecialty of bioethics that deals with matters in neuroscience, e.g. personality, consciousness, free will, or what it means to be human.[6]

In the following paragraph we provide a brief survey of the present pharmacological, technical and surgical options of neuro-enhancement before we will address controversial ethical positions in terms of individual, social and professional aspects.

---

[2]For a history of antidepressants see also the pharmaceutical industry-critic Healy 2004, 4 ff.

[3]In the aftermath of the described American Prozac® (generic name: fluoxetine) optimism a critical debate on the implications and consequences of such psychopharmacological modification commenced, see Breggin 1995, Wurtzel 1994.

[4]Ritalin® (generic name: methylphenidate) is normally used in the drug therapy of children and adolescents with attention-deficit-hyperactivity-disorder (ADHD).

[5]According to this German report, which is based on interviews with 3,000 randomly chosen employees aged 20–50, 5% of the subjects take medication without a medical reason to enhance their abilities, 2.2% of inquired people do so on a regular basis, 1–1.9% take prescription drugs.

[6]The topic is so popular that there have been numerous upshots in that field: important publications, e.g. Levy 2007, Marcus 2002; a neuroethics website (http://neuroethics.upenn.edu/), the foundation of a neuroethics society in 2006 and, just recently, a journal on the topic.

30.2 Foreseeable Neuro-Enhancement Options

30.2.1 Pharmacological Means

The neuro-enhancement possibility which is closest 'at hand' is the use of psychopharmacological drugs. Promising, effective and relatively safe drugs have recently been developed to treat diseases such as dementia, depression, attention deficit or sleep disorders. And there has been evidence that some of the substances have a positive effect even in healthy individuals. Of course it is neither easy nor trivial to get hold of these substances for an enhancement-driven off-label use, because – at least in Germany – they are only available on prescription and some of them even come under the Narcotics Law. But it is by no means impossible to get these drugs: utilizing grandpa's Alzheimer tablets, ordering the drugs online via laissez-faire internet pharmacies or taking the illegal paths of the black market seem to be viable options (DAK-Gesundheitsreport 2009, p. 59). To which degree physicians are 'helpful' in this regard by either frankly providing the substances for enhancement purposes or by stretching the boundaries of the term 'disease' beyond adequate limits[7] thus 'treating' normal people, remains unclear at this stage (see also Section 30.3.2.2).

30.2.1.1 Cognition: Wakefulness, Learning, Memory and Attention

The following substances often come under the name of 'smart drugs'.

Stimulants such as *amphetamines* with their positive effect on wakefulness have already been used as 'go-pills' for fighter pilots during the Second World War (Hall 2003, p. 40). Countless amphetamine derivates with fancy names (e.g. Speed, Ecstasy, Ice, Shabu, Crystal) are on the lifestyle-drug black market and enable youngsters to party despite sleep deprivation and to have a heightened sense of being 'bonded' with their peers. High doses of these amphetamine derivates may cause serious side effects such as dehydration and cardiovascular symptoms, e.g. tachycardia or arrhythmia.

The stimulant *methylphenidate* (e.g. Ritalin®) is used to treat attention-deficit-syndrome (ADS) in children.[8] Improving or maintaining one's attention and concentration is of course a key ability in several fields of 'normal' human life, especially for chronically sleep-deprived people like students before exams, managers and politicians during long negotiation sessions, etc. This explains why there seems to be a certain interest in the drug by healthy individuals.[9] Studies on the effectiveness of Ritalin® in healthy people are contradictory: some noted positive effects on cognitive tasks (Oken et al. 1995), others observed – apart from

---

[7]Pathologizing normal phenomena and labelling them with medical terminology is called 'disease mongering' (Blech 2003, Payer 1992).

[8]Interestingly, in Germany Ritalin® is not approved for therapeutic use in adults.

[9]Though it is difficult to pin down the utilization rate, see Babcock (2000) and McCabe (2005).

the general side-effects – negative effects on certain abilities (e.g. impulsivity) (Elliott et al. 1997) and still others noticed no effect at all (Turner et al. 2003). At the beginning of the methylphenidate intake, temporary alterations such as a loss of appetite, sleep disorders and abdominal pain are often observed (Rapport and Moffitt 2002). With regard to the application of this substance to children with attention deficit disorder, its addictive potential has often been discussed, but not yet convincingly proven (see Volkow and Insel 2003, 1307f). Nevertheless, the side-effects are regarded to be so serious that the drug – e.g. in Germany – goes under the Narcotics Act.

In contrast, since February 2008, the stimulant *modafinil* does no longer come under the German Narcotics Act. The drug is approved for the treatment of narcolepsia, severe shiftworker's syndrome and certain forms of sleep-apnoe-syndrome (SAS). It improves vigilance (Caldwell et al. 2000, Walsh et al. 2004), and counterbalances the effects of sleep deprivation (Chatterjee 2004, p. 969). In experiments with mice the drug had a positive influence on learning processes (Beracochea 2003). Unlike other stimulants modafinil was found to have only low addictive potential (Müller 2004), but it might cause headaches, nausea or nervousness, perhaps even death due to exhausting sleep deprivation if taken for an extended period of time (Miller 2005, p. 8). With regard to the effects of the substance the military seems interested in utilizing the drug in combat situations (Normann and Berger 2008, p. 112).

Donepezil, a newer *acetylcholine-inhibitor* and the older *piracetam*, are two antidementive drugs approved for treating dementia and other cognitive deficiencies.[10] In a randomized trial of healthy older pilots donepezil improved their performance in complex flight simulation tasks (Yesavage et al. 2002). In a study conducted in 1976 piracetam was found to significantly increase the verbal memory in healthy subjects (Dimond and Brouwers 1976). Possible side-effects of the drugs are – amongst others – sleeplessness (Jackson et al. 2004) and nausea.

Not yet in clinical use are so called *AMPAkines*, which have proven to augment memory and increase learning capacities in animal models (see for example Porrino et al. 2005). Findings in healthy volunteers are ambiguous (Wesensten et al. 2007, Wetzenberg et al. 2007).

Some *CREB-modulators* (e.g. *rolipram*) also dramatically increased specific learning abilities in mice, while it decreased others (Barad et al. 1998). Other CREB-modulators are applied to gain the opposite effect: to disrupt memory consolidation. CREB-modulators are at an early stage of clinical testing in humans (Tully et al. 2003).

---

[10]Though their beneficiary effect on cognitively impaired individuals is a matter of ongoing discussion, see http://www.cochrane.org/reviews/en/ab006104.html NICE as well as IQWIG judged donepezil as effective (even if only with small efficacy) whereas piracetam has no evidence-based efficacy so far.

30.2.1.2 Mood and Emotion

Third-generation antidepressants, the SSRIs, have fewer side-effects than the older drugs such as tricyclics or MAO-inhibitors. Nevertheless, they can still cause problems such as a dry mouth, sexual dysfunction, problems with micturition, constipation a reduction of the seizure threshold and others. Whether SSRIs increase the suicide risk in children and adolescents is a matter of great concern, but so far not decisively proven (see Elliott 2005, p. 21). It has already been mentioned above that in the mid-1990s the SSRI Prozac® (*fluoxetine*) had become popular in the context of cosmetic psychopharmacology – particularly in the US.[11] On the one hand consumers described to feel 'better than well' when taking the drug, on the other hand critical voices commented on the negative consequences of such mood enhancement and an artificially happy society (Elliott 2003, Kramer 1997, Wurtzel 1994).[12] The scientific basis for the effectiveness of SSRIs in healthy individuals is narrow (Harmer et al. 2004, Repantis et al. 2008), however, people seem to note beneficial effects or why else would they be so popular?

Another class of drugs can be utilized to modify the emotional status: beta-blockers. There have been reports that propranolol, one of the early beta-blockers, was popular among musicians who appreciated the 'calming' effect of the substance, fighting the symptoms of stage fright: less shivering, less sweating, less heart bumping (Slomka 1992). Therapeutically, beta-blockers are prescribed for high blood pressure or certain forms of anxiety disorders; side-effects include headaches, dizziness, fatigue, nightmares or erectile dysfunction.[13]

Particularly interesting are certain findings reported when using the neuropeptide *oxytocin*, which is applied via nasal spray to women after a caesarean section to help the uterus contract. When breast-feeding a baby or experiencing an orgasm oxytocin is 'naturally' being secreted from the pituitary. In a double-blind placebo controlled neuroeconomics experiment oxytocin nasal spray was administered to volunteers in a fake deal situation. The oxytocin group showed significantly increased trusting behaviour and was willing to accept riskier deals than the placebo group (Kosfeld 2005). Such an increase in trust has been considered to reflect the possible potency of oxytocin as a moral enhancer.

---

[11] In the context of treating depression the SSRI citalopram is the most frequent substance (DAK-Gesundheitsreport 2009, p. 50), the enhancement utilization of which is not yet known.

[12] In this context the happy-pill 'Soma' in Aldous Huxley's novel *Brave New World* is often cited (Huxley 1932).

[13] In how far other anxiolytics, such as benzodiazepines like *midazolam* are used for enhancement purposes is unclear, but apart from its calming effect this substance has the potential to disrupt memory: it causes a retrograde amnesia. In clinical contexts this is useful for patients awaiting operations: to protect them against the stressful events before they are being taken to the operating theatre, but situations where the attenuation of painful memories e.g. chagrin d'amour, is desired as life-style improvement even by healthy people are imaginable. See also the movie 'Eternal sunshine of the spotless mind'.

30.2.2 Technical and Surgical Means

Unlike the pharmaceutical means, which might have an increasing significance as enhancers, technical and surgical options which clearly and solely aim at enhancing people are still more or less utopian or at an experimental stage.[14]

The technique which is probably closest to being realized as a neuroenhancer is the *trancranial magnetic stimulation* (TMS). This is a technology where a magnetic field is produced via a coil held above the head, so that the respective person's skull is not touched. The magnetic field depolarizes neurons in a targeted region of the brain, thus inducing a transient, 'virtual lesion'. TMS has been tried to treat epilepsy or depression, however, so far, with ambiguous results. To date, TMS is mostly used in experimental settings to decipher the function of various brain regions. With regard to neuro-enhancement TMS can boost concentration while under stimulation (George 2003). Side-effects are relatively rare (see Daskalakis 2002).

Deep brain stimulation (DBS) is a procedure where an electrode is placed in a certain brain region, comparable to a pacemaker in the heart. DBS is approved for the treatment of patients with otherwise untreatable Parkinson's disease or other movement disorders. It is also used, but still experimental and not approved for severe obsessive-compulsive disorder or severe major depression. Other therapeutic applications are subject of research (Merkel et al. 2007). There are hypothetical considerations about the use of DBS for enhancement purposes. A possible target of such electrical stimulation is the nucleus accumbens, the major reward processing region of the brain. The idea of electrically stimulating nucleus accumbens is to elicit feelings of happiness. Side-effects of DBS include the risks of the surgical procedure (Maguire and McGee 1999, p. 10), misplacement of the electrode and psychiatric symptoms such as mania, depression, etc. (Appleby et al. 2007). Also, studying a small sample, one group of neuroscientists reported that DBS diminished socio-moral competency (see Brentrup et al. 2004).

30.3 Ethical Issues

Neuro-Enhancement raises a whole variety of ethical questions – partly well-known from rather different contexts, partly related to enhancement in general, and partly specific for manipulating one's mind and character, the 'kernel' of human beings. In the following, we will systematically, but briefly discuss the main points of ethical concern and controversy.

The area of enhancement-ethics that, so far, has been debated extensively, is 'doping' in sports, which commonly – though not unanimously (Savulescu and Foddy 2007) – is taken to be harmful, unfair, and deceptive. Not infrequently, people generalize these negative connotations to '*mind* doping', assuming close

[14]See also Chapter 19.

analogy between both categories. However, we caution our readers and ourselves not only against too hasty generalizations, but also against the careless employment of 'misuse', 'abuse', or other negative words when talking about neuro-enhancement.

30.3.1 Enhancing Oneself: Individual and Social Aspects

30.3.1.1 Risks and Benefits: Medical and Otherwise

As described above, neuro-enhancers have not yet been proven to be effective beyond doubt or to be safe if used chronically. Some experts believe this to be merely a matter of time, others of principle – the latter assuming that any artificially induced enhancement of brain function either ultimately results in down-regulation and thereby less 'effectiveness' (tachyphylaxia) or might have to be paid for with side-effects.[15] For the time being, these are open questions, since – despite remarkable progress in the neurosciences – we are far from comprehensively understanding mental processes. It is for instance empirically proven *that* serotonin – amongst a host of other neurotransmitters – influences mood, but we do not know *how* exactly it does; we do not even know what mood – in terms of brain processes – really is (Freedman 1998, p. 145). Hence, any kind of manipulation is supposedly imprecise, effects and side-effects being close by.[16]

Certainly, this lack of data and understanding provides good reasons for cautiousness and it is disquieting that a number of people give enhancement a try, nevertheless. Indeed, this might already indicate a latent social pressure to make oneself 'fitter' by any means, in order to endure perform outstandingly in exams, at the workplace, or in other social situations.

Nevertheless, lack of sufficient data on safety and effectiveness of neuroenhancers should not be taken as an excuse to dismiss neuro-enhancement and its genuine ethical evaluation altogether. It might well turn out that optimistic hopes become true or that consumers are prepared to accept some side-effects – although people's risk aversiveness will most likely be higher in neuro-enhancement contexts in comparison with therapeutic contexts.

A prominent warning with regard to neuro-enhancement is the suggestion of its possible psychological *addiction* potential. What if consumers become unable to give up smart or happy pills making their life so much smoother, their performance so much better? However, some consider this a problem only as long as unknown risks pose a somehow weird threat and thus require the discontinuation of neuro-enhancement (Kramer 1997, 311f), others take 'addiction' to undermine

---

[15] It has been suggested, for instance, that enhancing cognitive functions is to be paid for by loosing emotional depth or creativity. See Whitehouse et al. 1997, p. 20. Generally sceptical: Kass 2003, p. 15.

[16] Paul Wolpe summarizes this aspect when he says that from the fact that a memory is a good thing it does not necessarily follow that more memory is better (2002, p. 393).

one's self-esteem or even to be a matter of principle (President's Council on Bioethics 2004, Chapter III B).

According to a liberal position, lack of knowledge about long term effects of neuro-enhancement makes but a good argument in favour of information, differentiated evaluation, research, and innovation (Greely et al. 2008).

The same is true, of course, for matters of benefit. It remains to be seen in much more detail, to which extent and in which mental domains neuropharmaceutics can contribute to achieving the very goals for which healthy people otherwise turn to coaching, which is considered to be 'psychotherapy for the worried well, not the suffering sick'. Behind part of the scepticism against cosmetic neuropharmaceutics one finds the strong conviction that its effects are mainly superficial – much in contrast to the 'deep' effects of psychotherapy (Brock 1998, Freedman 1998, p. 63, Whitehouse et al. 1997, p. 19). This, however, might or might not turn out to be nothing more than a dualist prejudice.

30.3.1.2 Autonomous or (Latently) Coerced Use of Neuro-Enhancement?

Definitely nobody would opt for a society in which people might be pressed or manipulated to improve their mood or cognitive functions – be it with or without potential risks to the user. Hence, nobody would seriously deny that neuro-enhancement should in any case be restricted to the autonomous, that is informed and voluntary, consent of the enhancing individual (for the special case of younger children, who are by definition non-autonomous, see Section 30.3.2.1 below). Accordingly, there is no question that some legal framework will have to be developed to prevent coercive enhancement, even in those paradigm scenarios like that of the heart surgeon or pilot who might benefit their patients or passengers by drug induced enhancement of their alertness during long night shifts.

However, there is substantial disagreement as to whether it will be possible to prevent enhancement opponents from getting latently coerced to 'play the enhancement game' *contre coeur* (Kramer 1997, 269ff). Suppose, a substantial number of students take highly effective attention pills in preparation for their exams; suppose, managers or scientists perform much better when taking pills to maintain their level of alertness and concentration and thus increase expectations at work. How can we ensure that people who consider self-manipulation by enhancement pills dangerous, wrong, untrue or phoney for whatever reason, can stick to their 'no' without risking disadvantages? It has already been suggested to test students for Ritalin® before exams – much like anti-doping controls for athletes. But apart from operational difficulties, this would of course not inhibit the use of Ritalin® and similar substances during study and work times. Pragmatists might see a parallel to other technological inventions, such as cars, computers, or cell phones that have substantially changed our daily life and have in many social and work environments ultimately become a must-do for people who want to get along. But even if the uneasiness society might have felt with regard to trains and cell phones or computers had been considerable, it is hard to deny that the real or only sensed 'depth' of intervention is much higher with neuro-enhancement than with using any external devices to improve performances.[17]

In other words: it will inevitably turn out problematic to protect both enhancement opponents in their liberty to abstain, and enhancement proponents in their liberty to profit from neuro-enhancement. At least, one single social strategy in handling this dilemma might get wide support, i.e. the control of the pharmaceutical industry in their predictable attempts to induce – by aggressive advertisment or by euphemistic drug information – demands for neuro-enhancement.

30.3.1.3 Arguments from Naturalness, Virtue or the Good Life

A prominent objection to neuro-enhancement by biomedical means appeals to its 'unnaturalness': Whereas improvements by training, eating or sleeping properly seem innocent according to this benchmark, computer-chips and pills surely do not. Given that many people indeed assume a substantial moral difference between drinking a pot of tea ('natural') and swallowing a wakefulness-inducing pill ('unnatural'), there seems to be a point in this argument. However, normative arguments from naturalness are notoriously problematic in at least three respects: first, it is difficult to draw a clear line between the natural and the unnatural (Juengst 1998, 61f, Caplan 2002). Is the use of a particular substance 'natural' according to its provenience, e.g. from a naturally occurring plant? Does it loose naturalness by a modification like producing wine from grapes, or an unusual utilization like smoking tobacco to get the nicotine 'kick' or by chemically synthezising caffeine-tablets? Secondly, naturalness is an ambiguous condition: in overwhelmingly numerous contexts we do not hesitate to applaud unnatural interventions, e.g. eradicating epidemics and famines, domesticating nature by safety measures, culture, and civilization. Thirdly and most importantly, it is by no means obvious that and why naturalness should be of any distinct and inherent moral value – independent of the interests, needs or desires of those who live in and on nature (Heyd 2003; in the context of enhancement see Glover 2006, 81f). Even for the mere presumption of the moral superiority of naturalness over unnaturalness, one needs an additional justificatory metaphysical argument: that nature is prima facie good as a divine creation, or that nature has an inherent goal or *telos* – positions that are questionable, to say the least. Scepticism with regard to naturalness as a normative orientation does however not preclude the wisdom of pragmatic caution. 'Nature knows best' can be a reasonable maxim in contexts where we lack the full range of understanding natural complexities – e.g. when intervening in ecological balances or living organisms. This is certainly relevant for attempts to intervene in the brain and even more so for potential future attempts to *radically* transgress current 'normal' human functions,

---

[17]However, the moral relevance of the body boundaries has been contested (see Clark and Chalmers 1998 and Anderson 2009).

for instance adding the ability to hear ultrasound, or to memorize 100-fold faster than today's best. Here a 'heuristics of fear' seems reasonable.

Another standard line of critique against biomedical neuro-enhancement argues along the lines of an impending loss of *virtues*. We see at least two variants of this approach: one invoking the virtue of humility in living with one's natural endowments (a possible implication of the naturalness argument) (Sandel 2007), and a second one, sometimes labelled 'psychopharmacological Calvinism' (Klerman 1972), that incriminates an attitude of by-passing the arduous way, the hard work of mental training or psychological introspection, the path of human development *per aspera ad astra* Its underlying plausibility derives from suspecting the loss of 'secondary' virtues such as self-discipline, frustration tolerance, patience, or humility (President's Council on Bioethics 2004, Chapter 6. III). These are concerns that must be taken seriously.

However, it is questionable whether such grim prospects are likely to come true. Our limited collective experience does not cover such negative predictions; rather, it remains to be seen whether neuro-enhancers really do more than strengthen people's real aspirations and character. This is not to be misunderstood as a *carte blanche* for neuro-enhancement, but rather as a quest for a step-by-step-policy. From what we know, quite a number of clients stop psycho-enhancement because they feel self-alienated[18]; others emphasize that they have discovered their real selves and integrate the improved traits without difficulties.

A society in which neuro-enhancement is a common way of modifying one's way of life has been accused of fostering a profane hedonism where the individual is seduced '[...] into resting content with a shallow and factitious happiness' (President's Council on Bioethics 2004, p. 269). Others take a more optimistic stance (Kramer 1997, 264f, Buchanan 2008), emphasizing the possibility of using enhancement potentials for the sake of both: individual flourishing and social well-being. In this perspective neuro-enhancement could be utilized for promoting creativity and intelligence on behalf of important goals and values rather than for trivializing people's life-plans and quieting their social conscience.

In terms of the effects neuro-enhancement might have on society there are at least three other aspects which have to be taken seriously. Firstly – with regard especially to cognitive enhancement – the aim of using such a strategy might be to gain an edge in a competitive situation. Accordingly, neuro-enhancement has often been accused of promoting the wrong values, i.e. rivalry and ruthlessness, thus pushing society's already disquieting competitive pressure even further – an undesirable perspective. On the other hand, proponents of neuro-enhancement have argued that the means of enhancement is not intrinsically wrong just because it might be utilized to promote questionable values – nobody would call language courses or extracurricular classes for children into question, though they are also means to excel within a competitive context. Obviously this is an ongoing controversy about using means for either illicit or beneficial purposes. Secondly there is

---

[18] Peter Kramer gives the example of his client Philip (1997, 291f).

the phenomenon of self-defeating neuro-enhancement, i.e. enhancement which – in case of collective usage – no longer provides the competitive advantage it is meant to achieve. According to this view, a society of neuro-enhanced individuals might end up in a vicious circle where more and more sophisticated means become necessary so that the enhancement in question would still qualify as an improvement.[19]

Thirdly, there are critics of neuro-enhancement who fear for a social quietism (DeGrazia 2005, p. 217), i.e. lacking motivation to protest against questionable social conditions. Instead of reducing inhumanely long working hours for, say, doctors or pilots (Wolpe 2002, p. 392), the availability and the pressure to take pills that keep people functioning' or that dull their dissatisfaction could cause a social standstill. Enhancement in this notion would rather be an improvement for the functioning of society rather than for the individual (Brock 1998, p. 56). In this context the fear for homogeneity has been explicated. Due to this concern, individuals might mutually be forced to adapt to a certain standard approved by society. Natural variety and the acceptance of differences might diminish; others doubt this assumption (Caplan 2002, p. 112, Kamm 2005).

30.3.1.4 Unfairness in Access or Fairness in Compensation?

Concerns about fairness in access to neuro-enhancement, especially to cognitive enhancement, are shared by many, whether they take a permissive and prohibitive stance in evaluating single cases (see Brock 1998, Whitehouse et al. 1997). If smartness, creativity, concentration, or demand for sleep could indeed effectively and safely be manipulated, access to these means might turn into an important opening to access privileged jobs and positions. Given that enhancement is generally not expected to get reimbursed by health services that (in many countries) are publicly funded, the availability of e.g. smart pills will finally depend on people's ability to pay for them. Consequently, the availability of neuro-enhancers will once more increase the gap between the poor and the already rich and privileged – this is a widespread concern.

Opponents taking a more liberal stance, usually emphasize that we already have a 2-class-society like good nurseries, elite schools, summer camps and Ivy League universities which all contribute to enormous differences in opportunities in our society – not to mention the influence of social upbringing (Caplan 2002). So why should we get worried about the prospects of neuro-enhancement and its sequelae of opportunity-inequalities, if we have been tolerating such inequalities all along? Some agree with these arguments, others oppose them as just one step too far on the path of growing inequalities in options. Still others conciliate such fears by hoping that eventually neuro-enhancers might not be all that expensive, anyhow. However, for the future, it does not seem likely that one will be able to buy brain boosters on the cheap.

---

[19]Norman Daniels illustrates this with an example from the field of PID: in a society where having a baby boy is an economic advantage compared to having a girl, and parents 'produce' a gender imbalance in favour of boys by means of PID, the former advantage of having male offspring might no longer apply (see Daniels 2000, p. 321).

Some bioethicists have taken the opposite direction in evaluating the prospective fairness issues: they emphasize that cognitive enhancement might be a much more cost-effective way of compensating unfortunate bad 'lotteries' in terms of people's natural gifts and social upbringings as opposed to programs for education and psychosocial rehabilitation. Would it not be desirable and fair to make smart pills etc. available to the cognitively under-privileged (Brock et al. 2001)? Should society not even pay for that for reasons of fairness?

30.3.1.5 Authenticity, Self-Control, and Cheating?

Impairment of 'authenticity' by neuro-enhancement has become somewhat of a standard concern in philosophical debates (Bolt 2007, DeGrazia 2005, Elliott 2000) – although authenticity is a notoriously loose concept. With a core meaning of 'doing it one's own way'. Authenticity can be used to allude to a metaphysical idea of the real self, as a synonym for autonomy (Quante 2002) (another contested concept), non-opportunism (Feinberg 1986, p. 32) or as the idea of acting with self-controlled consistency.

Given these imprecise conceptual contours, concerns about a person's authenticity using neuro-enhancement could be mislabelled as it might simply refer to arguments from valuable naturalness or virtue (as discussed in Section 30.3.1.3). Authenticity would then be understood as living with one's natural given, with one's talents and capacities that ought to get detected and unfolded by learning and exercise, but not substituted or manipulated by 'artificial' means such as smart pills. Or an authentic life would be the human life governed by the *per aspera ad astra* idea of hard work instead of easy pills as the appropriate means for achievements. If our analysis is valid, such authenticity concerns should be addressed more appropriately in terms of those underlying assumptions.

However, at least one of the above ideas dealing with authenticity does deserve distinct analysis and consideration, namely the idea of acting with self-controlled consistency. Rather than dwelling on an exhaustive interpretation of this idea, we take a look only at self-controlled action as *felt* by the agent herself or by those around her. In this perspective, concerns about someone's authenticity under neuro-enhancement get a specific and relevant meaning: Does the student under Ritalin® consumption *feel* to give a proof of his authentic exam knowledge? Does a woman consider her lover's tenderness authentic when it occurs under the influence of an SSRI? And does he himself believe to talk authentically? Whatever the metaphysical content of the underlying idea, it seems also worth looking at the conditions of *ascribing* or rather *presuming* authenticity in the above sense. Love from 'tenderness pills' or professional excellence from brain boosters cut deep into our ideals and expectations of who we want to love and who we want to be. Important as it is, the search for the exact psychological conditions of questioning some one's authenticity from inside or outside (first or third person perspective) will remain incomplete at this point in time – due to our empirical ignorance. It seems, however, likely that several factors will contribute to the presumption of authentic action.

For beneficial neuro-enhancement it is probably most important that the user herself does not feel self-alienated by her pills. They should not give her the impression of acting or feeling 'outside herself', of being externally controlled. Rather, she should be 'in charge' of them, using them as a tool of some kind, without loosing authorship and merits of her achievements – just the way we commonly feel with regard to two cups of coffee. It might partly be a matter of neurobiological function and psychological adaptability, partly a matter of individual and social habituation whether enhancers get to be experienced as mere tools or as controlling devices. Second, the neuro-enhancement consumer should ideally not consider herself acting wrongly – neither against the standards of virtuous conduct nor of expected or desired social interaction. At this point both ethical concerns and third person expectations come into play. As long as neuro-enhancement is – truly or allegedly, for good or bad reasons – prohibited or disliked by relevant other persons, its user is forced into deceit, which in the long run, is an unpleasant consequence.

Now, whether enhancement means *cheating* depends on the context it is used in and on the specific rules governing this context. For instance, using Ritalin® when listening to classical music in one's leisure time, hence in a clearly non-competitive situation, to intensify the auditory pleasure, would not be considered deceptive. Whether Ritalin® on exams or erythropoetin before sport competitions is to be considered cheating, depends on the reglementations that govern the particular praxis (e.g. anti-doping laws in competitive sports) or on the values this praxis is meant to foster. As bioethicist Erik Juengst points out:

> Either the institutions must redesign the game (e.g. education or sports) to find new ways to evaluate excellence that are not affected by available enhancements, or they must prohibit the use of the enhancing shortcuts (Juengst 1998, p. 40).

30.3.2 Enhancing Others: Special Aspects

30.3.2.1 Enhancing Children

Until now we have only considered individual and social ethical aspects of 'first-person' enhancements, i.e. situations where the enhanced person decides for herself to do so. Other possible scenarios are of course those, where the enhanced person has not herself consented to the procedure. From the wide range of conceivable settings (children, mentally incompetent people, forensic usage and prenatal enhancement) we would like to select the first aspect of enhancing children.[20]

Given the lively philosophical discussion that has been led especially in the context of genetic modification of one's offspring (Brock et al. 2001, Feinberg 1980,

---

[20]Considering cognitive enhancement we explicitly do not want to join a debate on query cases of ADS, but rather concentrate on unambiguous cases of neuro-enhancement.

Habermas 2002) we only want to look at the most important arguments for the special case of neuro-enhancement, as we are aware of the fact that we cannot provide a detailed and sufficiently deep evaluation of the arguments.

On the one hand children, being in a process of development, have a huge learning potential, and their brains have extraordinary plasticity. According to the physiology of learning, children have sensitive phases for certain processes. Parents and other people in (moral) charge of the child's development face the burden/dilemma of supporting this development (which necessarily implies delimitating decisions on which abilities to develop) while at the same time having to keep the child's future open to a sufficient degree until the child can autonomously decide on her path in life (Feinberg 1980). Thus, in terms of enablement, a responsible parent might well consider to support, say, the sensitive phase of language learning in a child with a cognitive neuro-enhancer and stimulating language lessons. From this perspective there might be a moral demand or at least a moral 'okay' for neuro-enhancing one's child – just as there is such a demand to enhance the child's cognitive abilities by sending it to school (Glover 2006, 99ff).

On the other hand sceptics have pointed out the danger of over-ambitious parents who do not act in the child's best interest. From this point of view the neuro-enhancement of the child rather appears like a constraint of possibilities, where the child has to meet the mental/ cognitive demands of her parents and of society[21] and cannot develop according to her own aptitude. A more general objection to biotechnologically changing one's children derives from Juergen Habermas' thoughts on the alteration of human nature (Habermas 2002):[22] if there are individuals whose traits are planned or 'made' by other humans rather than naturally 'given', this will introduce a decisive asymmetry among human fellows: rather than all being equals they can be classified into 'makers' and 'products'. Habermas deems this asymmetry deleterious to mankind's common ethics (*Gattungsethik*), in particular to reciprocal respect for autonomy and to self-respect for one's human agency. This position has been both applauded and criticised as implausible (Harris 2007, 137f, Buchanan 2008, pp. 21–22).

At present neuro-enhancement for children should not be considered an available option, as the effects and especially the long-term effects in healthy individuals are far too poorly understood at the moment (see Section 30.3.3). The protection of children against potential harm should be in the centre of interest. It is probably not until neuro-enhancement for competent adults is a viable and accepted way of creating one's life that neuro-enhancement for children should be considered.

---

[21] Analogous is Michael Sandel's example of the ambitious father of the tennis players Venus and Serena Williams (2007, p. 52).

[22] Habermas explicitly refers to genetic enhancement, nevertheless his arguments seem extendable to the case of other forms of enhancement.

30.3.2.2 Neuro-Enhancement as a New Domain for Physicians?

Apart from ethical considerations concerning individual or social aspects of neuro-enhancement there is the question which role medical doctors should play in this field (Talbot 2009).

According to one position, doctors should abstain from it altogether. Rather, their appropriate sphere of action is seen within the goals of medicine (Hanson and Callahan 1999) which essentially comprises curing or preventing diseases or maladies, respectively. The reference to a core of professional ethics is essential in this view. This core is perceived as a duty to not harm the patient ('primum nil nocere') and to promote his beneficence ('salus aegroti suprema lex') – both under the premise of treating or preventing disease. According to this view, leaving the proper medical domain threatens professional integrity (Miller and Brody 2001) and contradicts physicians' professional ethics.

Others argue that balancing the maxims of not harming and benefitting the patient as described above is inadequately paternalistic, disrespecting the patient's or client's autonomy. Instead, well-understood professional ethics requires the doctor to provide sufficient information and to respect the patient's well-reflected value-judgement at the same time. Examples of 'enhancements' which are already realized by doctors are the distribution of contraceptive drugs to women and cosmetic surgery.

In reality, doctors already need to deal with neuro-enhancement requests. Some might frankly be brought forward by the 'patient'; and although – in a solidarity-based healthcare-system – it causes problems in terms of fairness and allocation if a doctor prescribes the desired medication, it is probably less problematic than hidden requests. It is less tricky because the doctor can turn down a direct neuro-enhancement request similarly frankly. But what, if the doctor and the patient take the bait and mutually define the enquirer's condition as pathologic and, hence, her problem as 'treatable'? The phenomenon of medicalizing normal conditions is not new and has been labelled 'disease mongering' (Moynihan and Cassels 2005, Payer 1992). So, if someone who has a time-consuming, intellectually demanding job is not able to concentrate for more than 8 h, it is not far fetched and probably guided by the 'patient's' best interest that one might attest him 'concentration problems' and prescribe an appropriate drug similar to the situation of shift-workers who have 'shift-workers syndrome' and get Modafinil when they cannot stay alert properly because their circadian rhythm is disrupted.

Erik Parens, a pioneer in the ethics of enhancement, has suggested to think about handing (neuro)enhancement-matters over to so-called 'schmocters', a group of non-medical enhancement specialists, if doctors feel uneasy about this matter. This would avoid a discussion about the interpretation of the proper goals of medicine and of professional ethics. But – as others like the neurologist Anjan Chatterjee think – this would be wrong. According to Chatterjee the medical profession should openly discuss the new role of the physician and possibly embrace the role as 'gatekeeper' in people's pursuit of well-being (Chatterjee 2004, 972f). Physicians' familiarity with handling the medical substances and devices and the resulting reduction of side-effects strongly argue in favour of such practice.

Nevertheless there remains the accusation of the (therapy-derived) trust set in doctors being wrongly taken advantage of for neuro-enhancement purposes; some fear a loss of reputation of and confidence in physicians if they engage in enhancement enterprises. Apart from that there remains the problem of fair access which does not put additional burden on a solidarity-based health insurance. But this fairness problem already exists, thus it needs to be addressed in any case. So, insisting on doctors' being responsible for the clear-cut domain of treating the sick and not for allegedly extravagant things like neuro-enhancement simply ignores the fact that the very concepts are already blurry at times and that there probably are already (hidden) requests to deal with.

30.3.3 Neuro-Enhancement as an Issue of Research

Much of the real-life use and evaluation of neuro-enhancement depends or should depend upon valid data on the effectiveness and side-effects of the different substances under discussion. All of the neuro-enhancement-substances described above (in Section 30.2.1) are licensed and have been tested for therapeutic applications in patients but not for short or long term use by healthy individuals. Since both safety and effectiveness might differ in these two contexts, it seems important to provide sound and systematic data on specific enhancement use – unless one wants to prohibit neuro-enhancement right away for ethical reasons. From any not a priori prohibitive perspective it is desirable to perform enhancement research according to best research standards. Trials should cover reasonably long usage periods. They should, moreover, not be disguised as therapeutic research, and should of course not 'exploit' patients as a particularly vulnerable group. Finally, probands must be vigorously and broadly informed about the fact that risks and side-effects are only known from therapeutic use of these drugs in patients and that near to nothing is known about their effects in healthy subjects. In the near future these ethical concerns should also be addressed at the level of international research regulations. Neuro-enhancement research should not be left solely to the pharmaceutical industry who has predictable financial interests in promoting enhancement use.

In addition, one should reflect society's engagement in research that aims at discovering and developing new substances *primarily* for neuro-enhancement purposes. Most people would agree that this should not get high priority on publically funded research agendas. On the other hand, even in the face of unsolved national and global medical problems, a possibly effective and extensive way of really improving mental capacity should be welcomed with open arms and promoted where possible, perhaps even with public funding.

30.4 Conclusions

Safe and effective biomedical strategies to enhance cognitive functions or psychological well-being in healthy people might be an option in the near or later future. At present, smart-pills, happiness-pills, or wakefulness pills have either not been scientifically proven to be effective or have not sufficiently been tested with regard to their long-term safety. Nevertheless, there seems to be an increasing demand for those substances which might indicate that there are in fact positive effects.

Besides matters of risk and effectiveness, neuro-enhancement raises a whole array of ethical questions which ought to be openly debated in society as well as among neuroscientists, lawyers, physicians or ethicists. Research on and the use and evaluation of neuro-enhancement should not be pushed underground In this regard we support the expert plea of Greely et al. in *Nature* in late 2008: we need a detailed interdisciplinary recognition of these new developments – neither ignoring its multidimensional dangers nor playing down its positive potential and thus its attractiveness for many people.

We dare to predict that neuro-enhancement cannot simply be disregarded as just another foolish peculiarity of what some consider the *US-American Way of Life* with a naïve belief in technical fix and superficially gearing to make life easy going. Rather, Europeans, too, do have reasons to ponder on the promises and perils of neuro-enhancement, – their bioethicists have taken up the lead with an increasing interest in the subject.

Acknowledgments We would like to thank our research partners Thorsten Galert, Reinhard Merkel, Christoph Bublitz, Isabella Heuser and Dimitris Repantis with whom we cooperate in the project 'Potentiale und Risiken des pharmazeutischen Enhancements psychischer Eigenschaften' (*Potentials and risks of pharmacologically enhancing psychological capacities*) which is funded by the Federal Ministry of Education and Research (2007–2009). We are particularly grateful for Isabella Heuser's brilliant suggestions on an earlier version of the manuscript. Further, we would like to thank our two unknown reviewers for their critical comments, and Silke Tandetzki for her help with the manuscript.

References

Anderson J (2009) Neuro-Prothetik, der erweiterte Geist und die Achtung vor Personen mit Behinderung. In: Schöne-Seifert B et al (eds) Neuro-enhancement – Ethik vor neunen Herausforderungen. Mentis, Paderborn

Appleby BS et al (2007) Psychiatric and neuropsychiatric adverse events associated with deep brain stimulation: a meta-analysis of ten years' experience. Mov Disord 22(12):1722–1728, Sep 15

Babcock Q et al (2000) Student perceptions of methylphenidate abuse at a public liberal arts college. J ACH 49:143–145

Barad M, Bourtchouladze R, Winder DG et al (1998) Rolipram, a type IV-specific phosphodiesterase inhibitor, facilitates the establishment of long-lasting long-term potentiation and improves memory. Proc Natl Acad Sci USA 95(25):15020–15025

Beracochea D et al (2003) Enhancement of learning processes following an acute modafinil injection in mice. Pharmacol Biochem Behav 76(3–4):473–479

Blech J (2003) Die Krankheitserfinder: Wie wir zu Patienten gemacht werden. Fischer, Frankfurt

Bolt LLE (2007) True to oneself? Broad and narrow ideas on authenticity in the enhancement debate. Theor Med Bioeth 28:285–300

Brentrup A et al (2004) Alterations of sociomoral judgement and glucose utilization in the frontomedial cortex induced by electrical stimulation of the subthalamic nucleus (STN) in Parkinsonian patients. In: 55. Jahrestagung der Deutschen Gesellschaft für Neurochirurgie e.V. (DGNC), German Medical Science. Düsseldorf, Köln, Doc DI.06.06. http://www.egms.de/en/meetings/dgnc2004/04dgnc0207.shtml

Breggin PR (1995) Talking back to prozac: what doctors aren't telling you about today's most controversial drug. Saint Martin's Press, New York, NY

Brock DW (1998) Enhancements of human function: some distinctions for policymakers. In: Parens E (ed) Enhancing human traits: ethical and social implications. Georgetown University Press, Washington, DC

Brock DW et al (2001) From chance to choice – genetics and justice. Cambridge University Press, Cambridge

Buchanan A (2008) Enhancement and the ethics of development. Kennedy Inst Ethics J 18(4): 1–34

Caldwell JA, Caldwell JL, Smythe NK, Hall KK (2000) A double-blind, placebo-controlled investigation of the efficacy of modafinil for sustaining the alertness and performance of aviators: a helicopter simulator study. Psychopharmacologia 150(3):272–282

Caplan AL (2002) No-brainer: can we cope with the ethical ramifications of new knowledge of the human brain? In: Marcus SJ (ed) Neuroethics: mapping the field. Conference proceedings. Dana Press, San Francisco, CA, pp 95–131, 13–14 May 2002

Chatterjee A (2004) Cosmetic neurology. Neurology 63(6):968–974

Clark A, Chalmers D (1998) The extended mind. Analysis 58:7–19

DAK-Gesundheitsreport (2009) http://www.presse.dak.de/ps.nsf/Show/A9C1DFD99A0104BAC 1257551005472DE/$File/DAK_Gesundheitsreport_2009.pdf. Accessed 20 Apr 2009

Daniels N (2000) Normal functioning and the treatment-enhancement distinction. Camb Q Healthc Ethics 9(3):309–322

Daskalakis ZJ et al (2002) Transcranial magnetic stimulation: a new investigational and treatment tool in psychiatry. J Neuropsychiatry Clin Neurosci 14(4):406–415

DeGrazia D (2005) Enhancement technologies and self-creation. In: Human identity and bioethics. Cambridge University Press, Washington, DC, pp 203–243

Dimond SJ, Brouwers EM (1976) Increase in the power of human memory in normal man through the use of drugs. Psychopharmacology (Berl) 49(3):307–309

Elliott C (2000) Pursued by happiness and beaten senseless: prozac and the American dream. Hastings Cent Rep 30(2):7–12

Elliott C (2003) Better than well. American medicine meets the American dream. Norton & Company, New York, NY

Elliott C (2005) Medicine goes to the mall: enhancement technologies and quality of life. VM: Ethics J Am Med Assoc 7:2

Elliott R, Sahakian BJ, Matthews K, Bannerjea A, Rimmer J, Robbins TW (1997) Effects of methylphenidate on spatial working memory and planning in healthy young adults. Psychopharmacology (Berl) 131(2):196–206

Feinberg J (1980) The child's right to an open future. In: Aiken W, LaFollette H (eds) Whose child? Children's rights, parental authority, and state power. Rowman & Littlefield, Totowa, NJ

Feinberg J (1986) Autonomy. In: Feinberg J (ed) Harm to self: the moral limits of criminal law. Oxford University Press, Oxford

Freedman C (1998) Aspirin for the mind? Some ethical worries about psychopharmacology. In: Parens E (ed) Enhancing human traits. Ethical and social implications. Georgetown University Press, Washington, DC

George MS (2003) Stimulating the brain. Sci Am Sept 289:66–73

Glover J (1984) What sort of people should there be? Penguin, London

Glover J (2006) Choosing children: the ethical dilemmas of genetic intervention. Oxford University Press, Oxford

Gordijn B (2004) Medizinische Utopien – Eine ethische Betrachtung. Vandenhoeck & Ruprecht, Göttingen

Greely H et al (2008) Towards responsible use of cognitive-enhancing drugs by the healthy. Nature 456:702–705

Habermas J (2002) The future of human nature. Polity, Oxford

Hall SS (2003) The quest for a smart pill. Sci Am Sept 289:54–65

Hanson MJ, Callahan D (eds) (1999) The goals of medicine: The forgotten issues in health care reform. Georgetown University Press, Washington, DC

Harmer CJ et al (2004) Increased positive versus negative affective perception and memory in healthy volunteers following selective serotonin and norepinephrine reuptake inhibition. Am J Psychiatry 161(7):1256–1263

Harris J (2007) Enhancing evolution: the ethical case for making people better. Princeton University Press, Princeton, NJ

Heyd D (2003) Human nature: an oxymoron? J Med Philos 28:151–169

Healy D (2004) Let them eat prozac. New York University Press, New York, NY

Huxley A (1932) Brave new world. Petersen, Hamburg

Jackson D et al (2004) The safety and tolerability of donepezil in patients with Alzheimer's disease. Br J Clin Pharmacol 58(Suppl 1):1–8

Juengst ET (1998) What does enhancement mean? In: Parens E (ed) Enhancing human traits: ethical and social implications. Georgetown University Press, Washington, DC

Kamm F (2005) Is there a problem with enhancement? Am J Bio 5(3):5–14, May–June

Kass L (2003) Ageless bodies, happy souls: biotechnology and the pursuit of perfection. New Atlantis Spring 2003:9–28

Klerman GL (1972) Psychotropic hedonism vs. pharmacological Calvinism. Hastings Cent Rep 2(4):1–3

Kramer PD (1997) Listening to prozac: a psychiatrist explores antidepressant drugs and the remaking of the self. Penguin, New York, NY

Kosfeld M et al (2005) Oxytocin increases trust in humans. Nature 435:673–676

Lenk C (2002) Therapie und Enhancement: Ziele und Grenzen der modernen Medizin. LIT, Münster

Levy N (2007) Neuroethics. Oxford University Press, Oxford

Maguire GQ, McGee EM (1999) Implantable brain chips? Time for debate. Hastings Cent Rep 29(1):7–13

Marcus SJ (ed) (2002) Neuroethics: mapping the field. Dana Press, New York, NY

McCabe SE et al (2005) Non-medical use of prescription stimulants among US college students: prevalence and correlates from a national survey. Addiction 99:96–106

Merkel R et al (2007) Intervening in the brain. Springer, Berlin

Miller MC (2005) What is modafinil? Harv Ment Health Lett 21(7):8

Miller FG, Brody H (2001) The internal morality of medicine: an evolutionary perspective. J Med Phil 26(6): 581–599

Moynihan R, Cassels A (2005) Selling sickness: how the world's biggest pharmaceutical companies are turning us all into patients. Nation Books, New York, NY

Müller U et al (2004) Effects of modafinil on working memory processes in humans. Psychopharmacology (Berl) 177(1–2):161–169

Normann C, Berger M (2008) Neuroenhancement: status quo and perspectives. Eur Arch Psychiatry Clin Neurosci 258(Suppl 5):110–114

Oken BS, Kishiyama SS, Salinsky MC (1995) Pharmacologically induced changes in arousal: effects on behavioral and electrophysiologic measures of alertness and attention. Electroencephalogr Clin Neurophysiol 95(5):359–371

Parens E (ed) (1998) Enhancing human traits: ethical and social implications. Georgetown University Press, Washington, DC

Payer L (1992) Disease-mongers: how doctors, drug companies and insurers are making you feel sick. Wiley, New York, NY

Porrino LJ, Daunais JB, Rogers GA et al (2005) Facilitation of task performance and removal of the effects of sleep deprivation by an ampakine (CX717) in nonhuman primates. PLoS Biol 3(9):e299

President's Council on Bioethics (ed) (2004) Beyond therapy: biotechnology and the pursuit of happiness. Dana Press, New York, NY

Quante M (2002) Personales Leben und menschlicher Tod: personale Identität als Prinzip der biomedizinischen Ethik. Suhrkamp, Frankfurt am Main

Rapport MD, Moffitt C (2002) Attention deficit/hyperactivity disorder and methylphenidate. A review of height/weight, cardiovascular, and somatic complaint side effects. Clin Psychol Rev 22(8):1107–1131

Repantis D et al (2008) Antidepressants for neuroenhancement in healthy individuals: a systematic review. Poiesis Prax: Int J Tech Assess Ethics Sci (27 Nov):1–36. http://www.springerlink.com/content/175414g2916w1322. Accessed 30 May 2009

Sandel M (2007) The case against perfection: ethics in the age of genetic engineering. Harvard University Press, Cambridge, MA

Savulescu J, Foddy B (2007) Ethics of performance enhancement in sport: drugs and gene doping. Principles of health care ethics. In: Ashcroft RE et al (eds) Principles of healthcare ethics. Wiley, London, pp 511–520

Savulescu J, Bostrom N (2009) Human enhancement. Oxford University Press, Oxford

Schöne-Seifert B, Talbot D (2009) Enhancement: Die ethische Debatte. Mentis, Paderborn

Schöne-Seifert B et al (2009) Neuro-enhancement: Ethik vor neuen Herausforderungen. Mentis, Paderborn

Slomka J (1992) Playing with propranolol. Hastings Cent Rep 22(4):13–17

Talbot D (2009) Ist Neuro-Enhancement keine ärztliche Angelegenheit? In: Schöne-Seifert B et al (eds) Neuro-enhancement: Ethik vor neuen Herausforderungen. Mentis, Paderborn, pp 321–346

Tully T, Bourtchouladze R, Scott R, Tallman J (2003) Targeting the CREB pathway for memory enhancers. Nat Rev Drug Discov 2(4):267–277

Turner DC, Robbins TW, Clark L, Aron AR, Dowson J, Sahakian BJ (2003) Relative lack of cognitive effects of methylphenidate in elderly male volunteers. Psychopharmacology (Berl) 168(4):455–464

Volkow ND, Insel TR (2003) What are the long-term effects of methylphenidate treatment? Biol Psychiatry 54:1307–1309

Walsh JK et al (2004) Modafinil improves alertness, vigilance, and executive function during simulated night shifts. Sleep 27(3):434–439

Wesensten NJ, Reichardt RM, Balkin TJ (2007) Ampakine (CX717) effects on performance and alertness during simulated night shift work. Aviat Space Environ Med 78(10):937–943

Wezenberg E, Verkes RJ, Ruigt GS et al (2007) Acute effects of the ampakine farampator on memory and information processing in healthy elderly volunteers. Neuropsychopharmacology 32(6):1272–1283

Whitehouse PJ, Juengst E, Mehlman M, Murray TH (1997) Enhancing cognition in the intellectually intact. Hastings Cent Rep 27(3):14–22

Wolpe PR (2002) Treatment, enhancement and the ethics of neurotherapeutics. Brain Cogn 50(3):387–395

Wurtzel E (1994) Prozac nation. Quartet Books, London

Yesavage JA, Mumenthaler MS, Taylor JL et al (2002) Donepezil and flight simulator performance: effects on retention of complex skills. Neurology 59(1):123–125, July 9

Part V
Teaching Ethics in Psychiatry

Chapter 31
Teaching Ethics in Psychiatry

Deborah Bowman

*Setting an example is not the main means of influencing another,
it is the only way*

(Albert Einstein).

Contents

D. Bowman (✉)
Centre for Medical and Healthcare Education, St George's University of London,
London SW17 0RE, UK
e-mail: dbowman@sgul.ac.uk

H. Helmchen, N. Sartorius (eds.), *Ethics in Psychiatry*, International Library
of Ethics, Law, and the New Medicine 45, DOI 10.1007/978-90-481-8721-8_31,
© Springer Science+Business Media B.V. 2010

31.1 Introduction

Those involved in teaching psychiatry undertake significant responsibility which is commonly balanced with other obligations e.g. clinical care, research and professional development. Developing the moral awareness of medical students and junior doctors is as integral to education in psychiatry as sharing knowledge, diagnostic and clinical skills. Yet, unlike other 'tools of the trade', many psychiatrists will have received little explicit training in the effective teaching and learning of ethics. However, the absence of a formal training in ethics should not preclude psychiatrists from incorporating ethics into their clinical teaching. Clinical staff who provide an integrated approach to ethics as part of good patient care are powerful role models with the potential to reach students more effectively than many ethics specialists who may sometimes be charged with teaching in something of an academic vacuum (McKneally and Singer 2008). Furthermore, clinicians are well-placed to ensure that ethics education is embedded in a professional context which will enhance considerably the impact and value of ethics teaching and learning (Siegler 2001).

Learning in ethics will, inevitably, be informed by its location, both geographically and temporally. 'Geographically', much of the most influential education in ethics occurs implicitly in the clinical environment and 'back stage' via role modelling, coffee room discussions and observation. 'Temporally', the needs, priorities and responses of learners who are medical students and those who are junior, but qualified, practising clinicians will differ (Malek et al. 2000, Weiss Roberts et al. 2004). It is the author's view that ethics must be integrated throughout a medical training from the very earliest days as an undergraduate through to specialty qualification and beyond. Whilst most curricula make variable claims to a 'core' body of knowledge that covers common ethical topics, it is argued that education in ethics is a far more subtle and complex exercise than merely ticking 'issues' off the list defined by a core curriculum.

31.2 The Teacher as Moral Agent

31.2.1 Self-Awareness and Personal Reflection

Teaching and learning is a human endeavour and it is therefore influenced by personality, connectedness and emotion. For a teacher, self-awareness and personal reflection is essential. As Fish and Coles (2005) note all educators are working to assumptions, values and preferences whether or not they do so explicitly. The starting point for education in ethics is awareness of one's own educational assumptions, professional preferences and potential ethical 'blind spots'. Arguably, in psychiatry there is particular scope for personal predilection to prevail and as such 'values testing' has been said to be an integral part of ethical practice (Bloch and Green 2009). To become self-aware is challenging and an ongoing process (Rokeach 1973), but at the very least teachers of ethics should consider their own experiences, preferences and practices in psychiatry. There are multiple ways in which one can elucidate one's own assumptions, biases and blind spots including the use of self-assessment tools (Dowie and Elstein 1988, Karel 2000) and drawing on empirical analyses of professionals' ethical beliefs (Bruhn 1991, Pomerantz and Grice 2001, Sigmon 1995) to evaluate better one's own 'position' within the profession. For scholarly analyses of the place of values in psychiatry, one should turn to the excellent work of Bill Fulford and his colleagues (Fulford 1989, 2004, 2008, Fulford and Williams 2003, Fulford et al. 2002, Woodbridge and Fulford 2004). However, for those with neither the time nor the inclination for lengthy research and introspective self-analysis, a simple commitment to discuss ideas with others, to seek out those who disagree and think differently from oneself and to listen carefully when an opposing or challenging argument is presented will enable the teacher to offer learners multiple perspectives and avoid imposing only parochial and unhelpfully skewed approaches to ethical practice.

31.2.2 Conflicts of Interest in Education

The influence of the teacher is further demonstrated when considering the effect of conflicts of interest in education. Slowly, and perhaps, painfully, the biomedical community has come to accept that most, if not, all researchers have conflicts of interest. However, conflicts of interest in education, have received considerably less attention, but are no less significant for those charged with teaching. It is awareness of, and responses to, those conflicts of interest that are required not an artificially 'neutral' state achieved via the unrealistic elimination of inevitable conflicts of interest.

Just as is the case in research and clinical practice, conflicts of interest abound in ethics education. Some may be relatively obvious, for example the tension between patient care and student learning or the sponsorship of educational events and materials by pharmaceutical companies. The latter conflict being an area that has prompted some in the profession to seek greater attention to, and collective reflection on, the nature of philanthropic donation in psychiatry as a specialty (Weiss Roberts and Coverdale 2006). Practical conflicts of interest regarding patient care and student learning warrant careful attention, but are relatively easily addressed by explicitly discussing, as part of 'ethics education' the tensions that can arise when seeking to meet the patient's interests, the clinician's goals and the learner's priorities (Hoop 2004, Sayer et al. 2002). Requiring students to reflect on the impact of their desire and need to learn, to seek consent properly and considerately from patients, to attend to the dignity and feelings of patients, to remember the value of learning from patients, rather than on patients, will do much to ensure that the business of 'ethics education' is itself ethical.

Other conflicts of interest in education are less readily revealed. Consider, for instance, the interests of a practising psychiatrist who has, in the course of her career pursued further psychotherapeutic training and developed a strong preference for a psychodynamic approach to depression rendering her less likely than

many of her peers to prescribe medication. Such a preference is not unacceptable
– clinical judgement and a range of expertise characterise psychiatric practice and
medicine. However, this psychiatrist does have a strong preference that potentially
skews her teaching practice. Unless she is aware of, and acknowledges, that she differs
from some of her peers in her prescribing habits and treatment approach, she
risks giving the students only a partial education. Furthermore, the mere admission
of her preference for a therapeutic model does not, itself, guarantee that she is acting
ethically, for she may introduce difference, consciously or otherwise, as a mechanism
for denigrating or undermining her pharmacologically minded colleagues. As
a teacher therefore, she must both be aware of, and acknowledge, her professional
preferences and ensure students have the chance to experience other perspectives
whilst avoiding defensiveness and encouraging learners to explore the range of professional
choices made and the implications of those choices when treating one of
the commonest psychiatric conditions.

31.2.3 Transference and Counter-Transference

Emotional or interpersonal conflicts of interest, potential and actual, can occur in
ethics education. One would hope that psychiatrists are alert to the phenomena of
transference and counter-transference, whilst being better placed than many clinicians
to understand the value of professional boundaries. The feelings that arise
between teacher and learner may not be explicitly acknowledged or considered, but
nonetheless exist and can influence powerfully the effectiveness of the education
(Brown 1985, Cahn 1986, Wilson 1982). Most experienced teachers have students
or juniors whom they remember fondly – perhaps the brilliant and able, but as yet
undeveloped, junior psychiatrist or the learner whose lack of confidence reminds
the teacher of him or herself years before. Equally, but perhaps less commonly discussed
in public, most experienced educators can readily recall the student or junior
whom they found to be challenging or problematic – perhaps the 'arrogant' junior
who lacked insight into his limitations or the student with a seemingly insurmountable
lack of interest in the specialty or apparent 'contempt' for the subject at hand.
Likewise, learners too will feel differently about, and respond variably, to teachers
who may be teaching identical subject matter because of the feelings those teachers
evoke in them as individual learners. Such emotional responses and differential
experiences of 'teaching', particularly perhaps in a subject such as ethics where
the focus is commonly on multiple perspectives and diverse ways of understanding
a problem, are inevitable. The trick is for teachers to remain alert to their own
emotions and to consider how they may enhance or inhibit teaching (Bowman and
Hughes 2005).

31.2.4 Teachers as Role-Models

Teachers in medical education, particularly clinical teachers, have long been identified
as role models for students and junior doctors. Ethnographic accounts of
professional socialisation have memorably captured the ways in which medical

education and training is much more than a process whereby knowledge and skills are conveyed (Becker et al. 1961, Atkinson 1981, Konner 1987, Sinclair 1997). Rather, medical education, and specifically the powerful effects of role models, communicates the values and norms shared by doctors (Ashcroft 2000), thereby indicating the so-called 'hidden curriculum' (Hafferty and Franks 1994, Wear 1998). Clearly, when role models are sound, learners will have a valuable insight into ideal professional practice. Regrettably when role models are questionable or simply negative, their influence will be equally enduring but considerably less desirable. As such, teachers have a significant responsibility to be good role models because it is one of the most powerful ways of teaching (Wright et al. 1997, Paice et al. 2002, Cruess et al. 2008). It is also important to remember that role modelling is unlike other forms of education in that one does not 'do it' intermittently in a pre-determined format as is the case with lectures or bed-side teaching. Role modelling occurs constantly, whenever the learner is able to observe his or her teacher. Teachers are role models to their students and juniors irrespective of whether they are interacting and whether the teacher realises he or she is being observed.

31.3 Content, Process and Ethics Education

31.3.1 Contents of Ethics Education

It is not difficult to find consensus-based content lists that describe an ideal curriculum in medical ethics (Doyal and Gillon 1998, UNESCO 2005a, b). There have been some attempts to adapt the notion of the 'core' to the specialty of psychiatry leading to debate about whether there is, or should be, a common approach to ethics education in psychiatry (Moffic et al. 1991, Coverdale et al. 1992, Coverdale 1996). Weiss Roberts et al. (1996, 2006) suggest that psychiatric trainees working in North America value ethics education and have particular interest in specific areas of ethics including relationships with colleagues, responding to error, cultural determinants of care and resource allocation. Core curricula and consensus statements provide a useful resource for anyone interested in teaching ethics and, furthermore, may act as a symbolic catalyst for improving the claims of ethics in a busy medical training timetable (Mattick and Bligh 2006).

Whatever the content of a curriculum, ethics education incorporates knowledge, cognitive skills such as reasoning, critique and logical analysis and clinical skills whereby abstract ethical learning is integrated with other clinical learning and applied appropriately in practice. Indeed, as Singer et al. (2001) argue, unless clinicians are able to draw on learning in ethics to enhance daily practice and better serve their patients, it is difficult not to conclude that ethics education has ultimately failed.

It is the author's view that flexibility should be a characteristic of all ethics teaching both in terms of the content and process of teaching. A teacher's openness to differing learning styles, student preferences, variable interests and contrasting priorities makes the oft-recommended notion of 'student centred learning' a meaningful reality rather than an abstracted ideal. Teaching ethics in the clinical setting allows for a rich and responsive approach in which generic principles or virtues can be demonstrated 'in action' whilst modelling best practice in seeking consent for patient participation, remaining alert to, and respectful of, patient dignity and encouraging learners to participate in care rather than remain passive recipients of 'teaching'. Weiss Roberts et al. (2003) describe well how the nuanced world of clinical psychiatry provides rich opportunities to incorporate, model and develop effective pedagogical approaches to learning ethics. Psychiatry naturally demands a case-based, narrative-oriented, evaluative, inclusive approach which mirrors the skills commonly used in ethical analysis, particularly what Ashcroft (2000) describes as 'patient-centred ethics'.

31.3.2 The Significance of Emotion in Ethics Education

Whatever the fine print of an ethics curriculum or teaching plan, it is argued that there is likely to be a common process in which students develop their engagement with, and approach to, moral problems (Bowman 2005). For most students, an ethical education will involve emotion, discomfort and perhaps irritation as personal responses are interrogated, constructively challenged and set alongside contrasting perspectives. Notwithstanding the historical emphasis that ethics has placed upon rationality, it is argued that teachers of ethics have to be aware of the significance of emotion and its effects on a learner's willingness to participate in, and apply, even the 'best' education (Bowman and Hughes 2005, Leget 2004). Emotion is, after all, experienced by all clinicians and yet is rarely explicitly discussed in relation to ethical decision-making. Instead, emotion commonly remains implicit in clinical decision-making, and is frequently repackaged as 'empathic neutrality'. It is argued that rather than being an 'irrational', undesirable or obfuscatory response, emotion is significant to ethical education because it indicates one is making a judgement about an important issue, suggests that the issue is complex, uncertain and yet, in some way, intrinsic to being 'a good clinician', and reveals the nature of ethical decision making to be related to values, discretion and multi-layered responses rather than a purportedly 'neutral' endeavour.

Ethics are shaped by emotions as is clinical practice (as is life). Awareness of emotion reveals values, differences and hierarchy of priorities. However, there is no inherent moral worth in an emotion. Different emotions will reveal different, and sometimes conflicting, values and not all will be positive. Attention to emotion, and its diminution as medical training progresses (Berry 2007), will enhance the worth and effectiveness of not only the ethical education of those who learn, but also the daily practice of those who teach.

31.4 Multiple Methods in Ethics Education

31.4.1 Adoption of a Multiplicity of Methods

Some writers have suggested that particular methods of teaching ethics, predominantly case analysis, narrative and problem-based learning approaches, are well-suited to psychiatry, both generically and in psychiatric sub-specialties (Parker 1995, Sondheimer 1996, Sondheimer and Martucci 1992, Schnapp et al. 1996, Garfield et al. 2002, Sondheimer 1998, Weiss-Roberts et al. 2004). However, the author believes that irrespective of specialty, the most effective ethics teaching is likely to occur when a multiplicity of methods is adopted. Didactic, large group lectures are appropriate for sharing information, for example outlining the details of Mental Health legislation or the provisions for non-voluntary treatment in psychiatry. Small group tutorials or seminars enable students and tutors to discuss ideas in depth, to form and to challenge arguments and to explore intuition (Goldie et al. 2002). Within methods for teaching ethics reside hotly debated challenges about theoretical preferences with 'Principalism' (Beauchamp and Childress 2001) continuing to dominate (albeit often in bastardised form) whilst other theoretical approaches jostle for position (Cowley 2005). For the author's part, it is the responsibility of the ethics educator both to be explicit about one's own preferences and to offer diverse theories and contrasting analytical frameworks to learners, for example inviting multiple analyses of a clinical problem using a range of theoretical approaches and decision-making tools.

31.4.2 Clarity of Textbooks Versus Confusing Reality

Clinical learning provides the learner with a window onto 'ethics in practice' as the neat, ordered and illusory clarity of textbook psychiatry yields to the uncertain, sometimes confusing and often memorable reality of daily practice (Calton et al. 2008). Individual mentoring allows students and their teachers the time to reflect on, discuss and explain the 'back story' to decisions, choices and practice. The supervisory relationships common to psychiatric, psychological and psychotherapeutic training offer a particularly valuable opportunity to incorporate ethics into a trainee's educational development (Weiss Roberts et al. 2003, Whitman 2001).

31.4.3 The 'Ethics Road Show'

As well as sharing information, ethics education should provide learners the opportunity to discuss their own experiences of making ethical choices. At the author's own institution, the 'ethics road show' provides a safe environment for students to share their experiences of the clinical setting (Bowman 2006). There are no preset learning objectives nor lesson plans: it is nothing more complex than a small group of students and a trusted tutor discussing the dilemmas that clinical students experience daily but are rarely found in ethics textbooks. In nearly a decade of facilitating the 'ethics road shows', particular themes recur. The question of if and how students are introduced to patients, the difference between learning from or on patients, the dissonance between 'classroom ethics' and observed behaviour in the clinical setting and the effects of professional hierarchy, competition and insecurity on remaining true to ethical ideals have been discussed by every group irrespective of location or specialty. In the armoury of strategies available to those seeking to avoid or correct the process of 'ethical erosion' (Feudtner et al. 1994, Baldwin et al. 1998, Satterwhite and Satterwhite 2000, Jagsi and Lehmann 2004) amongst students and doctors, the 'ethics road show' has proved to be an invaluable means of revisiting core values, sharing experiences, and empowering learners.

31.4.4 Humanities and the Arts

Another approach to effective education in ethics is to draw on the humanities and arts. Although the discipline of 'medical humanities' is well-established in some areas and growing in others, there remains scope for its incorporation into main stream ethics teaching and learning (Bowman 2003). The narrative of stories, art, drama and music offers diverse perspectives on health, illness and moral choices whilst reflecting the way in which medicine in general, and psychiatry in particular (Rudin et al. 1998), are based on individual accounts, stories and experiences from which the 'history' and symptoms are delineated via consultation and examination (Carson 2001, Hudson-Jones 1999). In the arts and humanities, the complexity of the human condition is captured, challenged and explored, often from multiple perspectives. The world is not represented as linear, context is all and uncertainty abounds in fiction, art and drama resonating with those teachers and learners who grapple with the enormity of applying biomedical 'truths' to messy human fallibility. Kidd and Connor (2008) describe the humanities as providing a mirror for students. Continuing with the reflective theme, perhaps the more apt metaphor is that the humanities allow both students and their teachers to follow in Alice's shoes and 'go through the looking glass' in that the best stories enable us to experience not merely a reflection of ourselves, but to experience ourselves hearing and responding to the varied narratives of others. Through the 'looking glass' of humanities, moral dilemmas and ethical practice are endowed with the rich, pluralistic fabric of multiple perspectives, unspoken but significant thoughts and personal experiences that are rarely, if ever, present as comprehensively in the 'clinical interview' however skilled the psychiatrist.

31.5 Conclusion

Those invited to teach ethics and facilitate the ethical development of future clinicians bear both educational privilege and power. To teach ethics is to learn ethics – the process inevitably requires the 'teacher' to reflect on, develop and grow his or her own expertise, both in ethics and in clinical practice. Ethics teaching and learning goes, as I have argued in this chapter, well beyond the boundaries of a 'core curriculum'. Messages about ethics, its content and its importance, are conveyed both deliberately in 'teaching sessions' but also unwittingly via role modelling, silently observed behaviour and responses to student concerns (Baldwin et al. 1998). As such, it is not only formally designated 'teachers of ethics' who have the power to foster professionalism, moral imagination, and courage in learners, but everyone whom a learner encounters in their clinical education. It is hoped that this chapter has been thought-provoking for others to enjoy and acknowledge individual and collective responsibilities in creating a culture of constructive, engaging and relevant education in ethics.

References

Ashcroft R (2000) Teaching for patient-centred ethics. Med Health Care Philos 2000(3):287–295

Atkinson P (1981) The clinical experience: the construction and reconstruction of medical reality, 2nd edn. Gower, Farnborough

Baldwin DC, Daugherty SR, Rowley BD (1998) Unethical and unprofessional conduct observed by residents during their first year of training. Acad Med 73(11):1195–1200

Beauchamp TL, Childress JF (2001) Principles of biomedical ethics, 5th edn. Oxford University Press, Oxford

Becker HS, Geer B, Hughes EC, Strauss AL (1961) Boys in white: student culture in medical school. University of Chicago Press, Chicago, IL

Berry PA (2007) The absence of sadness: darker reflections on the doctor–patient relationship. J Med Ethics 22:266–268

Bloch S, Green SA (2009) Psychiatric ethics. Oxford University Press, Oxford

Bowman D (2003) The ethicist's tale: using the humanities to facilitate teaching and learning in ethics. In: Illingworth, S (ed) Ethics teaching highlighted in contextualised scenarios. LTSN, Newcastle

Bowman D (2005) To irritate, to soothe, to scrutinise and to humanise: the challenges of an ethical education in European medical schools. Die Psychiatrie 2(3):158–164

Bowman D (2006) The Road Less Travelled: the 'Ethics Road Show' and its place in Preventing Ethical Erosion. Conference Proceedings of the 11th International Ottawa Medical Education Conference, 2006

Bowman D, Hughes P (2005) Emotional responses of tutors and students in problem-based learning: lessons for staff development. Med Educ 39(2):145–153

Brown RD (1985) Ethical issues in graduate education: faculty and student responsibilities. J High Educ 56:403–418

Bruhn JG (1991) Values in health care: choices and conflicts. Charles C. Thomas, Springfield, IL

Cahn SM (1986) Saints and scamps: ethics in academia. Rowman & Kamp, Totowa, NJ

Calton L, Essex J, Bowman D, Barrett C (2008) Ethics teaching for clinical practice: a student perspective. Clin Teach 5(4):222–226

Carson AM (2001) That's another story: narrative methods and ethical practice. J Med Ethics 27:198–202

Coverdale JH (1996) The status of ethics education in Australasian psychiatry. Aust N Z J Psychiatry 30(6):813–818

Coverdale JH, Bayer T, Isbell P, Moffic S (1992) Are we teaching psychiatrists to be ethical? Acad Psychiatry 16:199–205

Cowley C (2005) The dangers of medical ethics. J Med Ethics 31(12):739–742

Cruess SR, Cruess RL, Steinert Y (2008) Role modelling: making the most of a powerful teaching strategy. Br Med J 336:718–721

Dowie, J, Elstein, A (eds) (1988) Professional judgment: a reader in clinical decision making. Cambridge University Press, Cambridge

Doyal L, Gillon R (1998) Medical ethics and law as a core subject in medical education. Br Med J 316:1623–1624

Feudtner C, Christakis DA, Christakis NA (1994) Do clinical clerks suffer ethical erosion? Students' perceptions of their ethical environment and professional development. Acad Med 69(8):670–679

Fish D, Coles C (2005) Medical education: developing a curriculum for practice. Open University Press, Maidenhead

Fulford KWM (1989) Moral theory and medical practice. Cambridge University Press, Cambridge

Fulford KWM (2004) Ten principles of values-based medicine. In: Radden, J (ed) The philosophy of psychiatry: a companion. Oxford University Press, New York, NY

Fulford KWM (2008) Values-based practice: a new partner to evidence-based practice and a first for psychiatry? Mens Sana Monogr 6(1):10–21

Fulford KWM, Dickenson D, Murray TH (eds) (2002) Healthcare ethics and human values: an introductory text with readings and case studies. Blackwell Publishers, Oxford

Fulford KWM, Williams R (2003) Values based child and adolescent mental health services? Curr Opin Psychiatry 16(4):369–376

Garfield D, Atre-Vaidya N, Sierles F (2002) Teaching the American psychiatric association practice guidelines to psychiatry residents. Acad Psychiatry 26(2):70–75

Goldie J, Schwartz L, McConnachie A, Morrision J (2002) The impact of three years' ethics teaching, in an integrated medical curriculum, on students' proposed behaviour on meeting ethical dilemmas. Med Educ 36:489–497

Hafferty F, Franks R (1994) The hidden curriculum, ethics teaching and the structure of medical education. Acad Med 69:861–871

Hoop JG (2004) Hidden ethical dilemmas in psychiatric residency training: the psychiatry resident as dual agent. Acad Psychiatry 28(3):183–189

Hudson-Jones A (1999) Narrative in medical ethics. Br Med J 318:253–256

Jagsi R, Lehmann LS (2004) The ethics of medical education. Br Med J 329:332–334

Karel MJ (2000) The assessment of values in medical decision-making. J Aging Stud 14(4):403–422

Kidd MG, Connor JTH (2008) Striving to do good things: teaching humanities in Canadian medical schools. J Med Humanit 29(1):45–54

Konner M (1987) Becoming a doctor: a journey of initiation in medical school. Viking, New York, NY

Leget C (2004) Avoiding evasion: medical ethics and emotion theory. J Med Ethics 30:490–493

Malek JI, Geller G, Sugarman J (2000) Talking about cases in bioethics: the effect of an intensive course on health care professionals. J Med Ethics 26:131–136

Mattick K, Bligh J (2006) Undergraduate ethics teaching: revisiting the consensus statement. Med Educ 40:329–332

McKneally MF, Singer PA (2008) 'Teaching bioethics to medical students and postgraduate trainees' in the clinical setting. In: Singer, PA and Viens, AM (eds) The Cambridge textbook of bioethics. Cambridge University Press, Cambridge

Moffic HS, Coverdale J, Bayer TL (1991) Ethics education for psychiatry. J Clin Ethics 2:161–166

Paice E, Heard S, Moss F (2002) How important are role models in making good doctors? Br Med J 325:707–710

Parker M (1995) Autonomy, problem-based learning and the teaching of medical ethics. J Med Ethics 21:305–310

Pomerantz AM, Grice JW (2001) Ethical beliefs of mental health professionals and undergraduates regarding therapist practices. J Clin Psychol 57(6):737–748

Rokeach M (1973) The nature of human values. The Free Press, New York, NY

Rudin E, Edelson R, Servis M (1998) Literature as an introduction to psychiatric ethics. Acad Psychiatry 22:41–46

Satterwhite RC, Satterwhite WM (2000) An ethical paradox: the effect of unethical conduct on medical students' values. J Med Ethics 26:462–465

Sayer M, Bowman D, Evans D, Wessier A, Wood D (2002) Use of patients in professional medical examinations: current UK practice and the ethico-legal implications for medical education. Br Med J 324:404–407

Schnapp WB, Stone S, Van Norman J, Ruiz P (1996) Teaching ethics in psychiatry: a problem based approach. Acad Psychiatry 20:144–149

Siegler M (2001) Lessons from 30 years of teaching clinical ethics. Virtual Mentor 3:10, October

Sigmon ST (1995) Ethical practices and beliefs of psychopathology researchers. Ethics Behav 5(4):295–309

Sinclair S (1997) Making doctors: an institutional apprenticeship. Berg, Oxford

Singer P, Siegler M, Pellegrino E (2001) Clinical ethics revisited. BMC Med Ethics 2:1

Sondheimer AN (1996) Ethics and child and adolescent psychiatry: curricular design and clinical teaching. Acad Psychiatry 20:150–157

Sondheimer A (1998) Teaching ethics and forensic psychiatry: a national survey of child and adolescent psychiatrists. Acad Psychiatry 22:240–252

Sondheimer A, Martucci CL (1992) An approach to teaching ethics in child and adolescent psychiatry. J Am Acad Child Adolesc Psychiatry 31(3):415–422

UNESCO (2005a) Preliminary Draft Declaration on Universal Norms on Bioethics, 9 February 2005

UNESCO (2005b) Explanatory Memorandum on the Elaboration of the Preliminary Draft Declaration on Universal Norms on Bioethics, Paris, 21 February 2005

Wear D (1998) On white coats and professional development: the formal and the hidden curricula. Ann Intern Med 129:734–737

Weiss Roberts L, Coverdale J (2006) Philanthropy, ethics and leadership in academic psychiatry. Acad Psychiatry 30(4):269–272

Weiss Roberts L, Green Hammond KA, Geppert CMA, Warner TD (2004) The positive role of professionalism and ethics training in medical education: a comparison of medical student and resident perspectives. Acad Psychiatry 28:170–182

Weiss Roberts L, Johnson ME, Brems C, Warner TD (2006) Preferences of Alaskan and New Mexico psychiatrists regarding professionalism and ethics training. Acad Psychiatry 30:200–204

Weiss Roberts L, McCarty T, Lyketsos C (1996) What and how psychiatry residents at ten programs wish to learn about ethics. Acad Psychiatry 20(3):131–143

Weiss Roberts L, McCarty T, Roberts BB, Morrison N, Belitz J, Berenson C, Siegler M (2003) Clinical ethics teaching in psychiatric supervision. J Lifelong Learn Psychiatry 1(4):436–444, Fall

Whitman S (2001) Teaching residents to use supervision effectively. Acad Psychiatry 25:143–147

Wilson EK (1982) Power, pretence and piggybacking: some ethical issues in teaching. J Higher Educ 53:268–281

Woodbridge K, Fulford KWM (2004) Whose values? A workbook for values-based practice in mental health care. Sainsbury Centre for Mental Health, London

Wright S, Wong A, Newill C (1997) The impact of role models on medical students. J Gen Intern Med 12:53–56

Part VI
Conclusions and Summary

Chapter 32
Summary and Conclusions

Norman Sartorius and Hanfried Helmchen

1

The debate on ethical implications of dealing with mentally ill people has been intensified among psychiatrists but also among other professionals and members of the general public particularly those who have or had a mental illness. In part this is due to a greater awareness of issues related to human rights in medical practice and research that has marked the late decades of the twentieth century. In part however the debate has been fuelled by developments of particular relevance to psychiatry, including the

- the increasing stigmatisation of the mentally ill and of psychiatric institutions (see Chapter 2) – despite the progress of knowledge and improvements of mental health services – and frequent use of more subtle forms of coercive measures (see Chapter 20);
- Problems arising in the psychiatric treatment of persons of a different cultural background – now that there are so many migrants with mental illness all over Europe (see Chapters 2 and 13);
- the processes of de- and re-institutionalisation of mentally ill people (see Chapter 21);
- the new economic restraints of psychiatric care (see Chapter 3) which intensify questions of prioritising and rationing (and thus of justice as equal access and distribution of resources see Chapter 12);
- the continuing and aggravating tension between standardisation of services as well as of interventions and individualisation of the physician–patient-relationship (see Chapters 1 and 15);
- the exponential growth of the need to deal with demented patients and end-of life-decisions including the chances and limitations of advance directives (see

N. Sartorius (✉)
Association for the Improvement of Mental Health Programmes, CH-1209 Geneva, Switzerland
e-mail: sartorius@normansartorius.com

H. Helmchen, N. Sartorius (eds.), *Ethics in Psychiatry*, International Library
of Ethics, Law, and the New Medicine 45, DOI 10.1007/978-90-481-8721-8_32,
© Springer Science+Business Media B.V. 2010

Chapter 10) as well as with the question of euthanasia and assisted suicide (see Chapter 24);

- the ever more frequent need to provide psychiatric treatment of patients with co-morbid physical illness treated by another specialist – now that methods of treatment of physical illness have made it possible to save life and maintain patients with chronic illness much longer (see Chapters 25 and 24);
- development of new biological treatments such as neuromodulation, e.g. deep brain stimulation (see Chapter 19);
- the use of psychiatric interventions for non-medical use, e.g. for neuroenhancement (see Chapter 30) or (still) for political misuse (see Chapter 29);
- research involving mentally ill patients and specifically those who have lost their capacity to consent (see Chapter 25);
- increased need to inform the concerned persons, e.g. in genetic counselling (see Chapter 27); when proposing measures of prevention and early detection of mental illnesses (see Chapter 16); concerning ways of safeguarding of their data (see Chapters 11 and 26) and relevant mechanisms of control (see Chapter 6);
- changes of the normative context of psychiatric acting, i.e. ethics principles, codes (see Chapter 8) and laws (see Chapter 5) as well as methods of their translation into psychiatric practice (see Chapter 7);
- significant increase of knowledge about side-effects and negative consequences of established psychiatric treatments – psychotherapy (see Chapter 18), social interventions (see Chapter 21), psychopharmacotherapy (see Chapter 17) and fundamental changes of aims of treatment of drug dependence (see Chapter 23);
- the tendency of the ethically dangerous reductionism of giving priority to a single ethical requirement;
- the ethical implications of the transformation of psychiatry into an empirically evidence-based quantitative science (see Chapters 13, 14, and 22).

The ethical implications of these developments are discussed in the literature, but in a fairly dispersed manner so that psychiatrists and others engaged in the care for the mentally ill (and interested in improving it) do not have easy access to the relevant materials. This book brings the essence of these materials together in chapters written by Europe's leading authorities in the matter. Morals are embedded in societies and depend on social and cultural influences in the settings in which they are used: it follows that a coherent perspective on matters of ethics can best be presented by experts who live and work in a group of countries that share many of their socio-cultural traits. Thus, the book presents a predominantly European perspective.

The fact that we were fortunate in winning leading authorities to contribute to the volume ensures that the relevant arguments are comprehensively exposed; this however does not mean that all the points of view espoused by the contributors are shared by the editors.

2

The review of the contributions assembled in this book makes it clear that the generally accepted ethical principles (such as respect for both the autonomy and welfare of the patient; avoidance of harm; justice; confidentiality) must be applied in conjunction with each other. To make only one of them absolute or give it priority is dangerous for the patient. It is in such a balanced framework that the ethical analysis of specific clinical cases can take place and can use the criteria of benefits and risks for the evaluation of the psychiatric intervention in question. If the empirical evidence provides an acceptable balance between expected efficiency and the potential unwanted effects then patients should be informed about these benefits and risks as well as about the benefit-risk-ratio and should be helped to find their own benefit-risk-ratio. The goal of this process is to arrive at a shared decision, a decision that is satisfactory for the patient and the physician. It goes without saying that the psychiatrist must take any impairment of the patient's capacity to consent into account: if this is not the specific issue in question may either overload patients with a responsibility they cannot bear or discriminate against them (see Chapter 9).

Ethical analysis must clarify what is the ethical problem, which relevant ethical principles are in question; what are the interests of the patient and others involved in the solution of the problem, and what procedure should be followed to reach an ethically reasoned decision which will be accepted by the patient and at its best by all involved persons such as the psychiatrist, the carers, family members, the legal guardian. It is important that all concerned also know that the clarity about these issues will not necessarily eliminate ambivalences and uncertainties that will accompany concrete decisions in the individual case.

Ethical problems in psychiatric everyday practice are related both to the patient–psychiatrist-relationship and to context factors. Psychiatrists and their patients may use different definitions and interpretations of benefits and risks, may differently understand the risk-benefit-ratio, and the patients may have difficulties to understand what a particular line of action might mean for them. The decisions may be influenced by contextual factors, particularly by economic (e.g. insufficient resources) and cultural factors (e.g. stigmatizing attitudes).

3

All these factors contribute to a loosening of the borders of professional psychiatric practice acting and challenge the professional identity of psychiatrists:

- the traditional obligation of physicians/psychiatrists to act in the best interest and welfare of the patient has been confronted by need to respect the autonomy and self-determination of their patients thereby confronting the psychiatrist with conflicts between the principles of welfare and wish of the patient;

- the psychiatrist has to learn to share decisions with the informed patient even if that patient is severely disabled (see Chapters 9, 10, 20, and 21);
- the clarity of the psychiatrist's obligation to prevent suicides is challenged by the demand of severely and hopelessly ill people, by the acceptance of advanced directive, to get help with assisted suicide (see Chapter 10 and 24);
- the psychiatrists' obligation to prevent the development of addiction by restricting (even legally demanded) prescription of drugs with addictive potential is challenged by the emergence of neuroenhancement (see Chapter 30). Practising psychiatrists worry about the mental health consequences of a liberal attitude towards neuroenhancement procedures because they see the potentially negative consequences of use and misuse of drugs e.g. the habituation and addiction. At the present stage of our knowledge and social development psychiatrists are bound to be against neuroenhancement; but the treatment of addictions with the aim of abstinence, valid for a long time, is challenged by new empirical findings with substitution treatments which aim primarily to lessen the individual suffering and delinquency (see Chapter 23).

4

The issues that are presented in this book and summarized above indicate that the ethical principles that have been formulated in recent years are broad guidelines and not a manual of operations. This is all the more valid because by definition there are no simple solutions for ethical dilemmas. Their application in clinical practice and their use in the conduct of research can be useful if they are understood as being essential parts of a framework in which a careful ethical analysis of each individual situation has to take place. The challenge before medical educators (see Chapters 6, 7, and 31) is to ensure that their students at undergraduate and postgraduate level understand this and accept the ethical analysis as being just as important (or even more important) as other decisions in which the life and welfare of their patients are at stake. The challenge before all of us is to remember – or to learn – this way of practicing psychiatry and of conducting research in our field.

We hope that this book will be helpful in learning how to overcome this challenge and how to be useful to our patients and to the development of psychiatry in ethically acceptable ways.

Index

Note: The locators follwed by 'f', 'n' and 't' refers to figures, note numbers and tables cited in the text

H. Helmchen, N. Sartorius (eds.), *Ethics in Psychiatry*, International Library
of Ethics, Law, and the New Medicine 45, DOI 10.1007/978-90-481-8721-8,
© Springer Science+Business Media B.V. 2010

Lightning Source UK Ltd.
Milton Keynes UK
October 2010